REFRIGERATION AND AIR CONDITIONING

SECOND EDITION

S.N. Sapali

Professor of Mechanical Engineering
College of Engineering, Pune

Delhi-110092
2018

₹ 475.00

REFRIGERATION AND AIR CONDITIONING, Second Edition
S.N. Sapali

© 2014 by PHI Learning Private Limited, Delhi. All rights reserved. No part of this book may be reproduced in any form, by mimeograph or any other means, without permission in writing from the publisher.

ISBN-978-81-203-4872-1

The export rights of this book are vested solely with the publisher.

Fourth Printing (Second Edition) ... ... **February, 2018**

Published by Asoke K. Ghosh, PHI Learning Private Limited, Rimjhim House, 111, Patparganj Industrial Estate, Delhi-110092 and Printed by Rajkamal Electric Press, Plot No. 2, Phase IV, HSIDC, Kundli-131028, Sonepat, Haryana.

Contents

Foreword .... *xiii*
Preface .... *xv*
Preface to the First Edition .... *xvii*

1. INTRODUCTION   1–34

     1.1   Introduction 1
     1.2   SI Units 2
     1.3   Terms Used in Thermodynamics 7
     1.4   Types of Energy 8
     1.5   Work 10
     1.6   Heat 11
     1.7   Work Done during a Quasi-static Process 12
     1.8   First Law of Thermodynamics 15
     1.9   Processes for Ideal Gases 17
                   1.9.1   Constant Volume Process 17
                   1.9.2   Constant Pressure Process 18
                   1.9.3   Constant Temperature Process 18
                   1.9.4   Adiabatic Process 20
                   1.9.5   Polytropic Process 21
     1.10   Control Volume 23
     1.11   Limitations of the First Law of Thermodynamics 25
     1.12   Second Law of Thermodynamics 27
     1.13   Pure Substance 28
     1.14   Other Phase Diagrams 31
     Numericals 33

2. METHODS OF REFRIGERATION   35–52

     2.1   Introduction 35
     2.2   Applications of Refrigeration 35
     2.3   Refrigeration Systems 39
                   2.3.1   Ice Refrigeration 39

	2.3.2	Evaporative Refrigeration	40
	2.3.3	Refrigeration by Dry Ice	41
	2.3.4	Liquid Gas Refrigeration	42
	2.3.5	Steam Jet Refrigeration	43
	2.3.6	Thermoelectric Refrigeration	44
	2.3.7	Vortex Tube	44
	2.3.8	Solar Refrigeration	46
	2.3.9	Magnetic Refrigeration	48
	2.3.10	Air Refrigeration Cycle	49
	2.3.11	Vapour Compression Refrigeration Cycle	50
	Exercises		51

3. AIR REFRIGERATION SYSTEMS   53–114

3.1	Introduction		53
3.2	Definitions		53
3.3	Refrigeration Load		54
3.4	Heating Load		54
3.5	Concept of Heat Engine, Refrigerator and Heat Pump		55
3.6	Air Refrigeration Systems		57
	3.6.1	Carnot Refrigerator	57
	3.6.2	Limitations of Reversed Carnot Cycle	61
	3.6.3	Modified Reversed Carnot Cycle	61
	3.6.4	Reversed Carnot Cycle with Vapour as Refrigerant	62
	3.6.5	Bell–Colemann or Reversed Brayton or Joule Cycle	63
	3.6.6	Actual Bell–Colemann Cycle	67
	3.6.7	Application of Aircraft Refrigeration	68
3.7	Methods of Air Refrigeration Systems		68
	3.7.1	Simple Air-cooling System	68
	3.7.2	Simple Air Evaporative Cooling System	84
	3.7.3	Boot-strap Air Cooling System	88
	3.7.4	Boot-strap Air Evaporative Cooling System	100
	3.7.5	Reduced Ambient Air Cooling System	101
	3.7.6	Regenerative Air Cooling System	106
3.8	Comparison of Various Air Cooling Systems Used for Aircraft		111
Exercises			112
Numericals			112

4. SIMPLE VAPOUR COMPRESSION REFRIGERATION SYSTEMS   115–177

4.1	Introduction		115
4.2	Advantages and Limitations of Air Refrigeration		115
	4.2.1	Advantages of Air Refrigeration	115
	4.2.2	Limitations of Air Refrigeration	115
4.3	Vapour Compression Refrigeration System (VC Cycle)		116
4.4	Vapour Compression Cycle when Vapour is Dry Saturated at the End of Compression		131
4.5	Vapour Compression Cycle when Vapour is Wet at the End of Compression		134

4.6	Deviations from the Simple Compression Cycle	141
	4.6.1 Suction Gas Superheating without Cooling	143
	4.6.2 Suction Gas Superheating with Cooling Effect	145
4.7	Effect of Subcooling the Liquid	146
	4.7.1 Liquid-suction Heat Exchanger	151
	4.7.2 Effect of Evaporator Pressure	152
	4.7.3 Effect of Condenser Pressure	154
4.8	Losses in Vapour Compression Refrigeration System	156

Exercises 174
Numericals 175

5. REFRIGERANTS 178–205

5.1	Introduction	178
5.2	Classification of Refrigerants	179
	5.2.1 Primary Refrigerants	179
	5.2.2 Secondary Refrigerants	180
5.3	Designation of Refrigerants	180
5.4	Desirable Properties of a Good Refrigerant	181
	5.4.1 Thermodynamic Properties of Refrigerants	181
	5.4.2 Chemical Properties of Refrigerants	183
	5.4.3 Physical Properties of Refrigerants	184
5.5	Properties of an Ideal Refrigerant	185
5.6	Properties of Important Refrigerants	185
5.7	Selection of a Refrigerant	187
5.8	New Refrigerants	189
5.9	Secondary Refrigerants	191
	5.9.1 Substances Used as Secondary Refrigerants	193
5.10	Toxicity and Safe Handling of Refrigerants	194
5.11	Oil and Refrigerant Relationship	195
5.12	Lubricating Oils	195
5.13	Effect of CFC on Ozone Depletion and Global Warming	197
	5.13.1 Ozone Depletion	197
	5.13.2 Global Warming	199
5.14	Montreal Protocol	200
5.15	Alternatives to CFC Refrigerants	201
	5.15.1 Certain Mixtures of Refrigerants as Replacement for HCFC R22	202
	5.15.2 Substitutes for CFC, R12 Refrigerant	202
5.16	Kyoto Protocol and TEWI	204

Exercises 204

6. MULTIPRESSURE SYSTEMS 206–259

6.1	Introduction	206
	6.1.1 Limitations and Drawbacks of the Simple Vapour Compression Refrigeration Cycle	207
6.2	Multistage Vapour Compression System	208
	6.2.1 Multistage Compression with Intercooling between the Stages	208
	6.2.2 Intermediate Pressure for Minimum Work	210

- 6.3 Types of Multistage Vapour Compression System with Intercooler 211
  - 6.3.1 Two-stage Compression with Flash Gas Removal 211
  - 6.3.2 Two-stage Compression with Flash Intercooling 212
  - 6.3.3 Two-stage Compression with Flash Gas Removal and Additional Gas Cooler 215
- 6.4 Multiple Evaporators and Compressors Systems 224
  - 6.4.1 Multiple Evaporators at the Same Temperature and a Single Compressor System 225
  - 6.4.2 Multiple Evaporators at Different Temperatures with a Single Compressor, Individual Expansion Valves and a Back Pressure Valve 228
  - 6.4.3 Multiple Evaporators at Different Temperatures with a Single Compressor, Multiple Expansion Valves and a Back Pressure Valve 232
  - 6.4.4 Multiple Evaporators with Individual Compressors and Individual Expansion Valves 235
  - 6.4.5 Multiple Evaporators with Individual Compressors and Multiple Expansion Valves 239
  - 6.4.6 Multiple Evaporators with Compound Compression and Individual Expansion Valves 243
  - 6.4.7 Multiple Evaporator System with Compound Compression, Individual Expansion Valves and Flash Intercoolers 246
  - 6.4.8 Refrigeration Unit with Multiple Evaporators with Compound Compression Multiple Expansion Valves and Flash Intercooler 250
- 6.5 Cascade System 253
- *Exercises* 257
- *Numericals* 257

7. VAPOUR ABSORPTION REFRIGERATION SYSTEMS 260–274

- 7.1 Introduction 260
  - 7.1.1 Refrigerant–Solvent Properties 261
- 7.2 Simple Vapour Absorption System 262
- 7.3 Practical Vapour Absorption System 263
- 7.4 Vapour Absorption Refrigeration System vs Vapour Compression Refrigeration System 265
- 7.5 COP of an Ideal Vapour Absorption Refrigeration System 265
- 7.6 Domestic Electrolux (Ammonia–Hydrogen) Refrigerator 269
- 7.7 Lithium Bromide Absorption Refrigeration System 271
- *Exercises* 274
- *Numericals* 274

8. PSYCHROMETRY 275–319

- 8.1 Introduction 275
- 8.2 Psychrometry 277

	8.3	Psychrometric Chart	296
	8.4	Typical Air Conditioning Processes	298
		8.4.1 Sensible Heating of Air	298
		8.4.2 Sensible Cooling of Moist Air	301
		8.4.3 Cooling and Dehumidification of Moist Air	302
	8.5	Adiabatic Cooling or Cooling with Humidification Process	305
	8.6	Heating and Humidification	307
		8.6.1 Heating and Humidification by Steam Injection	308
	8.7	Adiabatic Mixing of Air Streams	309
	8.8	Air Washer	312
	8.9	Chemical Dehumidification or Sorbent Dehumidification	313
	Exercises		317
	Numericals		318

9. COOLING LOAD ESTIMATION AND PSYCHROMETRIC ANALYSIS   320–358

	9.1	Introduction	320
	9.2	Thermodynamics of Human Body and Mathematical Model	321
	9.3	Effective Temperature	322
	9.4	Human Comfort Chart	323
	9.5	Outside Design Conditions	324
	9.6	Sources of Heat Load	325
	9.7	Conduction through Exterior Structures	326
		9.7.1 Overall Heat Transfer Coefficient	326
		9.7.2 Cooling Load Estimation by CLTD Method	327
	9.8	Heat Gain through Glass	327
		9.8.1 Factors Affecting Solar Radiation at a Place	328
		9.8.2 Method of Estimation	328
	9.9	Infiltration	329
		9.9.1 Sensible Heat Loss Effect of Infiltration Air	329
		9.9.2 Latent Heat Loss Effect of Infiltration Air	329
	9.10	Ventilation	330
	9.11	Outside Air Load	331
	9.12	Heat Load from People	332
	9.13	Lighting	333
	9.14	Heat Gain from Equipment and Appliances	333
	9.15	System Heat Gain	334
	9.16	Room Cooling Load	335
	9.17	Cooling Coil Load	335
	9.18	Psychrometric Analysis of the Air Conditioning System	335
		9.18.1 Determining Supply Air Conditions	335
		9.18.2 Room Sensible Heat Factor (RSHF)	337
	9.19	Summer Air Conditioning System Provided with Ventilation Air (Zero Bypass Factor)	338
		9.19.1 Grand Sensible Heat Factor (GSHF)	339
		9.19.2 Winter Air Conditioning	340

		9.20	Effective Room Sensible Heat Factor (ERSHF) 342

       Exercises 357
       Numericals 357

10. AIR CONDITIONING SYSTEMS AND EQUIPMENT   359–382

    10.1   Introduction 359
    10.2   Classification of Air Conditioning Systems 360
    10.3   Unitary System 361
              10.3.1   Window Air Conditioner 361
              10.3.2   Split Air Conditioner 362
    10.4   Central Air Conditioning Systems 362
    10.5   Reheat System 365
    10.6   Multizone System 365
    10.7   Dual Duct System 366
    10.8   Variable Air Volume (VAV) System 367
    10.9   All-water System 368
    10.10   Air-water Systems 369
    10.11   Unitary vs. Central Systems 369
    10.12   Air Conditioning Equipment 370
    10.13   Cooling Coil 370
              10.13.1 Coil Selection 371
    10.14   Heating Coils 371
    10.15   Air Cleaning Devices (Filters) 372
              10.15.1 Types of Filters/Cleaners 373
              10.15.2 Types of Media 373
              10.15.3 Electronic Air Cleaners 374
              10.15.4 Choice of Filter 375
    10.16   Humidifiers 375
    10.17   Fan 376
              10.17.1 Types of Fans 376
              10.17.2 Performance Characteristics of Fans 377
              10.17.3 Fan Selection 379
              10.17.4 Fan Ratings 379
              10.17.5 System Characteristics 380
              10.17.6 Fan–system Interaction 380
              10.17.7 Selection of Optimum Fan Conditions 381
              10.17.8 Fan Laws 381
       Exercises 382

11. COMPRESSORS   383–414

    11.1   Introduction 383
    11.2   Compression Process 384
    11.3   Indicator Diagrams 387
              11.3.1   Comparison of Indicator Diagrams 388
    11.4   Overall Volumetric Efficiency 388
              11.4.1   Clearance Volumetric Efficiency (η_{cv}) 390

	11.4.2	Effect of Heat Exchange Loss	390
	11.4.3	Effect of Valve Pressure Drops	391
	11.4.4	Leakage Loss	391
11.5	Design Features of a Reciprocating Compressor		394
11.6	Determination of Compressor Motor Power		395
11.7	Performance of a Reciprocating Compressor		395
	11.7.1	Performance of Ideal Compressor	396
	11.7.2	Actual Performance	397
	11.7.3	Capacity of a Compressor	398
11.8	Capacity Control of Reciprocating Compressor		399
11.9	Rotary Compressors		401
	11.9.1	Rolling Piston Compressor	401
	11.9.2	Rotary Vane Type Compressor	402
	11.9.3	Screw Compressors	403
	11.9.4	Single Screw Compressor	403
	11.9.5	Oil-injected Compressor	404
	11.9.6	Oil-injection Free Compressor	405
11.10	Centrifugal Compressor		409
	11.10.1	Work Done by Impeller	410
	11.10.2	Power Input	411
	11.10.3	Performance of Centrifugal Compressor	411
11.11	Hermetically Sealed Compressor		413
Exercises			414

12. EVAPORATORS AND CONDENSERS   415–442

12.1	Introduction		415
12.2	Conduction		416
	12.2.1	Steady State Heat Conduction through a Slab	416
	12.2.2	Steady State Heat Flow through a Cylindrical Wall	417
	12.2.3	Steady State Conduction through a Composite Wall	417
	12.2.4	Steady State Conduction through a Composite Cylinder	418
	12.2.5	Overall Heat Transfer Coefficient	419
12.3	Convection Heat Transfer		420
12.4	Evaporator		423
	12.4.1	Dry and Flooded Evaporators	423
12.5	Frosting and Defrosting of Coolers		427
	12.5.1	Methods of Defrost	427
	12.5.2	Latent Heat Defrost	428
	12.5.3	Electric Defrost	428
12.6	Selection from Manufacturer's Data Sheets		429
12.7	Liquid Coolers (Chillers)		430
	12.7.1	Direct Expansion Type	430
	12.7.2	Flooded Shell-and-Tube Coolers	431
	12.7.3	Baudelot Cooler	431
	12.7.4	Shell-and-Coil Cooler	431

12.8	Condensers		432
	12.8.1	Water-cooled Condensers	433
	12.8.2	Air-cooled Condensers	437
	12.8.3	Evaporative Condensers	440
Exercises			442

13. EXPANSION DEVICES — 443–458

13.1	Introduction		443
13.2	Hand-operated Expansion Valve		444
13.3	Automatic Expansion Valve (AEV)		444
13.4	Thermostatic Expansion Valve (TEV)		446
	13.4.1	External Equalizer	448
	13.4.2	Gas-charged TEV	450
	13.4.3	Refrigerant-charged Expansion Valves	450
	13.4.4	Behaviour of Charge in the Bulb	452
	13.4.5	Rating and Selection of TEV	453
13.5	Capillary Tube		453
	13.5.1	Selection of Capillary Tube	456
13.6	Float Valves		456
	13.6.1	Low Pressure Float Valve	457
	13.6.2	High Pressure Float Valve	457
Exercises			458

14. REFRIGERANT PIPING, ACCESSORIES AND SYSTEM PRACTICES — 459–480

14.1	Introduction		459
14.2	Materials		459
14.3	System Practice for HCFC Systems Refrigerant Line Sizing		460
	14.3.1	Pressure Drop	460
	14.3.2	Sizing of Liquid Lines	460
	14.3.3	Sizing of Suction Lines	463
	14.3.4	Sizing of Discharge Lines	464
14.4	Piping Layout		464
	14.4.1	Piping Layout for Suction Lines	464
	14.4.2	Piping Layout for Hot Gas Lines	465
	14.4.3	Condenser-to-Receiver Lines	466
14.5	Multiple System Practices for HCFC Systems		467
	14.5.1	Hot Gas Lines	467
	14.5.2	Suction Lines	467
14.6	System Practices for Ammonia		468
	14.6.1	Pipe Sizing	469
	14.6.2	Ammonia Piping	471
	14.6.3	Compressor Piping	472
	14.6.4	Condenser and Receiver Piping	472
	14.6.5	Suction Traps	473
	14.6.6	Oil Separators	473

14.7	Installation Arrangement		474
	14.7.1	Location of Equipment	474
	14.7.2	Piping Arrangement	475
14.8	Dehydration		475
	14.8.1	Dehydration by Heating	475
	14.8.2	Dehydration by Vacuum	476
	14.8.3	Dehydration by Dry Air	476
	14.8.4	Heat and Vacuum Method	476
	14.8.5	Vacuum and Heat Method	476
	14.8.6	Heat and Dry Air Method	476
14.9	Charging		477
	14.9.1	Charging through Suction Valve	477
	14.9.2	Charging through Charging Valve	478
	14.9.3	Checking the Charge	479
14.10	Testing for Leaks		479
	14.10.1	Water Submersion Test	480
	14.10.2	Pressure Testing	480
	14.10.3	Halide Leak Detector	480
Exercises			480

15. AIR DISTRIBUTION SYSTEM AND DUCT DESIGN   481–514

15.1	Introduction		481
15.2	Classification of Ducts		482
15.3	Duct Material		482
15.4	Continuity Equation		483
15.5	Energy Equation for a Pipe Flow		484
15.6	Total, Static and Velocity Pressure		486
15.7	Static Regain		487
15.8	Pressure Loss in the Duct		487
	15.8.1	Pressure Loss from Friction in Piping and Ducts	487
	15.8.2	Friction Factor 'f'	488
15.9	Rectangular Sections Equivalent to Circular Sections		490
15.10	Dynamic Losses in Duct		494
	15.10.1	Pressure Loss due to Sudden Enlargement	494
	15.10.2	Pressure Loss Due to Contraction	495
	15.10.3	Pressure Loss at Entry or Exit from Duct	495
	15.10.4	Pressure Loss in Bends, Tees and Branch Offs	495
	15.10.5	Pressure Loss in Fittings	495
15.11	Methods of Duct Design		496
	15.11.1	Equal Friction Method	497
	15.11.2	Static Regain Method	498
	15.11.3	Velocity Reduction Method	500
15.12	Duct Arrangement Systems		502
	15.12.1	The Perimeter System	503
	15.12.2	Extended Plenum System	504
15.13	Air Distribution Systems		508
Exercises			514

16. CRYOGENICS   515–529

  16.1   Introduction 515
  16.2   Applications of Nitrogen 516
  16.3   Applications of Oxygen 517
  16.4   Applications of Argon 518
  16.5   Cryogenic Air Separation Plant 518
  16.6   Cooling Methods 519
  16.7   Air Liquefaction System 521
         16.7.1   System Performance Parameters 521
  16.8   Simple Linde Cycle 523
  16.9   Claude Cycle 525
  16.10 Small-to-Medium-sized Hydrogen Liquefiers 526
  16.11 Simon Helium Liquefier 527
  Exercises 529

17. FOOD PRESERVATION   530–537

  17.1   Introduction 530
  17.2   Food Deterioration and Spoilage 530
  17.3   Factors of Food Deterioration and Spoilage 531
         17.3.1   Enzymes 531
         17.3.2   Micro-organisms 532
         17.3.3   Bacteria 532
         17.3.4   Yeasts 533
         17.3.5   Moulds 533
  17.4   Food Preservation Processes 533
         17.4.1   Drying 533
         17.4.2   Refrigerated Storage 533
  17.5   Mixed Storage 534
  17.6   Freezing and Frozen Storage 535
         17.6.1   Advantages of Food Preservation 535
         17.6.2   Disadvantages of Food Preservation 535
  17.7   Cold Storage 536
         17.7.1   Cold Chain 536
         17.7.2   Construction of Cold Stores 536
  Exercises 537

Appendices *539–566*
Index *567–574*

Foreword

Focusing on the need of a first level textbook for the undergraduates, postgraduates and a professional reference book for practicing engineers, the author of this work Dr. S.N. Sapali has brought forth this edition using his extensive teaching and research experience in the field of Refrigeration and Air Conditioning. It is a great pleasure to write a foreword to such a book which satisfies a long felt requirement.

For me, this would have been just the required textbook for my young engineering students and engineers in the field of Refrigeration and Air Conditioning. The style of the book reflects the teaching culture of premier engineering institutions like IITs and COEP, since a vast topic has to be covered in a comprehensive way in a limited time. Each chapter is presented in an elegant simplicity requiring no special prerequisite knowledge of supporting subjects. Self-explanatory sketches, graphs, line-schematics of processes have been generously used to curtail long and wordy explanations. Numerous illustrated examples, exercises and problems at the end of each chapter are a good source of material to practice the application of the basic principles presented in the text. SI system of units has been used throughout the book.

It is not a simple task to bring out a comprehensive book on all-encompassing subject like Refrigeration and Air Conditioning. As our knowledge of refrigeration processes has advanced, these machines have continued to develop on a scientific basis. The present day refrigeration systems have to satisfy the energy constraints and standards in addition to meeting the competitiveness of the world market. To understand the working of Refrigeration and Air Conditioning system fully, one requires the basic knowledge of many disciplines—Thermodynamics, Fluid Flow, Heat Transfer as applied to a system with both special and temporal variation in a state of non-equilibrium.

This book includes vast information on Vapour Compression Refrigeration systems, the properties of various refrigerants, and multi-pressure and multi-evaporator systems. The air conditioning part not only explains the psychrometry but also the various air conditioning processes. It also includes the theory of Refrigeration system components like compressor, condenser and expansion devices. The basic principles and applications of cryogenic gases and air liquefaction systems are also included in the text to a limited extent.

I congratulate the author, Dr. S.N. Sapali on bringing out this excellent book for the benefit of students in R.A.C. While many a student will find it rewarding to follow this book for their class work. Also, I hope that it will motivate a few of them to specialize in some key areas and take up Refrigeration Research as a career. With great enthusiasm I recommend to the students and practicing engineers.

Dr. B.B. Ahuja
Deputy Director
College of Engineering, Pune

Preface

It is a great pleasure on the part of the author to bring out the second edition. The author is extremely thankful to the students and teachers for their love and affection shown for this reference book for the last 4 years. The teachers and students directly or indirectly helped the author to elevate the standard of this book by providing their suggestions for corrections and modifications.

The aim of bringing this edition remained unchanged and only the contents of the book has been thoroughly revised and rearranged. The book is conveniently written in seventeen chapters covering Refrigeration and Air Conditioning fundamentals, which are also useful for commercial and industrial purposes. Chapter 17 on Food Preservation is new to this edition.

The content of this book wherever necessary is amplified including recent advances on Refrigeration and Air Conditioning. Another important feature of the book which helps students to understand concepts easily is its theoretical and practical approach.

The outstanding features of this edition are:

1. All the printing and calculation mistakes are removed keeping eagle eye during proof reading.
2. The quality of text is considerably improved by including computer drawings.
3. Many chapters are modified and amplified to meet the requirement of the undergraduate students.
4. Selected questions from different competitive examinations are also included at the end of each chapter.

Author extends his thanks to all the distinguished professors and researchers engaged in this field for their constructive suggestions which helped the author to update this text to the present level. He also extend his heartfelt gratitude to Mr Ajai Kumar Lal Das for bringing this edition in a short period.

Irrespective of all cares taken by the author, I do not claim for perfection. Therefore, teachers and students are requested to bring errors, if any, to the author's notice for further improvement of the text.

S.N. Sapali

Preface to the First Edition

This book is designed for a first course in Refrigeration and Air Conditioning that can be covered in one semester. Refrigeration and Air Conditioning is part of the curriculam of engineering disciplines such as Mechanical, Automobile, and Chemical Engineering. This book has evolved from the author's teaching of the subject matter to senior undergraduate and postgraduate students. The object of this book is to present the topics in the most precise, compact and lucid manner. The book has been developed in a logical and coherent manner with neat illustrations, along with a fairly large number of solved examples and unsolved problems. Answers to many unsolved numerical problems are also given.

The following goals have been pursued in the preparation of this text:

- To cover the basic principles of refrigeration.
- To develop a very good understanding of the subject matter, systematically and in a step-by-step approach.
- To give practice in solving numerical examples to help the students develop their knowledge and skills in a progressive manner.

The book contains in all sixteen chapters. Chapter 1 includes introduction to thermodynamics while Chapter 2 is on various refrigeration methods. Chapter 3 discusses the concepts of air refrigeration cycles. Chapter 4 deals with simple vapour compression refrigeration systems. The properties of various refrigerants and environmental aspects are covered in Chapter 5. Multi-pressure, multistage and multi-evaporator systems are included in Chapter 6. Chapter 7 is an introduction to vapour absorption refrigeration systems. Chapter 8 introduces psychrometry while cooling load estimation and psychrometric analysis are carried out in Chapter 9. Air conditioning systems and equipment are discussed in Chapter 10. The compressor, the main component of refrigeration and air conditioning systems, is fully discussed in Chapter 11. The various evaporators and condensers are presented in Chapter 12. Chapter 13 is dedicated to the expansion devices. The information related to refrigerant piping, accessories and system practices is included in Chapter 14. Fundamentals related to air distribution system and duct design are elaborated in Chapter 15. The basic principles and applications of cryogenic gases and air liquefaction systems are included in Chapter 16 to a limited extent.

The appendices include property tables and charts of some refrigerants, psychrometric charts, and properties of some insulating, building and other materials.

Although every care has been taken to check mistakes and errors, it is difficult to attain perfection. Any errors, mistakes and suggestions for the improvement of this book, brought to my notice will be thankfully acknowledged and incorporated in the next edition.

I take this opportunity to thank PHI Learning, for publishing this book. I am also thankful to the staff of PHI Learning, especially Darshan Kumar and Pankaj Manohar for their endless efforts to make this book as best as it could be. I am deeply indebted to all my family members, especially my elder brother late **(Prof.) A.N. Sapali** for providing me with moral support. Last but not the least, I am most grateful to my wife **Vijaylaxmi**, our daughter **Rashmi** and son **Kiran**, as without their patience and understanding this work would not have seen the light of the day.

S.N. Sapali

Chapter 1

Introduction

1.1 INTRODUCTION

Refrigeration is a process which cools a closed space by removing heat from it. In other words, refrigeration is a phenomenon by virtue of which the temperature of a body is reduced compared with its surroundings. Likewise, the term refrigeration can be used in many ways.

For an in-depth understanding of the principles of refrigeration, it is important at this stage to refresh our memory about the SI units of fundamental quantities such as density, specific volume, force, pressure, energy, power, and so on. It is also equally important to study the basic laws of thermodynamics. Therefore, these topics are briefly discussed in the following text.

All things in nature are classified into solids, liquids and gases. The measurement of these quantities is based on their mass and density.

Mass: The mass of a body is expressed in kilograms (kg), and its volume in cubic metres (m^3). The other measurements of fluid volume are litre (L) or millilitre (mL). One litre is 1/1000th part of 1 m^3 and 1 mL is equal to 1 cm^3.

Density (ρ): Density is defined as mass per unit volume (kg/m^3 or kg/L) and specific volume (v) is defined as volume per unit mass (m^3/kg or L/kg), i.e.

$$\text{Density} = \frac{m}{V} \qquad (1.1)$$

$$\text{Specific volume} = \frac{V}{m} \qquad (1.2)$$

The density and specific volume of a substance are not constant but vary with temperature of the substance. The density of water is taken as 1000 kg/m^3 for most of the calculations for the temperature ranges of refrigeration, which gives fairly accurate results. For higher accuracy, its value should be taken from standard property tables.

Specific gravity: The specific gravity of a substance is the ratio of the density of the substance (ρ) to that of a standard substance. In the case of liquids, the standard substance is water at its maximum density (1000 kg/m³). If ρ_w is the density of water, then the specific gravity (ρ_r) of any substance is given by

$$\rho_r = \frac{\rho}{\rho_w} \tag{1.3}$$

Since specific gravity is a ratio, it has no dimensions.

Mass and volume flow rates: The mass flow rate is measured in kg/h or kg/min. Similarly, the volume flow rate is measured in L/s, m³/s, m³/min or m³/h.

Velocity: A body moving in a fixed straight line direction without changing its speed is said to have a constant velocity. It is a vector quantity and the units of velocity should properly indicate direction as well as magnitude. It is measured in m/s or km/h.

Acceleration: The rate of change of velocity is the acceleration which is measured in m/s². The bodies in motion always experience a change in velocity and so they experience an acceleration. Acceleration may be either positive or negative, depending on whether the velocity is increasing or decreasing. The simplest form of accelerated motion is a uniformly accelerated motion where the motion occurs in a single direction and the speed changes at a constant rate.

Acceleration due to gravity: A falling body experiences a uniformly accelerated motion. The action of gravity accelerates a falling body at the rate of 9.807 m/s. This value is known as *standard acceleration of gravity* (g) at sea level.

1.2 SI UNITS

The SI or the International System of Units is the simplest and an extension and refinement of the metric system. There are six basic SI units, and the units of other thermodynamic quantities are derived from these basic units. The basic SI units are as follows:

Length: The metre (m) is defined as the length of the path travelled by light in vacuum during a time interval of $\frac{1}{299792458}$ of a second.

Mass: The kilogram (kg) mass is equal to the mass of the international prototype of the kilogram. This international prototype is made of platinum–iridium metal and has been kept at the International Bureau of Weights and Measures, Severes, France.

Time: Time second (s) is the duration of 9,192,631,770 periods of the radiation corresponding to the transition between the two hyper line levels of the ground state of the caesium 133 atoms.

Electric current: The electric current is that quantity of current which, if maintained in two straight parallel conductors placed 1 metre apart in vacuum, would produce between these conductors a force equal to 2×10^{-7} newton per metre of length.

Temperature: The temperature in kelvin (K) is $\dfrac{1}{273.6}$ times the thermodynamic temperature of the triple point of water.

Luminous intensity: Luminous intensity in candela (cd) in a given direction of source is that which emits monochromatic radiation frequency of 540×10^{12} Hz and has a radiant intensity in that direction of $\dfrac{1}{683}$ watts per sterdian.

Force: The unit of force is newton (N). One newton force is that force which, when applied to a body having a mass of one kilogram, gives it an acceleration of one metre per second per second (1 m/s^2). Mathematically, force can be expressed as

$$F = (C)(m)(a) \tag{1.4}$$

where F = force in newtons (N), m = mass in kilograms (kg), a = acceleration in metre per second per second (m/s^2) and C = proportionality constant. The SI unit of force, newton, is derived assuming this constant as unity, i.e.

$$1 \text{ N} = (1 \text{ kg}) \times \left(\dfrac{1 \text{ m}}{\text{s}^2}\right) = 1\, \dfrac{\text{kg-m}}{\text{s}^2}$$

The cgs unit of force is dyne, which is the force exerted on 1 gram mass for 1 cm/s^2 acceleration, i.e.

$$1 \text{ dyne} = 1\, \dfrac{\text{g-cm}}{\text{s}^2}$$

or

$$1 \text{ N} = 10^5\, \dfrac{\text{g-cm}}{\text{s}^2} = 10^5 \text{ dyne}$$

Pressure: Pressure is the force exerted per unit area. Thus

$$p = \dfrac{F}{a} = \dfrac{\text{Force in newton (N)}}{\text{Area in square metre (m}^2\text{)}}$$

Fluid pressure: It is the rate of change of momentum of the fluid particles per unit time per unit area.

EXAMPLE 1.1 A rectangular tank with base 3 m × 4 m is filled with 20,000 kg of water. Determine (a) the gravitational force acting on the base of the tank and (b) the pressure exerted on the base of the tank. Also write the units.

Solution: (a) The gravitational force, $F = m \times g$

$$= 20{,}000 \times 9.80665$$
$$= 196{,}133 \text{ N} \qquad \textbf{Ans.}$$

(b) The area of the base of the tank is 12 m^2.

∴ $\qquad$ Pressure $= \dfrac{F}{a} = \dfrac{196{,}133}{12} = 16{,}344.4 \text{ N/m}^2$ or Pa $\qquad$ **Ans.**

Another common unit of pressure is the bar. One bar is equal to 100,000 Pa (or 10^5 Pa). Pressure is also measured in terms of a column of fluid (like mm of mercury column or mm of water column).

Atmospheric pressure: The earth is surrounded by an envelope of atmosphere or air that extends upwards from the earth's surface to a height of 80 km or more. Since air has mass, subject to the action of gravity, it exerts pressure known as *atmospheric pressure*.

Suppose there is an air column with its cross section 1 m², starting from the surface of the earth at sea level and extending to the upper limit of atmosphere. Such an air column is supposed to have a mass such that the gravitational force exerted at sea level would be 101,325 N. This is the force exerted on 1 m², therefore the pressure exerted by the atmosphere at sea level is 101,325 N/m². This atmospheric pressure is considered as standard and is used in all the calculations.

The atmospheric pressure changes somewhat with temperature, humidity and altitude.

It can be seen that one standard atmosphere is given by

$$1 \text{ atm} = 1.01325 \text{ bar} = 1.033 \text{ kgf/cm}^2$$
$$= 760 \text{ mm Hg} = 10.336 \text{ metre of water column}$$

The pressure of any fluid or system measured with the help of instruments such as Bourdon's gauge or diaphragm gauge is known as *gauge pressure*. But thermodynamic investigations are concerned with absolute pressure. The relation between the gauge pressure and the absolute pressure is given by the equation,

$$p_{abs} = p_{gauge} + p_{atm} \tag{1.5}$$

This relationship is shown schematically in Figure 1.1.

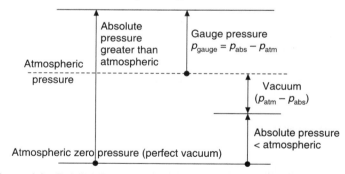

Figure 1.1 Relation between absolute, atmospheric and gauge pressure.

Energy and work: Energy is an idea, central to the development of all branches of science and engineering. A fundamental postulate of thermodynamics is that matter possesses energy (which can be in several forms) and energy is conserved.

Energy: Energy is nothing but the ability to do work. It is always said that energy exists in transition only. The unit of work or energy is obtained from the product of force and distance. The SI unit of work or energy is newton-metre (N-m) or joule (J).

$$1 \text{ N-m} = 1 \text{ J} = 10^7 \text{ erg} = 10^7 \text{ dyne-cm}$$

The conversion of MKS unit of energy to SI unit is shown below:

$$1 \text{ kcal} = 4186.8 \text{ N-m} = 4.1868 \text{ kJ} = 3.968 \text{ Btu}$$

or

$$1 \text{ kJ} = 0.239 \text{ kcal}$$
$$= 0.948 \text{ Btu}$$

Power: Power is the rate of doing work. The unit of power is watt (W). One watt power is defined as doing work at the rate of one joule per second (J/s). In electrical engineering, one watt power is defined as:

$$1 \text{ W} = 1 \text{ (volt)} \times 1 \text{ (ampere)} = 1 \text{ J/s}$$

The relationship between horsepower and watt is shown as follows:

$$1 \text{ hp (imperical)} = 746 \text{ N-m/s or J/s or W}$$

$$1 \text{ hp (metric)} = 75 \frac{\text{kg/m}}{\text{s}} = 75 \times 9.80665 \text{ N-m/s}$$

$$= 736 \text{ N-m/s or W}$$

Further, the units of energy can be obtained from that of power. Thus,

$$1 \text{ J} = 1 \text{ W-s}$$
$$1 \text{ kWh} = 1000 \times 3600 \text{ J} = 3600 \text{ kJ}$$

Enthalpy: Enthalpy is nothing but the sum of internal energy and the product of pressure and volume. It is denoted by H and measured in joules. Mathematically,

$$H = U + pV \tag{1.6}$$

where
U = internal energy (J)
p = pressure (N/m^2)
V = volume (m^3).

The specific enthalpy is denoted by h and measured in J/kg.

Entropy: Entropy is difficult to define and can be said to be the property of the system such that for any reversible process between the state points 1 and 2, its change is given by

$$\delta s = \int_1^2 \left(\frac{\delta Q}{T}\right)_{rev} \tag{1.7}$$

The SI unit of specific entropy is kJ/kg-K. It can also be expressed as 1 kcal/kg-°C = 4.186 kJ/kg-K.

Specific heat: It is the amount of heat required to raise the temperature of a substance by one degree. The specific heat may be at constant volume (c_v) or it may be at constant pressure (c_p). The SI unit of specific heat is kJ/kg-K.

Refrigerating effect: The refrigerating effect or capacity is measured in terms of ton refrigeration or simply ton denoted by the symbol TR. One TR is equivalent to the production of cold at the rate at which heat is to be removed from one US tonne of water at 0°C to freeze it to ice at 0°C in 24 hours.

$$1 \text{ TR} = \frac{(1 \times 2000 \text{ lb}) \times 144 \text{ Btu/lb}}{24 \text{ h}} = 12{,}000 \text{ Btu/h}$$

Here the latent heat of fusion of ice has been taken as 144 Btu/lb. In general, 1 TR is always equal to 12,000 Btu of heat removal per hour irrespective of the working substance used and its temperature.

Also,

$$1 \text{ TR} = 50.4 \text{ kcal/min} = 211 \text{ kJ/min}$$
$$= 3.5167 \text{ kW} = 12{,}660 \text{ kJ/h}$$

The refrigerating or cooling effect is produced by the refrigerating machine. The conventional refrigerating machine or system working on vapour compression cycle mainly consists of four components, namely evaporator, compressor, condenser and expansion device.

The evaporator is practically a heat exchanger placed in a confined space where cold is generated. To produce cold, power in the form of electrical energy is supplied to the compressor unit. So the compressor is one which produces the refrigerating effect. The performance of the refrigerating unit is evaluated with the help of a term called the *coefficient of performance* (COP). It is defined as

$$\text{COP} = \frac{\text{Refrigerating effect (watts)}}{\text{Work input to compressor (watts)}}$$

Small capacity refrigerating units are normally fitted with a hermetic compressor. A hermetic compressor is one where the assembly of the compressor and motor is kept in a sealed vessel. It means that the work input to the compressor has to include the motor efficiency as well. For such a unit the performance is evaluated with the *energy efficiency ratio*.

Energy Efficiency Ratio (EER)

Energy efficiency ratio is the ratio of refrigeration effect produced in kJ/h to the power input to the motor of a compressor.

$$\text{EER} = \frac{\text{Refrigerating effect (kJ/h)}}{\text{Work input (kW)}}$$

The EER takes into account the combined operating efficiency of the motor and the compressor.

Normally, this term is used by the manufacturers of hermetic compressors while evaluating the performance of the compressor through experiment in a calorimeter test set-up.

The performance of a refrigerating or air conditioning unit varies as per the climatic conditions, i.e. seasons. The condenser has to dissipate heat to the atmospheric air in the case of an air-cooled condenser. In summer the ambient air temperature is higher so the compressor consumes more power, hence lowering the performance of the refrigerator while in winter the process reverses.

Seasonal Energy Efficiency Ratio (SEER)

The anticipated performance of a refrigerating unit over an average season is called the seasonal energy efficiency ratio.

1.3 TERMS USED IN THERMODYNAMICS

System: A thermodynamic system or simply a system is defined as a quantity of matter or a region chosen for the study. The region outside the system is called *surroundings*. The real or imaginary surface that separates the system from its surroundings is termed *boundary*.

The various systems that are studied in thermodynamics include closed system, open system, isolated system and adiabatic system.

Property: Any characteristic of a system is known as its property. These properties are measurable followed by units of measurements. A few familiar properties are pressure, temperature, volume, enthalpy, etc. Thermodynamic properties are classified into two groups: (a) intensive and (b) extensive.

Intensive properties are those which are independent of the mass of a system. For example, temperature, pressure and density.

Extensive properties vary directly with the mass (or size) of a system. For example, mass, volume and total energy.

State: It is the condition of a system. This condition of the system is described with the help of thermodynamic properties such as pressure, temperature and volume.

The state of a pure substance is defined by any two independent properties. In thermodynamics, one assumes the equilibrium state (i.e. no unbalanced potentials within the system).

Thermodynamic process: Any change that a system undergoes from one equilibrium state to another state is because of a process. The series of states through which a system passes during a process is called the *path* of the process as shown in Figure 1.2.

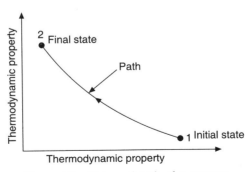

Figure 1.2 State and path of a process.

To understand a process thoroughly, it is necessary to specify the initial and final states of the process, the path followed and energy interactions with the surroundings.

Thermodynamics processes are either reversible or irreversible. In a reversible process, if the process is reversed, then the system follows the exact path of the initial process and also returns both the system and the surroundings to their initial states. All real processes are irreversible.

There are a number of factors, both internal and external, that cause irreversibility in the fluid processed. Internal irreversibility is due to internal fluid friction resulting from intermolecular forces and turbulence in the fluid.

For example, assume a high pressure gas confined behind a piston in a cylinder. If the piston is moved rapidly, a portion of the gas adjacent to the piston immediately expands into the space created by the receding piston, while another portion of the gas tends to remain at the rear. This causes pressure and temperature differentials in the fluid, with the resulting turbulence and fluid friction accounting for some of the energy that otherwise would be delivered as useful work. For a process to be internally reversible, it must employ an ideal fluid (no intermolecular forces of attraction in the fluid) and the process should be very slow. All the ideal processes discussed here are treated to be internally reversible.

External irreversibility is mainly on account of (a) mechanical friction encountered at rubbing surfaces such as bearings and cylinder walls, and (b) another reason is heat transfer which by its very nature can occur in only one direction, from higher temperature to lower temperature. The *frictionless adiabatic* (isotropic) process is an externally reversible process and it has particular significance in the analysis of a vapour compression refrigeration cycle.

In fact all the processes encountered in thermodynamics used for the analysis of different thermodynamic cycles are assumed to be ideally reversible.

Cyclic process or cycle: A closed system is said to undergo a cyclic process or cycle, when it passes through a series of states in such a way that its final state is equal in all respect to its initial state. This implies that all its properties have regained their initial values. The system is then in a position to be put through the same cycle of events again, and the procedure may be repeated indefinitely. Work can be transferred to or from the system continuously by devising a machine which undergoes a cyclic process.

For example, in a refrigerating machine, the working fluid (refrigerant) undergoes different processes such as isentropic compression (in compressor), condensation (in condenser), expansion (in expansion valve or capillary) and evaporation (to extract heat in the evaporator), and the cycle repeats. If we consider this whole machine as a system it cools the evaporator at the cost of external work supplied to the compressor. Such a system will be dealt with later in more detail in this book.

1.4 TYPES OF ENERGY

We have already defined the term *energy* as the ability to do work. Now we shall discuss the types of energy. Broadly, energy can be classified into two groups, namely (a) stored energy which is contained within the system boundaries, e.g. potential energy, kinetic energy and internal energy and (b) energy which crosses the boundary (energy in transition, e.g. heat and work).

Potential energy (PE): This is the stored energy in a system. This energy topic can be studied in two ways—(a) microscopic way and (b) macroscopic way.

The *internal potential energy* is the energy of molecular separation or configuration. It is the energy that molecules have as a result of their positions in relation to one another. The potential energy of a system at a molecular level is better explained in statistical thermodynamics. The energy of a system at macroscopic level is discussed in the following text.

The energy stored in the system, as a whole, by virtue of its elevation with reference to an arbitrary chosen datum level is known as potential energy.

Consider a system of mass m (kg) at height z (m) from a certain datum level. Then the work done in bringing this mass m to the datum level is given by

$$\text{PE or } W = m \cdot g \cdot z \text{ (J)} \tag{1.8}$$

where g = gravitational acceleration, 9.80685 m/s²

EXAMPLE 1.2 A tank located at a height of 500 metres contains 250 cubic metres of water. Determine the gravitational potential energy with reference to the ground.

Solution: Assuming the density of water to be 1000 kg/m³, the total mass of water is 250,000 kg.
Applying Eq. (1.8),

$$\text{PE} = (250{,}000) \times (9.80665) \times 500$$
$$= 1.22583 \times 10^9 \text{ J}$$
$$= 1.22583 \times 10^6 \text{ kJ} \qquad \textbf{Ans.}$$

Kinetic energy (KE): This can also be dealt with in two ways—at microscopic and macroscopic levels. The energy of molecular motion or velocity is called *internal energy*. When energy is supplied to a substance, it increases the motion or velocity of the molecules, hence the internal KE of the substance is increased, and this increase is reflected by an increase in the temperature of the substance. Conversely, if the internal KE of the substance is diminished by the loss of energy, the motion of the molecules will decrease and the temperature will decrease accordingly. This KE at microscopic level is better dealt with in statistical thermodynamics. The KE of the system at macroscopic level is discussed in the following text.

Kinetic energy is the energy that a body (system) possesses by virtue of its motion or velocity. For example, a flowing fluid, a falling body and the moving parts of a piece of machinery all have kinetic energy because of their motion. The amount of kinetic energy a body possesses is dependent on its mass m and its velocity v as shown in Eq. (1.9).

$$\text{KE} = \frac{mv^2}{2} \tag{1.9}$$

where
$\quad$ KE = kinetic energy (J)
$\quad m$ = mass (kg)
$\quad v$ = velocity (m/s).

EXAMPLE 1.3 A car having a mass of 1565 kg is moving with a velocity of 60 km/h. What is the kinetic energy?

Solution: The velocity of the car is 60 km/h, i.e. 16.67 m/s. Applying Eq. (1.9),

$$\text{KE} = \frac{(1565)(16.67)^2}{2} = 217361 \text{ J} = 217.361 \text{ kJ} \qquad \textbf{Ans.}$$

Internal energy (U): The molecules of any system may possess potential energy (PE), kinetic energy (KE) and nuclear energy (NE), etc. The total internal energy of a system (U) is the sum of its internal kinetic and potential energy. This relationship is shown by Eq. (1.10), i.e.

$$U = KE + PE + NE \tag{1.10}$$

1.5 WORK

In mechanics, *work* (W) is defined as the product of force (F) and distance (s), while the direction of application of force on the body is in the direction of motion. This is shown in Figure 1.3.

$$W = F \times s \tag{1.11}$$

where force F is in newtons (N), distance s is in metres (m) and work W in joules (J).

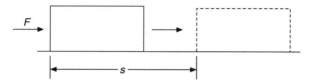

Figure 1.3 Work as in mechanics.

Work is considered as one of the basic modes of energy transfer in thermodynamics and work transfer occurs between the system and its surroundings. Work is said to be done by the system if the sole effect external to the system can be reduced to the raising of weight. Thus in thermodynamics,

  (a) Work is either done on a system or it is done by the system.
  (b) The weight may not be raised actually, but the net effect of work can be converted to raise the weight.

Suppose a battery drives a motor as shown in Figure 1.4. Here the motor drives a fan. If we limit the system boundary for the battery and motor as shown, then work is done by the system (battery and motor) on the surroundings (fan). It means work crosses the boundary.

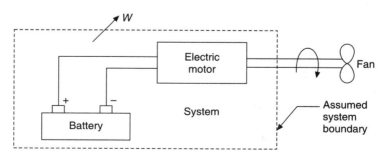

Figure 1.4 Battery-motor system driving a fan.

Now, replace the fan with a pulley-and-weight arrangement as represented in Figure 1.5. The weight will be raised with the help of the pulley and motor arrangement. Thus, the sole effect external to the system is to raise the load.

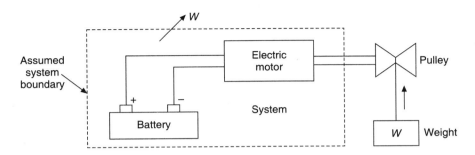

Figure 1.5 Work transfer from a system.

Comments on work

(i) Work is nothing but energy in transition. It appears only when it crosses the boundary.

(ii) The amount of work performed by a system on the surroundings and vice versa is dependent on the path it follows. Therefore, it is a path function and not a property of the system.

(iii) The work is an inexact differential, i.e. $\int \delta W \neq W_2 - W_1$.

1.6 HEAT

Heat is denoted by Q or q and is measured in kJ. Heat is something which appears at the boundary when a system changes its state due to a difference in temperature between the system and the surroundings. Here 'something' is a form of energy transfer and this energy transfer is due to temperature difference. Heat, like work, is a transient quantity, which only appears at the boundary when a change is taking place within the system. It is apparent that neither δQ nor δW is an exact differential, and therefore any integration of the elemental quantities of work or heat which appear during a change from state 1 to state 2 must be written as in Eq. (1.12).

$$\int_1^2 \delta W = W_{12} \text{ or } W \quad \text{and} \quad \int_1^2 \delta q = q_{12} = q \quad (1.12)$$

In thermodynamics, loosely speaking, heat is considered to flow across the boundary. Strictly speaking, it is the energy which is transferred. Heat is therefore defined as a form of energy that is transferred across a boundary by virtue of temperature difference.

Heat is an interaction, which may occur between two systems (at different temperatures), when they are brought into communication. The concept of heat is related with the temperature difference between two systems or between a system and surroundings. Heat is not stored in the system; it is the energy in transit. It is not the property of the system, rather it is a path function. It is represented by an inexact differential, i.e.

$$\int_1^2 \delta q = q_{12} \text{ or } q$$

Comments on heat

(i) Heat does not inevitably cause temperature rise. For example, boiling of water at 100°C to convert into steam.

(ii) Heat is not always present when a temperature rise occurs. For example, compression of gas in an adiabatic insulated cylinder.

According to Rutherford

(i) Heat is not a conserved fluid that can be transferred from one body to another. Heat exists in transition phase only.

(ii) Heat should not be confused with temperature.

Difference between heat and work

If a system is in a stable equilibrium state, then no work interaction between the system and its surroundings can take place, whereas there is no such restriction for heat interaction. Consider, for example, a gas contained in a rigid container at high pressure and temperature. The rigidity of the container provides an upper limit to the volume of the system. In this case, no work interaction will occur. But due to temperature difference between the system and the surroundings, heat interaction would take place.

Secondly, for heat interaction between the system and its surroundings, the temperature potential difference should exist between them, but no temperature difference is required for work interaction.

After studying the concept of work and heat, it is also important to understand the other forms of work.

1.7 WORK DONE DURING A QUASI-STATIC PROCESS

There are numerous ways to obtain work from a system such as by rotating shafts, electrical work, the displacement of the piston in a cylinder–piston arrangement.

In this section, a piston–cylinder arrangement is considered and the work done at the moving piston (boundary) during a quasi-static process is assumed. *Quasi-static process* is one which is infinitely slow. Such a process passes through a number of equilibrium states and is therefore so slow that at any instant it will be in an equilibrium state. Such a process is also called *reversible process*.

A technically important phenomenon in many engineering processes is the one in which work is obtained when a system expands through the piston–cylinder arrangement and work is performed on the system during a compression process.

In a quasi-static process from state 1 to state 2, assume that the piston moves a small distance dx from left to right. This process is assumed as quasi-static.

$$\text{Total force acting on the piston, } F = pA \tag{1.13}$$

where p = pressure of the system (N/m²), A = cross-sectional area of piston (m²).

The small work done on the piston

$$\delta W = pA\delta x \tag{1.14}$$

or

$$\delta W = p\delta V \tag{1.15}$$

where δV = change in volume (m³).

The net work done by the system in moving the piston from state 1 to state 2 can be worked out by integrating Eq. (1.15). But this is only possible if the relationship between p and V is known for the process. The relationship between p and V can be known from the graph shown in Figure 1.6.

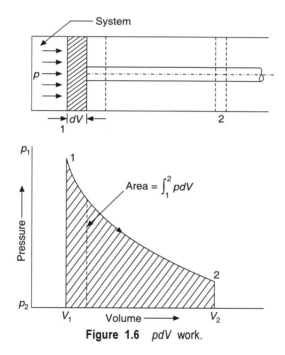

Figure 1.6 pdV work.

The work done W_{1-2} is

$$W_{1-2} = \int_1^2 \delta W$$

$$= \int_1^2 p\delta V \tag{1.16}$$

This integration indicates that it is an area under the curve. If the process is carried out from state 2 to state 1, it is a compression process. The piston will do the work which is considered negative, and the work done by the system is positive. It is also important to note that the system considered is a closed non-flow process. The $\int p\delta V$ is the work done during a reversible process for only a non-flow system.

Flow work

In the preceding text, we saw the mechanical work, the $\int p\delta V$ work in a thermodynamic reversible process. Now we shall see another important form of work, i.e. flow work.

Flow work is invariably associated with the maintenance of flow of fluid through a channel or conduit. This concept of flow work is explained with the help of Figure 1.7.

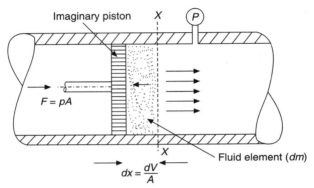

Figure 1.7 Flow work.

Consider a fluid flowing through a tube (cross-sectional area) with a uniform velocity. Let us assume an imaginary line X–X. Some work is required to push the mass past this imaginary line from left to right. This work is known as flow work or flow energy. This flow work is necessary for maintaining a continuous flow through this tube. Let δm be a small quantity of mass with its volume δV on the left side of line X–X.

The specific volume of the fluid is

$$v = \frac{dV}{dm} \tag{1.17}$$

Let p be the absolute pressure at section X–X. The fluid immediately upstream (left of dm) will act like a piston and will force the fluid element (δm) to cross section X–X. In order to push the fluid mass δm across the section X–X, a force pA must act through a distance δx.

Therefore,
$$dx = \frac{dV}{A} \tag{1.18}$$

And the flow work per unit mass $= (pA)\dfrac{dx}{dm}$

$$= (pA)\dfrac{dV/A}{dm}$$

$$= p\dfrac{dV}{dm} = pv \qquad (1.19)$$

where p is the pressure of fluid (N/m²) and v is the specific volume (m³/kg).

1.8 FIRST LAW OF THERMODYNAMICS

Heat and work, the two forms of energy, are related by the first law of thermodynamics. It is a law of conservation of energy which states that energy can neither be created nor be destroyed. This law cannot be proved mathematically, but no exception has been observed.

Scientist Joule had shown that Q_{1-2} is proportional to W_{1-2} ($Q_{1-2} \propto W_{1-2}$) for a system undergoing a cyclic process. This constant of proportionality is known as *Joule's equivalent* or the *mechanical equivalent of heat*.

If the cycle shown in Figure 1.8 involves many more heat and work transfers, the same conclusion will be found and when expressed mathematically, we have

$$(\Sigma W)_{\text{cycle}} = J(\Sigma Q)_{\text{cycle}}$$

As $J = 1$,

$$\oint dW = \oint dQ \qquad (1.20)$$

where the symbol $\oint$ denotes the cyclic integral for the closed path. This is the first law applied to a closed system undergoing a cyclic process.

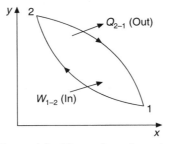

Figure 1.8 Thermodynamic cycle.

Other statements of first law of thermodynamics

It is said that energy can neither be created nor be destroyed. This implies that the sum of the energies of a system at microscopic and macroscopic levels is fixed, unless there is an interaction with the surroundings involving an energy exchange.

This can be simply stated as: *The total energy of an isolated system, measured with respect to any given frame of reference, remains constant.* Mathematically, for an isolated system,

$$E(\text{total}) = U + KE + PE + \text{Chemical energy} + \cdots = \text{constant}$$

The expression $(\Sigma W)_{\text{cycle}} = (\Sigma Q)_{\text{cycle}}$ applies only to a system undergoing a cyclic process. But in practice, a system may undergo a non-cyclic process which produces a change in state in the system.

Let us consider a system interacting with its surroundings, which involves work and heat transfer (Figure 1.9).

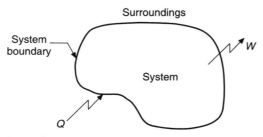

Figure 1.9 System interacting with the surroundings which involves work and heat transfer.

If Q is the amount of heat transferred to the system and W is the work obtained from it, then $(Q - W)$ is the energy stored in the system. This stored energy in the system is not the heat or work but is referred as internal energy or simply *energy of the system*, i.e.

$$\Delta U = Q - W \tag{1.21}$$

where ΔU is the increase in internal energy of the system.

If more energy transfers are involved in the process, as shown in Figure 1.10, the application of first law gives

$$(Q_2 + Q_3 - Q_1) = \Delta U + (W_1 + W_3 - W_2) \tag{1.22}$$

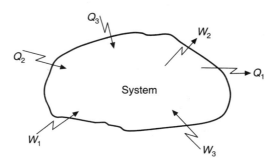

Figure 1.10 System interacting with the surroundings involving more energy exchanges.

1.9 PROCESSES FOR IDEAL GASES

A gas is said to undergo a process when it passes from one condition (initial) to another (final) condition. This change in the condition of an ideal gas may occur in a number of ways. Only the following non-flow processes are explained here. A non-flow process is the one in which the gas (mass) of the system does not cross the boundary. The non-flow processes are:

(a) Constant volume or isochoric process ($V = C$)
(b) Constant pressure or isobaric process ($p = C$)
(c) Constant temperature or isothermal process ($T = C$)
(d) Reversible adiabatic or isentropic process ($s = C$)
(e) Polytropic process.

1.9.1 Constant Volume Process

This is illustrated in Figure 1.11(a) and (b). A gas is heated in a rigid container so that its volume remains constant.

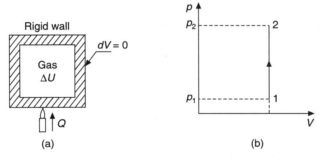

Figure 1.11 (a) Heating of gas in a closed vessel and (b) p–V diagram for a constant volume process.

Work done
$$dW = p \cdot dV = 0 \quad \text{(since } dV = 0\text{)}$$

Heat supplied
$$dQ = mc_v dT$$

From the first law of thermodynamics,
$$dQ - dW = dU$$

Therefore,
$$mc_v dT - 0 = dU$$

For unit mass
$$dU = c_v \, dT$$

$\therefore$
$$c_v = \left(\frac{dU}{dT}\right)_v \tag{1.23}$$

Since
$$\int_1^2 dU = \int_1^2 mc_v \cdot dT$$

∴ $\quad U_2 - U_1 = mc_v(T_2 - T_1)$ (1.24)

1.9.2 Constant Pressure Process

Consider a gas in the cylinder with a frictionless piston arrangement represented in Figure 1.12. When this gas is heated, its temperature increases and the piston displaces due to the expansion of gas in such away that the pressure of the gas remains constant. Therefore, the volume of the gas increases in accordance with Charle's law,

$$T_1 V_2 = T_2 V_1 \quad (1.25)$$

where T_1 and T_2 are the initial and final absolute temperatures and V_1 and V_2 are the initial and final volumes of the gas respectively. Since the volume of the gas increases during the process, work is done by the gas on surroundings and also at the same time its internal energy increases. For this process, the energy equation can be written as

$$\Delta Q = \Delta U + \Delta W \quad (1.26)$$

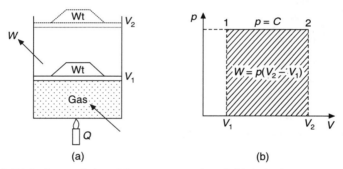

Figure 1.12 (a) Frictionless cylinder–piston arrangement and (b) isobaric process on p–V diagram.

The equation for work done will be

$$W_{1-2} = \int_1^2 p\,dV = p(V_2 - V_1) \quad (1.27)$$

= Area under the curve on p–V plot

And the internal energy equation becomes

$$\Delta U_{1-2} = \Delta Q_{1-2} - \Delta W_{1-2} \quad (1.28)$$

1.9.3 Constant Temperature Process

According to Boyle's law, when a gas is compressed or expanded at a constant temperature, the pressure varies inversely with volume. It means that the pressure increases as the gas is compressed and decreases as the gas is expanded. When the gas expands, it performs work. If the temperature of the gas is to remain constant, then the external energy supplied to it must be equivalent to the amount of work being obtained during the process.

During the compression process, work is performed on the gas, and if the gas is not cooled during compression, the internal energy of the gas increases by an amount equal to the work of the compression process. Therefore, if the temperature of the gas is to be maintained constant during the compression process, then the gas must reject heat to the surroundings by an amount equal to the amount of work done on it. As the temperature remains constant, there is no change in the internal energy of the gas, i.e. $\Delta U = 0$. Therefore, the energy equation reduces to

$$\Delta Q_{1-2} = \Delta W_{1-2} \tag{1.29}$$

A p–V diagram of an isothermal process is shown in Figure 1.13. In this process, both pressure and volume change according to Boyle's law. The path followed by an isothermal process is shown by the curve 1–2 and the work by the area under the curve.

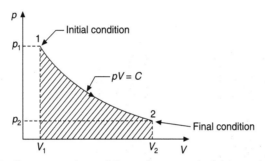

Figure 1.13 Pressure–volume diagram for a constant temperature process.

Therefore,
$$W_{1-2} = \int_1^2 p\,dV$$

$$= \int_1^2 \frac{C}{V}\,dV \quad (\because pV = C)$$

$$= p_1 V_1 \ln \frac{V_2}{V_1} \tag{1.30}$$

$$= p_1 V_1 \ln \frac{p_1}{p_2} \tag{1.31}$$

$$= mRT_1 \ln \frac{p_1}{p_2} \quad (\because pV = mRT) \tag{1.32}$$

As $\Delta U = 0$ for the isothermal process, the heat transferred or supplied is

$$Q_{1-2} = p_1 V_1 \ln \frac{p_1}{p_2} \tag{1.33}$$

1.9.4 Adiabatic Process

The properties, pressure, temperature and volume of the system will vary during the process and none will remain constant. In an adiabatic process, the system (gas) changes its state without transfer of heat to or from the surroundings. During an adiabatic expansion process, the gas does external work and energy is required to do this work. But no transfer (exchange) of heat to or from the surroundings occurs. The gas must perform external work at the cost of its stored energy. The adiabatic expansion process is always accompanied by decrease in gas temperature since external work is done at the expense of the stored (internal) energy.

Conversely, when a gas is compressed adiabatically, work is performed on the gas. Therefore, the internal energy of the gas increases followed by an increase in its temperature as no heat is transferred to the surroundings. The adiabatic process is represented mathematically as

$$pV^\gamma = \text{constant} \tag{1.34}$$

where p = pressure of gas in N/m², V = volume in m³ and γ = ratio of specific heats.

Since no heat, as such, is transferred to or from the gas during an adiabatic process, ΔQ is always zero and the energy equation for an adiabatic process becomes:

$$\Delta U + \Delta W = 0 \tag{1.35}$$

Therefore, $\Delta W = -\Delta U$

The work done for the reversible adiabatic expansion process is represented by the area under the curve of a p–V diagram as shown in Figure 1.14.

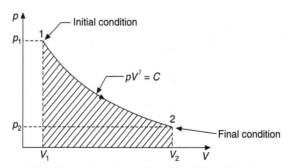

Figure 1.14 Pressure–volume diagram of an adiabatic process.

The work done of an adiabatic process may be calculated as:

$$W_{1-2} = \frac{p_1 V_1 - p_2 V_2}{\gamma - 1} \tag{1.36}$$

The change in internal energy for the adiabatic expansion process can be calculated by simply putting negative sign to Eq. (1.36).

After having discussed the concept of isothermal process in Section 1.9.3, let us at this stage carry out a comparison of isothermal and adiabatic processes.

In an adiabatic expansion process, there is no exchange of heat between the gas and the surroundings and all the work of expansion is done at the expense of internal energy of the gas. Therefore, the internal energy of the gas decreases at the same rate as that of doing external work and the temperature of the gas decreases accordingly. But in an isothermal expansion process, the temperature of the gas remains constant, and hence the internal energy of the gas remains constant too. Therefore, all the energy to do work is supplied to the gas from an external source. This external energy supplied during the process produces work.

Now a question arises as to by which expansion process do we get more work from the gas. Of course, the answer is isothermal process. This is illustrated graphically on p–V plots in Figure 1.15.

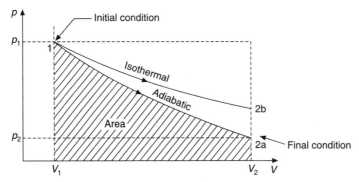

Figure 1.15 Pressure–volume diagrams of adiabatic and isothermal expansion processes.

Now consider the isothermal and adiabatic compression processes. In any compression process, work is done on the gas by a piston and a quantity of energy equal to the quantity of work done on the gas is transferred to the gas. In an isothermal compression, energy is transferred to the surroundings by an amount equal to the amount of work done on the gas. Therefore, the internal energy of the gas remains constant and the temperature also remains at the same level. On the other hand, in adiabatic compression, there is no transfer of energy as heat from the gas to the surroundings. Therefore, an amount of energy equal to the amount of work being done on the gas during compression remains with the system in the form of internal energy. Hence the temperature of the gas increases. Also, the work required for the compression of gas adiabatically will be more compared to the isothermal compression for the same pressure limits.

1.9.5 Polytropic Process

A polytropic process can be defined by the comparison of isothermal and adiabatic processes. An isothermal expansion process is the process in which the energy to do work is supplied from an external source, whereas in an adiabatic process, this energy is supplied entirely from the gas itself. Therefore, these two processes are of extreme limits while all other expansion processes fall in between these two limits. Therefore, a polytropic process is the one in which the energy to do the work of expansion is supplied partly from an external source and partly from the gas itself. Therefore, the polytropic process is a process that can be represented by the equation

$$pV^n = \text{constant} \quad (1.37)$$

where n is called the *polytropic index* whose value lies between 1 and 1.4. For $n = 1$, $pV = C$. It means that the process becomes identical to isothermal and the energy to do the work is supplied entirely by the surroundings. When $n = 1.4$, the process becomes adiabatic, indicating that the energy to do work comes from the gas itself. The work done in adiabatic, polytropic and isothermal processes can be compared by representing these expansion processes as respective p-V plots as shown in Figure 1.16.

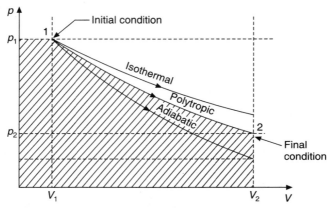

Figure 1.16 Pressure–volume diagrams of polytropic, adiabatic, and isothermal expansion processes.

During a polytropic compression process, the gas loses heat not at a rate which is sufficient to maintain the temperature constant. If the rate of loss of heat is greater and greater, the compression process approaches the isothermal stage. On the other hand, if the rate of loss of heat is smaller and smaller, the compression process approaches the adiabatic stage.

Let us consider the practical situation of compression of gas in the compressor. Such a practical process is very close to the adiabatic compression. This is on account of very short time available for the compression process which is based on the speed of the compressor, i.e. the number of cycles it completes per second. Due to this very short period, hardly any heat is transferred to the surroundings. If the compressor cylinders are cooled by water (by providing jackets), then the compression process approaches the isothermal.

Work done in a polytropic process

$$\text{Isotropic work, } W = \frac{mR(T_2 - T_1)}{1 - n} = \frac{p_2 V_2 - p_1 V_1}{1 - n} \quad (1.38)$$

The following relationships for a polytropic process may also be obtained.

$$T_2 = T_1 \left(\frac{V_1}{V_2}\right)^{n-1} \quad (1.39)$$

$$\frac{T_2}{T_1} = \left(\frac{p_2}{p_1}\right)^{(n-1)/n} \tag{1.40}$$

$$\frac{p_2}{p_1} = \left(\frac{V_1}{V_2}\right)^{n} \tag{1.41}$$

$$p_2 = p_1 \left(\frac{T_2}{T_1}\right)^{n/(n-1)} \tag{1.42}$$

1.10 CONTROL VOLUME

A control volume is a region in space to be observed with respect to the matter and energy which crosses its boundaries. To study an open system, the concept of a control volume as shown in Figure 1.17 is considered.

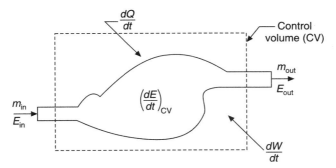

Figure 1.17 Control volume–energy balance.

The conservation of mass principle may be written as:

Mass flow into control volume = mass flow out of control volume + increase in mass inside the control volume

Mathematically,

$$m_{in} = \left(\frac{dm}{dt}\right)_{CV} + m_{out} \tag{1.43}$$

where
 m_{in} = mass flow per unit time into the control volume
 m_{out} = mass flow per unit time out of the control volume and the subscript CV indicates the control volume.

If multiple inlet and exit streams are involved, then we must take the summation of all such flows to determine the mass balance, i.e.

$$\sum_{in} m_{in} = \left(\frac{dm}{dt}\right)_{CV} + \sum_{out} m_{out} \tag{1.44}$$

The mass entering and leaving the control volume may experience a change in pressure and internal energy. Besides, heat transfer and work transfer across the control volume also occur. We therefore consider that the mass flow in and out of the control volume transports internal energy across the boundaries of the control volume. Thus, the energy-conservation principle for such a system is as shown in Figure 1.17 and can be written as:

$$\begin{pmatrix} \text{Transport of total energy into} \\ \text{control volume + heat added} \\ \text{to control volume} \end{pmatrix} = \begin{pmatrix} \text{Increase in total energy of control volume} \\ \text{+ transport of internal energy out of control} \\ \text{volume + work done by all elements as they} \\ \text{pass control volume} \end{pmatrix}$$

Mathematically, it can be written as

$$E_{in} + \frac{dQ}{dt} = \frac{dW}{dt} + \left(\frac{dE}{dt}\right)_{CV} + E_{out} \tag{1.45}$$

Where E_{in} and E_{out} represent the total energy transport per unit time at entrance and exit respectively and the subscript CV represents the change within the control volume.

The total energy, E_{in}, entering the control volume, can include kinetic energy, potential energy and internal energy, and flow work. Therefore,

$$E_{in} = m_{in}\left(u_1 + \frac{V_1^2}{2} + gz_1 + pv_1\right) \tag{1.46}$$

The terms, internal energy (u) and flow work (pv) are combined as specific enthalpy ($h = u + pv$). Therefore, Eq. (1.46) can be written as:

$$E_{in} = m_{in}\left(h_1 + \frac{V_1^2}{2} + gz_1\right) \tag{1.47}$$

Similarly, the total energy leaving the control volume becomes

$$E_{out} = m_{out}\left(h_2 + \frac{V_2^2}{2} + gz_2\right) \tag{1.48}$$

where
p_1, p_2 = absolute pressure (N/m²)
v_1, v_2 = specific volume (m³/kg)
u_1, u_2 = specific internal energy (J/kg)
V_1, V_2 = velocity of fluid (m/s)
z_1, z_2 = elevation above an arbitrary datum (m).

Subscripts 1 and 2 refer to the inlet and outlet sections. For simplification, let us assume steady state conditions. Therefore,

$$\frac{dQ}{dt} = Q \text{ and } \frac{dW}{dt} = W, \text{ and } \left(\frac{dE}{dt}\right)_{CV} = 0$$

and
$$m_{in} = m_{out} = m$$

Hence Eq. (1.45) reduces to:

$$Q + m\left(h_1 + \frac{V_1^2}{2} + gz_1\right) = W + m\left(h_2 + \frac{V_2^2}{2} + gz_2\right) \quad (1.49)$$

or
$$Q - W = m\left[(h_2 - h_1) + \left(\frac{V_2^2 - V_1^2}{2}\right) + g(z_2 - z_1)\right] \quad (1.50)$$

or
$$Q - W = m(\Delta h + \Delta KE + \Delta PE) \quad (1.51)$$

This is the Steady Flow Energy Equation (SFEE).

1.11 LIMITATIONS OF THE FIRST LAW OF THERMODYNAMICS

Before discussing the second law of thermodynamics on which the refrigeration effect is based, we shall discuss here terms like PMM-I, thermal energy reservoir, heat engine, thermal efficiency and also analyse the limitations of the first law of thermodynamics.

PMM-I: PMM-I stands for Perpetual Motion Machine of first kind. The first law states that energy can neither be created nor be destroyed but can be transformed from one form to another. A device which violates this first law, is called PMM-I. A device, which continuously produces work without requiring any energy, is a PMM-I. This is illustrated in Figure 1.18.

Figure 1.18 PMM-I (Such a device is impossible).

The converse of PMM-I is that there can be no machine that would continuously consume work without some other form of energy appearing simultaneously (Figure 1.19).

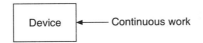

Figure 1.19 Converse of PMM-I (This is also not possible).

The limitations of the first law of thermodynamics are:

(i) The first law of thermodynamics states that energy can be transformed from one form to another, but it does not tell us how much energy can be converted from one form to another. Thus it is a qualitative statement in this sense, but not quantitative.

(ii) Energy of an isolated system remains constant, as stated by the first law. However, it does not give information as to whether a system undergoes a process or not.

Let us consider the following examples. Let a room be heated by an electric resistor (Figure 1.20). The first law of thermodynamics dictates that the amount of electrical energy supplied to the resistance wire will be equal to the amount of energy transferred to the room air as heat. Now, attempt to reverse this process. If the same amount of heat is supplied to the resistance wire, the process will not generate electric energy and still not violate the first law. Again, consider a paddle-wheel mechanism that is operated to cause the fall of a mass (Figure 1.21). As the weight falls, the paddle-wheel rotates, stirring the fluid. Therefore, the fluid

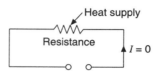

Figure 1.20 Supply of heat to the wire will not generate electricity.

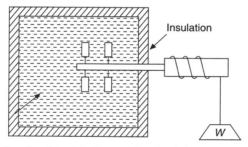

Figure 1.21 Supply of heat to the paddle wheel does not cause it to rotate.

gets heated. Now, try reversing the process, i.e. transferring heat from the fluid to the paddle-wheel. This does not make the paddle-wheel to rotate in the reverse direction raising the weight from the lower level to the higher level. Still the first law is not violated.

It is clear from the above discussion that though processes will proceed naturally in a certain direction but not in the reverse direction. The first law places no restriction on the direction of a process; but satisfying the first law does not ensure that the process will actually occur in the reverse direction as well. These limitations of the first law make it necessary to study the "second law of thermodynamics".

Thermal energy reservoir

A hypothetical body with a large thermal capacity (mass × specific heat) that can supply or absorb a finite amount of heat energy without undergoing a change in temperature is termed *thermal energy reservoir*. In practice, large bodies of water such as oceans, lakes and rivers as well as the atmospheric air are considered to be thermal reservoirs.

A reservoir that supplies energy in the form of heat is called a *source* and one that absorbs energy in the form of heat is called a *sink*.

Cyclic heat engine

Work is a very useful form of energy which can be adapted to practical applications in a variety of ways. Electrical work can power numerous devices, and mechanical work can drive automobiles, machines, etc. Heat is not necessarily so useful. The purpose of a heat engine is to convert heat into work, working as per a thermodynamic cycle. The internal combustion engine is a cyclic engine, which sequentially takes fuel and air, compresses and burns the mixture, and produces work output while emitting the products of combustion into the surroundings.

1.12 SECOND LAW OF THERMODYNAMICS

It has been discussed that a heat engine must reject some heat to a low temperature reservoir to complete the cycle, i.e. no heat engine working in a cycle can convert all the heat received by it into useful work. This limitation on the thermal efficiency of a heat engine forms the basis for *Kelvin–Planck* statement.

Kelvin–Planck statement: *It is impossible to construct a device that operates in a cycle and produces no effect other than withdrawal of energy as heat from a single reservoir and converting all of it into work.*

It can be simply put as: It is impossible for any device that operates in a cycle to receive heat from a single reservoir and produce an equivalent amount of work.

It is also stated as: No engine can have a thermal efficiency of one hundred per cent as shown in Figure 1.22.

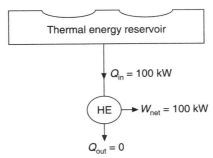

Figure 1.22 A heat engine that violates the Kelvin–Planck statement of the second law (PMM-II).

A device that violates the second law of thermodynamics is referred to as Perpetual Motion Machine of second kind (PMM-II).

Clausius statement: *It is impossible to construct a device that operates in a cycle and produces no effect other than the transfer of heat from a low temperature body to a high temperature body without external aid of work.*

A vapour compression refrigeration system will have the primary components, say evaporator, compressor, condenser and expansion valve. The compressor is required to raise the pressure of the refrigerant vapour from evaporating to condensing pressure. Unless this compressor unit is driven by some external power, the refrigeration system will not work. It indicates that this machine requires external energy to transfer heat from a body kept at low temperature to a body at higher temperature.

A device violating the Clausius statement is indicated in Figure 1.23.

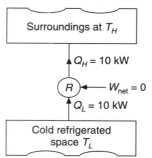

Figure 1.23   A refrigerator that violates the Clausius statement on the second law of thermodynamics (PMM-II).

1.13 PURE SUBSTANCE

A pure substance is one with homogeneous and invariable chemical composition, even if it undergoes a phase change. Water is a pure substance. A refrigerant is also a pure substance, which is the working media in the vapour compression refrigeration cycle, where it experiences different phase changes. Therefore, it is necessary to study the behaviour of this substance at different pressure and temperature conditions.

The knowledge of two independent properties suffices to determine the thermodynamic state of a fluid when it is in equilibrium and any other thermodynamic property is a function of the chosen pair of independent properties. Pressure p, specific volume v and temperature T are considered to be the primary properties. The equation expressing their relationship is called the equation of state. Here we have three variables to study the variation of one with respect to the other while the third is kept constant. As an example, consider a pure substance like water. Imagine the unit mass of ice below freezing point (at $-10°C$) enclosed in a cylinder by a piston under a constant load such that it exerts one atmospheric pressure on ice as shown in Figure 1.24. Let us follow the events which occur when heat at a uniform rate is supplied to the cylinder keeping the pressure constant on the ice.

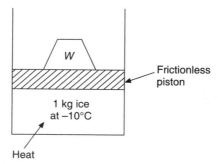

Figure 1.24   At 1 atm pressure and $-10°C$, water is in the solid phase.

(i) As the temperature rises from –10°C up to 0°C, ice expands as is indicated by the line AB (Figure 1.25). The heat that is supplied to raise this temperature is nothing but a sensible heat.

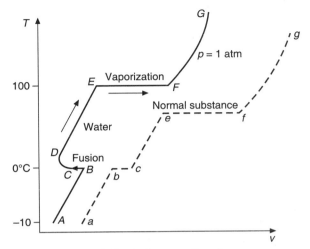

Figure 1.25 Isobars on *T–v* diagram.

(ii) Further addition of heat does not continue to raise the temperature of ice, but causes melting of ice to the liquid phase (line BC). The phase change from solid to liquid occurs at constant temperature and is accompanied by a reduction in specific volume. The heat added for the constant temperature melting of ice is called the *latent heat of fusion* or *latent enthalpy of fusion* because the pressure on ice is kept constant.

(iii) Further heating of liquid water at 0°C results in a rise in temperature, a contraction in volume until the temperature is about 4°C (point D) and subsequent expansion until a temperature of 100°C is reached (point E).

(iv) Further heat addition after point E causes a second phase change to occur at constant temperature, i.e. the liquid changes to vapour with a large increase in volume until the liquid is completely vaporised (point F). The heat supplied in this process is called the *latent heat of vaporisation*.

(v) Once the vaporisation is complete, further heating again causes the rise in temperature (line FG).

Water behaves in a different way compared with other substances. It contracts on melting and also contracts before it expands when its temperature is raised above the melting point. The curve shown with dotted lines in Figure 1.25 shows the behaviour of normal pure substances. Constant pressure lines of Figure 1.25 are called *isobars*.

The isobars for water at lower pressures are shown in Figure 1.26 which are obtained in the same manner as discussed above.

First, there is a slight rise in the melting point. Secondly, there is a remarkable drop in the boiling point and an increase in volume and enthalpy during evaporation. At further reduction in

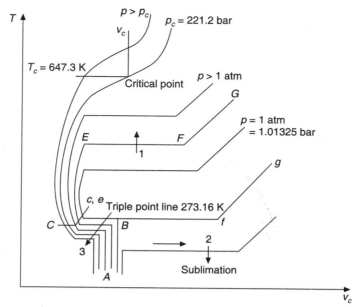

Figure 1.26 Families of isobars on T–v plot (water).

pressure to say 0.006112 bar, the melting and boiling temperatures become equal, and the change in phase, ice–water–vapour is represented by a single horizontal line.

The temperature at which all the three phases, solid, liquid and vapour, exist in equilibrium is called the *triple point*. Pure water at temperature 273.16 K and a pressure of 0.006112 bar, exists in all three phases as ice, water and vapour in a thermodynamic equilibrium condition, in a closed vessel. If the pressure is further reduced (below 0.006112 bar) the ice, instead of melting, sublimates directly into vapour.

Consider heating of water above the atmospheric pressure. The isobar curve is similar to that of the atmospheric isobar, but its change in volume reduces during evaporation. At very high pressure of the order of 221.2 bar, this change in volume falls to zero, and the horizontal portion of the curve (evaporation line) reduces to a point. This point is referred to as *critical point* and the properties at this point as critical temperature T_c, critical pressure p_c and critical volume v_c. These values for water are p_c = 221.2 bar, T_c = 647.3 K and v_c = 0.00317 m³/kg.

At a pressure above the critical point, the two phases, liquid and vapour cannot be distinguished, i.e. there is no definite transition from liquid to vapour. The latent heat of enthalpy of vaporisation falls to zero at the critical pressure.

Figure 1.27 shows the corresponding behaviour of normal substances such as R12, R134a, CO_2, etc. The only difference is in the behaviour near the triple point line. Only the left side of the curve is shown. Also, both normal boiling and melting points rise with an increase in pressure. The liquid expands on heating at all pressures and temperatures. This figure is constructed with the help of Figure 1.26. For detailed construction, the readers may refer to any standard book on thermodynamics.

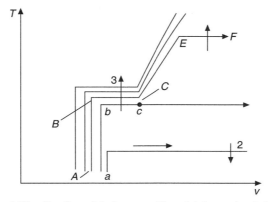

Figure 1.27 Families of isobars on T–v plot (normal substance).

1.14 OTHER PHASE DIAGRAMS

Many a time, one requires representing the thermodynamic process and cycles on the following phase diagrams: (a) p–v diagram, (b) T–s diagram, (c) h–s diagram, and (d) p–h diagrams. The topics and subject matter discussed within the scope of these diagrams deal with the properties of substances above the triple point. Also, the condition involving the mixture of solid and liquid phases is not dealt with in respect of any process here. Therefore, most of the charts show properties above the triple point with the states such as (i) liquid phase saturated and sub-cooled, (ii) liquid plus vapour phase and (iii) vapour phase in saturated and superheat condition.

The above four phase diagrams are shown from sub-cooled liquid state to superheated vapour state. The phase diagrams also show the change in state accompanying a constant pressure process above the triple point line.

Figure 1.28 shows a p–v diagram. The constant pressure line having points v_1, v_f, v_g and v_{sup} indicate how the specific volume of a pure substance changes in different phases, where subscript 'f' indicates saturated liquid, the subscripts 'g' stands for saturated vapour, the subscript 'sub' means superheated state and the subscript '1' is for subcooled liquid.

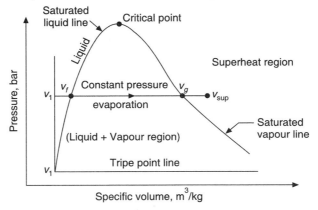

Figure 1.28 p–v diagram of pure substance.

The *T–s* and *h–s* diagrams are shown in Figure 1.29 and Figure 1.30 respectively. The changes in specific entropy s_1, s_f, s_g and s_{sup} at constant pressure are shown in these two diagrams.

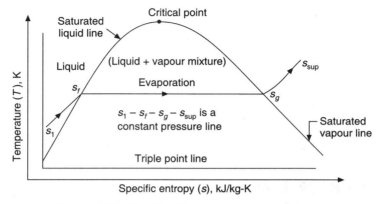

Figure 1.29 The *T–s* diagram of a pure substance.

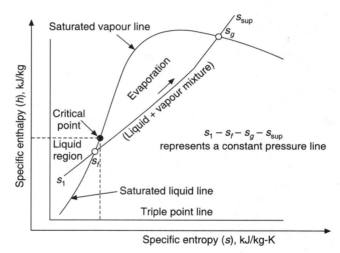

Figure 1.30 The *h–s* diagram of a pure susbtance.

The *p–h* diagram for a pure substance is separately discussed in the chapter on vapour compression refrigeration.

NUMERICALS

1. A tank 0.5 m × 2 m at the base is filled to a depth of 0.8 m with lubricating oil. If the quantity of oil in the tank is 1300 kg, then determine (a) the density, (b) the specific volume and (c) the specific gravity of the oil.
2. A tank 1.5 m × 0.7 m is filled to a depth of 0.8 m with a 65% solution of ethylene glycol having specific gravity of 1.1. What is (a) density, (b) total mass and (c) specific volume of the solution?
3. A force of 1 kN acts on a mass of 10 kg. What is the resulting acceleration?
4. What is the gravitational force acting on the mass of 10 kg?
5. A cylindrical tank, 1.2 m high and 0.4 m in diameter, contains water. If the pressure at the base of the tank is 1.5 bar, what is the total force exerted at the base?
6. A gauge on a refrigeration condenser reads 830 kPa while a barometer reads 772 mm of Hg. What is the absolute pressure of the refrigerant in the condenser?
7. A gauge on the suction side of a refrigeration compressor indicates –30 kPa while a gauge on the discharge side of the compressor reads 675 kPa. What is the increase in pressure of the refrigerant vapour during the compression process?
8. Two kilograms of air having a volume of 1.2 m³ and temperature of 290 K is heated so that the air expands at a constant atmospheric pressure of 1.01325 bar to another volume of 1.6 m³. Compute the external work done by the gas and the total energy added to the air.
9. A gas contained within a piston–cylinder assembly expands in a constant pressure process from $v_1 = 0.3$ m³/kg to $v_2 = 0.72$ m³/kg. The mass of the gas is 0.5 kg. For the gas as a system, the amount of energy transferred by doing the external work is +80 kJ. Calculate the pressure in bar.
10. An ideal gas is assumed to expand from an initial state where $p_1 = 500$ kPa and $V = 0.1$ m³ to a final state where $p_2 = 100$ kPa. For the process, p/V = constant is valid. Sketch the process on a p–V diagram and determine the work in kJ.
11. In Problem 10, if the expansion process is according to $pV^\gamma = C$, i.e. reversible adiabatic, what is the work done in kJ?
12. In Problem 10, if the expansion process is according to $pV^n = C$, $n = 1.3$, then what would be the work done in kJ?
13. Air enters a compressor operating at a steady state at a pressure of 1 bar, and temperature of 290 K and velocity of 6 m/s through an inlet with an area of 0.1 m². At the exit, the pressure, temperature and velocity are 7 bar, 450 K and 2 m/s respectively. Heat transfer from the compressor to its surroundings occurs at the rate of 3 kJ/s. Employing the ideal gas model, calculate the power input to the compressor in kW.

[**Ans.** –119.4 kW]

14. The summer and winter ambient air temperatures in a particular locality are 45°C and 20°C respectively. Find the value of Carnot COP of an air conditioner (i) when used for cooling and (ii) when used for heating.

Assume the refrigeration temperature of 5°C in summer and 50°C in winter. Take a temperature difference of 5°C in a heat exchanger (evaporator and condenser side).

[Ans. COP_R = 6.1, COP_{HP} = 9.2]

15. 50 kg of ice at −5°C is placed in a trolley to cool some vegetables. After 12 hours it was found that the ice had melted into water which had the temperature 10°C. What is the average rate of cooling in kJ/h provided by the ice and the refrigeration effect in TR? Assume, that the specific heat of ice = 194 kJ/kg-K, and specific heat of water = 4.1868 kJ/kg-K, latent heat of fusion of ice at 0°C = 353 kJ/kg.

[Ans. 1610.8, 0.127]

16. An inventor claims that he has developed a refrigerating machine which operates between −20°C and +30°C and consumes 1 kW power. The machine gives a refrigerating effect of 21.6 MJ/h. Verify the validity of his claim.

17. A domestic food freezer maintains a temperature of −15°C. The outdoor temperature is 30°C. If the heat load on the freezer is 1.75 kW, what is the minimum power required to pump out this heat? If the actual COP of the freezer is half that of the ideal COP, what is the actual power required?

[Ans. Minimum power = 0.305 kW. Actual power = 0.61 kW]

18. A reversible heat engine runs a reversible heat pump. The heat rejected by the heat pump and the heat engine is used to warm up a building. If the thermal efficiency of the heat engine is 27% and the COP of the heat pump is 4, find the ratio of the heat supplied to the building to the heat supplied to the heat engine.

[Ans. 2.08]

19. Determine the specific volume, enthalpy and entropy of 1 kg of R22 having a saturation temperature of −5°C and a quality of 32%.

Chapter 2

Methods of Refrigeration

2.1 INTRODUCTION

The Chinese were the first in the world to use ice in the food in 2000 BC. Naturally harvested ice was used for storage transportation of flesh during middle of the nineteenth century. Around 1855 the first machine was built to manufacture artificial ice through a refrigeration unit.

Refrigeration is the process of removal of heat from the confined (closed) space so as to reduce its temperature below the surrounding temperature and maintain it at that temperature. The temperature at which refrigeration is to be produced depends on the particular application. Let us discuss a few applications of refrigeration at this stage.

2.2 APPLICATIONS OF REFRIGERATION

1. A refrigeration system may be used to chill water to about 5–7°C, which is then used to cool the air in comfort air-conditioning systems of auditoriums, theatres, hospitals, etc. The capacity of these air-conditioning units may range from 150 TR to 250 TR.
2. Refrigeration systems with open ammonia compressors are used to manufacture ice. The capacity of the ice plants may be 20 tonnes per day or more.
3. Domestic refrigeration includes household refrigerators and home freezers. Domestic units are usually small in size, producing refrigerating effect in the range 300–500 W at about –5°C evaporator temperature.
4. Refrigeration finds widespread applications in food processing and in storage units.
   (a) **Milk and milk products:** Whole milk for human consumption is pasteurised at 75°C for a short time, and then recooled to 4°C immediately. This cooling is done in a counterflow plate heat exchanger with the use of chilled water at 2°C. The pasteurisation of milk is carried out in the following stages:

(i) Raw milk at 4°C is heated by the outgoing milk up to about 25°C.
(ii) This milk is finally heated by hot water up to the pasteurizing temperature of 75°C and held for a few seconds.
(iii) The milk is cooled by the incoming milk down to about 10°C.
(iv) The final stage of cooling from 10°C to 4°C is by chilled water at 2°C.

Long-term storage of butter is at −25°C. Milk is treated at low temperature for other products as well.

(b) **Soft drinks:** Most of the soft drinks are carbonated, i.e. they have a proportion of dissolved carbon dioxide, which causes the bubbles and also the typical effervescent taste. Each litre of water will have dissolved in it 3.5–5 litre of carbon dioxide. The solubility of CO_2 in water depends on the pressure and temperature. The relationship between pressure and temperature for 3.5 and 5 volumes is shown in Figure 2.1.

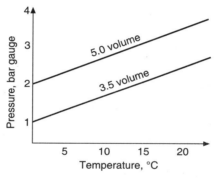

Figure 2.1 Solubility of carbon dioxide in water.

(c) **Storage of vegetables and food products:** Table 2.1 shows the storage temperature of a few vegetables and food products and their approximate storage life.

Table 2.1 Storage conditions for food products

Product	Temperature, °C	Relative humidity, %	Approximate storage life
Mangoes	10	85–90	2–3 weeks
Potatoes	3 to 4.5	85–90	5–8 months
Tomatoes	14 to 21	85–90	2–4 weeks
Bananas	14.5	95	8–10 days for ripening
Fish fresh	0.5 to 1.5	90–96	5–15 days
Butter	−18 to −23	80–85	2 months
Pasteurised milk	0.5	–	7 days
Eggs	−1.5 to −0.5	80–85	6–9 months
Grapes	−0.5	85–90	3–6 months

5. The progress in technology has facilitated better living conditions. Therefore, air conditioning is not only provided in auditoriums, hotels, etc. but also during transportation. The comfort air-conditioning systems are fitted in railway wagons and passenger cars.
6. Air conditioning improves the productivity by creating the required ambient conditions. For example:
   (a) While spinning cotton threads to the required quality one would require to maintain RH at more than 90% along with temperature control.
   (b) Printing industries achieve better print quality by maintaining the RH at less than 25% in the printing press.
   (c) Manufacturing industries can improve the life of various cutting tools, slip gauges, etc. by storing these in a tool room having RH less than 30% and temperature of about 20°C.
7. The cryogens (liquid N_2, O_2 and Ar) need to be produced and maintained at temperatures below 100 K and so require refrigeration systems.
8. Liquid nitrogen is used to store bull semen for cattle industry. It is also used in transport refrigeration systems of ice creams which need to be maintained at temperature below –25°C.
9. Gases like nitrogen, oxygen and argon are produced in their purest form through cryogenic distillation in cryogenic air separation plants. These gases have widespread applications as enumerated in the following sub-paragraphs:

Applications of nitrogen

Air is the main source of nitrogen. Nitrogen in its purest form can be economically produced in large scale by cryogenic air separation technology that finds widespread applications.

The quality of nitrogen product required differs significantly from one industry to another. Though electronics industry demands the highest purity of 1 ppm O_2, dust level 1/std ft^3 and moisture level at 0.1 ppm, the purity of 2 to 100 ppm O_2 is sufficient for aluminium, rubber, glass, textile, chemicals and steel industries. The purity of 10 ppm O_2 at a pressure of 300–400 bar (30–40 MPa) will meet its requirements in petroleum industry. However, the product pressure at POU depends on the specific applications.

Liquid nitrogen (LIN) is a useful source of cold and finds diverse applications. Some of these applications are mentioned here:

(a) With liquid nitrogen, food can be frozen in a few seconds thus preserving much of its original taste, colour and texture. It is reported that weight losses can be reduced considerably when food is frozen cryogenically rather than by any other means. The purity of LIN from 95–98% is sufficient for freezing the food.
(b) Cryosurgery is a technique that destroys cancer cells by freezing. It has been used in some top medical centres for the treatment of tumors of prostate, liver, lung, breast and brain as well as for cataracts, gynaecological problems and other diseases.
(c) LIN is used for storing biological specimens, especially bull semen for the cattle industry.
(d) In ground freezing, LIN enables tunnelling and excavation operations performed in wet and unstable soils.

- (e) In heat treatment of metals, LIN is known to transform metallurgical properties, which improve the wear resistances of carbon tool steels.
- (f) The automobile tyres have been one of the most difficult items to recycle or even worse—to discard. Cryogenic provides the necessary technology for the effective recovery, separation, and reuse of all materials used in the tyre. In fact, the use of LIN is the only known way to recover rubber from the steel radial tyre.
- (g) Mechanical breakdown of solids into smaller particles is known as grinding. Cryogenic grinding is a method of powdering materials at sub-zero temperatures. The materials are frozen with LIN when they are ground.

Applications of oxygen

Oxygen was one of the first atmospheric gases liquefied by Cailletet and Pictet in 1877. Later the Polish scientists Olzewski and Wroblewski at Cracow in 1883 produced stable liquid oxygen in a U-tube whose properties could be studied. Henceforth, oxygen began its useful life in industry early in the twentieth century.

- (a) A huge quantity of oxygen is consumed in steelmaking industry following LD or BOF process. These processes need 99.5% (conventional standard grade) purity of oxygen to accelerate the oxidation and conversion of iron to steel. The daily consumption amounts to several thousand tonnes, and all the modern steel plants, therefore, have sufficient tonnage oxygen plants.
- (b) One of the most common usages of oxygen is in the fabrication and cutting of metals using an oxy-acetylene torch.
- (c) Another major use of oxygen is in the field of medicine. The purity of oxygen needed is 99.999%.
- (d) Oxygen is used in the preparation of chemicals. For example, the manufacture of ethylene oxide requires 40% oxygen while acetylene consumes almost 20%. Titanium dioxide, propylene oxide and vinyl acetate need 10–15% O_2 for their manufacture.

In glass manufacturing, oxygen is added to enrich the combustion air in glass-melting furnaces. Jet aircraft for high altitude missions are equipped with oxygen systems for breathing purposes. Coal gasification is also one of the large consumers of gaseous oxygen.

Applications of argon

Usually argon is obtained from air which contains 0.0093% of argon by volume. It is highly inert and finds its applications in a wide range of conditions, both at cryogenic and at high temperatures. Argon is relatively expensive and its use is limited to applications where its highly inert properties are essential.

- (a) The largest consumption of argon worldwide is in the argon–oxygen decarburisation process for producing low carbon stainless steels.
- (b) MIG welding developed by Airco in the 1940s and TIG welding represent large markets for argon.
- (c) The light bulb industry uses argon to fill light bulbs. This gives longer life to the filament because argon does not react, even at high temperatures.

2.3 REFRIGERATION SYSTEMS

Refrigeration systems are grouped into the following three systems:

(i) **Non-cyclic refrigeration systems:** Such system include refrigeration using ice, refrigeration by evaporation, and refrigeration by dry ice. These systems were used before the invention of cyclic refrigeration systems. For example, natural ice was used to preserve flesh during its shipment from one country to another.

(ii) **Cyclic refrigeration systems:** These include the air refrigeration cycle, the vapour compression refrigeration cycle, and the vapour absorption cycle.

(iii) **Other refrigeration systems:** These are thermoelectric refrigeration cycle, steam-jet refrigeration cycle, vortex tube refrigeration system, magnetic refrigeration system, solar refrigeration systems, and so on.

A few of the above systems are discussed in the following sub-sections.

2.3.1 Ice Refrigeration

The quantity of heat required to convert one kg of ice from and at 0°C is 335 kJ. (Latent heat of fusion, h_L = 335 kJ). Therefore, ice can be used to produce cold in a confined chamber, which can be used to store vegetables or food. The mechanism of cooling the vegetables and food products using ice is shown in Figure 2.2 and Figure 2.3.

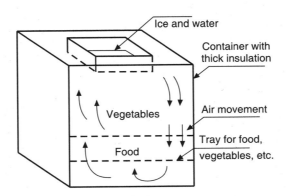

Figure 2.2 Ice refrigeration for food and vegetables storage.

Ice is kept in a small cabin at the top level of the insulated container. The arrangement is such that air comes in contact with the metallic surface in which ice is kept and gets cooled. Due to this phenomenon, its density increases and it starts moving downwards and the low density air from the low level moves upwards creating convection currents. Therefore, food, vegetables, etc. kept in the trays at different levels are cooled down and are preserved. The cooling of the commodities would continue till ice melts completely. The melted ice water is drained out.

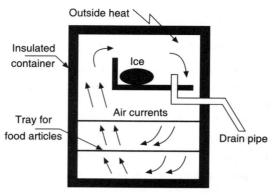

Figure 2.3 Ice refrigeration (cut section).

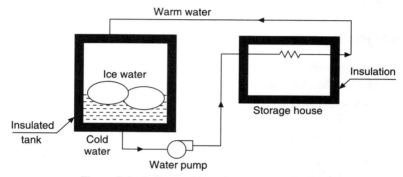

Figure 2.4 Indirect refrigeration system using ice.

Ice is also used to produce refrigeration on a large scale as shown in Figure 2.4. This system of refrigeration is suitable for transport refrigeration, and for the storage of food and vegetables for a limited period in case one does not want to invest in purchasing a refrigeration system.

2.3.2 Evaporative Refrigeration

This method of generating cold or refrigeration is very old and used by common people. It can be explained with a simple example of cooling water in a porous pot, shown in Figure 2.5.

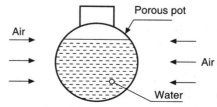

Figure 2.5 Evaporative cooling of water in a porous pot.

Water in the pot percolates through the porous pot. The air surrounding the pot comes in contact with this water and cools down. This is due to the evaporation of water. The heat required for the evaporation of water is partially from water and air. In turn, both air and water cool down. Since air is free to move, one does not feel the cold air but the cold water in the pot can be felt. The temperature of water in the porous pot may be reduced by 5–15°C compared to the temperature of ambient air. Based on this principle, one can answer as to why the flowing water is always cold? Why does one feel cool breeze near lakes and oceans? Human body is kept cold due to evaporative cooling in hot climate conditions as well as during physical exertion. The use of desert bag is another example of evaporative cooling to keep drinking water cold.

Evaporative refrigeration phenomenon can also be employed to produce artificial ice. Such a system can be explained with the help of a sketch shown in Figure 2.6.

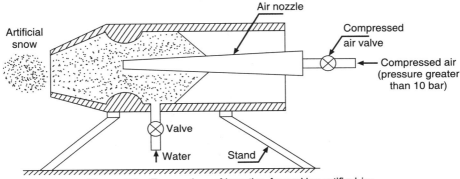

Figure 2.6 Evaporative refrigeration for making artifical ice.

The arrangement consists of an air expansion nozzle and water supply line. When air is expanded from a pressure of the order of 5 bar and temperature 25°C, to atmospheric pressure (1.0 bar), its temperature would even fall to sub-zero. Therefore, air is expanded from its high pressure to ambient pressure producing cold and simultaneously water is supplied to the chamber to mix with air. The high velocity air forces the water and causes it to break up into tiny droplets (fog). Since the surrounding air temperature is below sub-zero, the water droplets may condense and freeze forming small particles of ice. In this method, artificial snow is produced when the temperature of the surrounding air is 0°C or below 0°C.

Evaporative cooling phenomenon is employed in cooling towers and water-cooled condensers.

2.3.3 Refrigeration by Dry Ice

Solid carbon dioxide is known as 'dry ice'. The critical point of CO_2 is 31°C, whereas its critical pressure is 73.8 bar. To manufacture solid gaseous CO_2, carbon dioxide is compressed to a pressure of about 69 bar and is cooled at constant pressure (69 bar) to a temperature of 28°C at which it condenses to liquid completely. Expansion of liquid CO_2 to 1 bar pressure would result into solid CO_2 (–78.5°C) in a good insulated container.

The speciality of the solid CO_2 is that it sublimates into vapour directly, producing cold. Therefore, it is used to preserve the eye-ball during the eye surgery.

The drawbacks of ice refrigeration systems are:
1. There is no working substance that can be be utilized to produce a refrigeration effect at the required temperature.
2. The temperature of the refrigerated space cannot be controlled accurately.
3. One has to replenish the ice periodically to produce cold, otherwise it may not be possible to maintain the refrigeration temperature.

2.3.4 Liquid Gas Refrigeration

Liquid nitrogen (boiling point 77 K) boils at atmospheric temperature. It requires about 200 kJ to convert 1 kg liquid into dry vapour. If used, it produces a cooling effect of 200 kJ/kg liquid nitrogen. Nitrogen is an inert gas so it can be used for producing the refrigeration effect. (The cost of liquid nitrogen is very low as it is produced in cryogenic air separation plants.). Liquid nitrogen finds its application in providing cooling effect for transporting ice-creams, frozen foods, etc.

The block diagram of a liquid nitrogen cooling arrangement is shown in Figure 2.7.

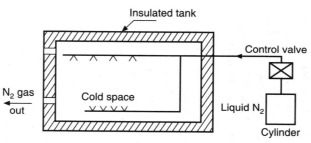

Figure 2.7 Liquid nitrogen cooling.

The control valve of the liquid nitrogen is shown outside the tank but in practice, it may be placed inside the tank. Liquid nitrogen is allowed to flow through the coil of tubes inside the tank and while absorbing heat from the cold space, it evaporates and produces cold. The temperature inside the tank is sensed by an electronic sensor, which gives feedback to the control valve so that the regulation of liquid nitrogen is possible. The nitrogen gas passes to the atmosphere through the vents provided.

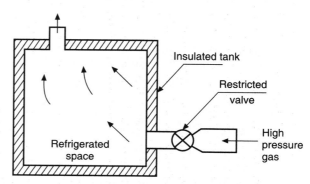

Figure 2.8 Refrigeration by throttling of gas.

Liquid nitrogen cylinders are available in the capacity of 50 litre, 100 litre, and 200 litre commercially.

Another way of producing cold in a refrigerated space is shown in Figure 2.8.

2.3.5 Steam Jet Refrigeration

It is a well-known fact that the boiling point of water changes as per the pressure. Water boils at 100°C at 1 atm pressure. It boils at 150°C if the pressure is 4.758 bar, but it boils at 10°C if its pressure is kept at 1.2276 kPa. Such a low pressure (vacuum) on the surface water is maintained by throttling the steam through nozzles.

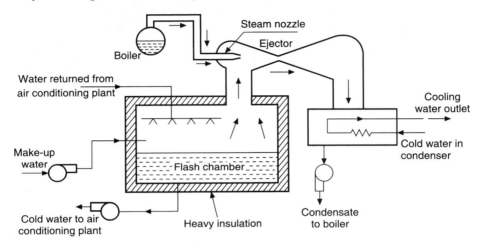

Figure 2.9 Steam jet refrigeration system.

Suppose a pressure of 5.593 kPa (vacuum) is maintained in a flash chamber which contains 100 kg of water. If 1 kg of water is allowed to flash at such a low pressure, it absorbs heat of about 2394 kJ from the remaining water of 99 kg. The fall in the temperature of water in the chamber is

$$1 = \frac{2394}{(4.186)(99)} = 5.8°C$$

If 2 kg of water evaporates, then decrease in temperature of the chamber water is

$$= \frac{(2)(2394)}{(4.186)(99)} = 11.67°C$$

On evaporation of a definite quantity of water, the temperature of the remaining water in the chamber may reach even 0°C. In that case the ice formed is pumped out.

Steam from the boiler is expanded through a nozzle (Figure 2.9), this helps to keep away the vapour that is formed due to flashing of water in the flash chamber.

The condensate formed flows through the ejector alongwith the steam into the condenser and the collected condensate is supplied back to the boiler. One would be required to supply the make-up water.

2.3.6 Thermoelectric Refrigeration

It is one of the non-conventional refrigeration methods used for producing low temperature on the basis of the reverse Seebeck effect. When the junctions of two dissimilar conductors are maintained at two different temperatures T_1 and T_2, an electromotive force (emf) E is generated. This phenomenon is called *Seebeck effect*. This principle is used in thermocouples for measuring temperatures.

When a battery is connected between the junctions of two dissimilar conductors which are initially maintained at the same temperature and a current is made to flow through the circuit, it is observed that the junction temperatures are different, one junction becoming hot (T_1) and the other becoming cold (T_2). This principle is reverse of the Seebeck effect and called *Peltier effect*. In this case the refrigeration effect is obtained at the cold junction and heat is rejected to the surrounding at the hot junction. This principle is the basis for thermoelectric refrigeration systems. The position of cold and hot junctions can be reversed by reversing the direction of current through the conductor. The principles of Seebeck effect and Peltier effect are illustrated in Figure 2.10(a) and (b), respectively.

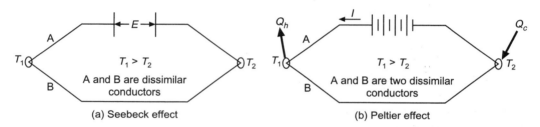

Figure 2.10 Principle of thermoelectric refrigeration.

The heat transfer rate at each junction is given by

$$Q = \Phi I$$

where Φ is the Peltier coefficient in volts and I is the current in amperes.

2.3.7 Vortex Tube

The vortex tube was invented by a French physicist named Georges J. Ranque in 1931, when he was studying processes in a dust separated cyclone. The vortex tube, also known as the Ranque-Hilsch vortex tube, is a mechanical device that separates a compressed gas into hot and cold streams. It has no moving parts. It is one of the non-conventional types of refrigerating systems for the production of refrigeration. The schematic diagram of the vortex tube is shown in Figure 2.11.

It consists of a nozzle, a diaphragm, a valve, a hot-air side, and a cold-air side. The nozzle is of converging or diverging type or of converging-diverging type as per the design. An efficient nozzle is the one, which is designed to have higher velocity, greater mass flow and minimum inlet losses. The chamber is a portion of the nozzle and provides facility for the tangential entry of high velocity air-stream into the hot side. Generally the chambers are not of circular form, but of the gradually converted spiral form. The hot-side is cylindrical in cross section and is of different lengths as per design. A throttle valve obstructs the flow of air through the hot-side and it also controls the quantity of hot air passing through the vortex tube. The diaphragm is a cylindrical piece of small thickness and has a small hole of specific diameter at its centre. The air stream travelling through the core of the hot-side is emitted through the diaphragm hole. The cold-side is a cylindrical portion through which the cold air passes.

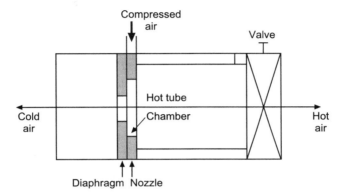

Figure 2.11 Vortex tube.

In this vortex tube, compressed air is passed through the nozzle as shown in the figure. Air expands and acquires a high velocity due to the particular shape of the nozzle. A vortex flow is created in the chamber and air travels in a spiral like motion along the periphery of the hot-side. This flow is restricted by the valve. When the pressure of the air near the valve is made more than the outside pressure by partly closing the valve, a reversed axial flow through the core of the hot-side starts from the high-pressure region to the low-pressure region. During this process, heat transfer takes place between the reversed stream and the forward stream. Therefore, the air stream through the core gets cooled below the inlet temperature of the air in the vortex tube, while the air stream in the forward direction gets heated up. The cold stream escapes through the diaphragm hole into the cold-side, while the hot stream is passed through the opening of the valve. By controlling the opening of the valve, the quantity of the cold air and its temperature can be varied. The vortex tube has the following advantages:

- It uses air as refrigerant, so there is no leakage problem.
- It is simple in design and does not require any control system to operate it.
- There are no moving parts in the vortex tube.
- It is light in weight and occupies much less space.
- The initial cost is low and its working expenses are also less, whereas compressed air is readily available.
- Maintenance is simple and no skilled labour is required.

On the disadvantage side, the vortex tube has low COP, limited capacity and only a small portion of the compressed air appearing as the cold air limits its wide use in practice.

2.3.8 Solar Refrigeration

Solar energy can be used to provide refrigeration at temperatures below 0°C. There are three approaches to use solar energy for refrigeration:

1. Photovoltaic operated refrigeration system
2. Mechanical solar refrigeration system
3. Solar vapour absorption system

Photovoltaic operated refrigeration system

Photovoltaic (PV) involves the direct conversion of solar radiation to direct current (dc) electricity using semiconducting materials. The PV powered solar refrigeration cycle is simple in operation. Solar photovoltaic panels produce dc electrical power that can be used to operate a dc motor, which is coupled to the compressor of a vapour compression refrigeration system. The major considerations in designing the PV refrigeration cycle involve appropriately matching the electrical characteristics of the motor driving the compressor with the available current and voltage being produced by the PV array.

Mechanical solar refrigeration system

Solar mechanical refrigeration uses a conventional vapour compression system driven by mechanical power that is produced with a solar-driven heat power cycle. The heat power cycle usually considered for this application is a Rankine cycle in which a fluid is vapourised at an elevated pressure by heat exchange with a fluid heated by solar collectors. A storage tank can be included to provide some high temperature thermal storage. The vapour flows through a turbine or piston expander to produce mechanical power, as shown in Figure 2.12. The fluid exiting the expander is condensed and pumped back to the boiler pressure where it is again vapourised.

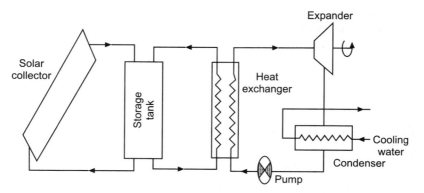

Figure 2.12 Mechanical solar refrigeration system.

Solar vapour absorption system

Unlike the PV and solar mechanical refrigeration options, the absorption refrigeration system is considered a "heat driven" system. It replaces the energy-intensive compressor in a vapour compression system with a set of devices such as absorber, generator, pump and pressure reducing valve. A schematic of a single-stage absorption system using ammonia as the refrigerant and water as the absorbent is shown in Figure 2.13. Absorption cooling systems that use lithium bromide-water absorption-refrigerant working fluids cannot be used at temperatures below 0°C. The condenser, the expansion device and the evaporator operate in the exactly same manner as for the vapour compression system. Ammonia vapours exiting the evaporator (State 1) are absorbed in a liquid solution of water-ammonia in the absorber. The absorption of ammonia vapours into the water-ammonia solution is analogous to a condensation process. The process is exothermic and so cooling water is required to carry away the heat of absorption. The principle governing this phase of the operation is that a vapour is more readily absorbed into a liquid solution as the temperature of the liquid solution is reduced. The ammonia-rich liquid solution leaving the absorber (State 2) is pumped to a higher pressure, passed through a heat exchanger and delivered to the generator (State 4). In the generator, the liquid solution is heated, which promotes desorption of the refrigerant (ammonia) from the solution. Unfortunately, some water also is desorbed with the ammonia, and it must be separated from the ammonia using the rectifier. Without the use of a rectifier, water exits at State 5 with the ammonia and travels to the evaporator, where it increases the temperature at which refrigeration can be provided. This solution temperature needed to drive the desorption process with ammonia-water is in the range between 120°C and 130°C. Temperatures in this range can be obtained using the low-cost non-tracking solar collectors. At these temperatures, the evacuated tubular collectors may be more suitable than the flat-plate collectors as their efficiency is less sensitive to the operating temperature.

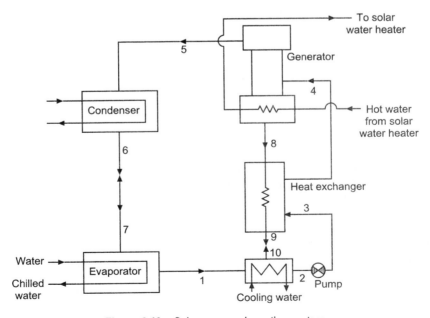

Figure 2.13 Solar vapour absorption system.

2.3.9 Magnetic Refrigeration

The magnetic refrigeration technique is used to attain extremely low temperatures of the order of 1 K or below 1 K. This technique can also be used to produce cold in common refrigerators depending on the design of the system. Magnetic refrigeration was the first method developed for cooling below about 0.3 K.

This cooling technology is based on *magnetocaloric effect*. Here a suitable material is subjected to a storage charging magnetic field. The decrease in the strength of an externally applied magnetic field allows the magnetic domains of a chosen (magnetocaloric) material to become disoriented from the magnetic field by the agitating action of the thermal energy present in the material. If this material is isolated from the surroundings, i.e. no energy enters into the material during this time (i.e. adiabatic process), the temperature of the material drops as the domains absorb thermal energy to perform their reorientation.

The magnetocaloric effect is observed in the chemical element gadolinium and some of its alloys. The temperature of the gadolinium increases when subjected to a magnetic field and its temperature decreases when the magnetic field leaves. This is as shown in Figure 2.14.

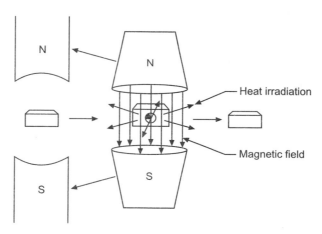

Figure 2.14 Magnetic effect.

The magnetic refrigeration cycle is analogous to the vapour compression cycle as illustrated in the following stages with the help of Figure 2.15.

1. **Adiabatic magnetisation:** According to law of thermodynamics, the entropy of the system/material does not decrease unless it loses energy. A magnetic substance is placed in an insulated environment. Applying the magnetic field ($+H$) to the substance causes the magnetic diapoles of the atoms to align; thereby decreasing its entropy and heat capacity. Since no energy (heat) is lost, therefore the total entropy remains constant. The net effect is that it heats up the magnetic substance.

2. **Isomagnetic enthalpic heat transfer:** During this process the magnetic field is held constant to prevent the diapoles from reabsorbing the heat. The added heat is removed by a fluid or gas. Once the material is cooled sufficiently, the magnetocaloric substance and coolant are separated.

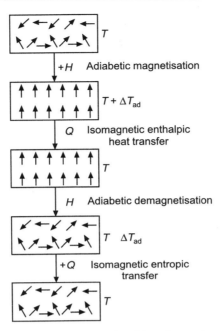

Figure 2.15 Magnetic refrigeration cycle.

3. **Adiabatic demagnetisation:** The substance is subjected to another adiabatic condition so that the total entropy remains constant. Here, the magnetic field is decreased, the thermal energy causes the magnetic moments to overcome the field and thus the sample cools.
4. **Isomagnetic entropic transfer:** The magnetic field is held constant to prevent the material from heating up again. The material is now in the space where the cooling effect is created. The working material is cooler than the refrigerated space, the energy enters into the working material ($+Q$). The cycle is now complete and gets repeated.

2.3.10 Air Refrigeration Cycle

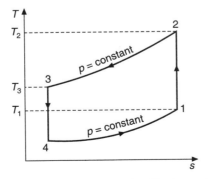

Figure 2.16 Reversed Carnot cycle with air as refrigerant.

Reversed Carnot cycle with air as a working fluid is the most efficient cycle. The T–s diagram of such a cycle is a rectangular one as shown in Figure 2.16.

If air is used as a refrigerant, this cycle would appear like that in Figure 2.16. The isentropic compression and expansion are the processes 1–2 and 3–4 respectively. Processes 2–3 and 4–1 are the constant pressure cooling and heating processes respectively.

At point 4, the temperature of air must be lower than that of cold space so that the air absorbs heat from the cold space and experiences an increase in temperature up to T_1. Similarly, at point 2, the temperature of air must be higher than that of the surroundings so that heat can be rejected. These processes are briefly discussed as follows:

Process 1-2: For isentropic compression, $p_1 V_1^\gamma = C$,

Specific work,
$$W_{1-2} = \frac{p_1 V_1 - p_2 V_2}{\gamma - 1} \tag{2.1}$$

and
$$Q_{1-2} = 0$$

Process 2-3: Isothermal heat rejection, $pV = C$

$$Q_{2-3} = W_{2-3} = -RT_2 \log p_3/p_2 \tag{2.2}$$

Process 3-4: Isentropic expansion, $p_3 V_3^\gamma = C$

Specific work
$$W_{3-4} = \frac{p_3 V_3 - p_4 V_4}{\gamma - 1} \tag{2.3}$$

Process 4-1: Isothermal heat absorption, $p_4 V_4 = C$

$$Q_{4-1} = RT_1 \ln p_4/p_1 \tag{2.4}$$

$$\text{COP} = \frac{\text{Heat absorbed from the system}}{\text{Work done on the system}}$$

$$= \frac{R T_1 \log_e (p_4/p_1)}{R (T_2 - T_1) \log_e (p_4/p_1)}$$

$$= \frac{T_1}{T_2 - T_1} = \frac{T_L}{T_H - T_L} \tag{2.5}$$

$$\frac{V_1}{V_2} = \frac{V_4}{V_3} = \gamma \quad \text{(compression ratio)}$$

2.3.11 Vapour Compression Refrigeration Cycle

A vapour compression refrigeration plant is shown in Figure 2.17. It mainly consists of the components such as a compressor, a condenser, an expansion device (valve or capillary tube) and an evaporator. In this figure, one can see that the existence of the refrigerant in different phases is at different cycle conditions. If we trace the path from point 2 (outlet of compressor) through the condenser, receiver and up to the inlet of the expansion valve, the refrigerant is in pressurized state. Therefore, this part of the flow diagram forms the high-pressure side. The remaining part, i.e. from point 3 through evaporator and up to the inlet of compressor, is the low-pressure side. To explain the working of this plant, the cycle is represented on a T–s diagram (Figure 2.17). This has been explained in Chapter 3.

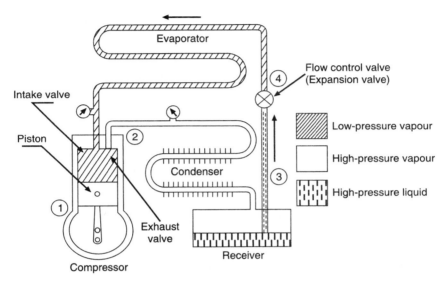

Figure 2.17 Vapour compression system with its components and the condition of refrigerant in the flow circuit.

EXERCISES

1. Define the terms refrigeration, refrigerant and refrigerating equipment.
2. Write a short note on the systems of refrigeration.
3. Briefly explain the applications of a refrigeration system.
4. Explain the working of ice refrigeration with the help of a simple schematic diagram. What are its drawbacks?
5. What do you understand by evaporative refrigeration? Explain with the help of examples.
6. What is dry ice? How can it be used for the refrigeration purpose? List its drawbacks.
7. What do you understand by refrigeration by expansion of air? In which cycle is this concept used?
8. Explain the working of a vapour compression refrigeration system with the help of a neat sketch. Name some of the common refrigerants used in this system.

9. Draw a neat diagram of a simple steam jet refrigeration system and explain its working.
10. Explain the concept of vortex tube. What are its advantages and disadvantages?
11. Explain the concept of thermoelectric refrigeration.
12. Explain the working of a solar vapour absorption system with the help of a diagram.
13. Explain the working principle of magnetic refrigeration with the help of a diagram.
14. Explain the concept of liquid gas refrigeration with the help of a diagram. Where is such refrigeration used in practice?
15. Describe the various methods of refrigeration.

Chapter 3

Air Refrigeration Systems

3.1 INTRODUCTION

Refrigeration is an application of thermodynamics. Refrigeration means transfer of heat from a lower temperature region to a higher temperature region with the help of an external aid. Devices that produce refrigeration effect are called *refrigerators* (or heat pumps) and the cycles on which they operate are called *refrigeration* cycles. The various refrigeration cycles are—reversed Carnot cycle, Bell–Colemann air refrigeration cycle, vapour compression refrigeration cycle, etc. Air refrigeration cycles are dealt with in this chapter.

3.2 DEFINITIONS

Refrigeration is a process of removing heat from a confined space so that its temperature is first lowered and then maintained at that low temperature compared to that of the surroundings, i.e. it is a phenomenon by virtue of which one reduces the temperature of a confined space compared with that of the surroundings.

The device used to produce cold or refrigeration effect is called a *refrigeration system* or *refrigerator*. The basic components of a refrigeration system are evaporator, compressor, condenser and expansion valve. In addition to these, there may be a *refrigerant* accumulator, temperature controller, etc. The refrigerant is a working substance, which circulates through the refrigeration system during its operation.

The R134a is the refrigerant used in domestic refrigerators and low-temperature refrigeration systems. The R22 is the refrigerant used in window air-conditioning units while ammonia is preferred in large air-conditioning systems. Air is the refrigerant in air-refrigeration cycles.

Unit of refrigeration or rating for refrigeration

The definition of refrigeration indicates that refrigeration is nothing but the rate of removal of heat. The SI unit of heat is joule, the time rate of which is watt. The unit of refrigeration effect is watt (W) or kilowatt (kW).

The standard unit of refrigeration is ton of refrigeration or simply ton denoted by the symbol TR. One TR is equivalent to the production of cold at the rate at which heat is to be removed from one US ton of water at 0°C in 24 hours. One US ton = 2000 lb, and latent heat of fusion of ice at 0°C = 144 Btu/lb.

Thus,
$$1 \text{ TR} = \frac{1 \times 2000 \text{ lb} \times 144 \text{ Btu/lb}}{24 \text{ h}}$$
$$= 12{,}000 \text{ Btu/h} = 200 \text{ Btu/min}$$

But 1 Btu = 1.055 kJ

∴ 1 TR = 211 kJ/min
     = 3.516 kW

In the above definition of 1 TR, one ton equals 2000 lb instead of 2240 lb as per the conversion factor. In US, one ton is equal to 2000 lb.

3.3 REFRIGERATION LOAD

The refrigeration effect or cooling effect is produced in a refrigeration cycle by the refrigerating equipment. The average rate at which heat is removed from the cold space by the equipment is known as the cooling load. It is expressed in kW or TR as stated above.

The cooling load on refrigerating equipment results from several different sources. Some of the common sources of heat that contribute to the cooling load on the refrigerating equipment are as follows:

1. Ingress of heat into the refrigerated space from outside by conduction through the insulated walls.
2. Solar radiations that enter the refrigerated space through transparent glass or other transparent materials.
3. Heat on account of warm outside air entering the refrigerated space through open doors or through cracks around windows and doors.
4. Heat emitted by warm products whose temperature is to be lowered to the refrigerated space temperature.
5. Heat emitted by people occupying the refrigerated space. For example, people present in an air-conditioned space or the people working in the cold storages during loading and unloading the goods.
6. Heat emitted by any heat generating equipment installed in the refrigerated space such as lamps, motors, electronic devices, etc.

It is to be noted here that all these sources of heat are not present in every application. The significance of any one heat source with relation to the total cooling load will vary considerably with each application.

This topic is discussed in more detail in this chapter.

3.4 HEATING LOAD

In those regions where the atmospheric temperature falls considerably (below 10°C), especially during winters, heating is needed to keep the rooms warm. The rate of heat to be supplied to such a conditioned space is known as heating load. In Western countries, the houses are facilitated with solar heating systems.

3.5 CONCEPT OF HEAT ENGINE, REFRIGERATOR AND HEAT PUMP

It is a well-known fact that heat flows in the direction of decreasing temperature, i.e. from a high temperature body to a low temperature body. Such heat transfer occurs in nature without any external aid or device.

Heat engine

It is a prime mover that generates heat from the fossil fuel. It works according to the second law of thermodynamics stated by Kelvin and Planck—*It is impossible to construct a device that operates continuously in a cycle and produces no effect other than the withdrawal of heat energy from a single reservoir and converts all the heat into useful work*. This means that the heat engine (HE) rejects part of the heat available from the heat source while converting it to useful work. This is shown in Figure 3.1. It takes heat at the rate of Q_1 from the heat source and generates the work at the rate of W while rejecting heat at the rate of Q_2 to the heat sink. The performance of a heat engine is known as 'thermal efficiency' or 'Carnot efficiency' and is mathematically represented as

$$\eta_{\text{Carnot}} = \frac{Q_1 - Q_2}{Q_1} = \frac{T_1 - T_2}{T_1}$$

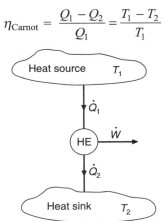

Figure 3.1 A heat engine.

Reverse heat transfer, i.e. from a body at low temperature to a body at high temperature, is possible only with a special device called *refrigerator*.

Refrigerator

It works according to the second law of thermodynamics stated by Clausius—*It is impossible to construct a device that operates in a cycle and produces no effect other than the transfer of heat from a body at low temperature to a body at high temperature without any external aid*.

Refrigerators are cyclic devices having working substances called *refrigerants* used in refrigeration cycles. The working principle of the refrigerator (R) is shown in Figure 3.2.

Here Q_2 is the amount of heat removed from the cold space at temperature T_2. Q_1 is the amount of heat rejected to the environment at temperature T_1 and W is the net work input to the refrigerator.

We know that the term 'efficiency' is used to indicate the performance of a heat engine. The performance of a device means its ability to carry out the assigned task.

Efficiency is the ratio of the output (desired quantity) to the input (reference quantity). The output is always less than the input. The output and input quantities are taken in the same units.

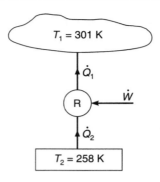

Figure 3.2 A refrigerator.

Coefficient of Performance (COP)

The objective of the refrigerator is to remove heat (Q_2) from a closed space. To accomplish this, it needs W_{net} as work input. Therefore, COP of a refrigerator is:

$$\text{COP}_R = \frac{\text{Desired effect}}{\text{Required input}} = \frac{Q_2}{W_{net}} \tag{3.1}$$

But

$$W_{net} = Q_1 - Q_2$$

∴

$$\text{COP}_R = \frac{Q_2}{Q_1 - Q_2} = \frac{1}{(Q_1/Q_2) - 1} \tag{3.2}$$

The COP may be greater than one in many cases.

Heat pump

A heat pump (HP) transfers heat from a low temperature space to a higher temperature space. The objective of a heat pump is to supply heat Q_1 to warm a space as shown in Figure 3.3.

The COP of a heat pump is expressed as the ratio of heat supplied (Q_1) to the work input (W_{net}). Mathematically,

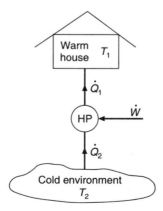

Figure 3.3 A heat pump.

$$\text{COP}_{\text{HP}} = \frac{\text{Desired effect}}{\text{Work input}} = \frac{\text{Heating effect}}{\text{Work input}} = \frac{Q_1}{W_{\text{net}}} \qquad (3.3)$$

Comparing Eqs. (3.2) and (3.3),

$$\text{COP}_{\text{HP}} = \text{COP}_{\text{R}} + 1 \qquad (3.4)$$

For fixed rates of Q_2 and Q_1, Eq. (3.4) shows that the COP_{HP} is always greater than one since the COP_{R} is a positive value. In reality, however, part of Q_1 is lost to the outside air through piping and other devices, and COP_{HP} may drop below one when the outside air temperature is too low.

3.6 AIR REFRIGERATION SYSTEMS

Air is used as a refrigerant (working media) in air refrigeration systems. Air absorbs heat from the low temperature space and rejects heat to the high temperature surroundings while undergoing an air refrigeration cycle. Air does not change its phase while undergoing a cyclic process. Therefore, the heat carrying capacity per unit mass of air is very small in comparison to that of a refrigerant of a vapour compression refrigeration cycle. To obtain a required refrigeration effect, a large quantity of air needs to be handled, requiring a bigger-sized compressor, heat exchanger and expansion device. In an aircraft, rammed air is available and hence used for its air conditioning. Therefore, such systems are also discussed in this chapter.

3.6.1 Carnot Refrigerator

The Carnot refrigeration system works on reversed Carnot cycle. It is only a theoretical system in its conception but serves as an ideal cycle ever to be achieved in reality. The p–v and T–s diagrams of reversed Carnot cycle using air as a working medium are shown in Figure 3.4.

The following processes are imagined to take place in a reciprocating compressor:

Process (1–2) reversible adiabatic compression process: Air is compressed from initial pressure p_1 to pressure p_2 till the temperature rises from T_1 to T_2 in a reversible adiabatic compression process. The piston is assumed to move very fast.

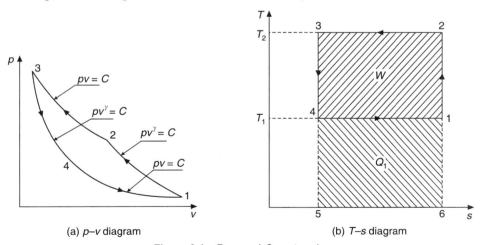

Figure 3.4 Reversed Carnot cycle.

Process (2–3) isothermal compression process: Air is compressed isothermally till its pressure rises from p_2 to p_3. It is assumed that at point 2, a body at temperature T_2 is brought in contact with cylinder. Heat Q_2 is rejected at constant temperature T_2 during isothermal compression. The piston is assumed to move at dead slow speed.

Process (3–4) reversible adiabatic expansion process: Reversible adiabatic expansion of air in the clearance volume takes place till its pressure and volume change from p_3 and v_3 to p_4 and v_4 during which the temperature falls to T_4. The piston is assumed to move at a great speed.

Process (4–1) isothermal expansion process: Air from p_4 and v_4 expands isothermally till the pressure and volume reach p_1 and v_1. It is assumed that when a cold body at temperature T_1 is brought in contact with the cylinder, heat is absorbed by the air. To satisfy the isothermal heat absorption at constant temperature, the piston has to move at a dead slow speed. The heat Q_1 is absorbed during the process. The air attains the original state on completing the cycle.

All the processes involved in the cycle are reversible and therefore the cycle is called a *reversible cycle*.

The heat absorbed from cold body at temperature T_1 (refrigeration effect)

$$Q_1 = \text{area } (1\text{--}4\text{--}5\text{--}6) \text{ on } T\text{--}s \text{ diagram}$$

$$= T_1(s_1 - s_4), \text{ in kJ/kg}$$

Work done during the cycle

$$W = \text{area } (1\text{--}2\text{--}3\text{--}4) \text{ on } T\text{--}s \text{ diagram}$$

$$= (T_2 - T_1)(s_1 - s_4), \text{ in kJ/kg}$$

Coefficient of performance of Carnot refrigerator,

$$\text{COP}_R = \frac{\text{Refrigeration effect } Q_1}{\text{Net work done } W}$$

$$= \frac{T_1(s_1 - s_4)}{(T_2 - T_1)(s_1 - s_4)}$$

$$= \frac{T_1}{T_2 - T_1} \tag{3.5}$$

$$= \frac{1}{(T_2/T_1) - 1} \tag{3.6}$$

The COP of heat pump working on reversed Carnot cycle,

$$\text{COP}_H = \frac{\text{Useful effect}}{\text{Work input}}$$

$$= \frac{\text{Heat rejected } Q_2}{\text{Work input } W}$$

$$Q_2 = \text{Area } (6\text{-}2\text{-}3\text{-}5)$$

$$\text{COP}_H = \frac{T_2 (s_1 - s_4)}{(T_2 - T_1)(s_1 - s_4)} \tag{3.7}$$

$$= \frac{T_2}{T_2 - T_1} \tag{3.8}$$

$$= \frac{1}{1 - (T_1/T_2)}$$

It follows that the Carnot COP depends on temperatures T_1 and T_2 only. In the sense, it does not depend upon the working substance (refrigerant) used.

The reversed Carnot cycle may be employed for cooling or heating purpose. For cooling, let T_1 be the refrigeration temperature and T_2 be the surrounding temperature. If we substitute $T_1 = 0$ (absolute zero), the minimum possible refrigeration temperature, in Eq. (3.6) then $\text{COP}_R = 0$. The upper limit to refrigeration temperature is $T_1 = T_2$, i.e. when the refrigeration temperature is equal to the temperature of the surroundings (ambient) at which $\text{COP}_R = \infty$. **Thus Carnot COP for cooling varies between 0 and ∞.**

From Eq. (3.8), it is clear that the Carnot COP for heating varies between 1 and ∞. Also,

(i) the refrigeration temperature T_1 should be as high as possible, and
(ii) the surrounding temperature T_2 should be as low as possible.

These two points are applicable to all refrigerating machines, whether theoretical or practical. A sub-consideration reveals that the refrigeration and surrounding temperature cannot be varied at our will.

The refrigeration temperature is fixed by the particular application, i.e. this must be less than the temperature of the substance in the refrigerated space to be cooled. For example, if the temperature of the refrigerated space would be say –5°C then the actual refrigerant temperature would be less than –10°C. This difference in temperature will be the base to decide the evaporator size for the required rate of cooling.

The effect of raising the refrigeration temperature on the Carnot COP for the fixed surroundings temperature 47°C is shown in Figure 3.5.

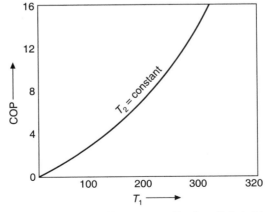

Figure 3.5 Carnot COP variation with respect to T_1 when T_2 is held constant (47°C).

Similarly, the temperature of the refrigerant must be higher than the surrounding temperature (cooling media in condenser) so that heat can be rejected. Thus, it can be said that if the temperature of the available cooling medium like water or air for heat rejection is lower, then Carnot COP for refrigeration will improve. For this reason, the COP of Carnot refrigerator will be higher in winter than in summer. This effect is shown in Figure 3.6.

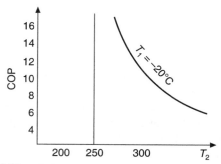

Figure 3.6 Carnot COP variation with respect to T_2, when T_1 is kept constant (–20°C).

EXAMPLE 3.1 A refrigeration system operates on reversed Carnot cycle between the temperature limits –23°C and 45°C. The refrigerating capacity is 10 TR. Determine (a) the COP, (b) the heat rejected from the system per hour, and (c) the power required.

Solution:

(a) $$COP_R = \frac{T_1}{T_2 - T_1}$$

$$= \frac{273 - 23}{(273 + 45) - (273 - 23)} = \frac{250}{318 - 250} = 3.67 \quad \textbf{Ans.}$$

(b) Also, $$COP_R = \frac{\text{Refrigerating effect}}{\text{Work input}}$$

∴ $$\text{Work input} = \frac{10 \times 211 \times 60}{3.67} = 34{,}332 \text{ kJ/h}$$

Heat rejected = refrigerating effect per hour + work input per hour
$$= (10 \times 211 \times 60) + 34{,}332$$
$$= 160{,}932 \text{ kJ/h} \quad \textbf{Ans.}$$

(c) $$\text{Power in kW} = \frac{34332}{3600} = 9.536 \text{ kW} \quad \textbf{Ans.}$$

EXAMPLE 3.2 A Carnot refrigerator works on a reversed Carnot cycle. This unit requires 1.5 kW power for every one TR of refrigeration at –23°C. Determine (a) the COP, (b) the higher temperature of cycle and (c) the heat rejected in kJ/min. Also calculate the heat removal rate and COP when this device is used as a heat pump.

Solution: $T_1 = 273 - 23 = 250$ K
(a) For refrigerator:

$$COP_R = \frac{\text{Refrigerating effect}}{\text{Work done}}$$

$$= \frac{1 \text{ TR}}{1.5 \text{ kW}} = \frac{3.516 \text{ kW}}{1.5 \text{ kW}} = 2.34 \qquad \textbf{Ans.}$$

(b) Also, $\qquad \text{COP}_R = \dfrac{T_1}{T_2 - T_1} = \dfrac{250}{T_2 - 250}$

$\therefore \qquad T_2 = 356.6 \text{ K} \qquad \textbf{Ans.}$

(c) For heat pump:

Heat rejected or supplied to a space at T_2,

$= \text{Heat absorbed} + \text{Work done}$

$= 3.516 + 1.5 = 5.01 \text{ kW} \qquad \textbf{Ans.}$

$\therefore \qquad \text{COP}_{HP} = \dfrac{\text{Heat rejected}}{\text{Work done}} = \dfrac{5.01}{1.5} = 3.3 \qquad \textbf{Ans.}$

3.6.2 Limitations of Reversed Carnot Cycle

We have seen in Figure 3.4 that the Carnot cycle consists of four processes, namely reversible adiabatic compression (1–2), isothermal compression (2–3), reversible adiabatic expansion (3–4), and isothermal expansion process (4–1).

During the reversible adiabatic compression process (1–2), the piston is supposed to move very fast and for the isothermal compression process, the piston has to move at a very slow speed. It means that the piston has to move at very high speed in the first part of its compression stroke and for the remaining part of the stroke, it has to move at very slow speed, which are both practically impossible situations.

During the isothermal process, it is assumed that a hot or cold imaginary body is brought in contact with cylinder walls so that heat exchange between the system and the surrounding takes place. This is also not practical.

The internal and external friction is assumed negligible in cycle processes, which is not a correct assumption. The working substance is the ideal gas. In practice there is no working substance that behaves as an ideal gas. Therefore, the reversed Carnot cycle is an ideal cycle which gives the upper limit of the COP for a particular refrigeration system.

In a practical refrigeration cycle working on air as a refrigerant, the constant pressure heat absorption/rejection process replaces the isothermal heat absorption/rejection process of the reversed Carnot cycle.

3.6.3 Modified Reversed Carnot Cycle

The reversed Carnot cycle with perfect gas as working fluid is the most efficient cycle. The *T–s* diagram of such a cycle is a rectangular one as shown Figure 3.7.

If air was used as a refrigerant, this cycle would appear like the one shown in Figure 3.7. The isentropic compression and expansion are the processes 1–2 and 3–4 respectively. The processes 2–3 and 4–1 are constant pressure cooling and heating processes respectively. This cycle differs from the Carnot cycle (Figure 3.4) in that it operates between the same two temperatures but with additional two areas *X* and *Y*.

At point 4 the temperature of air must be lower than that of cold space so that air absorbs

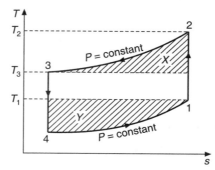

Figure 3.7 Reversed Carnot cycle with air as refrigerant

heat from the cold space and experiences increase in temperature up to T_1. The addition of area Y decreases not only the COP but also the refrigerating effect. Similarly, at point 2, the temperature of air must be higher than that of the surroundings so that heat can be rejected. This adds area X to the cycle. Therefore, this addition of area X increases not only the power but also decreases the COP.

3.6.4 Reversed Carnot Cycle with Vapour as Refrigerant

Instead of air, a refrigerant can be used that condenses during the heat rejection process and evaporates during the heat absorption (in evaporator) process. With such a refrigerant, the reversed Carnot cycle becomes as if it has been fitted with the saturated liquid and saturated vapour lines as shown in Figure 3.8.

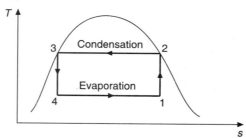

Figure 3.8 Reversed Carnot cycle when a condensable fluid is the refrigerant.

Process 2–3: It is a constant temperature condensation process.
Process 4–1: It is also a constant temperature process but an evaporation process. The isothermal and isobars lines are one and the same in the two-phase region.

At point 1, the refrigerant is a mixture of liquid and vapour. When this mixture is compressed (called *wet compression*) in a reciprocating compressor, then liquid refrigerant may be trapped in the cylinder head by the rising piston, which may damage the valves and the cylinder head.

Ideally, at point 2 the refrigerant should be in a saturated vapour state but practically the case is different as it contains liquid droplets. This so happens because in high-speed compressors, the liquid droplets that are to be vaporized by the internal heat transfer do not get much time to change the phase. Therefore, these droplets remain up to state point 2. It means that **state 2** does not reflect the true condition of the refrigerant. Hence, it is practically impossible to terminate the compression process exactly at state **point 2.**

Another drawback of wet compression is that the droplets of liquid may wash the lubricating oil from the cylinder walls, which leads to many undesirable effects like increased friction, wear, and blow-by losses, etc.

On account of these disadvantages, the dry compression is preferred. For this, the state point 1 should be on the saturation curve. The refrigerant must be in saturated vapour state at the suction of the compressor. This is as indicated in Figure 3.9.

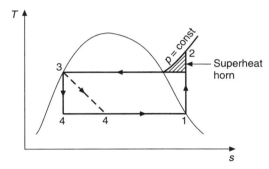

Figure 3.9 Modified reversed Carnot cycle using dry compression and throttling process.

Compression of dry vapour results in a high temperature at the end of compression, say, at point 2. This temperature is much higher than the condensing temperature. The refrigerant leaves the compressor in a superheated condition. The area of that part of the cycle, which is above the condensing temperature, is called the *superheat horn,* and represents the additional work required by dry compression.

In the reversed Carnot cycle, it is assumed that expansion occurs isentropically and the resulting work would be used to drive the compressor. But it is impractical because the work that is derived by the expansion device is very small. Another difficulty is that the expansion takes place in two phases and poses the lubrication problem. Therefore, the necessity is to reduce the pressure of the liquid from condenser to evaporator, which is certainly possible with a throttling device (a valve or a capillary tube). This throttling process 3–4 is isenthalpic and irreversible in nature. Therefore, **this is one more modification introduced in the original reversed Carnot cycle as indicated by the dashed line 3–4 in Figure 3.9.** Now the cycle 1–2–3–4 is the standard vapour compression refrigeration cycle which would be discussed in detail in the next chapter.

3.6.5 Bell–Colemann or Reversed Brayton or Joule Cycle

The Bell–Colemann refrigerator using air as a refrigerant is shown with the help of a block diagram in Figure 3.10. It consists of a compressor, a cooler, an expander and a refrigerator.

The Bell–Colemann air refrigeration cycle is the modification of the reversed Carnot cycle with air as a working medium. It can be operated as an open cycle shown in Figure 3.10(a), in which the cold air available at the outlet of the expander is used for refrigeration and is let out in the atmosphere. In the closed cycle [Figure 3.10(b)] the same air is circulated repeatedly. The cold air available at the outlet of the expander is employed to cool the other fluid in the heat exchanger. The other fluid acts as the secondary refrigerant in this cycle.

The cycle is represented on *p–v* and *T–s* diagrams as shown in Figure 3.11 and Figure 3.12 respectively.

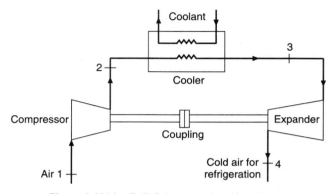

Figure 3.10(a) Bell–Colemann air refrigeration open cycle.

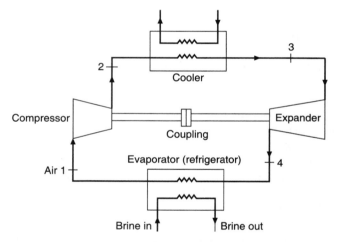

Figure 3.10(b) Bell–Colemann air refrigeration closed cycle (dense air cycle).

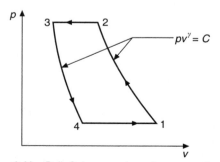

Figure 3.11 Bell–Colemann air cycle on *p–v* diagram.

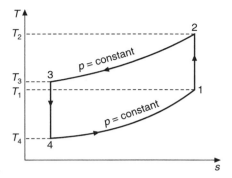

Figure 3.12 Bell–Colemann cycle on T–s diagram.

The four processes of the closed cycle are as follows:

Isentropic compression process (1–2): The air from the refrigerator is drawn into the compressor and compressed isentropically to state 2. During compression, both pressure and temperature increase while the specific volume decreases from v_1 to v_2. During the isentropic process, no heat is absorbed or rejected by the air.

Constant pressure cooling process (2–3): The high temperature air is cooled from temperature T_2 to T_3 in the cooler at constant pressure. Here the frictional pressure loss on account of the friction between the air and the heat exchanger is neglected. The specific volume decreases from v_2 to v_3. The heat rejected by the air during this process is

$$Q_{2-3} = c_p(T_2 - T_3)$$

Isentropic expansion process (3–4): The air is expanded isentropically from pressure p_2 ($p_3 = p_2$) to p_4 ($p_4 = p_1$ = atm pressure), while the temperature of air also decreases from T_3 to T_4. The specific volume of air increases from v_3 to v_4. During this no heat exchange takes place.

Constant pressure heat absorption process (4–1): The cold air from the expander is circulated through the refrigerator where it absorbs heat from brine and thus cools the brine. During this, the temperature of the air increases from T_4 to T_1 and the specific volume increases from v_4 to v_1. The heat absorbed from the refrigerator during constant pressure process per kg of air is

$$Q_{4-1} = c_p(T_1 - T_4) \tag{3.9}$$

Work done during the cycle per kg of air

$$= \text{Heat rejected} - \text{heat absorbed}$$
$$= c_p(T_2 - T_3) - c_p(T_1 - T_4) \tag{3.10}$$

Coefficient of performance, $\text{COP} = \dfrac{\text{Heat absorbed}}{\text{Work input}}$

$$= \frac{c_p(T_1 - T_4)}{c_p[(T_2 - T_3) - (T_1 - T_4)]} = \frac{T_1 - T_4}{(T_2 - T_3) - (T_1 - T_4)}$$

$$= \frac{T_4 (T_1/T_4 - 1)}{T_3 (T_2/T_3 - 1) - T_4 (T_1/T_4 - 1)} \qquad (3.11)$$

For the isentropic compression process 1–2,

$$\frac{T_2}{T_1} = \left(\frac{p_2}{p_1}\right)^{\frac{\gamma-1}{\gamma}}$$

Similarly, for the isentropic expansion process 3–4

$$\frac{T_3}{T_4} = \left(\frac{p_3}{p_4}\right)^{\frac{\gamma-1}{\gamma}}$$

Since $p_2 = p_3$ and $p_1 = p_4$, the equations would be:

$$\frac{T_2}{T_1} = \frac{T_3}{T_4}; \qquad \frac{T_2}{T_3} = \frac{T_1}{T_4}$$

Putting these values in Eq. (3.11), we get

$$\text{COP} = \frac{T_4}{T_3 - T_4} = \frac{1}{(T_3/T_4) - 1}$$

$$= \frac{1}{\left(\dfrac{p_3}{p_4}\right)^{(\gamma-1)/\gamma} - 1} = \frac{1}{\left(\dfrac{p_2}{p_1}\right)^{(\gamma-1)/\gamma} - 1}$$

$$= \frac{1}{r^{(\gamma-1)/\gamma} - 1} \qquad (3.12)$$

where r = compression or expansion ratio = $\dfrac{p_2}{p_1} = \dfrac{p_3}{p_4}$.

Practically, the reversible adiabatic process is not possible. The reversible adiabatic or isentropic compression and expansion processes in the above analysis are replaced by the polytropic process pv^n = constant. The COP of the cycle can be obtained as follows:

Compression work per kg air, $W_C = \dfrac{n}{n-1}(p_2 v_2 - p_1 v_1)$ kJ/kg

$$= \frac{n}{n-1} R(T_2 - T_1)$$

Expansion work per kg air, $W_E = \dfrac{n}{n-1} R(T_3 - T_4)$

The net work done, $W = W_C - W_E$

$$= \frac{n}{n-1} R[(T_2 - T_1) - (T_3 - T_4)]$$

$$= \frac{n}{n-1}\left(\frac{\gamma-1}{\gamma}\right) c_p[(T_2 - T_1) - (T_3 - T_4)]$$

since the characteristic gas constant, $R = \left(\frac{\gamma-1}{\gamma}\right) c_p$

From our previous analysis, $\quad \dfrac{T_4}{T_1} = \dfrac{T_3}{T_2}$

$$\text{COP} = \frac{\text{Refrigerating effect}}{\text{Work done}}$$

$$= \frac{c_p (T_1 - T_4)}{\dfrac{n}{n-1}\left(\dfrac{\gamma-1}{\gamma}\right) c_p [(T_2 - T_1) - (T_3 - T_4)]}$$

$$= \frac{c_p (T_1 - T_4)}{\dfrac{n}{n-1}\left(\dfrac{\gamma-1}{\gamma}\right) c_p [(T_2 - T_3) - (T_1 - T_4)]}$$

$$= \frac{n-1}{n} \times \frac{\gamma}{\gamma - 1} \times \frac{T_4}{T_3 - T_4} \tag{3.13}$$

3.6.6 Actual Bell–Colemann Cycle

The actual Bell–Colemann air refrigeration cycle shown in Figure 3.13 deviates in four ways from the cycle shown in Figure 3.12.

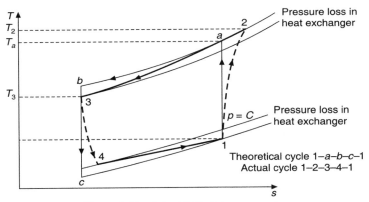

Figure 3.13 Actual Bell–Colemann cycle.

Process (1–2): The compression process in the compressor is non-isentropic due to internal and external friction between the air and the surface temperature T_2 at the end of compression is more than T_a.

Process (2–3): The cooling of air from temperature T_2 to T_3 takes place in the heat exchanger. Heat exchanger is a coil of tubes in which the air flows. Therefore, due to friction between the air and the metal tube surface a pressure loss occurs. Therefore, the cooling process is not truly at constant pressure.

Process (3–4): The expansion of air from pressure p_3 to p_4 instead of from b to c is on account of friction between the air and the expander surface.

Process (4–1): This process takes place in the heat exchanger, hence a pressure drop.

Practically, the actual cycle indicates that the actual compression work is higher compared with that of isentropic compression. The work obtained in the expander is less than that available from isentropic expansion.

3.6.7 Application of Aircraft Refrigeration

The air refrigeration cycle is exclusively used for air conditioning of all types of aircraft except cargo aircraft. Air is the working substance in such air-conditioning systems. The coefficient of performance of the air refrigeration system for aircraft is very poor compared to the vapour compression refrigeration cycle employed for the same purpose due to its many advantages.

One may ask that since at high altitudes both atmospheric temperature and pressure decrease, then why are aircraft air conditioned? The reasons are attributable to external and internal heat leakages. Also, temperature of the order of –5°C is needed to preserve food and cold drinks.

Following are the external heat sources which add the heat in the compartment of occupants.

External heat sources

1. Solar radiations enter the compartments through windows.
2. Due to solar radiations, the outer surface of the aircraft gets heated. Therefore, heat is conducted into the compartments.
3. The speed of aircraft is very high—of the order of 1500 km/h. Therefore, there is a skin friction between the outer aircraft surface and the air due to which the whole aircraft body gets heated.

Internal heat sources

1. A normal healthy person dissipates heat at the rate of 180 W (180 J/s). For the capacity of 100 persons in an aircraft, the total heat dissipated would be 18 kW. To compensate for such a huge heat load, one requires air conditioning.
2. Electrical and electronic components generate heat throughout their use.
3. The engine parts of aircraft generate heat which is conducted into the compartments.
4. Food products, cold drinks, etc. need to be stored at a low temperature.

3.7 METHODS OF AIR REFRIGERATION SYSTEMS

The various methods of air refrigeration systems used for aircraft are as follows:

- Simple air-cooling system
- Simple air-evaporative cooling system
- Boot-strap air cooling system
- Boot-strap air evaporative cooling system
- Reduced ambient air cooling system
- Regenerative air cooling system

3.7.1 Simple Air-cooling System

Simple air-cooling system of an aircraft is shown with the help of a block diagram in Figure 3.14

and on *T–s* diagram in Figure 3.15. The components of the system are a diffuser, a compressor coupled to a turbine, a heat exchanger, a cooling turbine, and an aircooling fan.

A part of the ram air bled off at state **point 3** is cooled in a heat exchanger with the help of rammed air to state **point 4**. It is further cooled due to expansion to cabin pressure in the cooling turbine and is then supplied to the cabin. The work of expansion of cooling turbine is used to drive the air cooling fan which draws the ram air through the heat exchanger.

The remaining air at a high pressure (p_3) and a high temperature (T_3) and the compressed air is supplied to gas turbine through the combustion chamber. Turbine power is used to drive the main compressor and other equipment of the aircraft.

Simple air-cooling system is useful for ground surface air cooling and for aircraft at low speeds.

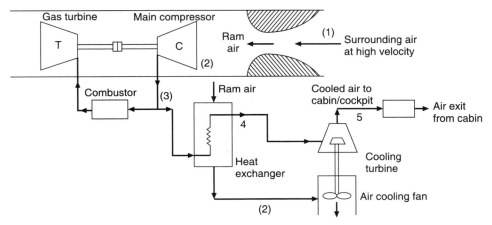

Figure 3.14 Simple air-cooling system.

Ramming process: The ambient air at p_1 and T_1 enters through a diffuser and due to ramming effect its pressure and temperatures are raised to p_2 and temperature T_2. This *ideal ramming action* is shown by the **vertical line 1–2** in Figure 3.15. In practice, because of internal friction and inherent irreversibilities, the temperature of the rammed air is more than T_2. Thus the actual ramming process is shown by the curve 1–2′. The pressure and temperature of the rammed air are now $p_{2'}$ and $T_{2'}$ respectively.

The diffuser efficiency η_d for ram compression is defined as:

$$\eta_d = \frac{\text{Actual pressure rise } (p_{2'} - p_1)}{\text{Ideal pressure rise } (p_2 - p_1)}$$

Let us apply the steady flow energy equation to the ramming process (diffuser). Let V_1 and V_2 be the relative velocities of air in m/s before and after diffuser respectively.

$$h_1 + \frac{V_1^2}{2} = h_2 + \frac{V_2^2}{2}$$

But $V_2 = 0$ after ram compression.

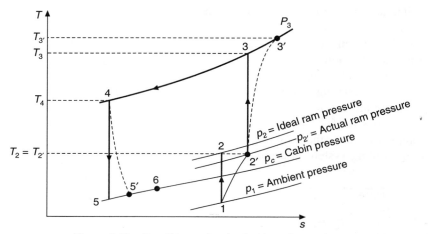

Figure 3.15 T–s diagram for simple air-cooling system.

$$\therefore \qquad h_2 - h_1 = \frac{V_1^2}{2}$$

$$c_p(T_2 - T_1) = \frac{V^2}{2} \quad \text{for unit mass of air.}$$

$$\therefore \qquad T_2 = T_1 + \frac{V^2}{2 c_p}$$

or
$$\frac{T_2}{T_1} = 1 + \frac{V^2}{2 c_p T_1} \qquad (3.14)$$

We know that $\quad c_p - c_v = R$

or $\quad c_p \left[1 - \dfrac{c_v}{c_p}\right] = R \quad$ or $\quad c_p \left[1 - \dfrac{1}{\gamma}\right] = R \quad (\because c_p / c_v = \gamma)$

$$\therefore \qquad c_p = \frac{\gamma R}{\gamma - 1}$$

$$\therefore \qquad T_2 = T_1 + \frac{1}{2}\left(\frac{\gamma - 1}{\gamma}\right)\frac{V_1^2}{R}$$

or
$$\frac{T_2}{T_1} = 1 + \frac{(\gamma - 1) V_1^2}{2 \cdot \gamma R T_1} \qquad (3.15)$$

Note that $\quad T_{2'} = T_2$

But sound velocity of ambient air, $c_1 = \sqrt{\gamma R T_1}$

$$\therefore \quad \frac{T_2}{T_1} = 1 + \left(\frac{\gamma-1}{2}\right)\frac{V_1^2}{c_1^2} = 1 + \left(\frac{\gamma-1}{2}\right)M^2 \qquad (3.16)$$

where Mach number, $\quad M = \dfrac{\text{Actual velocity, } V_1}{\text{Sound velocity, } c_1}$

and

$$\frac{p_{2'}}{p_1} = \left(\frac{T_2}{T_1}\right)^{\gamma/(\gamma-1)} \qquad (3.17)$$

where M is the mach number of the flight. It is defined as the ratio of aircraft velocity (V) to the local sound velocity (c).

The temperature $T_2 = T_{2'}$ is called the *stagnation temperature* of the ambient air entering the main compressor. The velocity of air at the outlet of the diffuser is assumed to be zero which is why the 'stagnation' is used. The properties corresponding to the point are stagnation properties. The stagnation pressure after isentropic compression (p_2) is determined with the help of diffuser efficiency.

Compression process: The isentropic compression of air in the main compressor is represented by the line 2′–3. In practice, because of friction, the compression process does not remain isentropic. So the actual compression is represented by the curve 2′–3′ on T–s diagram. The work done for compression process is given by

$$W_C = m_a c_p (T_{3'} - T_{2'}) \qquad (3.18)$$

where m_a = mass of air bled from the main compressor for refrigeration purposes since we are analyzing the refrigeration system.

Cooling process: The ram air is used to cool the compressed air in the heat exchanger. This process is a constant pressure cooling shown by the curve 3–4 in Figure 3.15. In practice, there is a pressure drop in the heat exchanger, which is not shown in the figure. The temperature of air decreases from $T_{3'}$ to T_4. The heat rejected in the heat exchanger is given by

$$Q_R = m_a c_p (T_{3'} - T_4) \qquad (3.19)$$

Expansion process: The cooled air is now expanded isentropically in the cooling turbine as shown by the curve 4–5. The actual expansion in the cooling turbine is shown by the curve 4–5′. The work obtained in cooling turbine due to expansion process is given by

$$W_R = m_a c_p (T_4 - T_{5'}) \qquad (3.20)$$

The work of this turbine is used to drive the cooling air fan, which draws cooling air from the heat exchanger.

Refrigeration process: Cool air from the outlet of cooling turbine (i.e. after expansion) is supplied to the cabin and cockpit where it gets heated by the heat of equipment and occupancy.

This process is shown by the curve 5′–6 in Figure 3.15. The refrigerating effect produced or heat absorbed is given by

$$= m_a c_p (T_6 - T_{5'}) = m_p (h_6 - h_5) \quad (3.21)$$

where T_6 = inside temperature of the cabin.

COP of the air cycle,

$$= \frac{\text{Refrigerating effect produced}}{\text{Work done}}$$

$$= \frac{m_a c_p (T_6 - T_{5'})}{m_a c_p (T_{3'} - T_{2'})} = \frac{T_6 - T_{5'}}{T_{3'} - T_{2'}} \quad (3.22)$$

If Q tons of refrigeration is the cooling load in the cabin, then the air required for the refrigeration purpose,

$$m_a = \frac{211 Q}{c_p (T_6 - T_{5'})} \text{ kg/min} \quad (3.23)$$

Power required for the refrigeration system,

$$P = \frac{m_a c_p (T_{3'} - T_{2'})}{60} \text{ kW} \quad (3.24)$$

And COP of the refrigerating system

$$= \frac{211 Q}{m_a c_p (T_{3'} - T_{2'})} = \frac{211 Q}{P \times 60} \quad (3.25)$$

EXAMPLE 3.3 An aircraft air conditioning needs refrigeration capacity of 10 TR. At the altitude of aircraft the atmospheric pressure and temperature are 0.9 bar and 10°C respectively. The pressure of air after ramming effect in a diffuser increases to 1.013 bar. The temperature of the air is reduced by 50°C in the heat exchanger. The pressure in the cabin is 1.01 bar and the temperature of air leaving the cabin is 25°C. The pressure of the compressed air is 3.5 bar. Assume that all the expansions and compressions are isentropic. Also carry out the calculation if the compression and expansion efficiencies are 90%.

Determine the following:

(a) Power required to take the load of cooling in the cabin.
(b) COP of the system.

If the efficiencies of the expansion and compression processes are 90%, analyse the problem for power requirement in both the cases.

Solution: Refer to Figure 3.16.

Given: $Q = 10$ TR; $p_1 = 0.9$ bar; $T_1 = 10°C = 10 + 273 = 283$ K;
$p_2 = 1.013$ bar; $p_5 = p_6 = 1.01$ bar; $T_6 = 25°C = 25 + 273 = 298$ K; $p_3 = 3.5$ bar.

Case 1: With 100% efficiency of compression and expansion processes:

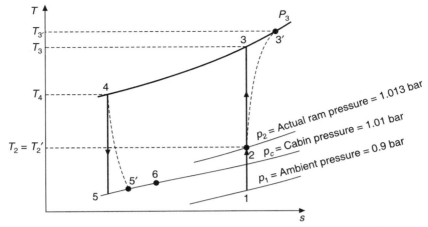

Figure 3.16 T–s diagram for simple air-cooling system—Example 3.3.

Temperature after ramming effect

$$\frac{T_2}{T_1} = \left(\frac{p_2}{p_1}\right)^{(\gamma-1)/\gamma} = \left(\frac{1.013}{0.9}\right)^{(1.4-1)/1.4} = (1.125)^{0.286} = 1.034$$

∴ $T_2 = T_1 \times 1.034 = 283 \times 1.034 = 292.6$ K

Similarly, $\dfrac{T_3}{T_2} = \left(\dfrac{p_3}{p_2}\right)^{(\gamma-1)/\gamma} = \left(\dfrac{3.5}{1.013}\right)^{(1.4-1)/1.4}$

$$= (3.45)^{0.286} = 1.425$$

∴ $T_3 = T_2 \times 1.425 = 292.6 \times 1.425 = 417$ K $= 144°C$

The temperature of air is reduced by 50°C in the heat exchanger, therefore, the temperature of air leaving the heat exchanger,

$$T_4 = 144 - 50 = 94°C = 367 \text{ K}$$

We know that $\dfrac{T_5}{T_4} = \left(\dfrac{p_5}{p_4}\right)^{(\gamma-1)/\gamma} = \left(\dfrac{1.01}{3.5}\right)^{(1.4-1)/1.4} = (0.288)^{0.286} = 0.7$

∴ $T_5 = T_4 \times 0.7 = 367 \times 0.7 = 257$ K

We know that the mass of air required for the refrigeration purpose is

$$m_a = \frac{211\,Q}{c_p(T_6 - T_5)} = \frac{211 \times 10}{1(298 - 257)} = 51.2 \text{ kg/min}$$

74 *Refrigeration and Air Conditioning*

(a) Power required to take the load of cooling in the cabin,

$$P = \frac{m_a c_p (T_3 - T_2)}{60} = \frac{51.2 \times 1(417 - 292.6)}{60} = 106 \text{ kW} \quad \textbf{Ans.}$$

(b) COP of the system:
We know that COP of the system is

$$= \frac{211 Q}{P \times 60} = \frac{211 \times 10}{106 \times 60} = 0.33 \quad \textbf{Ans.}$$

Case 2: When the efficiencies of compression and expansion are 90%:
Compression (2–3):

$$\eta_C = \frac{T_3 - T_2}{T_{3'} - T_2} = 0.9 = \frac{417 - 292.6}{T_{3'} - 292.6}$$

$$T_{3'} = 430.8 \text{ K} = 157.8°C$$

The temperature of air is reduced by 50°C in the heat exchanger, therefore, the temperature of air leaving the heat exchanger is

$$T_4 = 157.8°C - 50°C = 107.8°C = 380.8 \text{ K}$$

We know that

$$\frac{T_5}{T_4} = \left(\frac{p_5}{p_4}\right)^{(\gamma-1)/\gamma} = \left(\frac{1.01}{3.5}\right)^{(1.4-1)/1.4} = (0.288)^{0.286} = 0.7$$

∴

$$T_5 = T_4 \times 0.7 = 380.8 \times 0.7 = 266.5 \text{ K}$$

Turbine efficiency:

$$\eta_t = \frac{T_4 - T_{5'}}{T_4 - T_5} = 0.9 = \frac{380.8 - T_{5'}}{380.8 - 266.5}$$

∴

$$T_{5'} = 277.93 \text{ K}$$

Mass of air required for the refrigeration purposes,

$$m_a = \frac{211 Q}{c_p (T_6 - T_{5'})} = \frac{211 \times 10}{1(298 - 277.93)} = 104.6 \text{ kg/min}$$

(a) Power required to take the load of cooling in the cabin,

$$P = \frac{m_a c_p (T_{3'} - T_2)}{60} = \frac{104.6 \times 1(430.8 - 292.6)}{60} = 240.9 \text{ kW} \quad \textbf{Ans.}$$

(b) COP of the system:
We know that COP of the system is

$$= \frac{211 Q}{P \times 60} = \frac{211 \times 10}{240.9 \times 60} = 0.145 \quad \textbf{Ans.}$$

From the above calculation it is evident that a 10% reduction in the efficiency of both compression and expansion processes would increase the power requirement from 106 kW to 240.9 kW. In percentage, it is an increase of 127%. It is clear that utmost care should be taken to maintain the condition of both the compressor and turbine to the maximum possible.

EXAMPLE 3.4 An aircraft refrigeration plant has a capacity of 30 TR. The ambient temperature is 17°C. The atmospheric air is compressed to 0.95 bar and 30°C due to ram action. The air is then further compressed in a compressor to 4.75 bar and is then cooled in a heat exchanger to 67°C. It then expands in a turbine to 1 bar before it is supplied to the cabin. Air leaves the cabin at 27°C. The isentropic efficiencies of the compressor and the turbine are 0.9.
Determine the following:

(a) Mass flow rate of air circulated/second.
(b) COP.
(c) Specific power required. Take c_p = 1.004 kJ/kg-K and γ = 1.4 for air.

Sketch the cycle on T–s diagram.

Solution: Refer to Figures 3.17 and 3.18.

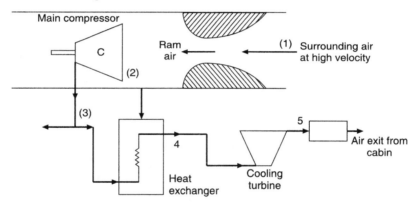

Figure 3.17 Simple air-cooling system—Example 3.4.

Given: Capacity = 30 TR = 30 × 3.517 = 105.51 kW

$T_1 = 17°C = 290$ K $\quad\quad p_2 = 0.95$ bar $\quad\quad T_2 = 30°C = 303$ K
$p_3 = 4.75$ bar $\quad\quad T_4 = 67°C = 340$ K $\quad\quad p_5 = 1$ bar
Cabin exit temperature = 27°C = 300 K; Also, $\eta_C = \eta_T = 0.9$

In compressor, $\quad\quad T_3 = T_2 \left(\dfrac{p_3}{p_2}\right)^{(\gamma-1)/\gamma} = 303\left(\dfrac{4.75}{0.95}\right)^{(1.4-1).0.4}$

∴ $\quad\quad T_3 = 479.89$ K

$$\eta_C = \dfrac{T_3 - T_2}{T_{3'} - T_2} = 0.9 = \dfrac{479.89 - 303}{T_{3'} - 303}$$

∴ $\quad\quad T_{3'} = 499.55$ K

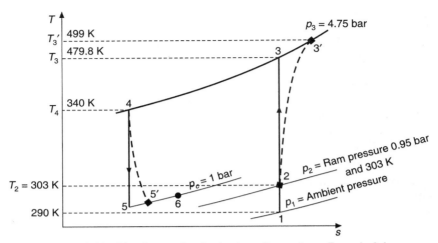

Figure 3.18 T–s diagram for simple air-cooling system—Example 3.4.

For turbine,
$$T_5 = T_4 \left(\frac{p_5}{p_4}\right)^{(\gamma-1)/\gamma} = 340\left(\frac{1}{4.75}\right)^{(1.4-1)/1.4}$$

∴ $T_5 = 217.84$ K

$$\eta_T = \frac{T_4 - T_{5'}}{T_4 - T_5} = 0.9 = \frac{340 - T_{5'}}{340 - 217.84}$$

∴ $T_{5'} = 230.0$ K

Refrigerating effect/kg $= c_p$(cabin temperature $- T_{5'}$)
$= 1.005(300 - 230.0) = 70.29$ kJ/kg
Capacity $= 105.51$ kW

(a) Mass flow rate in kg/s $= \dfrac{\text{Capacity}}{\text{Refrigerating effect}} = \dfrac{105.51}{70.29}$

∴ $m = 1.5$ kg/s **Ans.**

(b) Work done $W = W_C - W_T$
$W = c_p(T_{3'} - T_2) - c_p(T_4 - T_{5'})$
$= 1.005(499.55 - 303 - 340 + 230.06)$
$W = 87.04$ kJ/kg
$W = 87.04 \times 1.5 = 130.56$ kW

$$\text{COP} = \frac{\text{Capacity}}{W} = \frac{105.51}{130.56} = 0.81 \quad \textbf{Ans.}$$

(c) Specific power required (kW/TR) $= \dfrac{\text{Work in kW}}{\text{Capacity in TR}} = 4.352$ kW/TR **Ans.**

Air Refrigeration Systems   77

EXAMPLE 3.5 An aircraft is moving at a speed of 1000 km/h at an altitude of 8000 m, where the ambient pressure and temperature are 0.35 bar and –15°C respectively. The cabin of the plane is maintained at 25°C by using a simple air refrigeration system. The pressure ratio of compressor is 3. The air is passed through heat exchanger after compression and cooled to its original condition entering into the plane. A pressure loss of 0.1 bar takes place in the heat exchanger. The pressure of the air leaving the cooling turbine is 1.06 and the air pressure in the cabin is 1.013 bar. Considering the total cooling load of plane to be 70 kW, determine the following:

(a) Stagnation temperature and pressure.
(b) Mass flow rate of air circulated through the cabin.
(c) Volume handled by the compressor and expander.
(d) Net power delivered to the refrigeration system and COP of the system.

Solution: Refer to Figures 3.19(a) and (b).

Given: Plane speed, $C = 1000$ km/h $= 277.78$ m/s;
$p_1 = 0.35$ bar; $T_1 = -15°C = 258$ K;

$$\frac{p_2}{p_3} = 3 \quad \text{or} \quad p_2 = 3p_3;$$

$p_4 = (p_3 - 0.1)$ bar; cabin temperature $= 25°C = 298$ K
pressure at exit of turbine $= 1.06$ bar; pressure in the cabin $= 1.013$ bar;
cooling load $= 70$ kW

(a)   Stagnation temperature, $T_2 = T_1 + \dfrac{C^2}{2000\, c_p}$

$$= 258 + \frac{(277.78)^2}{2000 \times 1.005}$$

$T_2 = 296.6$ K   **Ans.**

Stagnation pressure,

$$\frac{p_2}{p_1} = \left(\frac{T_2}{T_1}\right)^{\gamma/(\gamma-1)}$$

$$p_2 = 0.35 \left(\frac{296.6}{258}\right)^{1.4/0.4} = 0.57 \text{ bar}$$ **Ans.**

Stagnation pressure, $p_2 = 0.57$ bar
$p_3 = 3p_2 = 1.71$ bar

$$T_3 = T_2 \left(\frac{p_3}{p_2}\right)^{(\gamma-1)/\gamma} = 405.97 \text{ K}$$

Pressure at turbine inlet, $p_4 = p_3 - 0.1 = 1.61$ bar

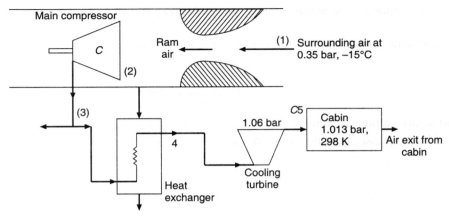

Figure 3.19(a) Simple air-cooling system—Example 3.5.

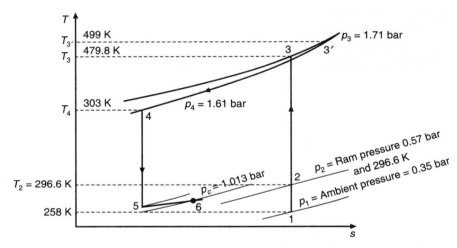

Figure 3.19(b) T–s diagram for simple air-cooling system—Example 3.5.

Turbine inlet temperature = original temperature of the air entering the plane
i.e. $T_4 = 258$ K

Turbine exit temp., $T_5 = T_4 \left(\dfrac{p_5}{p_4} \right)^{(\gamma-1)/\gamma}$

$$T_5 = 258 \left(\dfrac{1.06}{1.61} \right)^{(1.4-1)/1.4} = 264.25 \text{ K}$$

Refrigerating effect in cabin = $c_p(T_6 - T_5)$ = 1.005(298 − 264.25)
= 33.92 kJ/kg

(b) Mass flow rate $= \dfrac{\text{Cooling capacity in kg/s}}{\text{Refrigerating effect}}$

$= \dfrac{70}{33.92} = 2.06$ kg/s **Ans.**

(c) Volume through compressor $= \dfrac{mRT_2}{p_2} = \dfrac{2.06 \times 0.287 \times 296.6}{0.57 \times 10^2} = 1.66$ m³/s **Ans.**

Volume through expander $= \dfrac{mRT_5}{p_5} = \dfrac{2.06 \times 0.287 \times 229.86}{1.06 \times 10^2} = 1.282$ m³/s **Ans.**

(d) Net work $= W_C - W_T = mc_p(T_3 - T_2) - mc_p(T_4 - T_5)$
$= mc_p(T_3 - T_2 - T_4 + T_5)$

or
$W = 2.06 \times 1.005\,(405.97 - 296.6 - 258 + 229.86)$
$W = 167.33$ kW **Ans.**

and $\text{COP} = \dfrac{Q}{W} = \dfrac{70}{167.33} = 0.42$ **Ans.**

EXAMPLE 3.6 An air-cooling system for a jet plane cockpit operates on the simple cycle. The cockpit is to be maintained at 25°C. The ambient air pressure and temperature are 0.35 bar and –15°C respectively. The pressure ratio of the jet compressor is 3. The speed of the plane is 1000 km/h. The pressure drop through the cooler coil is 0.1 bar. The pressure of the air leaving the cooling turbine is 1.06 bar and that in the cockpit is 1.01325 bar. The cockpit cooling load is 58.05 kW.

Calculate the following:

(a) Stagnation temperature and pressure of the air entering the compressor.
(b) Mass flow rate of the air circulated.
(c) Volume handled by the compressor and expander.
(d) Net power delivered by the engine to the refrigeration unit.
(e) COP of the system.

Solution:

Given: Cockpit temperature = 25°C = 298 K
$p_1 = 0.35$ bar; $T_1 = -15°C = 258$ K

$\dfrac{p_3}{p_2} = 3$ or $p_3 = 3p_2$,

Plane speed, $C = 1000$ km/h = 277.78 m/s
Pressure drop through the cooler coil = 0.1 bar, Pressure of air leaving the turbine = 1.06 bar.
Pressure of air in the cockpit = 1.01325 bar, Cooling load = 58.05 kW

(a) Stagnation temperature, $T_2 = T_1 + \dfrac{C^2}{2c_p}$

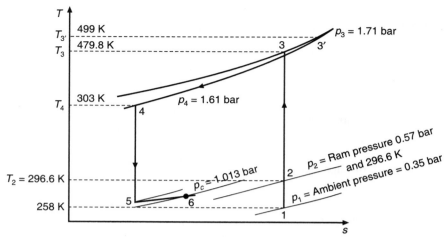

Figure 3.20 T–s diagram for simple air-cooling system—Example 3.6.

or
$$T_2 = 258 + \frac{(277.78)^2}{2 \times 1000} = 296.58 \text{ K}$$ **Ans.**

$$\text{Stagnation pressure, } p_2 = p_1 \left(\frac{T_2}{T_1}\right)^{\gamma/(\gamma-1)} = 0.35 \left(\frac{296.58}{258}\right)^{1.4/(1.4-1)} = 0.57 \text{ bar}$$

Ans.

(b) Compressor outlet pressure, $p_3 = 3p_2 = 1.71$ bar
= cooler inlet pressure

Cooler exit pressure = cooler inlet − pressure drop
= 1.71 − 0.1 = 1.61 bar
= turbine inlet pressure

Assuming cooling turbine inlet temperature is 30°C = 303 K

$$\frac{\text{Turbine exit temp}}{\text{Turbine inlet temp}} = \left(\frac{\text{Turbine exit pressure}}{\text{Turbine inlet pressure}}\right)^{(\gamma-1)/\gamma}$$

$$\text{Turbine exit temperature} = 303\left(\frac{1.06}{1.61}\right)^{(1.4-1)/1.4} = 294.1 \text{ K}$$

Refrigerating effect/kg = c_p(cockpit temperature − turbine exit temperature)
= 1.005(298 − 294.1) = 3.92 kJ/kg

$$\text{Mass flow rate} = \frac{\text{Cooling required in kJ/s}}{\text{Refrigerating effect kJ/kg}} = \frac{58.05}{2.5} = 23.2 \text{ kg/s} \quad \textbf{Ans.}$$

(c) $\qquad$ Compressor volume $= \dfrac{mRT_2}{p_2} = \dfrac{23.22 \times 0.287 \times 296.58}{0.57 \times 10^2}$

$\qquad\qquad\qquad\qquad\qquad = 34.67$ m³/s **Ans.**

$\qquad$ Expander volume $= \dfrac{mRT_5}{p_5} = \dfrac{23.22 \times 0.287 \times 295.51}{1.06 \times 10^2} = 18.13$ m³/s **Ans.**

(d) Assuming isentropic compression in compressor,

$$T_3 = T_2 \left(\dfrac{p_3}{p_2}\right)^{(\gamma-1)/\gamma} = 296.58 \left(\dfrac{1.71}{0.57}\right)^{(1.4-1)/1.4}$$

$\qquad\qquad T_3 = 405.9$ K

$\qquad$ Compressor work/kg $= c_p(T_3 - T_2) = 1.005(405.9 - 296.58)$
$\qquad\qquad W = 109.9$ kJ/kg
$\qquad$ Compressor work/s $=$ Mass flow $\times W = 23.22 \times 109.91 = 2552.0$ kJ/s
$\qquad$ Turbine work/s $= mc_p \Delta T$
$\qquad\qquad\qquad\qquad = 23.22 \times 1.005(333 - 295.51) = 874.87$ kJ/s
$\qquad$ Net work $= W_C - W_T = 2552 - 874.87 = 1677.13$ kJ/s **Ans.**

(e) $\qquad$ COP $= \dfrac{\text{Refrigerating effect in kJ/s}}{\text{Net work in kJ/s}} = \dfrac{58.05}{1677.13} = 0.035$ **Ans.**

EXAMPLE 3.7 A simple air refrigeration system is used for an aircraft to take a load of 20 TR. The ambient pressure and temperature are 0.9 bar and 22°C respectively. The pressure of air is increased to 1 bar due to isentropic ramming action. The air is further compressed in a compressor to 3.5 bar and then cooled in a heat exchanger to 72°C. Finally the air is passed through the cooling turbine and then supplied to the cabin at 1.03 bar. The air leaves the cabin at 25°C. Assuming the isentropic efficiency of compressor and turbine as 80% and 75% respectively, find

(a) the power required to take the cooling load in the cabin.
(b) the COP of the system.

Take $c_p = 1.005$ kJ/kg-K; $\gamma = 1.4$

Solution: The cycle of operations is represented on the T–s diagram shown in Figure 3.21.

Given:
$\qquad Q = 20$ TR $= 20 \times 3.517 = 70.34$ kJ/s; $\quad p_1 = 0.9$ bar; $T_1 = 22°C = 22 + 273 = 295$ K;
$\qquad p_2 = 1$ bar; $p_{3'} = p_3 = p_4 = 3.5$ bar; $\quad T_4 = 72°C = 72 + 273 = 345$ K;
$\qquad T_6 = 25°C = 25 + 273 = 298$ K; $\quad p_5 = p_{5'} = p_6 = 1.03$ bar;
$\qquad \eta_C = 80\%$; $\eta_T = 75\%$

Ram compression (1–2): For isentropic process,

$$T_2 = T_1 \left(\dfrac{p_2}{p_1}\right)^{(\gamma-1)/\gamma} = 295 \left(\dfrac{1}{0.9}\right)^{(1.4-1)/1.4} = 304 \text{ K}$$

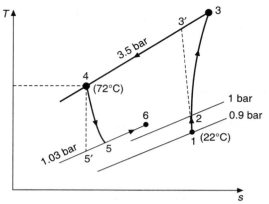

Figure 3.21 T–s diagram—Example 3.7.

Compressor: Consider the isentropic process (2–3′),

$$T_{3'} = T_2 \left(\frac{p_{3'}}{p_2}\right)^{(\gamma-1)/\gamma} = 304 \left(\frac{3.5}{1}\right)^{(1.4-1)/1.4} = 434.8 \text{ K}$$

$$\eta_C = \frac{T_{3'} - T_2}{T_3 - T_2} \text{ or } 0.8 = \frac{434.8 - 304}{T_3 - 304} \text{ or } T_3 = 467.5 \text{ K}$$

Cooling turbine: Consider the isentropic process (4–5′),

$$T_{5'} = T_4 \left(\frac{p_{5'}}{p_4}\right)^{(\gamma-1)/\gamma} = 345 \left(\frac{1.03}{3.5}\right)^{(1.4-1)/1.4} = 243.2 \text{ K}$$

$$\eta_T = \frac{T_4 - T_5}{T_4 - T_{5'}} \text{ or } 0.75 = \frac{345 - T_5}{345 - 243.2} \text{ or } T_5 = 268.7 \text{ K}$$

(a) Power required to take the cooling load in cabin

Let $\dot{m}_a$ be the mass flow rate circulated to cabin.

$$Q = \dot{m}_a c_p (T_6 - T_5)$$
$$70.34 = \dot{m}_a \times 1.005(298 - 268.7)$$
$$\therefore \quad \dot{m}_a = 2.3887 \text{ kg/s}$$

Net work (Power) = $\dot{m}_a c_p (T_3 - T_2) = 2.3887 \times 1.005 \times (467.5 - 304)$

= 392.5 kJ/s or kW **Ans.**

(b) COP of the system

$$\text{COP} = \frac{\text{Refrigerating effect in kW}}{\text{Net work in kW}} = \frac{70.34}{392.5} = 0.179 \quad \textbf{Ans.}$$

EXAMPLE 3.8 The following data refers to a simple air refrigeration cycle of 20 TR capacity.

Ambient air temperature and pressure = 20°C and 0.8 bar
Ram air pressure = 0.9 bar
Compressor outlet pressure = 3.6 bar
Temperature of air leaving H.E. = 60°C
Pressure of air leaving the turbine = 1 bar
Temperature of air leaving the cabin = 22°C
Compressor efficiency = 80%
Turbine efficiency = 75%

Assume no pressure drop in H.E. and isentropic ramming process.
Sketch the cycle on T–s and p–v diagrams.
Calculate the net power required and the COP of the system.

Solution: Refer to Figure 3.22.

Given: R_E = 20 TR = 20 × 3.517 = 70.34 kJ/s

Ambient pressure, p_1 = 0.8 bar; Ambient temperature, T_1 = 20°C = 293 K;
Ram pressure, p_2 = 0.9 bar; Compressor pressure, p_3 = 3.6 bar; η_C = 80%;
Temperature at exit of heat exchanger, T_4 = 60°C = 60 + 273 = 333 K;
Turbine exit pressure = cabin pressure = p_5 = 1 bar;
Turbine efficiency, η_T = 75%;
T_6 = 22°C = 22 + 273 = 295 K

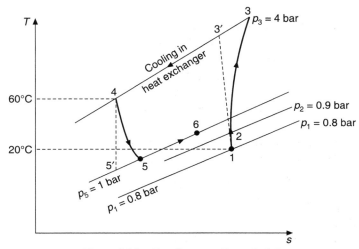

Figure 3.22 T–s diagram—Example 3.8.

Ram compression (1–2): $T_2 = T_1 \left(\dfrac{p_2}{p_1}\right)^{(\gamma-1)/\gamma} = 293\left(\dfrac{0.9}{0.8}\right)^{(1.4-1)/1.4} = 303$ K

Main compressor: For the isentropic compression process (2–3′),

$$T_{3'} = T_2 \left(\frac{p_3}{p_2}\right)^{(\gamma-1)/\gamma} = 303 \left(\frac{3.6}{0.9}\right)^{(1.4-1)/1.4} = 450.2 \text{ K}$$

$$\eta_C = \frac{T_{3'} - T_2}{T_3 - T_2} \text{ or } 0.8 = \frac{450.2 - 303}{T_3 - 303}$$

$$T_3 = 487.0 \text{ K}$$

Gas turbine: For the isentropic expansion process (4–5′)

$$T_{5'} = T_4 \left(\frac{p_5}{p_4}\right)^{(\gamma-1)/\gamma} = 333 \left(\frac{1}{3.6}\right)^{(1.4-1)/1.4} = 230.9 \text{ K}$$

$$\eta_T = \frac{T_4 - T_5}{T_4 - T_{5'}} \text{ or } 0.75 = \frac{333 - T_5}{333 - 230.9}$$

∴ $T_5 = 256.4$ K

Mass flow rate of air, $\dot{m}_a$

$$R_E = \dot{m}_a \times c_p(T_6 - T_5)$$
$$70.34 = \dot{m}_a \times 1.005(295 - 256.4)$$

∴ $\dot{m}_a = 1.813$ kg/s

(a) Net compressor power required, P

$$P = \dot{m}_a c_p (T_3 - T_2)$$
$$= 1.813 \times 1.005(487.0 - 303.0)$$
$$= 335.32 \text{ kJ/s or kW} \qquad \textbf{Ans.}$$

(b) COP of the system

$$\text{COP} = \frac{R_E}{P} = \frac{70.34}{335.3} = 0.2038 \qquad \textbf{Ans.}$$

3.7.2 Simple Air Evaporative Cooling System

The simple air evaporative cooling system is similar to the simple air cooling system with just one modification. *The system has an additional evaporative type heat exchanger to cool the air to a large extent before it is expanded in the cooling turbine.* A simple air evaporative cooling system is shown in Figure 3.23.

In the evaporative type heat exchanger, water evaporates and cools the air to a greater extent (i.e. from point 4 to point 4′). At high altitudes, the evaporative cooling may be obtained by using alcohol or ammonia. Water, alcohol and ammonia all have different refrigerating effects at different altitudes. At the altitude of the aircraft, atmospheric pressure is of the order of 0.8 to 0.9 bar. Water boils and provides a cooling effect to the air.

The various processes involved in the simple air evaporating refrigeration cycle are:

Process (1–2) → Ideal ram compression
Process (1–2′) → Actual ram compression
Process (2′–3) → Isentropic compression of air in main compressor
Process (2′–3′) → Actual compression of air in main compressor
Process (3′–4) → Constant pressure air cooling by ram air in heat exchanger
Process (4–4′) → Constant pressure air cooling by refrigerant in evaporator
Process (4′–5) → Isentropic expansion of air in cooling turbine
Process (4′–5′) → Actual expansion of air in cooling turbine
Process (5′–6) → Constant pressure heating of air to cabin temperature T_6

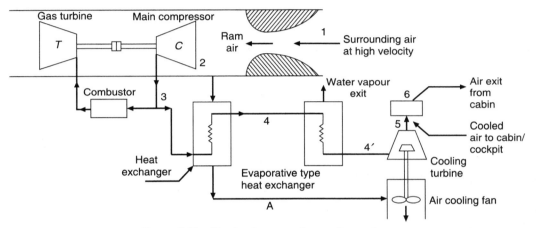

Figure 3.23 Simple air evaporative cooling system.

The T–s diagram for a simple evaporative cooling system is shown in Figure 3.24.

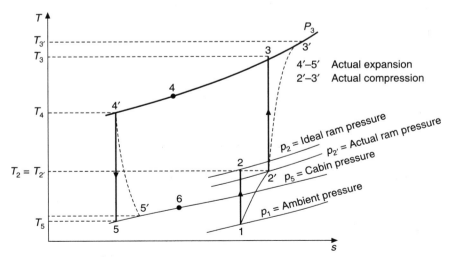

Figure 3.24 T–s diagram for a simple air evaporative cooling system of aircraft.

The cooling effect achieved in the aircraft refrigeration system

$$= \dot{m}_a \times c_p \times (T_6 - T_{5'})$$

where $\dot{m}_a$ is the mass flow rate of air in kg/min, T_6 and $T_{5'}$ are the cabin temperature and the air temperature at the outlet of the cooling turbine. If Q is the cooling load requirement of the cabin, then the air required for the refrigeration purpose,

$$\dot{m}_a = \frac{211 Q}{c_p (T_6 - T_5)} \text{ kg/min}$$

Power required for the refrigerating system,

$$P = \frac{\dot{m}_a c_p (T_{3'} - T_{2'})}{60} \text{ kW}$$

and COP of the refrigerating system

$$= \frac{211 Q}{\dot{m}_a c_p (T_{3'} - T_{2'})} = \frac{211 Q}{P \times 60}$$

EXAMPLE 3.9 A simple evaporative air refrigeration system is used for an aeroplane to take 20 TR of refrigeration load to maintain the cabin temperature at 25°C. The ambient air conditions are 20°C and 0.9 bar. The ambient air is rammed isentropically to a pressure of 1 bar. The air leaving the main compressor at pressure 4 bar is first cooled in the heat exchanger to a temperature of 100°C and then in the evaporator where its temperature is reduced by 10°C. The air from the evaporator is passed through the cooling turbine and then it is supplied to the cabin at a pressure of 1.05 bar. If the compression efficiency of the compressor is 80% and the expansion efficiency of cooling turbine is 75%, determine the following:

(a) Mass of air bled off the main compressor.
(b) Power required for the refrigerating system.
(c) COP of the refrigerating system.

Solution:

Given: $Q = 20$ TR; $\quad\quad\quad\quad T_1 = 20°C = 20 + 273 = 293$ K;
$p_1 = 0.9$ bar; $\quad\quad\quad\quad p_2 = 1$ bar; $p_3 = p_{3'} = 4$ bar;
$\eta_T = 75\%$ $\quad\quad\quad\quad\quad\quad \eta_C = 80\%$
$T_6 = 25°C = 25 + 273 = 298$ K; $\quad p_6 = 1.05$ bar

The T–s diagram for the simple evaporative air refrigeration system with the given conditions is shown in Figure 3.25.

Suppose T_2 = temperature of air entering the main compressor,
T_3 = temperature of air after isentropic compression in the main compressor,
$T_{3'}$ = actual temperature of air leaving the main compressor, and
T_4 = temperature of air entering the evaporative type heat exchanger.

We know that for an isentropic ramming process 1–2, taking ($\gamma = 1.4$)

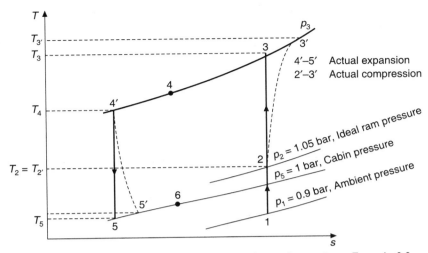

Figure 3.25 T–s diagram for a simple evaporative cooling system—Example 3.9.

$$\frac{T_2}{T_1} = \left(\frac{p_2}{p_1}\right)^{(\gamma-1)/\gamma} = \left(\frac{1}{0.9}\right)^{(1.4-1)/1.4} = (1.11)^{0.286} = 1.03$$

∴ $T_2 = T_1 \times 1.03 = 293 \times 1.03 = 301.8$ K

Now for the isentropic compression process 2–3,

$$\frac{T_3}{T_2} = \left(\frac{p_3}{p_2}\right)^{(\gamma-1)/\gamma} = \left(\frac{4}{1}\right)^{(1.4-1)/1.4} = (3.5)^{0.286} = 1.485$$

∴ $T_3 = T_2 \times 1.485 = 301.8 \times 1.485 = 448.5$ K

We know that the efficiency of the compressor,

$$\eta_C = \frac{\text{Isentropic increase in temperature}}{\text{Actual increase in temperature}} = \frac{T_3 - T_2}{T_{3'} - T_2}$$

or

$$0.8 = \frac{448.5 - 301.8}{T_{3'} - 301.8} = \frac{146.6}{T_{3'} - 301.8}$$

∴ $T_{3'} = 301.8 + 146.6/0.8 = 485.1$ K

The temperature of air leaving the first heat exchanger, given in the example, $T_4 = 100°C$
The temperature of air in the evaporative type HE is reduced by 10°C, therefore the temperature of air leaving the evaporator and entering the cooling turbine,

$$T_{4'} = T_4 - 10 = 100 - 10 = 90°C = 363 \text{ K}$$

Now for the isentropic expansion process 4′–5,

$$\frac{T_{4'}}{T_5} = \left(\frac{p_3}{p_6}\right)^{(\gamma-1)/\gamma} = \left(\frac{4}{1.05}\right)^{(1.4-1)/1.4} = (3.81)^{0.286} = 1.46$$

∴ $T_5 = T_{4'}/1.41 = 363/1.46 = 247.6$ K

Efficiency of the cooling turbine,

$$\eta_T = \frac{\text{Actual decrease in temperature}}{\text{Isotropic decrease in temperature}} = \frac{T_{4'} - T_{5'}}{T_{4'} - T_5}$$

or $\quad 0.75 = \dfrac{363 - T_{5'}}{363 - 247.6} = \dfrac{361.7 - T_{5'}}{115.4}$

∴ $T_{5'} = 363 - 0.75 \times 115.4 = 276.45$

(a) The mass of air bled off the main compressor,

$$\dot{m}_a = \frac{211\,Q}{c_p\,(T_6 - T_{5'})} = \frac{211 \times 20}{1\,(298 - 276.4)} = 195 \text{ kg/min} \quad \text{Ans.}$$

(b) Power required for the refrigerating system,

$$P = \frac{m_p c_p (T_{3'} - T_{2'})}{60} = \frac{195 \times 1\,(485.1 - 301.8)}{60} = 595.7 \text{ kW} \quad \text{Ans.}$$

(c) The COP of the refrigeration system

$$= \frac{210\,Q}{P \times 60} = \frac{210 \times 20}{595.7 \times 60} = 0.011 \quad \text{Ans.}$$

3.7.3 Boot-strap Air Cooling System

A boot-strap air cooling system is shown in Figure 3.26. The main advantage of this system is that the compression of air is carried out in two stages with intercooling. Cold atmospheric air is used as a coolant in both the heat exchangers. The air bled from the main compressor is first cooled in the first heat exchanger using cold ram air. This cooled air, after compression in the secondary compressor, leads to the second heat exchanger, where it is again cooled by the ram air before passing to the cooling turbine. This type of cooling system is mostly used in transport type aircraft.

The T–s diagram for a boot-strap air cycle cooling system is shown in Figure 3.27. The various processes are as follows:

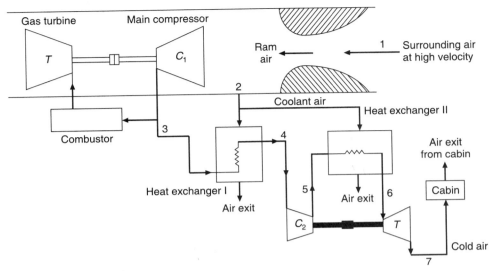

Figure 3.26 Boot-strap air refrigeration cycle for aircraft.

Process 1–2: It represents the isentropic ramming of ambient air from pressure p_1 and temperature T_1 to pressure p_2 and temperature T_2. The process 1–2′ represents the actual ramming process.

Process 2′–3: It represents the isentropic compression of air in the main compressor and the process 2′–3′ represents the actual compression of air.

Process 3′–4: The compressed air from the main compressor is cooled in heat exchanger I using ram air. The pressure drop in the heat exchanger is neglected.

Process 4–5: The isentropic compression of cooled air is from the first heat exchanger, in the second compressor. The actual compression is represented by the process 4–5′.

Process 5′–6: The compressed air from the secondary compressor is cooled in heat exchanger II using ram air. The pressure drop in the heat exchanger is neglected.

Process 6–7: It is isentropic expansion of cooled air in the cooling turbine up to the cabin pressure. The process 6–7′ represents the actual expansion of the cooled air in the cooling turbine.

Process 7′–8: It represents the heating of air up to the cabin temperature T_8.

The quantity of air required for the refrigeration purpose to meet the Q TR of refrigeration load in the cabin is

$$\dot{m}_a = \frac{211\, Q}{c_p (T_8 - T_{7'})} \text{ kg/min}$$

Power needed for the refrigerating system,

$$P = \frac{\dot{m}_a c_p (T_{3'} - T_{2'})}{60} \text{ kW}$$

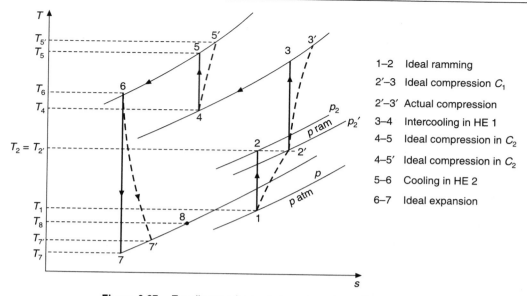

Figure 3.27 *T–s* diagram for boot-strap air cycle cooling system.

and COP of the refrigerating system

$$= \frac{211\,Q}{m_a c_p (T_{3'} - T_{2'})} = \frac{211}{P \times 60}$$

EXAMPLE 3.10 A boot-strap cooling system is used in an aeroplane for 15 TR capacity. The ambient air temperature and pressure are 20°C and 0.85 bar respectively. The pressure of air rises from 0.85 bar to 1 bar due to ramming action. The pressure of air at the outlet of the main compressor is 3 bar. The discharge pressure of air from the secondary compressor is 4.5 bar. The efficiency of each of the compressors is 85% and that of the turbine is 90%. 75% of the heat content of the air discharged from the main compressor is removed in the heat exchanger I and 30% of the heat of air discharged from the auxiliary compressor is removed in the heat exchanger II using rammed air. Assuming ramming action to be isentropic, the required cabin pressure to be 0.9 bar and the temperature of the air leaving the cabin to be not more than 20°C, find the following:

(a) The mass of air bled off the main compressor.
(b) The power required to operate the system.
(c) The COP of the system.

Take $\gamma = 1.4$ and $c_p = 1$ kJ/kg-K

Solution:
Given: $Q = 15$ TR; $T_1 = 20°C = 20 + 273 = 293$ K
  $p_1 = 0.85$ bar; $p_2 = 1$ bar
  $p_3 = p_{3'} = p_4 = 3$ bar; $p_5 = p_{5'} = p_6 = 4.5$ bar
  $\eta_{C1} = \eta_{C2} = 85\%$; $\eta_T = 90\%$
  $p_7 = p_{7'} = p_8 = 0.9$ bar; $T_8 = 20°C = 20 + 273$ K
  $\gamma = 1.4;\ c_p = 1$ kJ/kg-K;

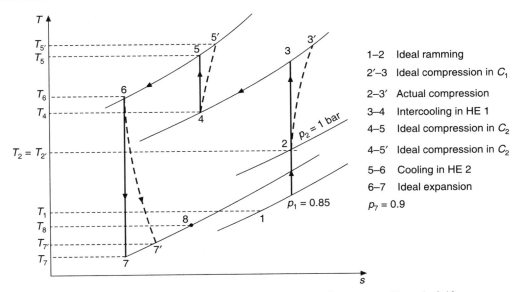

Figure 3.28 T–s diagram for boot-strap air cycle cooling system—Example 3.10.

The temperature entropy (T–s) diagram with the given conditions is shown in Figure 3.28. We know that for isentropic ramming process 1–2,

$$\frac{T_2}{T_1} = \left(\frac{p_2}{p_1}\right)^{(\gamma-1)/\gamma} = \left(\frac{1}{0.85}\right)^{(1.4-1)/1.4} = (1.176)^{0.286} = 1.047$$

∴ $T_2 = T_1 \times 1.047 = 293 \times 1.03 = 306.8$ K

Now for the isentropic compression process 2–3,

$$\frac{T_3}{T_2} = \left(\frac{p_3}{p_2}\right)^{(\gamma-1)/\gamma} = \left(\frac{3}{1}\right)^{(1.4-1)/1.4} = (3)^{0.286} = 1.37$$

∴ $T_3 = T_2 \times 1.37 = 306.8 \times 1.37 = 420.3$ K

Isentropic efficiency of the compressor,

$$\eta_{C1} = \frac{\text{Isentropic increase in temperature}}{\text{Actual increase in temperature}} = \frac{T_3 - T_2}{T_{3'} - T_2}$$

or $$0.8 = \frac{420.3 - 306.8}{T_{3'} - 306.8} = \frac{113.5}{T_{3'} - 306.8}$$

∴ $T_{3'} = 306.8 + 113.5/0.8 = 448.7$ K $= 175.7°$C

For perfect intercooling, the temperature of air at the outlet of heat exchanger I is equal to the temperature of air entering the main compressor. But in intercooling only 75% of the compression heat is removed and so the temperature of air at point 4 is

$$T_4 = T_2 + 0.25(T_3 - T_2) = 306.8 + 0.25 (420.3 - 306.8) = 335 \text{ K}$$

The isentropic process 4–5,

$$\frac{T_4}{T_5} = \left(\frac{p_3}{p_6}\right)^{(\gamma-1)/\gamma} = \left(\frac{4.5}{3}\right)^{(1.4-1)/1.4} = (1.5)^{0.286} = 1.122$$

∴ $T_5 = T_4 \times 1.122 = 335 \times 1.122 = 376 \text{ K} = 118.5°\text{C}$

Isentropic efficiency of the auxiliary (secondary) compressor,

$$\eta_{C2} = \frac{T_5 - T_4}{T_{5'} - T_4}$$

or $\quad 0.85 = \dfrac{376 - 335.85}{T_{5'} - 335} = \dfrac{41}{T_{5'} - 335}$

∴ $T_{5'} = 383.2 \text{ K} = 110°\text{C}$

For perfect after-cooling, the temperature of air at the outlet of heat exchanger II is equal to the temperature of air entering the auxiliary (secondary) compressor. But in after-cooling, only 30% of the compression heat is removed. Therefore, the temperature of air at point 6 is

$$T_6 = T_4 + 0.7(T_{5'} - T_4) = 335 + 0.7(383.2 - 335) = 368.7 \text{ K}$$

For the isentropic process 6–7,

$$\frac{T_7}{T_6} = \left(\frac{p_7}{p_6}\right)^{(\gamma-1)/\gamma} = \left(\frac{0.9}{4.5}\right)^{(1.4-1)/1.4} = (0.2)^{0.286} = 0.631$$

$$= (0.225)^{0.286} = 0.653$$

∴ $T_7 = T_6 \times 0.631 = 368.7 \times 0.631 = 232.6 \text{ K} = -40.3°\text{C}$

The turbine efficiency,

$$\eta_T = \frac{\text{Actual decrease in temperature}}{\text{Isentropic decrease in temperature}} = \frac{T_6 - T_{7'}}{T_6 - T_7}$$

or $\quad 0.9 = \dfrac{368.7 - T_{7'}}{368.7 - 232.6} = \dfrac{368.7 - T_{7'}}{136.1}$

∴ $T_{7'} = 246.2 \text{ K} = -26.8°\text{C}$

(a) Flow rate of air:
The mass of air bled off the main compressor

$$\dot{m}_a = \frac{211Q}{c_p(T_8 - T_{7'})} = \frac{211 \times 15}{1 \times (293 - 246.2)} = 67.3 \text{ kg/min} \qquad \textbf{Ans.}$$

(b) Power required to operate the system:

$$P = \frac{m_a\, c_p\, (T_{3'} - T_{2'})}{60} = \frac{67.3 \times 1\,(448.7 - 306.8)}{60} = 159 \text{ kW} \qquad \text{Ans.}$$

(c) COP of the system:
We know the COP of the system to be

$$= \frac{211\,Q}{m_a c_p\,(T_{3'} - T_2)} = \frac{211 \times 15}{67.3 \times 1\,(448.7 - 306.8)} = 0.33 \qquad \text{Ans.}$$

EXAMPLE 3.11 An aeroplane uses the boot-strap air cooling system. It requires 16 TR of refrigeration. Ambient temperature and pressure are –13°C and 0.6 bar respectively. The speed of the plane is 800 kmph. The ram air is compressed in the main compressor up to 3.6 bar pressure and this is further compressed in the secondary compressor up to 4.4 bar. The isentropic efficiency for each compressor is 85%. The air in the heat exchanger I is cooled up to 113°C by ram air while the air from the secondary compressor is cooled by ram air up to 87°C. This cooled air is then expanded in cooling turbine up to the cabin pressure of 1 bar with an isentropic efficiency of 90%. The cabin temperature is required to be maintained at 20°C. Find the following:

(a) Mass flow rate of air in kg/min.
(b) Power required to operate the system.
(c) COP of the system.

Assume, $\gamma = 1.4$ and $c_p = 1.01$ kJ/kg-K.

Solution:

Given:

$Q = 16$ TR
$p_1 = 0.6$ bar
$p_3 = p_{3'} = p_4 = 3.6$ bar
$\eta_C = \eta_{C2} = 85\%$
$T_6 = 87°C = 87 + 273 = 360$ K
$T_8 = 20°C = 20 + 273 = 293$ K

$C = 800$ kmph,
$T_1 = -13°C = -13 + 273 = 260$ K
$p_{5'} = p_5 = p_6 = 4.4$ bar
$\eta_T = 90\%$
$p_7 = p_{7'} = p_8 = 1$ bar
$T_4 = 113°C = 113 + 273 = 386$ K

The *T–s* diagram is shown in Figure 3.29 and system diagram in Figure 3.26.

Ram compression (1–2):

$$\text{Velocity of air, } C = 800 \text{ kmph} = \frac{800 \times 1000}{3600} = 222.2 \text{ m/s}$$

$$T_2 = T_1 + \frac{C^2}{2\,c_p} = 260 + \frac{(222.2)^2}{2 \times (1.01 \times 1000)} = 284.5 \text{ K}$$

$$p_2 = p_1 \left(\frac{T_2}{T_1}\right)^{\gamma/(\gamma-1)} = 0.6 \left(\frac{284.45}{260}\right)^{1.4/(1.4-1)} = 0.822 \text{ bar}$$

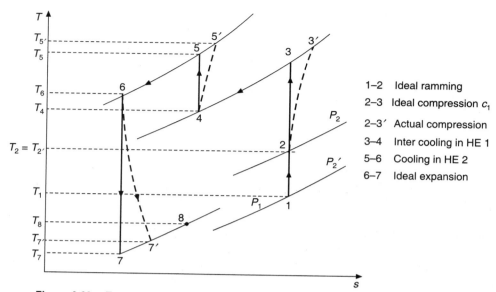

Figure 3.29 T–s diagram for boot-strap air evaporative cooling system—Example 3.11.

Process (2–3) in main compressor:

$$T_3 = T_2 \left(\frac{p_3}{p_2}\right)^{(\gamma-1)/\gamma} = 284.5 \left(\frac{3.6}{0.822}\right)^{(1.4-1)/1.4}$$

or
$$T_3 = 433.7 \text{ K}$$

$$\eta_{C1} = \frac{T_3 - T_2}{T_{3'} - T_2} \quad \text{i.e.} \quad T_{3'} = T_2 + \frac{T_{3'} - T_2}{\eta_{C1}}$$

∴
$$T_{3'} = 284.5 + \frac{(433.7 - 284.5)}{0.85} = 460.0 \text{ K}$$

Secondary compression process (4–5):

$$T_5 = T_4 \left(\frac{p_5}{p_4}\right)^{(\gamma-1)/\gamma} = 396 \left(\frac{4.4}{3.6}\right)^{(1.4-1)/1.4} = 413.0 \text{ K}$$

But
$$\eta_{C2} = \frac{T_5 - T_4}{T_{5'} - T_4}$$

or
$$T_{5'} = T_4 + \frac{T_5 - T_4}{\eta_{C2}} = 386 + \frac{(413.0 - 386)}{0.85} = 452.5 \text{ K}$$

Expansion in cooling turbine (6–7):

$$T_7 = T_6 \left(\frac{p_7}{p_6}\right)^{(\gamma-1)/\gamma} = 360 \left(\frac{1}{4.4}\right)^{(1.4-1)/1.4}$$

$$= 235.8 \text{ K}$$

$$\eta_T = 0.9 = \frac{360 - T_{7'}}{360 - 235.8}$$

or $\quad T_{7'} = 248.2$ K

(a) Refrigerating effect, i.e. heating of air in cabin in process (7–8)

$$Q = \dot{m}_a c_p (T_8 - T_{7'})$$

or $\quad 16 \times 211 = \dot{m}_a \times 1.01(293 - 248.2)$

∴ $\quad \dot{m}_a = 74.61$ kg/min **Ans.**

(b) Total compressor power required is to run the main compressor.

$$P_C = \dot{m}_a [c_p(T_{3'} - T_2)]$$

$$= \frac{74.61}{60} \times 1.01(460.0 - 284.5) = 220.60 \text{ kJ/s or kW} \quad \textbf{Ans.}$$

(c) $\quad$ Since 1 TR = 3.517 kW

$$\text{COP} = \frac{Q}{P_C} = \frac{16 \times 3.517}{220.6} = 0.255 \quad \textbf{Ans.}$$

EXAMPLE 3.12 A boot-strap air refrigeration system is used in an aeroplane for the 10 TR refrigeration load. The ambient air conditions are 15°C and 0.9 bar. This air is rammed isentropically to a pressure of 1.1 bar. The pressure of air bled off the main compressor is 3.5 bar and this is further compressed in the secondary compressor to a pressure of 4.5 bar. The isentropic efficiency of both the compressors is 90% and that of cooling turbine is 85%. The effectiveness of both the heat exchangers is 0.6. If the cabin is to be maintained at 25°C and the pressure in the cabin is 1 bar, find the following:

(a) Mass of air passing through the cabin.
(b) Power used for the refrigeration system.
(c) COP of the system

Draw the schematic and T–s diagram for the system, take $\gamma = 1.4$ and $c_p = 1$ kJ/kg-K

Solution: Figure 3.30 shows the T–s diagram.
Given:
Refrigeration effect = 10 TR = 10 × 3.517 = 35.17 kW;
$p_1 = 0.9$ bar; $\quad\quad\quad\quad T_1 = 15°C = 15 + 273 = 288$ K;
$p_2 = 1.1$ bar; $\quad\quad\quad\quad p_3 = p_{3'} = p_4 = 3.5$ bar;

96 Refrigeration and Air Conditioning

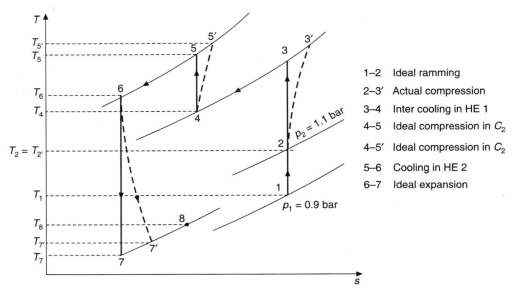

Figure 3.30 T–s diagram for boot-strap air evaporative cooling system—Example 3.12.

$p_5 = p_{5'} = p_6 = 4.5$ bar;
$\eta_C = \eta_{C1} = \eta_{C2} = 90\%$; $\qquad \eta_T = 85\%$;
$E_r = E_{r1} = E_{r2} = 0.6$
Cabin temperature, $T_8 = 25°C = 25 + 273 = 298$ K, $p_{7'} = p_7 = p_8 = 1$ bar

Isentropic ram compression (1–2):

$$T_2 = T_1 \left(\frac{p_2}{p_1}\right)^{(\gamma-1)/\gamma} = 288 \left(\frac{1.1}{0.9}\right)^{(1.4-1)/1.4} = 305 \text{ K}$$

Main compressor: Isentropic compression process (2–3):

$$T_3 = T_2 \left(\frac{p_3}{p_2}\right)^{(\gamma-1)/\gamma} = 305 \left(\frac{3.5}{1.1}\right)^{(1.4-1)/1.4} = 424.5 \text{ K}$$

$$\eta_{C1} = \frac{T_3 - T_2}{T_{3'} - T_2}; \quad 0.9 = \frac{424.5 - 305}{T_{3'} - 305}; \quad \therefore T_{3'} = 437.8 \text{ K}$$

Heat exchanger (HE–I): Its effectiveness is

$$E_r = \frac{T_{3'} - T_4}{T_{3'} - T_2}; \quad 0.6 = \frac{437.8 - T_4}{437.8 - 305}; \quad \therefore T_4 = 358.1 \text{ K}$$

Secondary compressor: Consider the isentropic process (4–5),

$$T_5 = T_4 \left(\frac{p_5}{p_4}\right)^{(\gamma-1)/\gamma} = 358.1 \left(\frac{4.5}{3.5}\right)^{(1.4-1)/1.4} = 384.8 \text{ K}$$

$$\eta_{C2} = 0.9 = \frac{384.8 - 358.1}{T_{5'} - 358.1}; \qquad \therefore T_5' = 387.8 \text{ K}$$

Heat exchanger II: Its effectiveness is

$$E_r = \frac{T_{5'} - T_6}{T_{5'} - T_2}$$

Hence, $\qquad 0.6 = \dfrac{387.8 - T_6}{387.8 - 305}; \qquad \therefore T_6 = 338.1 \text{ K}$

Cooling turbine: Consider the isentropic process (6–7'),

$$T_7 = T_6 \left(\frac{p_7}{p_6}\right)^{(\gamma-1)/\gamma} = 338.1 \left(\frac{1}{4.5}\right)^{(1.4-1)/1.4} = 220 \text{ K}$$

$$\eta_T = \frac{T_6 - T_{7'}}{T_6 - T_7}$$

or $\qquad 0.85 = \dfrac{338.1 - T_{7'}}{338.1 - 220} \qquad \therefore T_{7'} = 237.7 \text{ K}$

(a) Mass of air passing through the cabin, $\dot{m}_a$

$$\text{Refrigerating effect} = \dot{m}_a \times c_p(T_8 - T_{7'})$$

$$35.17 = \dot{m}_a \times 1 \times (298 - 237.7)$$

$\therefore \qquad \dot{m}_a = 0.58325 \text{ kg/s} = 35 \text{ kg/min} \qquad\qquad$ **Ans.**

(b) Power required for the refrigeration system, P_c

$$P_c = \dot{m}_a \times c_p(T_{3'} - T_2) = 0.58325 \times 1 \times (437.8 - 305)$$

$$= 77.46 \text{ kW} \qquad\qquad \textbf{Ans.}$$

(c) COP of the system

$$\text{COP} = \frac{\text{Refrigerating effect}}{\text{Power supplied}} = \frac{35.17}{77.46} = 0.454 \qquad\qquad \textbf{Ans.}$$

EXAMPLE 3.13 A boot-strap refrigeration system is used in an airplane. The following observations are made. The ambient air temperature is 15°C and the pressure is 0.85 bar. Due to ramming action the pressure increases to 1 bar. This ram air is used for heat exchangers. This air is then compressed in primary compressor to 3.25 bar. The discharge pressure of air from the secondary compressor is 4.25 bar. Assume compression efficiency of 0.9 and turbine efficiency of 0.85. Effectiveness for both the heat exchangers is 0.7. The cabin pressure is 0.9

bar and the temperature of air leaving the cabin is 22°C. Assume the remaining action to be isentropic. Calculate the COP and the power required per ton of refrigeration.

Solution: Refer to Figure 3.31 for the cycle and the *T–s* diagram.

Given : $T_1 = 15°C = 15 + 273 = 288$ K $p_1 = 0.85$ bar, $p_2 = 1$ bar
$p_3 = p_{3'} = 3.25$ bar $p_{5'} = P_5 = 4.25$ bar
$\eta_{C1} = \eta_{C2} = 0.9$ $\eta_T = 0.8$, $E_r = E_{r1} = E_{r2} = 0.7$
$p_7 = p_{7'} = p_8 = 0.9$ bar $T_8 = 22°C = 22 + 273 = 295$ K

Isentropic ram compression (1–2):

$$T_2 = T_1 \left(\frac{p_2}{p_1}\right)^{(\gamma-1)/\gamma} = 288 \left(\frac{1}{0.85}\right)^{(1.4-1)/1.4} = 301.7 \text{ K}$$

Main compressor: Considering isentropic compression,

$$T_3 = T_2 \left(\frac{p_3}{p_2}\right)^{(\gamma-1)/\gamma} = 301.7 \left(\frac{3.25}{1}\right)^{(1.4-1)/1.4} = 422.5 \text{ K}$$

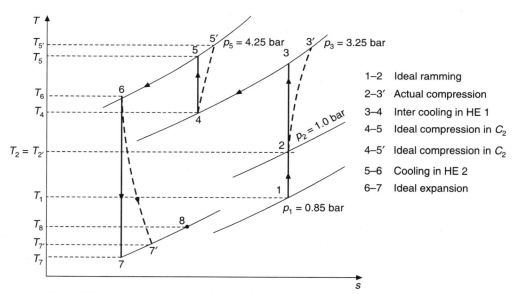

Figure 3.31 *T–s* diagram for boot-strap air evaporative cooling system—Example 3.13.

But the compression is not isentropic:

$$\eta_{C1} = 0.9 = \frac{T_3 - T_2}{T_{3'} - T_2} = \frac{422.5 - 301.7}{T_{3'} - 301.7}; \quad T_{3'} = 435.9 \text{ K}$$

Heat exchanger I: Ram air is employed in the heat exchangers for cooling the compressed air. Its effectiveness E_r is

$$E_r = \frac{T_{3'} - T_4}{T_{3'} - T_2}; \quad 0.7 = \frac{435.9 - T_4}{435.9 - 301.7}$$

$$\therefore \quad T_4 = 342.0 \text{ K}$$

Compression in secondary compressor: Considering isentropic compression (4–5'),

$$T_5 = T_4 \left(\frac{p_5}{p_4}\right)^{(\gamma-1)/\gamma} = 342 \left(\frac{4.25}{3.25}\right)^{(1.4-1)/1.4} = 369.3 \text{ K}$$

$$\eta_{C2} = 0.9 = \frac{369.2 - 342}{T_{5'} - 342}; \quad \therefore \quad T_{5'} = 372.2 \text{ K}$$

Heat exchanger II: Cold ram air is used for cooling the compressed air in HE2. Therefore,

$$E_r = \frac{T_{5'} - T_6}{T_{5'} - T_2}; \quad 0.7 = \frac{372.2 - T_6}{372.2 - 301.7}$$

$$\therefore \quad T_6 = 322.8 \text{ K}$$

Cooling turbine:

Considering isentropic expansion (6–7),

$$T_7 = T_6 \left(\frac{p_7}{p_6}\right)^{(\gamma-1)/\gamma} = 322.8 \left(\frac{0.9}{4.25}\right)^{(1.4-1)/1.4} = 207.2 \text{ K}$$

$$\eta_T = \frac{T_6 - T_{7'}}{T_6 - T_7}; \quad 0.80 = \frac{322.8 - T_{7'}}{322.8 - 207.2}$$

$$\therefore \quad T_{7'} = 230.3 \text{ K}$$

(a) Consider 1 kg of air.

$$\text{Compressor work} = c_p(T_{3'} - T_2) + c_p(T_5 - T_4)$$
$$= 1.005(435.9 - 301.7) + (372.2 - 342.0)$$
$$= 134.9 + 30.35 = 165.2 \text{ kJ/kg}$$

Specific refrigerating effect, $Q_2 = c_p(T_8 - T_{7'}) = 1.005(295.0 - 230.3)$

$$q = 65 \text{ kJ/kg}$$

$$\text{COP} = \frac{\text{Refrigerating effect}}{\text{Work input}} = \frac{65}{165.2} = 0.393 \quad \textbf{Ans.}$$

(b) Power required per ton of refrigeration, P

$$1 \text{ TR} = 3.517 \text{ kJ/s}$$

$$Q = \dot{m}_a \times q$$

∴
$$\dot{m}_a = \frac{1\ \text{TR}}{\text{Specific refrigerating effect}} = \frac{3.517}{65}\ \text{kg/s}$$

and
$$P = \dot{m}_a \times \text{compressor work} = \frac{3.517}{65} \times 165.2$$
$$= 8.94\ \text{kW} \qquad \textbf{Ans.}$$

3.7.4 Boot-strap Air Evaporative Cooling System

A boot-strap air evaporative cooling system is shown in Figure 3.32. The simple boot-strap air cooling system is modified introducing an evaporator between the second heat exchanger and the cooling turbine. Air is cooled to a low temperature in the evaporator before it is expanded in the cooling turbine. This improves the COP of the cycle or requires less compressor work for the same refrigeration effect.

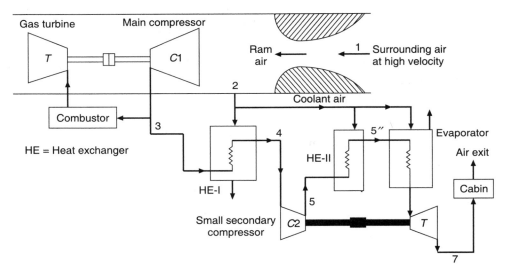

Figure 3.32 Boot-strap air evaporative cooling cycle for aircraft.

The T–s diagram for a boot-strap air evaporative cooling system is shown in Figure 3.33. The various processes of this cycle are same as those a simple boot-strap system and are indicated on the T–s diagram itself. Since the temperature of the air leaving the cooling turbine in the boot-strap evaporative system is lower than that of the simple boot-strap system, the mass of air (m_a) per TR of refrigeration is less in the boot-strap evaporative system.

The quantity of air required for the refrigeration load of Q TR in the cabin will be

$$m_a = \frac{211\,Q}{c_p(T_8 - T_{7'})}\ \text{kg/min}$$

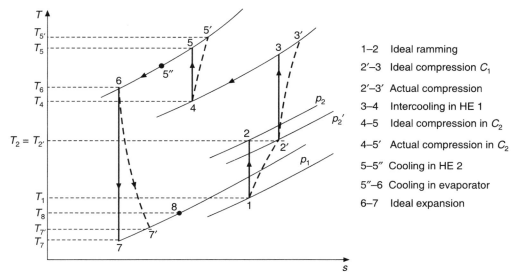

Figure 3.33 T–s diagram for a boot-strap air evaporative cooling system.

Power required for the refrigeration system is given by

$$P = \frac{m_p c_p (T_{3'} - T_{2'})}{60} \text{ kW}$$

and COP of the refrigeration system

$$= \frac{211 Q}{m_p c_p (T_{3'} - T_{2'})} = \frac{211 Q}{P \times 60}$$

3.7.5 Reduced Ambient Air Cooling System

The reduced ambient air cooling system is shown in Figure 3.34. It is used for air cooling of very high speed aircraft. Due to high speed of the aircraft, the temperature of the ram air is relatively high compared with other cooling systems seen earlier. Therefore, ram air cannot be employed directly as a cooling media but it is first expanded in the turbine T_1 and then used for cooling the compressed air in the heat exchanger.

This cooling system includes two cooling turbines and one heat exchanger. This high pressure and high temperature air bled off from the main compressor is cooled initially in the heat exchanger. The cooled air from the heat exchanger is passed through the second cooling turbine from where the air is supplied to the cabin. The work of the cooling turbine is used to drive the cooling fan (through reduction gears), which draws cooling air from the heat exchanger.

The T–s diagram for the reduced ambient air cooling system is shown in Figure 3.35. The various processes are as follows:

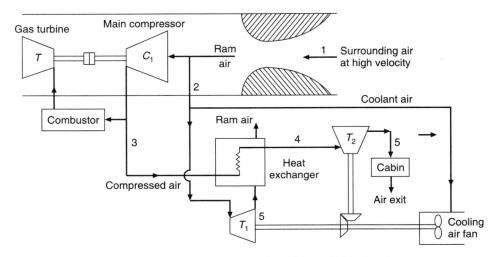

Figure 3.34 Reduced ambient air cooling cycle for aircraft.

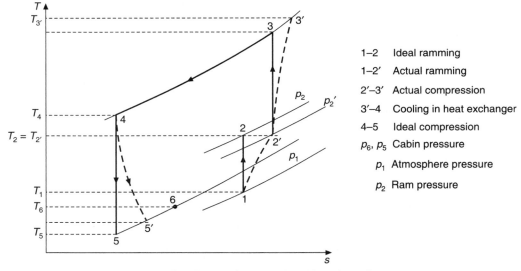

Figure 3.35 T–s diagram for reduced ambient air cooling system.

1–2	Ideal ramming
1–2'	Actual ramming
2'–3'	Actual compression
3–4	Cooling in heat exchanger
4–5	Ideal compression
p_6, p_5	Cabin pressure
p_1	Atmosphere pressure
p_2	Ram pressure

Process (1–2): Isentropic compression of ram air, actual compression shown by 1–2' line.

Process (2–3): Isentropic compression in main compressor, actual process by 2'–3'.

Process (3'–4): Constant pressure cooling of compressed air drawn from main compressor in heat exchanger.

Process (4–5): Isentropic expansion of cooled air in the second cooling turbine (T_2), and the actual expansion is represented by the process (4–5').

Process (5–6): Constant pressure heating of air up to temperature T_6 in cabin to produce refrigerating effect.

1. The quantity of air required to produce refrigeration effect of Q TR will be

$$m_a = \frac{211\,Q}{c_p(T_6 - T_{5'})} \text{ kg/min}$$

2. Power required for the refrigeration system is

$$P = \frac{m_a c_p (T_3 - T_{2'})}{60} \text{ kW}$$

3. COP of the system

$$= \frac{211\,Q}{m_a c_p (T_{3'} - T_{2'})} = \frac{211\,Q}{P \times 60}$$

where, P = power in kW, Q = refrigeration effect in TR.

EXAMPLE 3.14 The reduced ambient air refrigeration system used for an aircraft consists of two cooling turbines, one heat exchanger and one air cooling fan. The speed of the aircraft is 1500 km/h. The ambient air conditions are 0.8 bar and 10°C. The rammed air used for cooling is expanded in the first cooling turbine and leaves it at a pressure of 0.8 bar. The air bled from the main compressor at 6 bar is cooled in the heat exchanger and leaves it at 100°C. The cabin is to be maintained at 20°C and 1 bar. If the isentropic efficiency for the main compressor is 85% and both of the cooling turbines have 80% efficiency, find the following.

(a) Mass flow rate of air supplied to cabin to take a cabin load of 10 TR of refrigeration.
(b) Quantity of air passing through the heat exchanger if the temperature rise of ram air is limited to 80 K.
(c) Power used to drive the cooling fan.
(d) COP of the system.

Solution:
Given: $V = 1500$ km/h $= 417$ m/s; $p_1 = 0.8$ bar; $T_1 = 10°C = 10 + 273 = 283$ K; $p_3 = p_4 = 6$ bar; $T_4 = 100°C = 100 + 273 = 373$ K; $T_6 = 20°C = 20 + 273 = 293$ K; $p_6 = 1$ bar; $\eta_C = 85\%$; $\eta_{T1} = \eta_{T2} = 80\%$; Refrigeration effect $= Q = 10$ TR

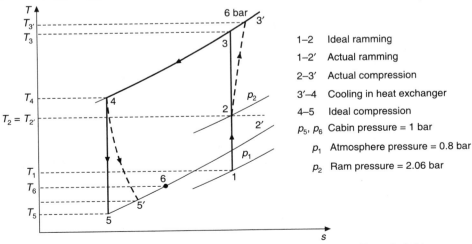

Figure 3.36 *T–s* diagram for reduced ambient cooling system—Example 3.14.

The T–s diagram for the reduced ambient air refrigeration system with the given conditions is shown in Figure 3.36.

Suppose T_2 = Stagnation temperature of ambient air entering the main compressor
p_2 = Pressure of air at the end of isentropic ramming, and

we know that $\quad T_2 = T_{2'} = T_1 + \dfrac{V^2}{2000 \, c_p} = 283 + \dfrac{(417)^2}{2000 \times 1} = 370$ K

Isentropic ramming process 1–2:

$$p_2 = p_1 \left(\dfrac{T_2}{T_1}\right)^{(\gamma-1)/\gamma} = 0.8 \left(\dfrac{370}{283}\right)^{1.4/(1.4-1)} = 2.06 \text{ bar}$$

The expansion of ram air in the first cooling turbine is shown in Figure 3.37. The vertical line 2–1 represents the isentropic expansion process and the curve 2–1′ represents the actual expansion process.

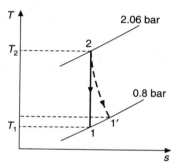

Figure 3.37 Expansion of ram air in first turbine on T–s diagram—Example 3.14.

Isentropic expansion of ram air in the first turbine,

$$T_2 = T_1 \left(\dfrac{p_2}{p_1}\right)^{(\gamma-1)/\gamma}$$

$$= 370 \Big/ \left(\dfrac{2.06}{0.8}\right)^{(1.4-1)/1.4} = 282 \text{ K}$$

Isentropic efficiency of the first cooling turbine,

$$\eta_{T1} = \dfrac{\text{Actual decrease in temperature}}{\text{Isentropic decrease in temperature}} = \dfrac{T_2 - T_{1'}}{T_2 - T_1}$$

$$0.8 = \dfrac{370 - T_{1'}}{370 - 282_1}$$

∴ $\quad T_{1'} = 299.6$ K

Main compressor: Considering isentropic compression (2–3),

$$T_3 = T_2 \left(\frac{p_3}{p_2}\right)^{(\gamma-1)/\gamma} = 370\left(\frac{6}{2.06}\right)^{(1.4-1)/1.4} = 502.6 \text{ K}$$

We know that isentropic efficiency of the compressor,

$$\eta_C = \frac{\text{Isentropic increase in temperature}}{\text{Actual increase in temperature}} = \frac{T_3 - T_2}{T_{3'} - T_2}$$

$$0.85 = \frac{502.3 - 370}{T_{3'} - 370}; \qquad \therefore T_{3'} = 525.5 \text{ K}$$

$$p_5 = p_{5'} = p_6 = 1.0 \text{ bar}$$

Now for the isentropic expansion of air in the second cooling turbine (process 4–5),

$$\frac{T_4}{T_5} = \left(\frac{p_4}{p_5}\right)^{(\gamma-1)/\gamma} = \left(\frac{6}{1.0}\right)^{(1.4-1)/1.4} = 1.669$$

$$T_5 = \frac{T_4}{1.669} = \frac{373}{1.669} = 223 \text{ K}$$

$$\eta_T = \frac{\text{Actual decrease in temperature}}{\text{Isentropic decrease in temperature}} = 0.8 = \frac{373 - T_{5'}}{373 - 223} = \frac{373 - T_{5'}}{149}$$

$$\therefore \qquad T_{5'} = 253.3 \text{ K}$$

(a) **Mass flow rate of air supplied to cabin:**
For mass flow rate of air supplied to cabin, apply the energy balance equation:
Total refrigeration load = heat carried away by the air

$$211 \times Q = m_a \times c_p \times (T_6 - T_{5'})$$

$$m_a = \frac{211 Q}{c_p(T_6 - T_{5'})} = \frac{211 \times 10}{1(293 - 253.9)} = 53.14 \text{ kg/min} \qquad \textbf{Ans.}$$

(b) **Quantity of ram air passing through the heat exchanger:**
Let m_R be the quantity of ram air passing through the heat exchanger.
Consider the heat balance for the heat exchanger (Refer Figure 3.36).
Heat given out by compressed air = Heat gained by ram air

$$m_a c_p (T_{3'} - T_4) = m_R \times c_p(\text{increase in temperature of ram air})$$

$$53.14 \times 1.0 \times (525.6 - 373) = m_R \times 1.0 \ (80)$$

$$\therefore \qquad m_R = 101.2 \text{ kg/s} \qquad \textbf{Ans.}$$

(c) **Power to drive the cooling fan:**
Work output from the first cooling turbine,

$$W_{T1} = m_R \times c_p(T_2 - T_{1'})$$

$$= 101.2 \times 1(370 - 299.6) = 7124.5 \text{ kJ/min}$$

and work output from the second cooling turbine,

$$W_{T2} = m_a \times c_p(T_4 - T_{5'})$$

$$= 53.14 \times 1(373 - 253.3) = 6360.8 \text{ kJ/min}$$

∴ Combined work output from both the cooling turbines,

$$W_T = W_{T1} + W_{T2} = 7124.5 + 6360.8 = 13484 \text{ kJ/min}$$

The power generated by both the turbines is combined and supplied to the cooling fan. Power used to drive the cooling fan = 13484 kJ/min = 224.7 kW **Ans.**

(d) COP of the system:

We know that COP of the system

$$= \frac{211 Q}{m_a c_p (T_{3'} - T_{2'})} = \frac{211 Q}{53.14 \times (525.60 - 370)} = 0.255 \qquad \textbf{Ans.}$$

3.7.6 Regenerative Air Cooling System

The regenerative air cooling system is shown in Figure 3.38 and its corresponding T–s diagram in Figure 3.39. The speciality of this system is the regenerative heat exchanger. The high pressure and high temperature air from the main compressor is first cooled by the ram air in the heat exchanger. This air is further cooled in the regenerative heat exchanger with a portion of the air bled after expansion in the cooling turbine. This type of cooling system is used for supersonic aircraft and rockets.

The various processes are as follows:

Process 1–2: Isentropic ramming of air and the curve 1–2′ represents the actual ramming process of air.

Process 2′–3: Isentropic compression of air in the main compressor and the process 2′–3′ represents the actual compression of air.

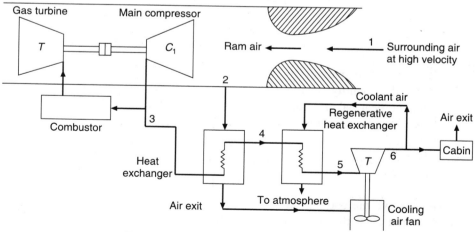

Figure 3.38 Regenerative air cooling system.

Process 3′–4: Cooling of compressed air by ram air in the heat exchanger.
Process 4–5: Cooling of air in the regenerative heat exchanger.
Process 5–6: Isentropic expansion of air in the cooling turbine up to the cabin pressure and 5–6′ represents the actual expansion of air.
Process 6′–7: Increase in air temperature while passing through the cabin.

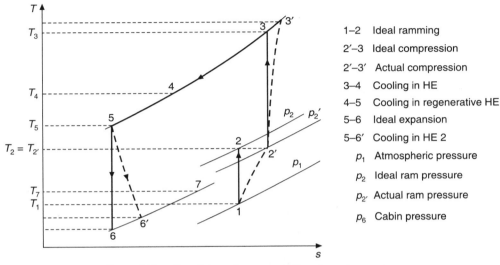

Figure 3.39 *T–s* diagram for regenerative air cooling system.

1. **Mass flow rate of air supplied to cabin**
   Let Q TR be the cooling load in the cabin.
   Mass flow rate of air supplied to cabin, apply energy balance equation.
   Total refrigeration load = heat carried away by the air

   $$211 \times Q = m_a \times c_p \times (T_7 - T_{6'})$$

   $$m_a, \text{ in kg/min} = \frac{211\,Q}{c_p\,(T_7 - T_{6'})}$$

   where, m_a = total mass of air bled from the main compressor,

2. **Mass flow rate of air bled off after cooling turbine and used for regenerative cooling.** From the energy balance of regenerative heat exchanger, we have

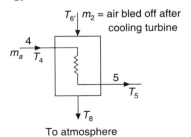

$$m_2 c_p(T_8 - T_{6'}) = m_a c_p(T_4 - T_5); \quad \therefore \quad m_2 = \frac{m_a(T_4 - T_5)}{(T_8 - T_{6'})}$$

T_8 = temperature of air escaping to atmosphere from the regenerative heat exchanger.
m_2 = mass of cold air bled from the cooling turbine for regenerative heat exchanger.

3. **Power required for the refrigeration system**

$$P = \frac{m_a c_p(T_{3'} - T_{2'})}{60} \text{ kW}$$

4. **COP of the refrigerating system**

$$= \frac{211 Q}{m_c c_p(T_{3'} - T_{2'})} = \frac{211 Q}{P \times 60}$$

where, P = power in kW, Q = refrigeration effect in TR.

EXAMPLE 3.15 A regenerative air cooling system is used in an aircraft to take 20 TR load. The ambient air at pressure 0.8 bar and temperature 10°C is rammed isentropically to a pressure of 1.2 bar. The air is further compressed in the main compressor to 4.75 bar and is cooled by the ram air in the heat exchanger whose effectiveness is 80%. The air from the heat exchanger is further cooled to 60°C in the regenerative heat exchanger with a portion of the air bled after expansion in the cooling turbine. The cabin is to be maintained at a temperature of 25°C and a pressure of 1 bar. If the isentropic efficiency for the compressor as well as that for the turbine is 90%, and the temperature of air escaping to atmosphere from the regenerative heat exchangers is 50°C, find the following.

(a) Mass of the air bled from cooling turbine to be used for regenerative cooling.
(b) Power required for maintaining the cabin at the required condition.
(c) COP of the system.

Solution:

Given: Q = 20 TR; p_1 = 0.8 bar $\quad\quad T_1$ = 10°C = 10 + 273 = 283 K
p_2 = 1.2 bar $\quad\quad p_3 = p_4 = p_5$ = 4.75 bar
effectiveness of HE = 0.8 $\quad\quad T_5$ = 60°C = 60 + 273 = 333 K
T_7 = 25°C = 25 + 273 = 298 K $\quad\quad p_7 = p_6 = p_{6'}$ = 1 bar
$\eta_C = \eta_T$ = 90% $\quad\quad T_8$ = 100°C = 100 + 273 = 373 K

The T–s diagram for the regenerative air-cooling system is shown in Figure 3.40. For the isentropic ramming of air (Process 1–2),

$$T_2 = T_1 \left(\frac{p_2}{p_1}\right)^{(\gamma-1)/\gamma} = 283 \left(\frac{1.2}{0.8}\right)^{(1.4-1)/1.4} = 317.8 \text{ K}$$

Main compressor: Considering isentropic compression,

$$T_3 = T_2 \left(\frac{p_3}{p_2}\right)^{(\gamma-1)/\gamma} = 317.8 \left(\frac{4.75}{1.2}\right)^{(1.4-1)/1.4} = 471 \text{ K}$$

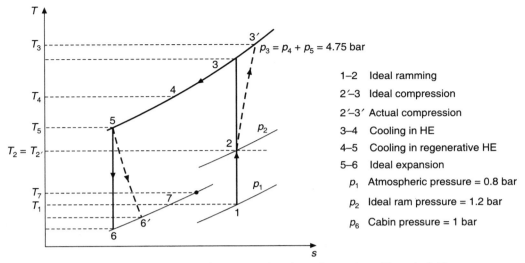

Figure 3.40 T–s diagram for regenerative air cooling system—Example 3.15.

Isentropic efficiency of the compressor,

$$\eta_C = \frac{\text{Isentropic increase in temperature}}{\text{Actual increase in temperature}} = \frac{T_3 - T_2}{T_{3'} - T_2}$$

i.e.
$$0.9 = \frac{471 - 317.8}{T_{3'} - 317.8} = \frac{153.2}{T_{3'} - 317.8}$$

∴
$$T_{3'} = 488 \text{ K}$$

Effectiveness of the heat exchanger (ε_H),

$$0.8 = \frac{T_{3'} - T_4}{T_{3'} - T_2} = \frac{488 - T_4}{488 - 317.8} = \frac{488 - T_4}{170.25}$$

∴
$$T_4 = 351.8 \text{ K}$$

Isentropic expansion in the cooling turbine (process 5–6),

$$\frac{T_5}{T_4} = \left(\frac{p_5}{p_6}\right)^{(\gamma-1)/\gamma} = \left(\frac{4.75}{1.0}\right)^{(1.4-1)/1.4} = 1.561$$

∴
$$T_6 = \frac{T_5}{1.561} = \frac{333}{1.561} = 213.3 \text{ K}$$

Isentropic efficiency of the cooling turbine,

$$\eta_T = \frac{\text{Actual decrease in temperature}}{\text{Isentropic decrease in temperature}} = \frac{T_5 - T_{6'}}{T_5 - T_6}$$

$$0.8 = \frac{333 - T_6'}{333 - 213} = \frac{333 - T_6'}{119.7}$$

$\therefore \quad T_6' = 225.2$ K

(a) Mass of air bled from the cooling turbine to be used for regenerative cooling:

Let m_2 = mass of air bled from the cooling turbine to be used for regenerative cooling,
m_a = total mass of air bled from the main compressor, and
m_s = mass of cold air supplied to the cabin.

For the energy balance of regenerative heat exchanger,

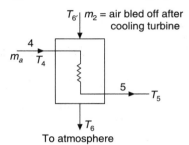

$$m_2 c_p (T_8 - T_{6'}) = m_a c_p (T_4 - T_5); \qquad \therefore m_2 = \frac{m_a (T_4 - T_5)}{(T_8 - T_{6'})}$$

$$m_2 = \frac{m_a (351.8 - 333)}{(323 - 225.2)} = 0.1923 m_a \qquad \text{(i)}$$

We know that the mass of air supplied to the cabin,

$$m_s = m_a - m_2$$
$$= \frac{211 \, Q}{c_p (T_7 - T_{6'})} = \frac{211 \times 20}{1 (298 - 225.2)} = 58.0 \text{ kg/min} \qquad \text{(ii)}$$

From Eq. (i), we find that

$$m_a - m_2 = 58.0 \quad \text{or} \quad m_a - 0.1923 m_a = 58.0$$

$\therefore \qquad m_a = \dfrac{58.0}{1 - 0.1923} = 71.8$ kg/min

and $\qquad m_2 = 0.1923 m_a = 0.1923 \times 71.8 = 13.8$ kg/min **Ans.**

(b) Power required for the compressor:
Power required for maintaining the cabin at the required condition is given as:

$$P = \frac{m_a c_p (T_{3'} - T_2)}{60} = \frac{71.8 \times 1 (488 - 317.8)}{60} = 203.6 \text{ kW} \qquad \textbf{Ans.}$$

(c) COP of the system:

$$\text{COP of the system} = \frac{211\,Q}{m_a c_p (T_{3'} - T_2)} = \frac{211 \times 20}{71.8 \times 1\,(488 - 317.8)} = 0.345 \qquad \textbf{Ans.}$$

3.8 COMPARISON OF VARIOUS AIR COOLING SYSTEMS USED FOR AIRCRAFT

The performance curves for the various air cooling systems used for aircraft are shown in Figure 3.41. These curves are plotted for various Mach numbers vs temperature of air at the outlet of cooling turbine:

1. One can observe that the simple air cooling system gives maximum cooling effect on the ground surface and decreases as the speed of the aircraft increases. It would be useful for the Mach numbers ranging between 0.4 and 1.5. This simple system can be employed with the evaporative system at high speeds.
2. The boot-strap system is better at low Mach number (1.4 and above) since it needs ram air for cooling in heat exchangers.
3. Boot-strap system is modified using evaporative system in addition to heat exchangers. At high Mach numbers, the temperature of ram air rises due to high speeds. Such high temperature ram air would be less effective in cooling the compressed air of the main compressor in the primary heat exchanger to the desired low temperature.
4. Regenerative air cooling systems are useful for low Mach numbers as can be seen from the graph.
5. Reduced ambient air cooling system is suitable to render low temperatures at high Mach numbers, hence it is only suitable for supersonic aircraft.
   The turbine discharge temperature of the air is variable. Therefore, in order to maintain the constant temperature of supply air to the cabin, it requires some control system.

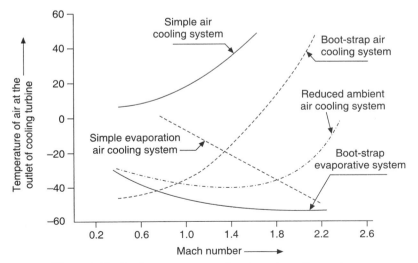

Figure 3.41 Performance curves of various air cooling systems.

EXERCISES

1. The COP of the heat pump is greater by unity compared to the reversed Carnot refrigerator system. Explain.
2. Discuss the working of a Carnot refrigerator with working substance as air as well as vapour. Derive an expression for its COP.
3. Differentiate between an engine, a refrigerator and a heat pump.
4. Define the term coefficient of performance as applied to refrigerators and heat pumps.
5. Discuss the Bell–Colemann air refrigeration cycle with the help of schematic, p–v and T–s diagrams. Deduce an expression for its COP. For what purpose is this cycle used in aircraft?
6. What are the advantages and disadvantages of Bell–Colemann air cycle?
7. Discuss the closed Bell–Colemann air cycle. Why are brine solutions used in these cycles?
8. State and explain the various air refrigeration cycles used for aircraft.
9. What are the advantages or benefits of using the air cycles for aircraft cooling?
10. Draw a neat diagram and explain the working of the simple air cooling cycle for aircraft. Also represent the cycle on T–s diagram and explain the various processes involved. Explain the procedure of obtaining the COP of the cycle.
11. Explain the working of boot-strap air refrigeration system.
12. Describe the working of the simple air evaporative cycle with the help of schematic and T–s diagrams.
13. Differentiate and enumerate the advantages of the boot-strap air evaporative system from the boot-strap air cooling system.
14. When can one prefer to have the reduced ambient air cooling system in aircraft? Explain its working with the help of a neat sketch.
15. Explain the working of the regenerative air cooling system with the help of schematic and T–s diagrams. Explain the various processes involved and write the expression for its COP.
16. Write short notes on the following:
    - Limitations of Carnot cycle with gas as refrigerant.
    - Refrigeration effect and its units.
    - Aircraft air conditioning.
17. Compare the different types of air cooling systems in terms of aircraft speed.

NUMERICALS

1. In a Bell–Colemann refrigeration plant, the air is drawn from cold chamber at 1 bar and 12°C, and compressed to 5 bar. The same is cooled to 25°C in the cooler before expanding in the expansion cylinder to cold chamber pressure of 1 bar.

(a) Determine the theoretical COP of the plant and the theoretical net refrigeration effect/kg of air. The compression and expansion be assumed isentropic. Assume $\gamma = 1.41$ and $c_p = 1.009$ kJ/kg-K.

(b) If the compression and expansion laws followed are $pv^{1.35} = C$ and $pv^{1.3} = C$ respectively, how will the result be modified?

2. A refrigeration unit working on Bell–Colemann cycle takes air from cold chamber at −10°C and compresses it from 1 bar with index of compression being 1.2. The compressed air is cooled to a temperature 10°C above the ambient temperature of 25°C before it is expanded in the expander where the index of expansion is 1.35.

   Determine the following:

   (i) COP.
   (ii) Quantity of air circulated per minute for the production of 2000 kg ice per day at 0°C from water at 20°C.
   (iii) Capacity of the plant in ton-refrigeration.
   Assume $c_p = 1$ kJ/kg-K for air.

3. An air refrigeration plant 25 TR capacity comprises a centrifugal compressor, a cooler heat exchanger and an air turbine. The compressor is coupled directly to the air turbine. The compressor also receives power from another prime mover. The processes in the compressor and the turbines are adiabatic but not isentropic. Air at temperature 21°C and 0.85 bar enters the compressor. It leaves the compressor at 90°C. The same air enters the turbine at 38°C and 1.5 bar. The turbine exit is at 0°C. Assuming no pressure drop in the cooler and the refrigerator (evaporator) section, and constant specific heats as $c_p = 1.004$ kJ/kg-K and $c_v = 0.712$ kJ/kg-K, determine the following.

   (i) The compressor efficiency.   (ii) The turbine efficiency.
   (iii) The flow rate of air.   (iv) The power input to the plant.
   (v) The coefficient of performance.

4. A high altitude flight aircraft is flying at an altitude of 1.5 km with a speed of 1.2 mach. The ambient atmospheric pressure and temperature are 0.2 bar and −40°C. The cabin is pressurised to 0.7 bar and has to be maintained at 25°C. The main compressor pressure ratio is 5 and the air enters the cooling turbine at 40°C. The exit from the cooling turbine is at 0.75 bar. The cockpit cooling load is 10 tonne. Assume internal efficiency of compressor as 85% and that of cooling turbine as 75%. Ram efficiency is 90%. Assume $\gamma = 1.4$ and $c_p = 1$ kJ/kg-K. Determine the following:

   (i) Stagnation temperature and pressure of air entering and leaving the main compressor.
   (ii) Mass flow rate of air.
   (iii) Ram air heat exchanger effectiveness.
   (iv) Volume handled by compressor and cooling turbine.
   (v) Net power delivered by engine to the refrigerating unit (Pressurisation + Refrigeration).
   (vi) COP of the system based on compressor work.
   (vii) Additional power only for refrigeration.

5. A boot-strap air refrigeration system is used for an airplane to take 20 TR of cooling load. The ambient conditions are 5°C and 0.85 bar. The air pressure increases to 1.1 bar due to ramming action which is considered ideal (isentropic). The pressure of air bled off the main compressor is 3.5 bar and this air is further compressed in the secondary compressor to 4.8 bar. The isentropic or internal efficiency of the main compressor as well as that of the secondary compressor is 90% and that of the turbine is 80%. Heat exchanger effectiveness of the primary heat exchanger is 0.6 and that of secondary heat exchanger is 0.6. Assuming c_p = 1.0 kJ/kg-K, determine the following:

   (a) The power required to take the cabin load and, (b) the COP of the system.

   The cabin may be maintained at 1 bar and 25°C. The cooling turbine runs the secondary compressor and uses its surplus power to run a fan for blowing in the ram air to waste.

6. A regenerative air refrigeration system of an aircraft with flight speed of 1500 km/h has 30 TR cooling load while the ambient conditions are 0.1 bar and –63°C. The ram efficiency is 90%. The pressure ratio of the main compressor is 5 with internal efficiency of 0.9. The air bled off the main compressor is cooled by ram air in a heat exchanger which is 60% effective. The air from the heat exchanger passes on to the cooling air turbine whose internal efficiency is 0.8. Some portion of air from the cooling turbine is led to the regenerative heat exchanger reducing the temperature to 30°C of the bled off compressed air. The cooling air gets heated to 92°C before discharging to atmosphere. The cabin is pressurized to 0.8 bar and maintained at 25°C. Determine the following:

   (a) The percentage of air extracted for regenerative cooling.
   (b) Power required to maintain the cabin at the required condition.
   (c) COP.

   Assume the cooling turbine power to be used for air exhaust fan.

7. The following data refer to the aircraft refrigeration system with evaporative cooling.

   Ram air pressure = 1.04 bar; ram air temperature = 25°C; compressor delivery pressure = 4 bar; pressure drop in ram air heat exchanger = 0.2 bar; ram air exchanger effectiveness = 0.8; cabin pressure = 1 bar; mass flow rate of air through cabin bled from main compressor = 30 kg/min; temperature of air leaving the cabin = 25°C and pressure = 1 bar; evaporative cooling effect = 30 kJ/kg of air flow. Determine the following:

   (a) Refrigeration capacity, (b) Power for refrigeration, and (c) COP.

Chapter 4

Simple Vapour Compression Refrigeration Systems

4.1 INTRODUCTION

The vapour compression system is the basis of operation for many of the refrigeration units. Therefore, one must understand the operation of the compression cycle accurately so as to be able to diagnose (identify) faults. Before studying the development of vapour compression cycle, let us review the advantages and disadvantages (limitations) of air refrigeration.

4.2 ADVANTAGES AND LIMITATIONS OF AIR REFRIGERATION

The advantages and limitations of air refrigeration are given in the following subsections:

4.2.1 Advantages of Air Refrigeration

(i) Air refrigeration systems use air as refrigerant which is available abundantly free of cost.
(ii) Air is an inert gas. It is inflammable, non-toxic and hence safe as a refrigerant.
(iii) Air refrigeration systems, which work on Bell–Colemann cycle, are suitable for aircraft on account of availability of ram air.

4.2.2 Limitations of Air Refrigeration

Various air refrigeration systems employed for cooling the cabins of aircraft were discussed in Chapter 3. In general, there are some limitations of using air as a refrigerant, which are as follows:

1. The heat carried away by air in air refrigeration systems is $\dot{m}_a \times c_p (T_{out} - T_{in})$. It depends on the mass flow rate of air ($\dot{m}_a$), specific heat c_p and temperature difference. The value of c_p is 1.00 kJ/kg which is very low. Therefore, for a fixed value of

temperature difference, it needs a large mass flow rate of air per TR (ton of refrigeration).
2. It requires a bigger size compressor to handle air. Therefore, the system becomes bulky.
3. Since the coefficient of performance is very poor, more power is required to cope the cooling load.
4. The system components like compressor and heat exchanger are bulky. Hence a large installation space per TR of refrigeration is required.
5. Air contains moisture, the condensation of which at low temperatures poses a maintenance problem.
6. Many a time air is contaminated with pollutants, hence requiring air filters which need to be cleaned regularly.

Today, refrigerants having better thermodynamic properties and low boiling points are used in refrigeration systems.

The following points should be carefully understood before studying the vapour compression cycle:

1. Fluids absorb heat while changing from liquid phase to vapour phase and release it while changing from vapour phase to liquid phase. Now the question arises, as to which fluids are these? Of course, these are refrigerants. So, it is necessary to have knowledge of refrigerants.
2. The temperature remains constant during the phase change, provided the pressure remains constant.
3. Heat flows only from a body at a higher temperature to a body at lower temperature.
4. The evaporating and condensing units are made from metals having high thermal conductivity.
5. Heat energy and other forms of energy are interchangeable. For example, electric energy may be converted to heat energy; heat energy to electrical energy and further to mechanical energy.

4.3 VAPOUR COMPRESSION REFRIGERATION SYSTEM (VC CYCLE)

A vapour compression refrigeration plant is shown in Figure 4.1. It mainly consists of components such as compressor, condenser, expansion device (valve or capillary tube) and evaporator. In this figure, one can see the existence of refrigerant in different phases at different cycle conditions. If we trace the path from point 2 (outlet of compressor) through the condenser, receiver and up to the inlet of the expansion valve, the refrigerant is in pressurized state. Therefore, this part of the flow diagram forms the high pressure side. The remaining part, i.e. from point 4 through the evaporator and up to the inlet of compressor is the low pressure side. To explain the working of this plant, the cycle is represented on T–s diagram (Figure 4.2), and p–v diagram (Figure 4.3). Here one more assumption is made, i.e. the refrigerant vapour leaves the evaporator in dry saturated condition and enters the compressor. The suction to the compressor is indicated by point 1 in all these diagrams.

Simple Vapour Compression Refrigeration Systems 117

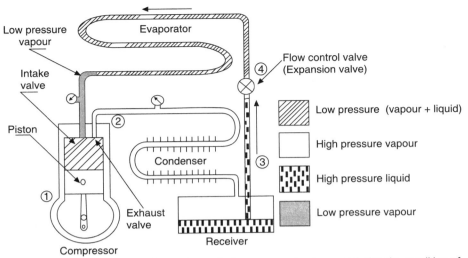

Figure 4.1 Vapour compression system with its components shown and also the condition of refrigerant in the flow circuit.

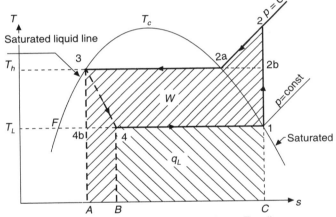

Figure 4.2 Vapour compression cycle on T–s diagram.

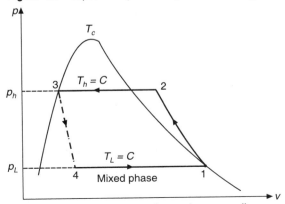

Figure 4.3 Vapour compression cycle on p–v diagram.

Assumptions made in theoretical vapour compression cycle

- All the processes involved in the cycle are assumed to be reversible.
- There is no internal and external friction between the refrigerant and the metallic surface of the heat exchangers (evaporator and condenser). Hence any pressure drop is neglected.
- There is no heat gain or loss except in evaporator and condenser.

Compression process (1–2)

Refer to Figures 4.1 through 4.4. The compressor takes the dry saturated refrigerant vapour during its suction stroke at pressure p_1. Then the vapour is compressed to pressure p_2 isentropically during its compression stroke. Therefore, the compressor needs external power. The vapour at pressure p_2 and temperature T_2 enters the condenser. The vapour at the discharge of the compressor is in superheated state as shown by point 2. Basically, the question arises as to why should the vapour be taken to higher pressure p_2. The answer is that the refrigerant at temperature T_2 corresponding to pressure p_2 will condense. This is all due to the properties of the refrigerant.

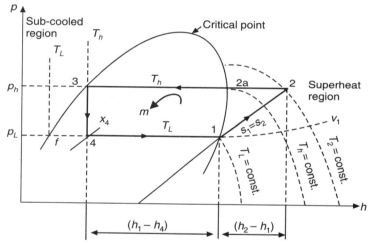

Figure 4.4 Vapour compression cycle on p–h diagram.

Just now, it has been stated that the compression process is isentropic. One can ask as to why is the compression process isentropic? Let us go through the following explanation:

The modern compressors work at high speed of about 2000 rpm, i.e. for every one cycle of operation the total time available is about 0.03 second, out of which for the compression process it is about 0.01 second. It means that during compression, the refrigerant vapour comes in contact with the cylinder walls for a very small period. Also, the temperature difference between the vapour and cylinder walls is very small. Therefore, heat is hardly transferred from vapour to these walls. It is also known that for good lubrication the friction is negligible. Hence, the process is considered reversible adiabatic or isentropic.

Condensation process (2–3)

The function of the condenser is to remove heat from vapour refrigerant so that it changes to liquid phase. In this unit, superheated vapour from state point 2 is first cooled so that it reaches the dry saturated vapour condition (point 2a), where the vapour starts condensing and further removal of heat would result in a liquid refrigerant at state point 3. The condenser may be either air- or water-cooled. Small capacity refrigeration units are air-cooled and large capacity refrigeration units are water-cooled.

Expansion process (3–4)

The high pressure liquid from state point 3 is expanded through an expansion valve, also called throttle valve, during which the enthalpy remains constant. This process is represented by line 3–4. At the end of this process (point 4), the refrigerant is again a mixture of liquid and vapour phases.

Evaporation process (4–1)

Evaporation of liquid refrigerant takes place in a heat exchanger called evaporator. The function of an evaporator is to transfer heat from its space to the refrigerant. The refrigerant at state point 4 enters the evaporator, takes the heat from space and slowly changes into saturated vapour as it flows towards the end of evaporator, i.e. state point 1. This point 1 indicates the starting and end points of the cycle.

Referring to Figure 4.2, we can compare the vapour compression cycle 1–2–3–4 with the reversed Carnot cycle 1–2–3–4b, as these operate in the same temperature limits of T_h and T_L.

(i) The area 4–4b–A–B, represents the loss of refrigerating effect due to throttling.

(ii) This area, 4–4b–A–B, also represents a loss of positive work due to throttling. It is possible to show that the area 4–4b–A–B equals the area 3–4–4b.

(iii) The area 2–2a–2b is called *superheat horn* which results in an increase of compression work.

Now we can summarize the following points for vapour compression cycle:

Net refrigerating effect, q_L = area 1–4–B–C.
Total heat rejected (kJ/kg), q_h = area 2–2a–3–A–C–2
Net work done (kJ/kg), w = area 1–2–2a–3–4–1
= $q_h - q_L$

From the above discussion, it is clear that for vapour compression cycle, the net work input is more and the refrigerating effect is less. Therefore, the theoretical COP of the vapour compression cycle is lower than that of the reversed Carnot cycle.

In vapour compression cycle, two processes, namely evaporation and condensation, are at constant pressure, and out of the remaining two, one is isentropic and the other is isenthalpic process. Therefore, it is advantageous to represent the cycle on the p–h diagram rather than with the combination of any other variables. The cycle is shown in Figure 4.4 using pressure-enthalpy and also all the process points of the simple vapour compression cycle are marked.

The analysis of the cycle has been carried out as given below assuming unit mass flow rate of refrigerant (i.e. $m = 1$ kg/s).

Process 1–2: Isentropic compression $s_1 = s_2$, $q = 0$.

$$\text{Work done, } w = (h_2 - h_1) \text{ kJ/kg} \tag{4.1}$$

where h_1 and h_2 are enthalpies in kJ/kg corresponding to points 1 and 2, respectively.

Process 2–3: Desuperheating and condensation at constant pressure, p_h

$$\text{Heat rejected, } q_h = (h_2 - h_3) \text{ kJ/kg} \tag{4.2}$$

h_3 = enthalpy at state point 3 in kJ/kg

Process 3–4: Isenthalpic expansion, $h_3 = h_4$

$$= h_{f4} + x(h_1 - h_{f4}) \text{ kJ/kg}$$

$$x = \frac{h_3 - h_{f4}}{h_1 - h_{f4}} \tag{4.3}$$

Process 4–1: Evaporation at constant pressure, p_L

$$\text{Refrigerating effect, } q_L = h_1 - h_4 \tag{4.4}$$

$$\text{For cooling, } COP_R = \frac{h_1 - h_4}{h_2 - h_1} \tag{4.5}$$

$$\text{For heating, } COP_H = \frac{h_2 - h_3}{h_2 - h_1} \tag{4.6}$$

$$\text{Refrigerant flow rate, } m = \frac{\text{Total refrigerating effect}}{\text{Refrigerating effect per unit mass}}$$

$$= Q_L / q_L \tag{4.7}$$

Let v_1 be the specific volume of the gas at the suction to compressor. The theoretical piston displacement of reciprocating compressor or volume of suction vapour

$$V = m v_1 \tag{4.8}$$

If η_V is the volumetric efficiency, then the actual piston displacement (V_p) of the compressor

$$V_p = m v_1 / \eta_V \tag{4.9}$$

$$\text{Power requirement, } W = m(h_2 - h_1) \tag{4.10}$$

$$\text{Heat rejected in the condenser, } Q_h = m q_L = m(h_2 - h_3) \tag{4.11}$$

If W^* is the power requirement per ton of refrigeration (TR), then the mass flow rate per ton of refrigerating effect,

$$\dot{m} = \frac{3.5164}{q_L} \text{ kg/s (TR)} \tag{4.12}$$

$$W^* = \dot{m} \times w = \frac{3.5164}{q_L} \times w$$

$$= \frac{3.5164(h_2 - h_1)}{(h_1 - h_4)} \qquad (4.13)$$

It is observed that
$$W^* \propto \frac{3.5164}{\text{COP}_R} \qquad (4.14)$$

Similarly, the suction volume requirement per ton is
$$V^* = \frac{3.5164}{q_L} \times v_1 \text{ (m}^3\text{/s)/(TR)} \qquad (4.15)$$

The temperature T_2 at the end of isentropic compression may be found by the following three methods:

(i) Using p–h diagram, trace the isentropic line from point 1 to pressure p_h
(ii) Using saturation properties and the specific heat of vapour:
$$s_1 = s_2 = s_{2a} + c_p \, l_n \, (T_2/T_{2a}) \qquad (4.16)$$
where, $s_{2a} = s_{g2}$ and $T_{2a} = T_h$, $T_2 = T_{\text{sup}}$

This equation is used for ideal gases. The use of this equation for refrigerant vapour is only to get the approximate value of T_2. This further helps in locating the isentropic curve in the superheat region so that we could get the enthalpy values which otherwise may include human errors.

(iii) Use superheat tables and interpolate for the degree of superheat $(T_2 - T_{2a})$ corresponding to the entropy difference $(s_2 - s_{2a})$.

EXAMPLE 4.1 Refrigerant R12 at –20°C enters the compressor as dry saturated vapour. It is compressed isentropically to the condenser pressure corresponding to its temperature 35°C. Find (a) the temperature and enthalpy at the end of compression, (b) the work of compression per unit mass flow rate of refrigerant.

Solution: (a) Referring to Figure 4.4, temperature T_2 can be obtained by the following methods:

(i) Using p–h chart:
Locate point 1 on the saturated vapour curve corresponding to –20°C. The saturation vapour pressure corresponding to 35°C is 8.47 bar from property table. Now from point 1, draw a line parallel to the isentropic lines to meet the pressure line 8.47 bar. (Line 1–2 in Figure 4.4). After this, note down the temperature T_2 = 48.5°C, enthalpy h_2 = 220 kJ/kg and h_1 = 185 kJ/kg. **Ans.**

(ii) From R12 property table, the properties at point 1 are
$T_1 = -20 + 273 = 253$ K, $p_1 = 1.5101$ bar, $h_1 = 185$ kJ/kg
$s_1 = 1.566$ kJ/kg-K
$s_{2a} = 1.542$ kJ/kg-K at $T_{2a} = 308$ K

Applying Eq. (4.16), this can be used for ideal gases. The use of this for refrigerant vapour is only to get an approximate value of T_2.
$s_1 = s_{2a} + c_p \log(T_2/T_{2a})$

$1.566 = 1.542 + 0.65 \log(T_2/308)$

$T_2 = 319.6$ K or $46.6°C$ **Ans.**

(iii) Using the superheat table

$s_2 - s_{2a} = 1.566 - 1.542 = 0.024$ kJ/kg-K

$T_2 = 47°C$ and enthalpy $h_2 = 372.6$ kJ/kg, which helps to locate isentropic curve in the superheat region so that we could get the enthalpy values which otherwise may include human errors. **Ans.**

(b) Work of compression per kg:

$w = (h_2 - h_1)$

$= 372.6 - 342.0 = 32$ kJ/kg using property chart of R12

$= 372.6 - 342.6 = 30$ kJ/kg using property superheat table of R12

A little deviation will always be observed while taking values of properties from the chart and table. Moreover, the reference point to construct chart and table are different. Therefore, only the difference in enthalpies would be considered while extracting the property values from charts or tables.

EXAMPLE 4.2 A deep freezer operates between the temperature limits $-20°C$ and $35°C$ and has the refrigerating capacity of 0.8 ton with R12 refrigerant. Calculate the compressor work assuming isentropic compression and refrigerant inlet to the compressor being dry saturated vapour.

Solution: From p–h chart:

At $-20°C$, $h_1 = 175$ kJ/kg dry saturated vapour, $h_2 = 220$ kJ/kg

As explained in Example 4.1, locate point 2 on p–h chart and get the value of $h_2 = 220$ kJ/kg

Work $= (h_2 - h_1) = 45$ kJ/kg

Mass flow rate of refrigerant:

First determine the refrigerant effect $(h_1 - h_4)$ kJ/kg

Draw the cycle on p–h chart and find $h_4 = 235$ kJ/kg

$h_1 - h_4 = 175 - 70 = 105$ kJ/kg

Mass flow rate $= \dfrac{Q_L}{h_1 - h_4} = \dfrac{0.8 \times 3.5167}{105}$

$= 0.02678$ kg/s

Compressor power $= m(h_2 - h_1)$

$= (0.02678)(45) = 1.2051$ kW **Ans.**

EXAMPLE 4.3 Determine the refrigerating effect per kilogram if the condenser temperature is $35°C$ and the temperature of evaporator is $-5°C$ for the R134a system.

Solution: From the property tables of R134a, enthalpy of R134a saturated vapour at $-5°C$

$h_1 = 402.25$ kJ/kg

Enthalpy of R134a liquid at $35°C$, $h_4 = 248.94$ kJ/kg

Refrigerating effect per kg $= (h_1 - h_4) = 402.25 - 248.94 = 153.31$ kJ/kg **Ans.**

EXAMPLE 4.4 For Example 4.3, if the temperature of evaporator is decreased to –10°C and the temperature of the liquid R134a approaching expansion valve is 35°C, what is the refrigerating effect per kilogram?

Solution: From the property table of R134a, enthalpy of saturated vapour at –20°C,
$$h_1 = 399.28 \text{ kJ/kg}$$
Enthalpy of liquid at 35°C, $h_4 = 248.94$ kJ/kg
$$\text{Refrigerating effect per kg} = (h_1 - h_4)$$
$$= (399.28 - 248.94)$$
$$= 150.34 \text{ kJ/kg} \quad \textbf{Ans.}$$

After studying Examples 4.3 and 4.4, one can conclude that as the temperature of the evaporator decreases, the refrigerating effect also decreases for constant condenser temperature.

EXAMPLE 4.5 In Example 4.3, if the evaporator temperature is same as –5°C and if the condenser temperature is raised to 40°C, find the specific refrigerating effect.

Solution: Specific enthalpy of liquid R134a at 40°C, $h_3 = h_4 = 256.35$ kJ/kg
Specific enthalpy of saturated vapour at –5°C, $h_1 = 402.25$ kJ/kg
Specific refrigerating effect $= (h_1 - h_4) = (402.25 - 256.35)$
$$= 145.9 \text{ kJ/kg} \quad \textbf{Ans.}$$

The comparison of this value with that in Example 4.3 shows that the specific refrigerating effect has decreased by 7.41 kJ/kg.

EXAMPLE 4.6 A vapour compression refrigerating system operates at –6°C evaporating temperature and condensing temperature of 36°C. If the volume flow rate of R22 through the system is 0.05 m³/s, determine the refrigerating capacity of the system.

Solution: From the property table of R22, for $v_1 = v_g = 0.05706$ m³/kg, $h_1 = h_g = 245$ kJ/kg and $h_{f3} = h_4 = 90$ kJ/kg, $\dot{m} = \dfrac{0.05}{0.05706} = 0.876 \text{ kg/s}$

Refrigerating effect $= (h_1 - h_4) = 155$ kJ/kg
Therefore, refrigerating capacity in kW $= \dot{m} q_L = (0.876)(155) = 135$ kW **Ans.**

EXAMPLE 4.7 A R134a system is working at evaporating temperature of –10°C and the condensing temperature of 40°C. Assuming that the system works as per the simple vapour compression cycle, find the following:
(a) The refrigerating effect per kg.
(b) The mass flow rate of refrigerant required to be circulated (in kg/s) to get one kW of refrigeration.
(c) The mass of refrigerant circulated per second for a 10 kW refrigeration unit.

Solution: (a) The refrigerating effect per kg:
Refer to Figure 4.4, from the property table of R134a
Enthalpy of saturated vapour at –10°C, $h_1 = 399.28$ kJ/kg
Enthalpy of saturated liquid at 40°C, $h_3 = h_4 = 256.35$ kJ/kg
Refrigerating effect per kg $= (h_1 - h_4) = 142.93$ kJ/kg **Ans.**

(b) Mass flow rate of refrigerant per kW = Q_L/q_L

$$= \frac{1 \text{ kW}}{142.93 \text{ kJ/kg}}$$

$$= 0.00699 \text{ kg/s} = 6.99 \text{ g/s} \quad \text{Ans.}$$

(c) Mass flow rate of refrigerant for a 10 kW system

$$= (10)(0.00699) = 0.0699 \text{ kg/s}$$
$$= 69.9 \text{ g/s} \quad \text{Ans.}$$

EXAMPLE 4.8 A saturated liquid refrigerant R22 at the condenser pressure corresponding to a saturation of 35°C is expanded through a capillary tube to an evaporator pressure corresponding to a temperature –20°C. Determine the condition of refrigerant at the end of the expansion (outlet of capillary tube) process.

Solution: The enthalpy of refrigerant before and after expansion is same for the isenthalpic process. Refer Figure 4.4. Enthalpy of saturated liquid refrigerant at 8.47 bar, i.e. corresponding to 35°C, h_{f3} = 243.14 kJ/kg. The refrigerant will be in mixed phase after expansion, $h_4 = h_{f4} + x(h_1 - h_{f4})$. At temperature –20°C corresponding to evaporator pressure 1.3268 bar, h_{f4} = 177 kJ/kg, h_1 = 396.97 kJ/kg

$$\therefore \quad x = \frac{h_4 - h_{f4}}{h_1 - h_{f4}} = \frac{90 - 20}{240 - 20} = 0.318 \quad \text{Ans.}$$

EXAMPLE 4.9 A compressor manufacturing company manufactures one model of compressor used for air conditioning applications in India. The compressor was tested in a refrigerating calorimeter keeping the evaporator temperature of 8°C and condensing temperature of 35°C using R22 refrigerant and it was found that the compressor produced a refrigerating effect of 5 TR. Using the *p–h* chart, determine the following:

(a) The mass flow rate of the refrigerant.
(b) Volume flow rate handled by the compressor, volume flow rate per ton of refrigeration.
(c) Power required by the compressor and power requirement of the compressor per ton of refrigeration.
(d) Heat rejected in the condenser.
(e) COP of the cycle. Assume a simple vapour compression cycle.

Solution: Required properties have been taken from the *p–h* chart as in Figure 4.5.

v_1 = 0.4 m³/kg, h_1 = 255 kJ/kg, h_2 = 275 kJ/kg, $h_3 = h_4$ = 90 kJ/kg

(a) Refrigerating effect: $q_L = (h_1 - h_4)$
$$= (255 - 90) = 165 \text{ kJ/kg}$$

Refrigerant mass flow rate, $\dot{m} = \dfrac{Q_L}{q_L} = \dfrac{5 \times 3.5167}{165} = 0.105 \text{ kg/s}$

$$= 6.3 \text{ kg/min} \quad \text{Ans.}$$

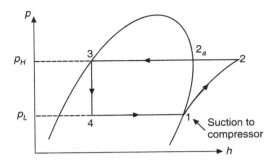

Figure 4.5 Flow diagram on p–h chart—Example 4.9.

(b) Refrigerant volume flow rate, $V = mv_1 = (6.3)(0.04) = 0.24$ m³/min **Ans.**

Refrigerant volume flow rate of compressor/ton $= V^* = \dfrac{0.23}{5} = 0.504$ (m³/min)/(TR) **Ans.**

(c) Power consumption,
$$W = m(h_2 - h_1)$$
$$= (0.105) \times (275 - 255)$$
$$= 2.1 \text{ kW} \quad \textbf{Ans.}$$

Power consumption per TR $= \dfrac{2.1}{5} = 0.42$ kW/TR **Ans.**

(d) Heat rejected, $Q_H = m(h_2 - h_3)$
$$= (0.105)(275 - 90) = 19.42 \text{ kW} \quad \textbf{Ans.}$$

(e) COP of the cycle, $\text{COP}_R = \dfrac{h_1 - h_4}{h_2 - h_1} = \dfrac{255 - 90}{275 - 255} = 8.2$ **Ans.**

EXAMPLE 4.10 An ammonia refrigeration system operates in a simple saturation vapour compression cycle to produce ice. The average temperature of condenser and evaporator are 35°C and –10°C, respectively. The plant produces 12 tons of ice everyday using available water at 30°C. The ice produced is at –5°C. Determine the following:

(a) Cooling capacity (refrigerating effect) of the plant in kW.
(b) Mass flow rate of refrigerant.
(c) Temperature of ammonia vapour at the outlet of compressor.
(d) The compressor cylinder diameter and stroke if its volumetric efficiency is 70%, speed = 1500 rpm and $L/D = 1.15$.
(e) Theoretical power requirement of the compressor.
(f) Theoretical COP.

Solution: This problem can be solved using the properties of ammonia table.

(a) Refrigerating capacity, Q_L:

The heat to be removed includes the following:
1. Sensible heat to cool water from 30°C to 0°C
2. Latent heat of fusion 335 kJ/kg for water
3. Sensible heat to cool ice from 0°C to –5°C.

$$Q_L = (12)(1000)[4.186(30-0)] + 335 + 1.95[0-(-5)] = 557.37 \times 10^4 \text{ kJ}$$

Rate of heat removed per second

$$Q_L = \frac{557.37 \times 10^4}{(24)(3600)} = 64.5 \text{ kW} \quad \text{Ans.}$$

(b) Mass flow rate:
From ammonia chart, at –10°C, $h_1 = 500$ kJ/kg and $v_1 = 0.4182$ m³/kg
At 35°C, $h_3 = h_4 = -580$ kJ/kg
Refrigerating effect, $q_L = (h_1 - h_4) = [500 - (-580)] = 1080$ kJ/kg
Mass flow rate = (64.5/1080) = 0.05956 kg/s = 214.4 kg/h **Ans.**

(c) Temperature at the end of compression
$T_2 = 90°C$ is obtained from the p–h chart by tracing the path of the isentropic line. **Ans.**

(d) Cylinder bore and stroke:

$$\text{Displacement volume} = \frac{mv_1}{\text{Volumetric efficiency}}$$

$$= \frac{(0.05956)(0.4518)}{0.70} = 0.03844 \text{ m}^3/\text{s}$$

$$\text{Displacement volume} = \left(\frac{\pi}{4}D^2\right)(L)(N)$$

$$0.03844 = \frac{\pi}{4}(D^2)(1.15D)\frac{(1500)}{60}$$

∴ $D = 0.119$ m, $L = 0.137$ m **Ans.**

(e) Power consumption (W):
Enthalpy at discharge, $h_2 = 700$ kJ/kg

$$W = m(h_2 - h_1)$$
$$= 0.05956(700 - 500) = 11.91 \text{ kW} \quad \text{Ans.}$$

(f) Theoretical COP = $\dfrac{h_1 - h_4}{h_2 - h_1} = \dfrac{(500-(-550))}{(700-500)} = 5.25$ **Ans.**

EXAMPLE 4.11 A Carnot refrigeration cycle absorbs heat at –20°C and reject it at 40°C. (a) Calculate the COP of this cycle. (b) If the cycle is absorbing 850 kJ/min at –20°C, how much work in kJ/min is required. (c) If the same cycle operates on simple vapour compression cycle with R134a refrigerant, what are the deviations in COP and the work input per kg of mass circulated.

Solution:

(a) $(COP)_C = \dfrac{T_L}{T_H - T_L} = \dfrac{273 - 20}{40 - (-20)} = 4.216$

(b) Also, $Q = 850$ kJ/min

∴ $(COP)_C = \dfrac{Q}{W}$

∴ $W = \dfrac{850}{4.216}$

or $W = 201.581$ kJ/min

(c) From the property table of R134a

$h_1 = h_g$ at –20°C

$h_2 = 425$ kJ/kg; $h_3 = h_4 = 257$ kJ/kg $= h_f$ at 40°C

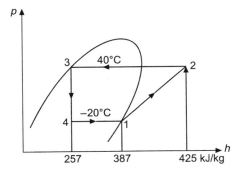

Figure 4.6 Vapour compression cycle on p–h diagram—Example 4.11.

Now, $COP = \dfrac{Q}{W}$

where $Q = \dot{m}(h_1 - h_4) = 387 - 257$

$= 130$ kJ/kg

and $\dot{W} = h_2 - h_1 = 425 - 387 = 38$ kJ/kg

∴ $COP = \dfrac{130}{38} = 3.42$

EXAMPLE 4.12 A Carnot heat pump cycle absorbs heat at −20°C and rejects it to a space maintained at 45°C. (a) Calculate the COP of the heat pump cycle. (b) If the cycle rejects heat at the rate of 850 kJ/min at 45°C, how much work in kJ/min is done in a simple vapour compression cycle using R134a refrigerant. What are the deviations in COP and the work supplied.

Solution:

(a) $(COP)_{HP} = \dfrac{T_H}{T_H - T_L} = \dfrac{273 + 45}{45 - (-20)} = 4.892$

(b) $(COP)_{HP} = \dfrac{Q_R}{W} = \dfrac{Q_a + w}{W} = \dfrac{850}{W} = 4.892$

∴ $W = \dfrac{850}{4.892} =$ kJ/kg

(c) Now from the property table of R134a, $h_1 = h_g$ at −20°C = 387 kJ/kg, $h_2 = 428$ kJ/kg, $h_3 = h_4 = h_f$ at 45°C = 264 kJ/kg

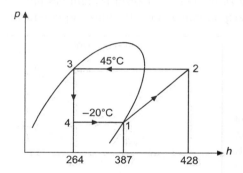

Figure 4.7 Vapour compression cycle on p–h diagram—Example 4.12.

Now,
$$Q_a = h_1 - h_4 = 387 - 264 = 123 \text{ kJ/kg}$$
and,
$$W = h_2 - h_1 = 428 - 387 = 41 \text{ kJ/kg}$$

$$COP = \dfrac{Q_R}{W} = \dfrac{Q_a + W}{W} = \dfrac{123 + 41}{41} = 4$$

EXAMPLE 4.13 A deep freezer works between the evaporating temperature and the condensing temperature. Determine the refrigerating effect per kg if the evaporating temperature is at −5°C and the condensing temperature at 35°C. This plant operates in a saturated vapour compression cycle with R134a. If the evaporating temperature is lowered to −20°C keeping the condensing temperature the same, what would be the % reduction in refrigerating effect/kg.

Solution: From the property table of R134a,

$$h_1 = h_g \text{ at } -5°C = 395 \text{ kJ/kg}, \quad h_2 = 421 \text{ kJ/kg}$$
$$h_3 = h_4 = h_f \text{ at } 35°C = 249 \text{ kJ/kg}$$

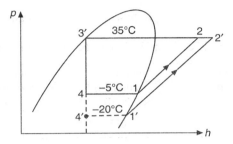

Figure 4.8 Vapour compression cycle on p–h diagram—Example 4.13.

and
$$h_1' = h_g \text{ at } -20°C = 387 \text{ kJ/kg}$$
$$Q = h_1 - h_4 = 395 - 249 = 146 \text{ kJ/kg}$$

and
$$Q' = h_1' - h_4 = 387 - 249 = 138 \text{ kJ/kg}$$

Now,
$$\% \text{ reduction} = 1 - \frac{138}{146} = 0.05479 \approx 5.5\%$$

EXAMPLE 4.14 A R22 refrigeration plant produces a refrigerating effect of 5 tons at the evaporating temperature –5°C and condensing temperature 30°C. If the refrigerating plant is shifted to Kharagpur where the atmospheric temperature is about 40°C, evaluate the plant in terms of (a) cooling effect, (b) power requirement of compressor and, (c) COP. Assume a simple vapour compression cycle.

Solution: Assuming 45°C as condenser temperature at Kharagpur, now, from the property table of R22, we get

$$h_1 = h_g \text{ at } 45°C = 404 \text{ kJ/kg}, \quad h_2 = 438 \text{ kJ/kg}, \quad h_3 = h_4 = h_f \text{ at } 45°C = 256 \text{ kJ/kg}$$

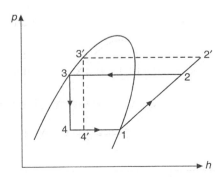

Figure 4.9 Vapour compression cycle on p–h diagram—Example 4.14.

(a) Cooling effect, $q = h_1 - h_4 = 404 - 256 = 148$ kJ/kg

(b) Work done, $W = h_2 - h_1 = 438 - 404 = 34$ kJ/kg

(c) $\text{COP} = \dfrac{Q}{W} = \dfrac{148}{34} = 4.358$

EXAMPLE 4.15 In Germany, a refrigerating unit with NH_3 as the refrigerant, operating with simple vapour compression cycle, produces 20 tons of ice per day, using the available water at 15°C. The evaporator temperature is –5°C. The company decides to sell this unit to a party in Rajasthan (India). Considering the climate condition of Rajasthan, evaluate the COP of the plant, so as to decide its cooling capacity, compressor power and COP.

Solution: Assuming 50°C as the condenser temperature in Rajasthan, heat to be removed from water at 15°C to convert it to ice at –1.5°C is given by

$$Q = \dfrac{20 * 1000}{24 * 3600}[4.187(15 - 0) + 335 + 2.1(0 \pm 1.5)]$$

$$= 94.515 \text{ kW (compressor power)}$$

where
specific heat for water = 4.187 kJ/kg-K
specific heat for ice = 2.1 kJ/kg-K
latent heat of fusion of ice = 335 kJ/kg

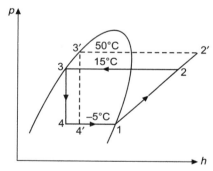

Figure 4.10 Vapour compression cycle on p–h diagram—Example 4.15

From the property table of NH_3,

$h_1 = h_g$ at –5°C = 1456 kJ/kg, h_2 = 1714 kJ/kg

$h_3 = h_4 = h_f$ at 50°C = 440 kJ/kg

$Q = h_1 - h_4 = 1456 - 440 = 1016$ kJ/kg

$W = h_2 - h_1 = 1714 - 1456 = 258$ kJ/kg

∴ $\text{COP} = \dfrac{Q}{W} = \dfrac{1016}{258} = 3.94$

EXAMPLE 4.16 A simple saturation R134a compression system has a high absolute pressure of 0.86 Mpa and a low pressure of 0.2 Mpa. Find (a) the refrigerating effect per kg, (b) the power required per kg, (c) the COP. Due to unavoidable reasons if the absolute pressure of the condenser increases to 1.13 Mpa, what will be its effect on the COP.

Solution: From the p–h chart of R134a,

$$h_1 = h_g \text{ at } 0.2 \text{ Mpa} = 394 \text{ kJ/kg},$$

$$h_2 = 420 \text{ kJ/kg}, \; h_3 = h_4 = h_f \text{ at } 0.86 \text{ Mpa} = 247 \text{ kJ/kg}$$

(a) Q = refrigerating effect/kg
$$= h_1 - h_4 = 394 - 247 = 147 \text{ kJ/kg}$$

(b) Power required/kg
$$W = h_2 - h_1 = 420 - 394 = 26 \text{ kJ/kg}$$

(c) $\therefore \; \text{COP} = \dfrac{Q}{W} = \dfrac{147}{26} = 5.65$

Now, if the condenser pressure is 1.13 Mpa, the properties are collected once again

(a) $h_1 = 394$ kJ/kg, $h_2 = 424$ kJ/kg, $h_3 = h_4 = 261$ kJ/kg

$$Q = h_1 - h_4 = 394 - 261 = 133 \text{ kJ/kg}$$

(b) Power required,
$$W = h_2 - h_1 = 424 - 394 = 30 \text{ kJ/kg}$$

(c) $\text{COP} = \dfrac{Q}{W} = \dfrac{133}{30} = 4.43$

4.4 VAPOUR COMPRESSION CYCLE WHEN VAPOUR IS DRY SATURATED AT THE END OF COMPRESSION

The various compressors used for pressurizing the vapour from the evaporator pressure to the condenser pressure are reciprocating hermetic compressors, screw compressors, centrifugal compressors, etc. The condition of vapour at the inlet to the compressor is either in dry saturated condition or in slightly superheat condition. The wet vapour entry into the reciprocating compressor is absolutely undesirable due to washing of the lubricating oil by the droplets of liquid refrigerant present in the wet vapour. The topics dealt with in this section as well as in Section 4.6 are only of academic interest.

Let us assume the condition of vapour refrigerant at the inlet to the compressor as wet and dry saturated at the end of the compression process as shown at points 1 and 2, respectively, on T–s diagram and p–h diagram in Figures 4.11(b) and (c).

The various processes are as follows:

Compression process (1–2): The wet condition of vapour is expressed by the term 'dryness fraction x'. Let wet vapour at pressure p_1 and saturation temperature with dryness fraction x_1

enter the compressor at state 1. It is compressed isentropically (reversible adiabatically) up to the discharge pressure p_2 and vapour is assumed to reach a dry saturated state. The vapour compression cycle is called *wet compression cycle*.

Condensation process (2–3): Dry saturated vapour at state 2 enters the condenser. The condenser is a heat exchanger which is either air-cooled or water-cooled. Let the refrigerant reject heat Q_1 at constant pressure ($p_2 = p_3$). It is assumed that the vapour is completely condensed and it becomes saturated liquid at state point 3.

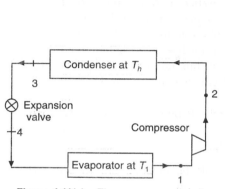

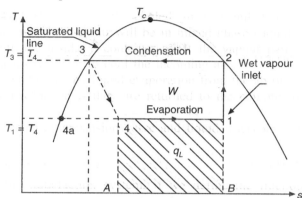

Figure 4.11(a) The vapour compression system shown in Figure 4.1 is represented here in a simple way.

Figure 4.11(b) T–s diagram when the vapour is in dry saturated state at the end of compression.

Expansion process (3–4): Saturated liquid at (p_3, T_3) represented by state 3 is now reduced to the evaporator pressure p_4 while passing through an expansion valve. It is an irreversible process hence shown by dotted curve (3–4). It is a constant enthalpy process, i.e. $h_3 = h_4$. During the expansion process, a part of liquid refrigerant vaporises and the refrigerant leaves the expansion valve as a vapour of low dryness fraction.

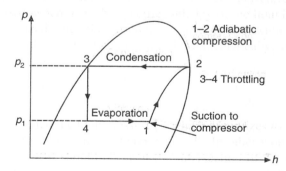

Figure 4.11(c) p – h diagram when the vapour is in dry saturated state at the end of compression.

Vaporization process (4–1): The refrigerant in mixed phase enters the evaporator at state 4. It absorbs heat from cold chamber producing a refrigerating effect. It leaves the evaporator at state 1. Thus, the cycle is completed.

Analysis of the cycle assuming 1 kg of refrigerant flow in the system

(a) **Work of compression (w):** Work is done in compressing the refrigerant during process (1–2) by the compressor. Many a time the work/power obtained during the expansion process is negligible. Therefore, the net work done during the cycle equals the compressor work. Hence,

$$w = (h_2 - h_1) \text{ kJ/kg} \tag{4.17}$$

$$= \text{area } (1\text{–}2\text{–}3\text{–}4\text{–}1) \text{ as shown in } T\text{–}s \text{ diagram}$$

(b) **Heat rejected at the condenser (q_R):** Dry saturated vapour refrigerant rejects heat at constant pressure (pressure loss due to friction neglected) to cooling medium during process (2–3). Therefore,

$$q_R = (h_2 - h_3) \text{ kJ/kg} \tag{4.18}$$

(c) **Expansion device:** The pressure of the refrigerant liquid is reduced to evaporator pressure as it passes through the expansion valve. The process is assumed to occur at constant enthalpy. The condition of vapour at the end of expansion is wet. Therefore, $h_3 = h_4$.

$$h_{f3} = h_{f4} + x_4 \times h_{fg4}$$

$$\therefore \quad x_4 = \frac{h_{f3} - h_{f4}}{h_{fg4}}$$

where h_{f3} = enthalpy of liquid at condenser pressure p_2, h_{f4} = enthalpy of liquid at evaporator pressure p_1, h_{fg4} = enthalpy of evaporation at evaporator pressure p_1, x_4 = dryness fraction of vapour after throttling.

(d) **Refrigerating effect (q_L):** Refrigerating effect is the heat absorbed by the refrigerant in the evaporator during the process (4–1), therefore the refrigerating effect is

$$q_L = (h_1 - h_4) \tag{4.19}$$

(e) **COP:** The coefficient of performance of the refrigerator is given as

$$= \frac{\text{Refrigerating effect, } q_L}{\text{Work of compression, } w} = \frac{h_1 - h_4}{h_2 - h_1} \tag{4.20}$$

(f) **Cooling water requirement in the condenser in the case of water-cooled condenser:** Under steady-state conditions, assuming that no heat is lost to the surroundings,

Heat given away by refrigerant = Heat absorbed by cooling water

$$\therefore \quad \dot{m}(h_2 - h_3) = \dot{m}_w \times c_{pw} \times (\Delta T)_w \tag{4.21}$$

where

$\dot{m}$ = mass flow rate of refrigerant, $\dot{m}_w$ = mass flow rate of cooling water in the condenser, $(\Delta T)_w$ = temperature rise of cooling water, c_{pw} = specific heat of water $\cong 4.19$ kJ/kg-K

(g) **Compressor power (P):**

$$P = \dot{m}(h_2 - h_1) \text{ kJ/s or kW} \quad (4.22)$$

(h) **Mass of refrigerant to be circulated, $\dot{m}$ per ton of refrigeration:**

Since $\quad 1 \text{ TR} = 3.5164 \text{ kJ/s}$

$$\therefore \quad \dot{m} = \frac{3.517 (\text{kJ/s})}{q_o (\text{kJ/kg})} = \frac{3.5164}{q_o} \text{ (kg/s)/(TR)} \quad (4.23)$$

(i) **Volume of refrigerant to be handled by compressor per TR:**

$$\dot{V} = \dot{m} \times v_1 \text{ m}^3/\text{s} \quad (4.24)$$

where v_1 represents the specific volume of refrigerant at inlet to compressor.

In case the volumetric efficiency η_V of compressor is considered,

$$\dot{V} = \frac{\dot{m} \times v_1}{\eta_V} \text{ m}^3/\text{s} \quad (4.25)$$

(j) **Relative COP:**

$$\text{Relative COP} = \frac{\text{Actual COP}}{\text{Ideal COP}} \quad (4.26)$$

Note: The compression process (1–2) in the compressor has to be terminated such that the condition of vapour at the end of compression would be dry saturated. This is practically very difficult. In an actual compression process, the state of vapour refrigerant would either be dry or be superheated.

4.5 VAPOUR COMPRESSION CYCLE WHEN VAPOUR IS WET AT THE END OF COMPRESSION

Let us assume that the condition of vapour refrigerant at the inlet to the compressor is wet and it is also wet at the end of compression process as shown at points 1 and 2, respectively, on T–s and p–h diagrams in Figures 4.12(a) and (b), respectively.

We discussed the wet compression process in Section 4.4 in detail. Therefore, only the following points are being recollected at this stage.

For unit mass flow of refrigerant,

$$\text{Compressor work, } w = (h_2 - h_1) \text{ kJ/kg}$$

$$\text{Refrigerating effect, } q_L = (h_1 - h_4) \text{ kJ/kg}$$

$$\text{COP} = \frac{w}{q_L} = \frac{h_2 - h_1}{h_1 - h_4} \quad (4.27)$$

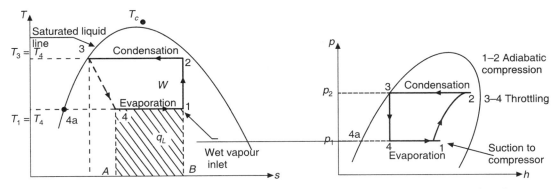

Figure 4.12(a) T–s diagram when the vapour is wet at the end of compression.

Figure 4.12(b) p–h diagram when the vapour is wet at the end of compression.

Wet compression vs dry compression

The liquid refrigerant absorbs heat from the cold space and changes to vapour state. The refrigerant has to dissipate this heat to the surroundings. To do so, its temperature has to be higher than the surrounding coolant. In order to increase the temperature of the refrigerant vapour greater than the surrounding, its pressure has to be increased. This is achieved with the help of a compressor.

In case the compression process involves the compression of dry saturated vapour or with slightly superheated vapour, it is called *dry compression*. In this case the complete compression process remains in superheated state.

In case the compression process involves the compression of wet refrigerant, the compression is called *wet compression*.

The refrigerant vapour would be compressed efficiently if there is a perfect sealing between the piston and the cylinder bore. This would be satisfied to a certain extent with better surface finish obtained in the manufacturing process. In a reciprocating compressor, the lubricating oil serves two purposes: (i) It reduces the friction between the rubbing surfaces, and (ii) it acts as a sealing agent between the piston and the cylinder.

In reciprocating compressors, wet compression is avoided due to the following reasons:

1. Liquid droplets present in the wet vapour wash away the lubricating oil from the cylinder walls of the compressor. Then there will not be any sealing agent between the piston and the cylinder bore. It results into a large friction between the piston and the cylinder. Subsequently the driving motor will be overloaded and may lead to burning of the same. Since there is no proper sealing between the piston and the cylinder bore, the compression efficiency will be very poor. The blow-by loss will increase leading to poor volumetric efficiency.

2. Liquid droplets in the refrigerant would enter the compressor and damage the valves and other moving parts.

Note: Dry compression is preferred over wet compression since it gives high volumetric efficiency and the mechanical efficiency of the compressor is increased with less chances of damage to it.

EXAMPLE 4.17 A refrigeration plant works between the temperature limits of –5°C and 25°C. The refrigerant CO_2 is wet at entry to the compressor and has dryness fraction of 0.6. The refrigerator has actual COP 70% of the theoretical COP. If there is no under-cooling, determine the ice formed during a period of 24 hours from water at 20°C. The mass of CO_2 circulated is 5 kg/min. Take enthalpy of fusion of ice as 336 kJ/kg.

The properties of CO_2 are as follows:

Saturation temperature (°C)	Specific enthalpy (kJ/kg)		Specific entropy (kJ/kg-K)
	h_f	h_g	s_f
25	81.25	202.75	0.2513
–5	–7.53	238.5	–0.04187

Solution:

Given: Actual $(COP)_a = 0.7 \times$ Ideal COP

Latent heat of fusion of ice, $h_L = 336$ kJ/kg, $\dot{m} = 5$ kg/min

$T_{s1} = T_1 = T_4 = -5°C = -5 + 273 = 268$ K

$T_{s2} = T_2 = T_3 = 25°C = 25 + 273 = 298$ K

Theoretical COP: Consider the isentropic process (1–2); $s_1 = s_2$

$$s_{f1} + x_1 \times s_{fg1} = s_{f2} + x_2 \times s_{fg2}$$

But

$$s_{fg} = \frac{h_{fg}}{T_s} = \left(\frac{h_g - h_f}{T_s}\right)$$

∴

$$s_{f1} + x_1 \times \frac{h_{g1} - h_{f1}}{T_{s1}} = s_{f2} + x_2 \times \frac{(h_{g2} - h_{f2})}{T_{s2}}$$

or

$$-0.04187 + 0.6 \times \frac{[238.5 - (-7.53)]}{268} = 0.2513 + x_2 \times \frac{(202.75 - 81.25)}{298}$$

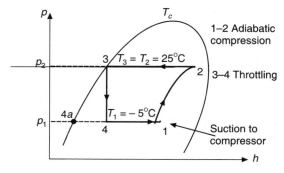

Figure 4.13 p–h diagram when the vapour is wet at the end of compression—Example 4.17.

∴ $\quad x_2 = 0.632$
$$h_1 = h_{f1} + x_1 \times h_{fg1} = h_{f1} + x_1(h_{g1} - h_{f1})$$
$$= -7.53 + 0.6[238.5 - (-7.53)] \approx 140 \text{ kJ/kg}$$
∴ $\quad h_2 = h_{f2} + x_2 \times h_{fg2} = h_{f2} + x_2(h_{g2} - h_{f2})$
$$= 81.25 + 0.632 \times (202.75 - 81.25) \approx 158 \text{ kJ/kg}$$
$$h_3 = h_{f3} = h_4 = 81.25 \text{ kJ/kg} \;(\because \text{ the process is isenthalpic})$$

Consider 1 kg of refrigerant flow:
$$\text{Work of compression, } w = (h_2 - h_1) = (158 - 140) = 18 \text{ kJ/kg}$$
$$\text{Refrigerating effect, } q_L = (h_1 - h_4) = (140 - 81.25) = 58.75 \text{ kJ/kg}$$
$$\text{Ideal COP} = \frac{q_L}{w} = \frac{58.75}{18} = 3.264$$

Mass of ice formed per 24 hours:
$$\text{Actual COP} = 0.7 \times 3.264 = 2.285$$
Actual refrigerating effect, $q_L = w \times \text{actual COP} = 18 \times 2.285 = 41.2$ kJ/kg
Total refrigerating effect produced,
$$Q_L = \dot{m} \times q_L = 5 \times 41.2 \text{ kJ/min} = 206 \text{ kJ/min} = 3.43 \text{ kW}$$

Let $\dot{m}_i$ be the ice formed/min from water at 20°C into ice at 0°C

∴ $\quad Q_L = \dot{m}_i \times [c_{pw} \times (20 - 0) + h_L]$
or $\quad 206 = \dot{m}_i [4.187 \times (20 - 0) + 336]$

∴ $\quad \dot{m}_i = \dfrac{206}{419.74} \text{ kg/min} = \dfrac{206}{419.74} \times 60 \times 24 \text{ kg/day}$
$$= 705.6 \text{ kg/day} \qquad \textbf{Ans.}$$

EXAMPLE 4.18 A F–12 refrigeration system operates between the temperature limits of −6°C and 36°C. The vapour is 95% dry at the beginning of isentropic compression and the liquid leaving the condenser is at 30°C. Assuming actual COP as 50% of the theoretical COP, calculate the amount of ice produced in kg per kWh at 0°C from water at 30°C. Latent heat of ice = 335 kJ/kg, specific heat of refrigerant in liquid stage = 1.24 kJ/kg-K.

Temperature	Enthalpy (kJ/kg)		Entropy (kJ/kg-K)	
°C	liquid	vapour	liquid	vapour
−6	30.53	184.94	0.1217	0.6996
36	70.55	201.80	0.2591	0.6836

Solution: The refrigeration cycle is shown in Figure 4.9 on T–s diagram.
$$x_1 = 0.95, \quad T_3 = 30°C$$
$$\text{Actual COP} = 0.5 \times \text{Theoretical COP}$$
Water is available at $T = 30°C$

$$L_i = 335 \text{ kJ/kg}, \quad c_{pf} = 1.24 \text{ kJ/kg-K}$$
$$h_1 = h_{f1} + x_1 \times h_{fg1} = h_{f1} + x_1(h_{g1} - h_{f1})$$
$$= 30.53 + 0.95(184.94 - 30.53) = 177.22 \text{ kJ/kg}$$

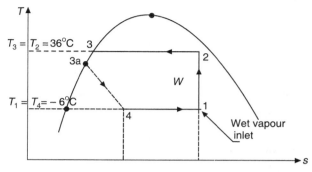

Figure 4.14 T–s diagram when the vapour is wet at the end of compression—Example 4.18.

Liquid refrigerant leaving the condenser is subcooled since the actual temperature of refrigerant at the outlet of the condenser is 30°C, which is less than the saturation temperature $T_{sa} = 36$°C.

$$h_{f3} = 70.55 \text{ kJ/kg}$$
$$\therefore \qquad = h_{f3} - c_{pf}(T_{sa} - T_3) = 70.55 - 1.24(36 - 30)$$
$$= 63.1 \text{ kJ/kg} = h_4 \text{ (since the process 3–4 is isenthalpic)}$$

Theoretical refrigerating effect/kg of refrigerant,
$$q_L = h_1 - h_4 = 177.22 - 63.1 = 114.1 \text{ kJ/kg}$$
$$s_{f1} + x_1(s_{g1} - s_{f1}) = s_{f2} + x_2(s_{g2} - s_{f2})$$
$$0.1217 + 0.95(0.6996 - 0.1217) = 0.2591 + x_2(0.6836 - 0.2591)$$
$$x_2 = 0.9696$$
$$h_2 = h_{f2} + x_2 \times (h_{g2} - h_{f2})$$
$$= 70.55 + 0.9696(201.8 - 70.55)$$
$$= 197.8 \text{ kJ/kg}$$

Theoretical work of compression,
$$w = h_2 - h_1 = 197.8 - 177.22 = 20.58 \text{ kJ/kg}$$
$$\text{Theoretical COP} = \frac{q_L}{w} = \frac{114.1}{20.58} = 5.54$$
$$\text{Actual COP} = 0.5 \times \text{Theoretical COP} = 0.5 \times 5.542 = 2.77$$

Work to be spent corresponding to 1 kWh, $W = 3600$ kJ/kWh
$\therefore$ Actual refrigerating effect/kWh
$$Q = W \times \text{Actual COP} = 3600 \times 2.77$$
$$= 9975.6 \text{ kJ/kWh}$$

Let $\dot{m}_i$ = ice formed/kWh

Ice is formed from water at $T = 30°C$

Then,
$$Q = \dot{m}_i \times [c_{pw}(T - 0) + L_i]$$
$$9975.6 = \dot{m}_i \times [4.187(30 - 0) + 335]$$
$$\dot{m}_i = 21.657 \text{ kg/kWh} \quad \textbf{Ans.}$$

EXAMPLE 4.19 Twenty-eight tonnes of ice from and at 0°C is produced per day in an ammonia refrigeration plant. The temperature range in the compressor is from +25°C to –15°C. The refrigerant is dry and saturated at the end of compression. If the actual COP is 60% of the theoretical COP, calculate the power supplied or required to drive the compressor.

Assume latent heat of ice = 335 kJ/kg. Use the properties of refrigerant given below:

Tempetature (°C)	h_f (kJ/kg)	h_g (kJ/kg)	Entropy (kJ/kg-K)	
			s_f	s_g
+25	+100.04	1319.22	+0.3473	4.4852
–15	–54.56	1304.99	–2.1338	5.0585

Solution: Given: $T_{s1} = T_1 = T_4 = -15°C = 258$ K
$T_{s2} = T_2 = T_3 = 25°C = 298$ K
Actual COP = 60% theoretical COP, $h_L = 335$ kJ/kg

Theoretical Coefficient of Performance (COP):
Consider isentropic compression process (1–2) from which $s_1 = s_2$;
The vapour is dry saturated at the end of compression, hence $x_2 = 1$ and $s_2 = s_g$ at T_2

$$s_1 = s_2$$
$$s_{f1} + x_1 s_{fg1} = s_{g2}$$
$$s_{f1} + x_1(s_{g1} - s_{f1}) = s_{g2}$$
$$-2.1338 + x_1(5.0585 + 2.1338) = 4.4852$$
$$\therefore \quad x_1 = 0.9203$$
$$h_1 = h_{f1} + x_1 h_{fg1} = h_{f1} + x_1(h_{g1} - h_{f1})$$
$$h_1 = -54.56 + 0.9203(1304.99 + 54.56)$$
$$\therefore \quad h_1 = 1196.63 \text{ kJ/kg}$$
$$h_2 = h_{g2} = 1319.22 \text{ kJ/kg}$$
$$h_3 = h_{f2} = h_4 = 100.04 \text{ kJ/kg (Isenthalpic process)}$$

For 1 kg of refrigerant:
Work of compression, $w = h_2 - h_1 = (1319.22 - 1196.63) = 122.59$ kJ/kg
Refrigerating effect, $q_L = (h_1 - h_4) = (1196.63 - 100.04)$
$$\therefore \quad q_L = 1096.57 \text{ kJ/kg}$$

Theoretical COP or Ideal COP = $\dfrac{q_L}{w} = \dfrac{1096.57}{122.59} = 8.95$

∴ Actual COP = 60% of Ideal COP = $\frac{60}{100}$ × 8.95 = 5.35

Refrigeration capacity = $\dot{m}_i [c_{pw}(0) + L_i]$ = $\frac{28 \times 1000}{24 \times 3600}$ (335)

= 108.565 kJ/s or kW

Actual COP = $\frac{\text{Capacity}}{\text{Power required}}$

Power required = $\frac{108.565}{5.35}$ = 20.29 kJ/s or kW   **Ans.**

EXAMPLE 4.20 A refrigeration plant is used for the production of ice. The saturation temperatures corresponding to condenser pressure and evaporator pressure are 25°C and –5°C, respectively. Refrigerant is 0.8 dry at the entry to the compressor. The quantity of refrigerant circulated through the cycle is 1 kg/s. Find the ice formed in kg per hour assuming the relative efficiency of the cycle to be 50%. The ice is formed at –5°C from water at 10°C.

Assume for water, c_{pw} = 4.187 kJ/kg-K
For ice, c_{pi} = 2.05 kJ/kg-K
Latent heat of fusion of ice = 335 kJ/kg
Use the following properties of refrigerant.

Temperature (°C)	h_f (kJ/kg)	h_{fg} (kJ/kg)	s_f (kJ/kg-K)
+25°C	81.5	121.5	0.2515
–5°C	–7.5	246.0	–0.0419

Solution: $T_1 = T_4 = 268$ K; $T_3 = 298$ K; $x_1 = 0.8$; $\dot{m}_i = 1$ kg/s

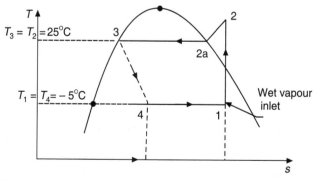

Figure 4.15 T–s diagram when the vapour is wet at the end of compression—Example 4.20.

Cycle efficiency = 50%

$$\therefore \quad s_1 = s_{f1} + x_1 s_{fg1} = s_{f1} + x_1 \frac{h_{fg1}}{T_1} = -0.0419 + \frac{0.8 \times 246}{268} = 0.692 \text{ kJ/kg-K}$$

Consider isentropic compression process.

$$s_1 = s_2$$

$$0.692 = s_{f2} + \frac{h_{fg2}}{T_2} + c_p \ln \frac{T_2}{298} = 0.2515 + \frac{121.5}{298} + 0.733 \ln \frac{T_2}{298}$$

$$\therefore \quad T_2 = 311.6 \text{ K}$$

$$h_1 = h_{f1} + x_1 h_{fg1} = -7.5 + 0.8 \times 246 = 189.3 \text{ kJ/kg}$$

$$h_2 = h_{fg2} + h_{f2} - c_p(T_{\text{sup}} - T_{\text{sat}}) = 81.5 + 121.5 + 0.733(311.6 - 298) = 212.991 \text{ kJ/kg}$$

$$h_3 = h_{f3} = h_4 = 81.5 \text{ kJ/kg}$$

Specific refrigerating effect = $q_L = h_1 - h_4 = 189.3 - 81.5 = 107.8$ kJ/kg

Total refrigerating effect, $Q = m \times q_L = 1 \times 107.8 = 107.8$ kW

Work done, $w = h_2 - h_1 = 212.99 - 189.3 = 23.69$ kJ/kg

$$\text{Ideal COP} = \frac{q_L}{w} = \frac{107.8}{23.69} = 4.55$$

Actual COP = $0.5 \times 4.55 = 2.275$

Heat to be removed from 1 kg of ice = $c_{p_w}(\Delta T_1) + h_L + c_{p_i} \times \Delta T_2$

$$= 4.187(10 - 0) + 335 + 2.05(0 + 5)$$

$$= 387.12 \text{ kJ/kg}$$

$$\text{Ice produced} = \frac{\text{Refrigerating effect}}{\text{Heat removed}} = \frac{107.8}{387.12} = 0.2785 \text{ kg/s}$$

Ice produced per hour = 1002.48 kg/h **Ans.**

$$\text{Actual COP} = \frac{Q}{\text{Work done}}$$

$$\text{Work done} = \frac{107.8}{2.275} = 47.28 \text{ kW} \quad \text{**Ans.**}$$

4.6 DEVIATIONS FROM THE SIMPLE COMPRESSION CYCLE

The following are the assumptions made for a simple saturated cycle:

1. The pressure losses in the flow lines of evaporator and condenser are neglected.
2. The compression process is an isentropic process.
3. No suction gas superheating and liquid subcooling is involved.
4. Evaporation and condensation occur only at constant temperature etc.

Practically, these assumptions are not true and if all these deviations are introduced in a simple saturated cycle, it becomes more and more complex and also more realistic.

The effect of superheating the suction vapour

The suction gas superheat phenomenon is shown in Figure 4.16. Here the pressure drop in the pipeline from evaporator to the compressor is small and neglected. The reasons for the suction gas superheating are as follows.

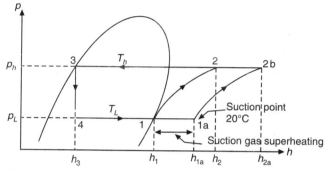

Figure 4.16 p–h diagram, 1-1a is suction gas superheating (ideal cycle 1-2-3-4).

Heating in the evaporator: The refrigerant vapour in the evaporator will be absorbing heat continuously and will continue to absorb heat even after the evaporator.

Heating between the evaporator and compressor: The tube connecting the evaporator and compressor is always exposed to atmosphere. Though the pipe is insulated, a certain heat transfer from surroundings to the vapour refrigerant occurs.

Heating in the cylinder head: Since this vapour reaching the compressor has to pass through the flow passages in the cylinder head, which is at higher temperature, further heating of vapour occurs in the cylinder head.

For hermetic compressors, the superheating is still more because of heat transfer from the motor to the suction gas inside the sealed unit.

In the following text, the effects of suction gas superheating on cycle performance are discussed:

(i) The heat of compression per kilogram for the superheated cycle (1a–2b–3–4) is slightly greater than that of the saturated cycle. This is because the isentropic lines become more flat as they go away from the saturated vapour curve.

(ii) For the same condensing temperature and pressure, the temperature of the discharge vapour leaving the head of the compressor is considerably higher (i.e. $T_{2b} > T_2$) than that of the saturated cycle.

(iii) For the superheated cycle, the vapour contains more heat at the compressor discharge ($T_{2b} > T_2$). Therefore, a greater quantity of heat must be rejected at the condenser per kilogram than that in the saturated cycle. For the rejection of this sensible heat to the surroundings, condenser is required to desuperheat it.

(iv) The specific volume of superheated vapour is always greater than that of the saturated vapour. Therefore, the specific volume of suction vapour at point 1a is greater than that at point 1. It means that for each kilogram of the refrigerant circulated, the compressor must compress a greater volume of vapour. Let us study these with the following numerical example.

EXAMPLE 4.21 A refrigeration system operates on R22 refrigerant. The evaporator and condenser temperatures are at –10°C and 40°C, respectively. If the actual suction to the compressor is at 20°C, determine the following in compression of saturated cycle.
(a) Percentage increase in heat of compression.
(b) Percentage increase in the rate of heat rejection in the condenser.

Solution: (Refer to Figure 4.11 to understand the solution. The properties are obtained from the *p–h* chart.)

Point 1: h_1 = 401.7 kJ/kg   Point 1a (superheat): h_{1a} = 422.0 kJ/kg
Point 2: h_2 = 437.5 kJ/kg   Point 2b: h_{2b} = 462.0 kJ/kg
Point 3: $h_{f3} = h_3$ = 249.7 kJ/kg

(a) Heat of compression: For saturated cycle
Heat of compression = $(h_2 - h_1)$ = (437.5 – 401.7) = 35.8 kJ/kg
For superheated cycle:
Heat of compression = $(h_2 - h_{1a})$ = (462 – 422) = 40 kJ/kg
Percentage change in heat of compression = ((40 – 35.8) / 35.8) × 100 = 11.7% increase
Ans.

(b) Heat rejection in condenser
For saturated cycle:
Heat rejected in condenser = $(h_2 - h_3)$ = (437.5 – 249.7) = 187.8 kJ/kg
For superheated cycle:
Heat rejected in condenser = $(h_2 - h_3)$ = (462 – 249.7) = 213 kJ/kg
Percentage increase in the rate of heat rejection
= [(213 – 187.8)/187.8] × 100 = 13.40%
Ans.

4.6.1 Suction Gas Superheating without Cooling

Here it is assumed that the superheating of suction vapour from evaporator temperature to the actual inlet temperature of the compressor occurs outside the evaporator. It is possible that refrigerant vapour after leaving the evaporator catches heat from atmosphere.

As such the refrigerating effect per unit mass of refrigerant circulated is same for both superheated and saturated cycles operating at the same evaporating and condensing temperatures. But the specific volume of refrigerant vapour at the suction to the compressor for superheated cycle is more (i.e. $v_{1a} > v_1$). Therefore, for the same refrigerating effect the compressor of superheated cycle has to handle a greater volume of refrigerant requiring greater power.

Moreover, the COP of the superheated cycle reduces in comparison to the saturated cycle. This is explained with the following example:

EXAMPLE 4.22 A refrigeration system operates with R12 refrigerant. The evaporator and condenser temperatures are at −5°C and 35°C, respectively. The actual suction to the compressor is at 15°C. If superheating of refrigerant vapour from −5°C to 15°C does not add any refrigerating effect,

(i) determine the percentage increase in volume flow rate per ton of refrigeration compared with the saturation cycle;
(ii) compare the COP for saturated and superheated cycles; and
(iii) determine the power required per TR.

Solution: For solution, refer to Figure 4.16.

Properties at
Point 1a: $p_{1a} = 2.61$ bar, $T_{1a} = 15°C$, $v_{1a} = 0.071$ m³/kg, $h_{1a} = 362$ kJ/kg
Point 2b: $p_{2b} = 9.61$ bar, $T_{2b} = 66.7°C$, $v_{2b} = 0.021$ m³/kg, $h_{2b} = 387.4$ kJ/kg
Point 1: $h_1 = 349.3$ kJ/kg, $h_2 = 372.4$ kJ/kg, $v_1 = 0.064$ m³/kg

(i) Volume flow rate:

For saturated cycle:
The refrigerating effect = $(h_1 - h_4)$
= (349.3 − 238.50) = 110.8 kJ/kg

Mass flow rate, $\dot{m} = \dfrac{3.516}{110.8} = 0.0317$ kg/s

The mass flow rate of refrigerant is same for both the superheated and saturated cycles. But the specific volumes of refrigerant vapour at point 1 and 1a are different.

Volume of vapour compressed per ton of refrigeration, = mv_1
= (0.0317)(0.064)
= 2.031 × 10⁻³ m³/TR

Therefore, the percentage increase in the volume flow rate of vapour for the superheated cycle is:

$$= \dfrac{(219 \times 10^{-3} - 2.031 \times 10^{-3}) \times 100}{2.031 \times 10^{-3}}$$

= 9.25%   **Ans.**

It means the compressor displacement must be higher by 9.25% for the superheated cycle.

(ii) Coefficient of performance:

For saturation cycle:

$$COP = \dfrac{h_1 - h_4}{h_2 - h_1} = \dfrac{349.3 - 238.5}{372.4 - 349.3} = 4.79 \quad \textbf{Ans.}$$

For superheated cycle,

$$COP = \dfrac{h_1 - h_4}{h_{2b} - h_{1a}} = \dfrac{349.3 - 238.5}{387.4 - 362} = 4.36 \quad \textbf{Ans.}$$

(iii) Power required per TR:
For saturated cycle:
$$\text{Power required/TR} = m(h_1 - h_2) = 0.0317(349.3 - 372.4)$$
$$= 0.73 \text{ kW} \qquad \textbf{Ans.}$$

For superheated cycle:
$$\text{Power required/TR} = m(h_{2b} - h_{1a})$$
$$= 0.0317(387.4 - 362) = 0.8 \text{ kW} \qquad \textbf{Ans.}$$

4.6.2 Suction Gas Superheating with Cooling Effect

It is assumed that the amount of heat received by the suction vapour produces useful cooling and in total it enhances the refrigeration effect that has been explained through the following example.

EXAMPLE 4.23 For the details given in Example 4.22 for the superheated cycle, calculate (i) the refrigerating effect, (ii) the mass flow rate, (iii) the volume of vapour compressed per TR, (iv) the power required per TR, and (v) the COP.

Solution: Refer to Figure 4.16. For the superheated cycle 1a–2b–3–4,

(i) The refrigerating effect $= h_{1a} - h_4$
$$= (362 - 238.5) = 124 \text{ kJ/kg} \qquad \textbf{Ans.}$$
This is higher compared to the saturated cycle which otherwise produces 110.8 kJ/kg.

(ii) Mass flow rate per TR
$$\dot{m} = \frac{3.5167}{124} = 0.0283 \text{ kg/s} \qquad \textbf{Ans.}$$
This value is less than 0.0317 kg/s of the saturated cycle.

(iii) Volume flow rate of vapour per TR
$$= \dot{m}.v_{1a} = (0.0283)(0.071) = 2.013 \times 10^{-3} \text{ m}^3/\text{s} \qquad \textbf{Ans.}$$
This is also less than 2.219×10^{-3} m³/s of the saturated cycle.

(iv) Power required per TR
$$= \dot{m}(h_{2b} - h_{1a}) = (0.0283)(387.4 - 362) \times 10^{-3} = 0.718 \text{ kW} \qquad \textbf{Ans.}$$
The power required in superheated cycle is 0.718 kW which is less than 0.73 kW of saturated cycle.

(v) Coefficient of performance
$$= \frac{h_{1s} - h_4}{h_{2b} - h_{1a}} = \frac{362 - 238.5}{387.4 - 362} = 4.86$$
which is higher than that of the saturated cycle. **Ans.**

Suction gas superheating without cooling effect and with cooling effect are the two extreme conditions, which are not practical. In actual cycles, the suction vapour superheating takes place in such a way that it gains heat partly from refrigerated space producing useful cooling and partly from outside of refrigerated space (after leaving the refrigerated space). In practical refrigerating units, it depends on the particular application.

Suction vapour superheating in many applications is unavoidable and also desirable. When suction vapour is directly drawn from the evaporator into the suction manifold of compressor, there is a possibility that small liquid droplets may enter into the cylinder. Compression of such vapour with small liquid droplets is called *wet compression*. As such, serious mechanical damage to the compressor may result. Since superheating the suction vapour eliminates the possibility of wet vapour entering the compressor, a certain amount of superheating is usually desirable.

4.7 EFFECT OF SUBCOOLING THE LIQUID

Subcooling means cooling the liquid refrigerant from state 3 to 3a as shown in Figure 4.12. From the figure, it is clear that subcooling of the liquid increases the refrigerating effect and also reduces the flashing of the liquid during the expansion process. Therefore, the performance of the cycle improves which means that the compressor power and its displacement will be smaller per ton of refrigeration.

Figure 4.17 shows the *p–h* diagram with the subcooling effect (cooling liquid from point 3 to 3a).

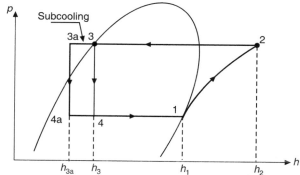

Figure 4.17 *p–h* diagram showing the subcooling effect (cooling liquid from point 3 to 3a).

As such, since the subcooling phenomenon improves the performance of the cycle, it is done by using a separate liquid subcooler. The use of the liquid subcooler is justified only if the gain in refrigerating effect and efficiency resulting from the liquid subcooler are sufficient to offset the additional cost of the subcooler.

Liquid refrigerant also becomes subcooled while stored in a receiver tank or while passing through the liquid line by dissipating heat to the surrounding air.

In large refrigeration plants, to achieve subcooling a water-cooled condenser and a separate heat exchanger are employed. This is shown in Figure 4.18.

In Figure 4.18 it is shown that the fresh cold water from cooling tower is supplied to the condenser and subcooler separately. This cold water quantity circulated should be sufficient to achieve the required subcooling.

In the case of aircooled condenser (domestic refrigerator or window air conditioner), the subcooling is achieved by providing extra condenser tubes.

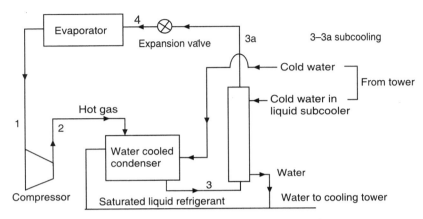

Figure 4.18 Flow diagram with liquid subcooler.

EXAMPLE 4.24 In a refrigeration plant working with R134 with 10 TR effect, the evaporator and condenser temperatures are maintained at $-5°C$ and $40°C$ respectively. If the liquid is subcooled from $40°C$ to $30°C$, in the condenser, then calculate for the single compressor cycle and for the subcooled cycle the following: (a) the specific refrigerating effect, (b) mass flow rate, (c) volume of vapour handled by compressor, (d) power required and, (e) COP.

Solution: $h_1 = h_g$ at $-5°C = 395$ kJ/kg; $h_2 = 421$ kJ/kg; $h_3 = h_4$ at $30°C = 242$ kJ/kg

(a) Specific refrigerating effect,
$$q = h_1 - h_4 = 395 - 242 = 153 \text{ kJ/kg}$$

(b) Mass flow rate,
$$\dot{m} = \frac{Q}{q} = \frac{10 \times 3.5167}{153} = 0.23 \text{ kg/s}$$

(c) Volume of vapour handled by compressor,
$$q_1 = 0.083 \text{ m}^3/\text{kg at } 5°C$$
$$\therefore \quad v_1 = mq_1 = 0.23 \times 0.83 = 0.019 \text{ m}^3/\text{s}$$

(d) Power required,
$$W = h_2 - h_1 = 421 - 395 = 6 \text{ kW}$$

(e) $$\text{COP} = \frac{Q}{W} = \frac{10 \times 3.5617}{6} = 5.861$$

EXAMPLE 4.25 A refrigeration plant working with R134a and simple vapour compression cycle, includes a liquid suction heat exchanger in the system. The heat exchanger cools the saturated liquid coming from the condenser from 35°C to 20°C. Compressor is isentropic with and without the heat exchanger. Calculate (a) the COP of system without the heat exchanger at the condensing temperature of 35°C and an evaporating temperature of –20°C and, (b) the COP of system with the heat exchanger.

Solution: Applying heat or enthalpy balance for the liquid–vapour heat exchanger

$$h_3 - h'_3 = h'_1 - h_1$$

From chart of R134a,

$$h_1 = h_g \text{ at } -2°C = 397 \text{ kJ/kg}; \quad h_2 = 420 \text{ kJ/kg}$$

$$h'_3 = h'_4 = 227 \text{ kJ/kg} = h_g \text{ at } 20°C$$

$$h_3 - h_4 = h_f \text{ at } 35°C = 249 \text{ kJ/kg}$$

∴ $\quad 249 - 227 = h'_1 - 397$

or $\quad h'_1 = 419 \text{ kJ/kg}$

∴ $\quad h'_2 = 447 \text{ kJ/kg} \quad$ (since the compression is isentropic)

(a) Refrigerating effect

$$Q = h_1 - h_4 = 397 - 249 = 148 \text{ kJ/kg}$$

and $\quad W = h_2 - h_1 = 420 - 397 = 23 \text{ kJ/kg}$

∴ $\quad \text{COP} = \dfrac{Q}{W} = \dfrac{148}{23} = 6.43 \quad$ without the heat exchanger

(b) Using the heat exchanger

$$Q' = h'_1 - h'_4 = 419 - 227 = 192 \text{ kJ/kg}$$

and $\quad W' = h'_2 - h'_4 = 447 - 419 = 28 \text{ kJ/kg}$

∴ $\quad \text{COP} = \dfrac{Q'}{W'} = \dfrac{192}{28} = 6.86 \quad$ with the heat exchanger

EXAMPLE 4.26 Assume a combination of superheated cycle and a subcooling cycle. A meat container requires a refrigerating system of 15 TR capacity at an evaporator temperature of –10°C and a condenser temperature of 40°C. The refrigerant R134a is subcooled by 8°C before entering the expansion valve and the vapour is superheated to 10°C before leaving the evaporator space. A three-cylinder compressor having stroke equal to 1.2 times the bore operates at 1200 rpm. Assume isentropic compression in the compressor and isenthalpic expansion in the expansion valve.

Determine the following:

(i) Refrigerating effect per kilogram.
(ii) Mass flow rate of refrigerant in kg/min.
(iii) Volume of vapour handled by the compressor per min.

(iv) Theoretical power.
(v) COP.
(vi) Bore and stroke of the compressor.

Solution: This problem can be solved either by using the *p–h* chart or by using the property table. Here it has been solved using the properties from the table:

Sat. temp. (0°C)	Absolute pressure (bar)	Specific volume of vapour (m³/kg)	Enthalpy of liquid (kJ/kg)	h of vapour (kJ/kg)	s of liquid (kJ/kg-K)	s of vapour (kJ/kg-K)
–10°C	2.0052	0.09963	186.78	392.75	0.9509	1.7337
40°C	10.165	0.01999	256.35	419.58	1.1903	1.7115

The cycle is represented on *p–h* diagram in Figure 4.19.

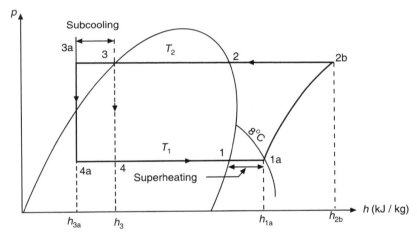

Figure 4.19 Superheating of suction gas—Example 4.26.

From ASHRAE data book, the specific heat of R134a vapour at 8°C is 0.92 kJ/kg-K, at –10°C it is 0.842 kJ/kg-K and for 50°C it is 1.218 kJ/kg-K.

Therefore, the average specific heat for the temperature range –10°C to 8°C is 0.881 kJ/kg-K. Also, the average specific heat for the temperature range 8°C to 50°C is 1.07 kJ/kg-K.

Now,
$$h_{1a} = h_1 + c_p(T_{1a} - T_1)$$
$$= 392.75 + (0.881)(18) = 408.5 \text{ kJ/kg}$$

This h_{1a} can also be obtained from the table 'Properties of Superheated Vapour' and it is 408.026 kJ/kg.

$$s_{1a} = s_1 + c_p \log (T_{1a}/T_1)$$
$$= 1.7337 + (0.881) \log 281/263 = 1.7920 \text{ kJ/kg-K}$$

This can also be obtained directly from the table of 'Properties of Superheated Vapour' which is 1.78996 kJ/kg-K.

For isentropic compression $1_{1a}-2_b$, $s_{1a} = s_{2b}$

$$s_{1a} = s_{2b} = s_{2g} + c_p \log \frac{T_{2b}}{T_2}$$

$$1.78996 = 1.7165 + 1.07 \log (T_{2b}/T_2)$$

$$T_{2b} = 335.2 \text{ K or } 62.2°\text{C}$$

Enthalpy at point 2b is given by

$$h_{2b} = h_2 + c_p(T_{2b} - T_2) = 419.58 + (62.2 - 40)(1.07) = 443.3 \text{ kJ/kg}$$

This is also obtained directly from the table 'Properties of Superheated Vapour' as 443.65 kJ/kg-K. The specific volume of vapour at point 1a is:

$$v_{1a} = \frac{T_{1a} \cdot v_1}{T_1} = \frac{(281)(0.09963)}{263}$$

$$= 0.106 \text{ m}^3/\text{kg} \; (0.108 \text{ m}^3/\text{kg from superheat table})$$

(i) Refrigerating effect/kg

$$q_L = (h_{1a} - h_{4a}) = (408.5 - 241.65)$$

$$= 166.85 \text{ kJ/kg} \qquad \text{Ans.}$$

(ii) Mass flow rate of refrigerant for 15 TR

$$\dot{m} = \frac{15 \times 3.5167}{166.85} = 0.316 \text{ kg/s}$$

$$= 18.96 \text{ kg/min} \qquad \text{Ans.}$$

(iii) Volume flow rate of compressor

$$= \dot{m} v_{1a} = (18.96)(0.108) = 2.04 \text{ m}^3/\text{min} \qquad \text{Ans.}$$

(iv) Theoretical power, W

$$W = \dot{m}(h_{2b} - h_{1a}) = (0.316) \times (443.3 - 408.5) = 10.9 \text{ kW} \qquad \text{Ans.}$$

(v) Coefficient of performance

$$\text{COP} = \frac{h_{1a} - h_{4a}}{h_{2b} - h_{1a}} = \frac{408.5 - 241.65}{443.3 - 408.5}$$

$$= 4.79 \qquad \text{Ans.}$$

(vi) Theoretical suction volume per cylinder per minute

$$= \frac{2.04}{3} = 0.68 \text{ m}^3/\text{min}$$

$$0.68 \text{ m}^3/\text{min} = \left(\frac{\pi}{4} D^2\right)(L) \times \text{rpm}$$

$$= \left(\frac{\pi}{4} D^2\right)(1.2D) \times 1200$$

$D = 0.084$ m $= 8.4$ cm **Ans.**

Stroke, $L = 1.2D = 1.2 \times 8.4 = 10.08$ cm **Ans.**

4.7.1 Liquid-suction Heat Exchanger

The refrigerant vapour at the outlet of the evaporator is at low temperature. Therefore, this cold vapour is employed to subcool the liquid refrigerant from the condenser outlet in a heat exchanger called *liquid-suction heat exchanger* and is shown in Figure 4.20.

The *p-h* diagram for this liquid-suction heat exchanger is same as shown in Figure 4.19. The use of liquid-suction heat exchanger is beneficial in two ways:

(i) It subcools the liquid.
(ii) It superheats the suction vapour up to a certain extent.

Therefore, the refrigerating effect increases per unit mass of refrigerant flow. Though the refrigerating effect increases, in total the coefficient of performance of a cycle employing a heat exchanger may be either greater than, less than, or same as that of the saturated cycle operating between the same pressure limits. The difference in COP is negligible and it is also true that the advantages accruing from the subcooling of the liquid in the heat exchanger are approximately offset by the disadvantages of superheating the vapour. Therefore, the use of a heat exchanger cannot be justified on the basis of an increase in system capacity and efficiency.

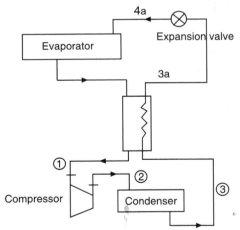

Figure 4.20 Vapour compression cycle with liquid–vapour heat exchanger.

So far we have discussed the impact of suction gas superheating and liquid subcooling on the refrigeration system.

The refrigerant experiences a drop in pressure while flowing through the piping, evaporator, condenser and receiver and through the valves and passages of the compressor. Understanding each one of these requires a separate study. But for the time being, we shall discuss the effect of evaporator pressure and condenser pressure on the saturation cycle separately.

4.7.2 Effect of Evaporator Pressure

Evaporator is a loop of tube coil through which the refrigerant flows in a mixed condition. While flowing, its pressure from evaporator inlet to its outlet decreases due to friction. The value of drop in pressure depends on the type of flow, evaporator tube diameter, length, number of bends, etc. In all, there is some pressure drop in the evaporator.

To understand the effect of drop in pressure on the performance, consider a simple saturated cycle as shown in Figure 4.21.

The line 4a–1a represents the real evaporation process in the evaporator while the refrigerant experiences a pressure drop of $(p_1 - p_{1a})$. As a result, the vapour leaves the evaporator at a lower pressure and saturation temperature and with a greater specific volume than in a system with no pressure drop.

For decreased evaporator pressure (temperature) for all the refrigerants, the following points w.r.t. Figure 4.21 are valid.

(i) Refrigerating effect decreases from $(h_1 - h_4)$ to $(h_{1a} - h_4)$.
(ii) Specific volume of vapour increases from v_1 to v_{1a}.
(iii) The specific volume of vapour at the suction to the compressor (v_{1a}) is higher. Therefore, for each suction stroke of the compressor, the mass of vapour pumped is less leading to poor volumetric efficiency of the compressor.
(iv) The work of compressor increases from $(h_2 - h_1)$ to $(h_{2b} - h_{1a})$ on account of increased pressure ratio (p_2/p_{1a}) and also the isentropic line becomes more flatter, i.e. the slope of line 1_a–2_b is less than the slope of line 1–2.
(v) In all, the performance of the system decreases, but the numerical values depend upon the working conditions of the plant and the range of pressure drop in the evaporator.

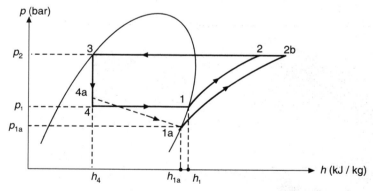

Figure 4.21 Saturation cycle with drop in evaporator pressure.

EXAMPLE 4.27 An R134a refrigeration system produces 10 ton of refrigerating effect at the evaporating temperature –10°C and the condensing temperature of 40°C. The absolute pressure measured at the inlet and outlet of evaporator are 2.0 bar and 1.7 bar, respectively. Considering there are no superheat and subcooling effects, calculate the following parameters with or without a drop of pressure in the evaporator:

(i) Refrigerating effect
(ii) Mass flow rate
(iii) Power requirement in kW/TR
(iv) COP

Assume specific heat c_p for vapour to be 1.07 kJ/kg-K.

Solution: Refer to Figure 4.21. The refrigerant properties are
At point 1:
Temperature at p_1 (2.0 bar) = $-10°C$
$$h_1 = 392.75 \text{ kJ/kg}, \quad s_1 = 1.7337 \text{ kJ/kg-K}, \quad v_1 = 0.09963 \text{ m}^3/\text{kg}$$
At point 1a:
Temperature at p_{1a} (1.7 bar) = $-14°C$
$$h_{1a} = 390.33 \text{ kJ/kg}, \quad s_{1a} = 1.7367 \text{ kJ/kg-K}, \quad v_{1a} = 0.11610 \text{ m}^3/\text{kg}$$
At point 2: $\quad s_{g2} = 1.7115$ kJ/kg-K, $\quad h_{g2} = 419.58$ kJ/kg
At point 3: $\quad h_3 = h_4 = 256.35$ kJ/kg

For isentropic compression 1–2,

$$s_1 = s_2 = s_{g2} + c_p \log \frac{T_2}{T_{2\,sat}} = 1.7337 = 1.7115 + 1.07 \log \left(\frac{T_2}{313}\right)$$

$$T_2 = 319.5 \text{ K}$$

Similarly, for isentropic compression, 1_a–2_b

$$s_{1a} = s_{2b} = s_{g2} + c_p \log \frac{T_{2b}}{T_{2\,sat}}$$

$$1.7367 = 1.7115 + 1.07 \log \frac{T_{2b}}{313}$$

$$T_{2b} = 320.45 \text{ K}$$

$$h_2 = h_{g2} + c_p \log \frac{T_{2b}}{T_{2\,sat}}$$

$$= 419.58 + 1.07 \log \left(\frac{319.5}{313}\right) = 419.58 \text{ kJ/kg}$$

Similarly,
$$h_{2b} = 420.6 \text{ kJ/kg}$$

(i) Refrigerating effect, q_L
For cycle 1–2–3–4:
$$q_L = (h_1 - h_4) = (392.75 - 256.35) = 136.4 \text{ kJ/kg} \textbf{ Ans.}$$

For cycle 1a–2b–3–4:
$$q_L = (h_{1a} - h_4) = (390.33 - 256.35) = 133.98 \text{ kJ/kg}$$
Ans.

(ii) Mass flow rate, $\dot{m}$

For cycle 1–2–3–4:
$$\dot{m} = \frac{(10 \times 3.5167)\ \text{kW}}{136.4\ \text{kJ/kg}} = 0.23\ \text{kg/s} \quad \textbf{Ans.}$$

For cycle 1a–2b–3–4:
$$\dot{m} = \frac{(10 \times 3.5767)\ \text{kW}}{133.98} = 0.26\ \text{kg/s} \quad \textbf{Ans.}$$

(iii) Power requirement, W

For cycle 1–2–3–4:
$$W = \dot{m}(h_2 - h_1)$$
$$= (0.23)(419.58 - 392.75) = 6.17\ \text{kW} \quad \textbf{Ans.}$$

For cycle 1a–2b–3–4:
$$W = \dot{m}(h_{2a} - h_{1a})$$
$$= (0.26)(420.6 - 390.33) = 7.9\ \text{kW} \quad \textbf{Ans.}$$

(iv) Coefficient of performance, COP

For cycle 1–2–3–4:
$$\text{COP} = \frac{h_1 - h_4}{h_2 - h_1} = \frac{392.75 - 256.35}{419.58 - 392.25} = 5.08 \quad \textbf{Ans.}$$

For cycle 1a–2b–3–4:
$$\text{COP} = \frac{h_{1a} - h_4}{h_{2a} - h_{1a}} = \frac{390.33 - 256.35}{420.6 - 390.3} = 4.42 \quad \textbf{Ans.}$$

4.7.3 Effect of Condenser Pressure

An increase in condenser pressure results in a decrease in refrigerating capacity and an increase in power consumption. This is represented by a p–h diagram using saturated cycle, in Figure 4.22. The condenser pressure increases due to the following factors.

To overcome the inertia of the spring-loaded valves

The pressure of the vapour refrigerant is raised considerably above the average condensing pressure by the compressor. This is necessary in order to discharge the vapour out of the cylinder through the discharge valves against the condensing pressure and the additional pressure encountered by the spring loading of the discharge valves. The average discharge pressure at the compressor outlet is 20–30% greater than the condenser pressure. In large refrigerating systems, the other reasons are due to non-condensable gases. The effect of non-condensable gases towards increasing the condenser pressure is negligible.

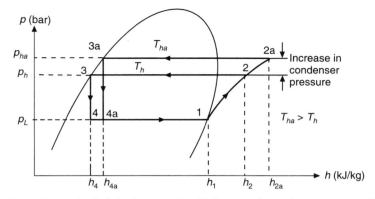

Figure 4.22 Saturation cycle (1–2–3–4) and cycle with increase in condenser pressure (1–2a–3a–4a).

Increase in temperature of cooling media

The condenser is either air- or water-cooled type. The medium that takes the heat from refrigerant vapour is water in the case of water-cooled condensers, and is given by $m\, c_p\, \Delta T$. The mass flow rate of water remains constant because of the constant capacity pump being used to circulate it and c_p also remains constant. In the field, water is cooled in the cooling tower. Due to deterioration in the performance of cooling tower, the water temperature may rise. In such a case ΔT will decrease and the heat carried away by the water decreases and ultimately the condenser temperature increases, accompanied by its pressure rise. The compressor pumps the refrigerant vapour continuously at a particular rate. If the rate of condensation of refrigerant is less, then more refrigerant vapour gets accumulated on the discharge side of the compressor, which leads to building of pressure.

EXAMPLE 4.28 A refrigeration system with R22 refrigerant produces a cooling effect of 1 TR. The evaporator is at –10°C (evaporator pressure 3.5 bar) and condenser temperature is 40°C (condenser pressure 15.3 bar).

While this system is in operation, the pressure measured at the condenser inlet is 18.5 bar, and no change is observed on the evaporator side. Assuming saturated cycle, evaluate the following parameters with an increase in the condenser pressure.
 (i) Refrigerating effect
 (ii) Mass flow rate
 (iii) Work of compressor
 (iv) COP

Solution: For the condenser pressure of 18.5 bar, the saturation temperature of R22 is 48°C. Refer to Figure 4.22. The following properties are collected from the *p–h* chart.

At point 1: $h_1 = 235$ kJ/kg $v_1 = 0.065$ m³/kg
At point 2: $h_2 = 262$ kJ/kg $T_2 = 62°C$
At point 2a: $h_{2a} = 285$ kJ/kg $T_{2a} = 82°C$
At point 3: $h_3 = h_f = 100$ kJ/kg, for point 3a: $h_{3a} = 112.5$ kJ/kg

(i) Refrigerating effect, q_L
For the cycle 1–2–3–4:
$$q_L = (h_1 - h_4) = (235 - 100) = 135 \text{ kJ/kg} \quad \text{Ans.}$$
For the cycle 1–2a–3a–4a:
$$q_L = (h_1 - h_{4a}) = (235 - 112.5) = 122.5 \text{ kJ/kg} \quad \text{Ans.}$$

(ii) Mass flow rate, $\dot{m}$
For the cycle 1–2–3–4:
$$\dot{m} = \frac{3.5167}{q_L} = \frac{3.5167}{135} = 0.026 \text{ kg/TR} \quad \text{Ans.}$$
For the cycle 1–2a–3a–4a:
$$\dot{m} = \frac{3.5167}{q_L} = \frac{3.5167}{122.5} = 0.0287 \text{ kg/TR} \quad \text{Ans.}$$

(iii) Work of Compression, W
For the cycle 1–2–3–4:
$$W = \dot{m}(h_2 - h_1) = 0.0287(262 - 235) = 0.77 \text{ kW} \quad \text{Ans.}$$
For the cycle 1–2a–3a–4a: $W = \dot{m}(h_{2a} - h_{1a})$
$$= 0.0250(285 - 235) = 1.43 \text{ kW} \quad \text{Ans.}$$

(iv) Coefficient of Performance (COP)
For the cycle 1–2–3–4:
$$\text{COP} = \frac{h_1 - h_4}{h_2 - h_1} = \frac{235 - 100}{262 - 235} = 5.0 \quad \text{Ans.}$$
For the cycle 1–2a–3a–4a:
$$\text{COP} = \frac{h_1 - h_{4a}}{h_{2a} - h_1} = \frac{235 - 112.5}{285 - 235} = 2.45 \quad \text{Ans.}$$

4.8 LOSSES IN VAPOUR COMPRESSION REFRIGERATION SYSTEM

A refrigerant flowing through the different sections of the vapour compression cycle mainly experiences two types of losses, i.e. thermal losses (gain) and pressure losses. The pressure loss is due to the friction between the fluid and the contacting surface and the thermal loss may be of heat gain type or heat loss type depending on the temperature difference between the fluid and the surroundings.

The vapour compression cycle with these losses is known as the *actual vapour compression cycle*. The above losses are listed as follows and also represented on *p–h* diagram and *T–s* plot in Figures 4.23 and 4.24, respectively.

(i) **Thermal losses (or gain)**
    (a) Superheating of vapour in the evaporator (1d–1c). This produces useful refrigerating effect.

(b) Heat gain and superheating of vapour in the pipeline between the evaporator and the compressor (1c–1b). Usually, this does not produce a useful refrigerating effect.

(c) Heat loss and desuperheating of vapour in the discharge line joining the compressor and the condenser (2b–2c).

(d) Heat gain in the liquid line (3a–3b). This is due to heat flow from the surroundings to the pipeline between the expansion valve and the condenser.

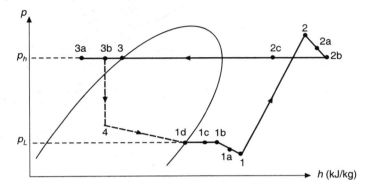

Figure 4.23 Actual vapour compression cycle on *p–h* diagram.

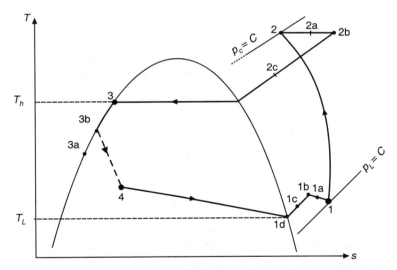

Figure 4.24 Actual vapour compression cycle on *T–s* diagram.

(ii) **Pressure losses**

(a) Pressure drops in the suction line (1b–1a).
(b) Pressure drops at the suction valve, suction port in the valve plate (1a–1).

(c) Polytropic compression of vapour which involves friction and heat transfer (1–2).
(d) Pressure drops at the compressor delivery valve (2–2a).
(e) Pressure drops in the delivery line, i.e., it includes discharge muffler, tube connection between the compressor and the condenser (2a–2b).
(f) Pressure drops in the condenser unit (2b–3).
(g) Pressure drops in the evaporator unit (4–1b).

The sum of all these pressure drops is ultimately borne by the compressor. It means the compressor has to raise the vapour pressure by this extra pressure drop, which increases the pressure ratio. This high pressure ratio has many disadvantages, like overheating of the compressor, lubricating problems, lesser volumetric efficiency and cooking of valve plate. Therefore, it is always desirable to minimise these losses as much as possible.

EXAMPLE 4.29 Suppose a 100 ton cooling capacity ammonia refrigeration plant is working at evaporating temperature –20°C and condensing temperature 34°C. The liquid is subcooled to 28°C in order to reduce the formation of flash gases. Ammonia vapour leaves the evaporator and enters the compressor at –6°C. There is a pressure loss of 0.1 bar at the suction valve and 0.5 bar at the discharge valves. Assume isentropic compression and take the volumetric efficiency of the compressor to be 80%.

Determine the following quantities:
(a) Refrigerating effect per kg
(b) Mass flow rate in kg/min
(c) Piston displacement in m³/min
(d) COP of the plant
(e) Power required (kW)

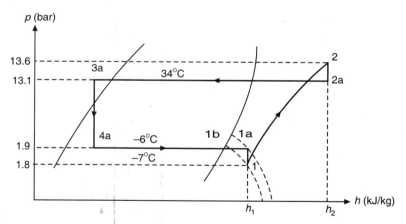

Figure 4.25 Saturation cycle 1b–2–3–4—Example 4.29.

Solution: The properties obtained from p–h chart are as follows:

Point 1a: $h_1 = h_{1a} = 495$ kJ/kg

Point 1: $p_1 = 1.8$ bar, $S_1 = 1.5$ kg/m³, $v_1 = 0.66$ m³/kg

Point 2: $p_2 = 13.6$ bar, $h_2 = 790$ kJ/kg
           $T_2 = 130°C$
Point 3a: $h_{3a} = h_4 = -615$ kJ/kg

(a) Refrigerating effect per kilogram, q_0
    Assuming that superheating takes place outside the evaporator,
    $$q_0 = h_1 - h_4 = (495 - (-615)) = 1110 \text{ kJ/kg} \quad \textbf{Ans.}$$

(b) Mass flow rate, $\dot{m}$
    $$\dot{m} = \frac{100 \times 211}{1110} = 19.0 \text{ kg/min} \quad \textbf{Ans.}$$

(c) Piston displacement, m³/min.
    Volume of vapour handled by compressor
    $$= \dot{m}v_1 = (19)(0.66) = 12.5 \text{ m}^3/\text{min}$$
    Therefore, piston displacement,
    $$= \frac{12.5}{\eta_V} = \frac{12.5}{0.8} = 15.68 \text{ m}^3/\text{min} \quad \textbf{Ans.}$$

(d) COP of the plant
    $$\text{COP} = \frac{h_1 - h_4}{h_2 - h_1} = \frac{495 - (-615)}{790 - 495}$$
    $$= 3.7 \quad \textbf{Ans.}$$

(e) Power required, kW
    $$= (\dot{m})(h_2 - h_1)$$
    $$= \left(\frac{19}{60}\right)(790 - 495)$$
    $$= 93.34 \text{ kW} \quad \textbf{Ans.}$$

EXAMPLE 4.30 Compare the COP of an ammonia refrigeration cycle that uses wet compression with one that uses dry compression and operates between $-20°C$ and $30°C$. Assume compression to be isentropic and no subcooling of liquid refrigerant. In wet compression, the vapour leaving the compressor is dry saturated whereas in dry compression, the vapour entering the compressor is dry saturated.

Solution: Refer to Figure 4.26 where the cycle (1–2–3–4) represents the ideal cycle having wet dry compression whereas in cycle (1a–2a–3–4), the process 1–2 represents dry compression with vapour dry saturated at entry to compression.

160 *Refrigeration and Air Conditioning*

Saturation temperature, t_s (°C)	Specific enthalpy (kJ/kg)			Specific entropy (kJ/kg-K)	
	h_f	h_{fg}	h_g	s_f	s_g
–20	89.8	1330.2	1420.0	0.3684	5.6244
30	323.1	1145.8	1468.9	1.2037	4.9842

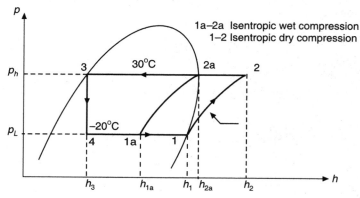

Figure 4.26 *p–h* diagram, ideal cycle 1–2–3–4—Example 4.30.

Specific heats: $\quad c_{pf} = 4.6$ kJ/kg-K, $\quad c_{pg} = 2.8$ kJ/kg-K

COP of wet compression cycle (1a–2–3–4)

Consider the isentropic compression process (1a–2a)

$$s_{f1a} + x_1(s_{g1} - s_{f1a}) = s_{g2}$$

$$0.3684 + x_1(5.6244 - 0.3684) = 4.9842$$

∴ $\quad x_1 = 0.878$

∴ $\quad h_1 = h_{f1a} + x_1 \cdot h_{fg1} = 89.8 + 0.878 \times 1330.2$

$\quad\quad = 1257.7$ kJ/kg

Compressor work, $\quad w = h_{2a} - h_{1a} = 1489.9 - 1257.7 = 232.2$ kJ/kg

Refrigerating effect, $\quad q_L = h_{1a} - h_4 = 1257.7 - 323.1 = 934.6$ kJ/kg

$$\text{COP} = \frac{q_L}{w} = \frac{934.6}{232.2} = 4.025$$

COP of dry compression cycle (1–2–3–4)

Consider the isentropic compression process (1–2); T_{2s} = saturation temperature

$$s_{g1} = s_{g2} + c_{pg} \log_e \frac{T_2}{T_{2s}}$$

$$5.6244 = 4.9842 + 2.8 \log_e \frac{T_2}{(273+30)}$$

∴ $T_2 = 380.8 \text{ K} = (380.8 - 273)°C = 107.8°C$

$h_2 = h_{g2} + c_{pg}(T_2 - T_{2s}) = 1468.9 + 2.8(107.8 - 30) = 1686.7 \text{ kJ/kg}$

Compressor work, $w = (h_2 - h_1) = 1686.7 - 1420.0 = 266.7 \text{ kJ/kg}$

Refrigerating effect,

$$q_L = h_1 - h_4 = 1420.0 - 323.1$$
$$= 1096.9 \text{ kJ/kg}$$

$$\text{COP} = \frac{q_L}{w} = \frac{1096.9}{266.7} = 4.113$$

COP of dry compression cycle is 4.113 and that of wet compression cycle is 4.025.

Ans.

EXAMPLE 4.31 A refrigeration system using NH_3 works on standard vapour compression cycle. The condensing temperature is 30°C. Investigate the performance of the system based on the following:

(i) COP
(ii) Mass of refrigerant circulated
(iii) Volume of refrigerant handled by compressor
(iv) Temperature of refrigerant at compressor delivery

The system operates at temperature in the evaporator of –10°C and 16°C, respectively. The capacity of the system is 10 TR.

Solution: Standard cycle is shown in Figure 4.27 with dry vapour at entry to compressor.

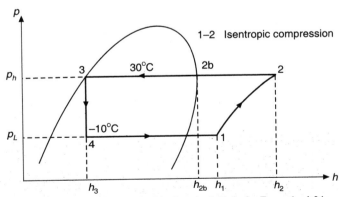

Figure 4.27 p–h diagram, ideal cycle 1-2-3-4—Example 4.31.

Given:

$$Q = 10 \text{ TR} = 10 \times 211 = 2110 \text{ kJ/min}$$

The properties of saturated refrigerant ammonia are as follows:

Saturation temperature, T_s (°C)	Specific volume, v_g (m³/kg)	Specific enthalpy (kJ/kg)			Specific entropy kJ/kg-K	
		h_f	h_{fg}	h_g	s_f	s_g
–10	0.4189	112.4	1314.2	1426.6	0.4572	5.5789
30	0.1107	323.1	1145.8	1468.9	1.2037	4.9842

Specific heat, c_{pg} = 2.8 kJ/kg-K

(i) COP of system: It is assumed that superheating of refrigerant at inlet to the compressor takes place in the evaporator and adds to the refrigeration effect.

For isentropic process (1–2):

$$s_{g1} + c_{pg} \log_e \frac{T_1}{T_{s1}} = s_{g2} + c_{pg} \log_e \frac{T_2}{T_{s2}}$$

$$5.5789 + 2.8 \log_e \frac{(16+273)}{(-10+273)} = 5.1421 + 2.8 \log_e \left(\frac{T_2}{273+30}\right)$$

$T_2 = 389.2$ K = (389.2 – 273) = 116.2°C

$h_1 = h_{g1} + c_{pg}(T_1 - T_{s1})$
    = 1426.6 + 2.8[(16 – (–10)] = 1499.4 kJ/kg

$h_3 = h_4 = h_{f3}$ = 256.12 kJ/kg

and

$h_2 = h_{g2} + c_{pg}(T_2 - T_{s2})$
    = 1459.4 + 2.8(116.2 – 30) = 1700.76 kJ/kg

Compressor work,

$w = h_2 - h_1$ = 1700.76 – 1499.4 = 200.36 kJ/kg

Refrigerating effect,

$q_L = h_1 - h_4$ = 1499.4 – 256.12 = 1243.28 kJ/kg

$$\text{COP} = \frac{q_L}{w} = \frac{1243.28}{200.36} = 5.02 \qquad \text{Ans.}$$

(ii) Mass of refrigerant circulated, $\dot{m}$

$$\dot{m} = \frac{Q_L}{q_L} = \frac{2110}{1243.28} = 1.697 \text{ kg/min} \qquad \text{Ans.}$$

(iii) Volume of refrigerant handled by compressor, v_1

$$v_1 = v_{g1} \times \frac{T_1}{T_{s1}} = 0.4189 \times \frac{(273+16)}{(273-10)} = 0.4603 \text{ m}^3/\text{kg}$$

$V_1 = \dot{m} \times v_1$ = 1.697 × 0.4603 = 0.781 m³/min   Ans.

(iv) Temperature of refrigerant at delivery, T_2 = 116.2°C   Ans.

EXAMPLE 4.32 An ammonia refrigeration system used in an ice-plant operates between the evaporator temperature of –20°C and condensing temperature of 30°C. It produces 10 tons of

ice per day from water at 25°C to ice at 0°C. Using the standard vapour compression cycle, determine the following.
  (i) Capacity
 (ii) Mass flow of refrigerant
(iii) Temperature at the outlet of compressor
(iv) Work done
 (v) COP
(vi) Ideal COP
(vii) Bore and stroke of the twin cylinder compressor, given $L/D = 1.3$, $\eta_V = 0.80$, and $N = 1000$ rpm
(viii) Mass flow of refrigerant, if the refrigerant is subcooled by 10°C before entering into the expansion valve.
Take $c_{pW} = 4.6$ kJ/kg-K, $L_{ice} = 335$ kJ/kg.

Solution: Mass of ice, $\dot{m}_i = 10$ tons/day $= 10 \times 10^3$ kg/day, $c_{pg} = 2.8$ kJ/kg-K

The properties of NH_3 refrigerant from property table are as follows:

Saturation, temperature, T_s (°C)	Specific volume, v_g (m³/kg)	Specific enthalpy (kJ/kg)			Specific entropy (kJ/kg-K)	
		h_f	h_{fg}	h_g	s_f	s_g
−20	0.6244	89.8	1330.2	1420.0	0.3684	5.6244
30	0.1107	323.1	1145.8	1468.9	1.2037	4.9842

(i) Capacity:

Total refrigerating effect, Q

$$Q = \dot{m}_i[c_{pw}(T - 0) + h_L]$$

$$= \frac{10 \times 10^3}{24 \times 60 \times 60} [4.6(25 - 0) + 335] = 50.93 \text{ kJ/s}$$

∴ Capacity $= \dfrac{50.93}{3.517} = 14.47$ TR **Ans.**

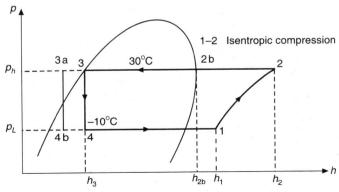

Figure 4.28 p–h diagram, ideal cycle 1–2–3–4—Example 4.32.

(ii) Mass flow rate of refrigerant, $\dot{m}$

$$h_1 = h_{g1} = 1420.0 \text{ kJ/kg}; \quad h_4 = h_3 = h_{f3} = 323.1 \text{ kJ/kg}$$

Refrigerating effect/kg,

$$q_L = h_1 - h_4 = 1420 - 323.1 = 1096.9 \text{ kJ/kg}$$

$$\dot{m} = \frac{\dot{Q}}{q_L} = \frac{50.93}{1096.9} \times 60 \text{ kg/min} = 2.786 \text{ kg/min} \quad \text{Ans.}$$

(iii) Temperature at the outlet of compressor, T_2. For isentropic (1–2):

$$s_1 = s_2, \text{ i.e. } s_{g1} = s_{g2} + c_{pg} \log_e \frac{T_2}{T_1}$$

$$5.6244 = 4.9842 + 2.8 \log_e \frac{T_2}{273 + 30}$$

∴ $T_2 = 379.2 \text{ K} = (379.2 - 273) = 106.2°C$ **Ans.**

(iv) Work done, w

$$h_2 = h_{g2} + c_{pg}(T_2 - T_{s2}) = 1468.9 + 2.8(106.2 - 30) = 1682.3 \text{ kJ/kg}$$

∴ $w = h_2 - h_1 = 1682.3 - 1420 = 262.3 \text{ kJ/kg}$ **Ans.**

(v) COP

$$\text{Actual COP} = \frac{\dot{Q}}{\dot{m} \times w} = \left(\frac{60}{2.786}\right) \times \frac{50.93}{262.3} = 4.182 \quad \text{Ans.}$$

(vi) Ideal COP

$$= \frac{q_L}{w} = \frac{1096.9}{262.3} = 4.182 \quad \text{Ans.}$$

(vii) Bore and stroke of the twin-cylinder compressor,
Given:

$$\frac{L}{D} = 1.3, \; \eta_V = 0.8, \; N = 1000 \text{ rpm}$$

Actual volume,

$$\dot{m} \times v_{g1} = 2 \times \frac{\pi}{4} D^2 \times L \times N$$

$$2.786 \times 0.6244 = 2 \times \frac{\pi}{4} D^2 \times 1.3D \times 1000$$

∴ $D = 0.0948 \text{ m} \quad \text{or} \quad 9.48 \text{ cm}$

$L = 1.3D = 1.3 \times 9.48 = 12.324 \text{ cm}$ **Ans.**

(viii) Mass flow of refrigerant if the refrigerant is subcooled by 10°C, $\dot{m}_a$
Subcooling is shown by process (3–3a)

$$\therefore \qquad T_3 - T_{3a} = 10°C$$
$$\therefore \qquad h_{3a} = h_3 - c_{pL}(T_{s3} - T_3) = 323.1 - 4.6 \times 10 = 277.1 \text{ kJ/kg}$$
$$h_{3a} = h_{4a} \qquad \text{since the process of throttling is isenthalpic.}$$

New refrigerating effect,
$$q_L = h_1 - h_{4a} = 1420 - 277.1 = 1142.9 \text{ kJ/kg}$$

$$\therefore \qquad \dot{m}_a = \frac{\dot{Q}}{q_L} = \frac{50.93}{1142.9} \times 60 \text{ kg/min} = 2.674 \text{ kg/min} \qquad \textbf{Ans.}$$

EXAMPLE 4.33 A food storage locker has a capacity of 12 TR and evaporates between the temperature of $-8°C$ and condensation temperature of $30°C$. The refrigerant R12 is subcooled by $5°C$ before throttling and the vapour is superheated by $2°C$ before leaving the evaporator. Assuming a twin-cylinder, single-acting compressor running at 1000 rpm with $L : D$ ratio of 1.5, calculate:

(i) COP
(ii) Specific power required
(iii) Diameter and stroke of compressor having volumetric efficiency of 0.945. Sketch the cycle on p–h diagram.

Solution: $Q = 12 \text{ TR} = 12 \times 211 = 2532 \text{ kJ/min}, \quad \dfrac{L}{D} = 1.5$

This example is solved using the p–h chart.
Since the refrigerant is subcooled by $5°C$, $T_3 = 30 - 5 = 25°C$. Since the refrigerant is superheated by $2°C$ leaving the evaporator, $T_1 = -8 + 2 = -6°C$.
The cycle is represented on p–h diagram in Figure 4.29.
From the p–h chart for refrigerant R12, we get:
$$h_1 = 186 \text{ kJ/kg}; \quad h_2 = 206 \text{ kJ/kg}; \quad h_3 = h_4 = 59 \text{ kJ/kg}$$
$$\therefore \qquad v_1 = 0.07244 \text{ m}^3/\text{kg}$$

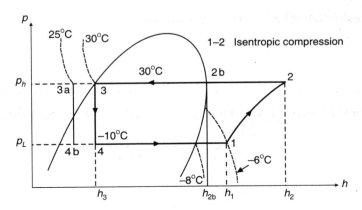

Figure 4.29 p–h diagram, ideal cycle 1–2–3–4—Example 4.33.

(i) COP: Consider 1 kg of refrigerant R12 flow.
Compressor work per kg,
$$w = h_2 - h_1 = 206 - 186 = 20 \text{ kJ/kg}$$
Refrigerating effect per kg,
$$q_L = h_1 - h_4 = 186 - 59 = 127 \text{ kJ/kg}$$
$$\text{COP} = \frac{q_L}{w} = \frac{127}{20} = 6.35 \quad \textbf{Ans.}$$

(ii) Specific power required, P
Mass flow of refrigerant,
$$\dot{m} = \frac{Q}{q_L} = \frac{2532}{127} = 19.94 \text{ kg/min}$$
$$P = \dot{m}w = \frac{19.94}{60} \times 20 \text{ kJ/s} = 6.647 \text{ kW} \quad \textbf{Ans.}$$

(iii) Diameter D and stroke L of compressor:
Given: $N = 1000$ rpm, twin cylinder $\eta_V = 0.945$
Volume flow rate of refrigerant,
$$\dot{V} = \dot{m} \times v_1 = 19.94 \times 0.07244 = 1.4444 \text{ m}^3/\text{min}$$

But $\dot{V}$ = No. of cylinders × Stroke volume × Volumetric efficiency × Speed

$\therefore \quad 1.4444 = 2 \times \left(\frac{\pi}{4} \times D^2 \times L\right) \times 0.945 \times 1000$

or $\quad 1.4444 = 2 \times \frac{\pi}{4} \times D^2 \times 1.5D \times 0.945 \times 1000$

$\therefore \quad D = 0.0866$ m and $L = 1.5D = 0.1299$ m $\quad \textbf{Ans.}$

EXAMPLE 4.34 An NH_3 refrigeration plant has a capacity of 20 TR. The condensation and evaporation temperatures are 35°C and −20°C, respectively. Refrigerant is dry and saturated at the entry to the compressor. There is no undercooling of the liquid refrigerant. If the actual COP is 0.7 times the theoretical COP, determine the following:
  (i) Mass flow rate of refrigerant.
  (ii) Power required to drive the compressor.
  (iii) Diameter and stroke of the compressor running at 4 rev/s and $L = D$. Its volumetric efficiency is 80% and is single acting.

Sketch the cycle on p–h diagram. You may use the p–h chart provided.

Solution: Cooling capacity of plant = Actual refrigerating effect = 20 TR
$$= 20 \times 3.517 = 70.34 \text{ kJ/s}$$
$$T_1 = T_4 = -20°C = 253 \text{ K}; \quad T_3 = T_4 = 35°C = 308 \text{ K}$$

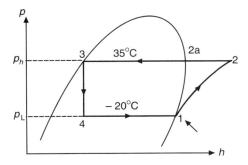

Figure 4.30 Flow diagram on p–h chart—Example 4.34.

Actual COP = 0.7 times the theoretical COP
From the p–h chart (ammonia chart), $h_1 = 1413$ kJ/kg; $h_2 = 1691$ kJ/kg
$h_3 = 342$ kJ/kg $= h_4$ (isenthalpic process)
Work done on compressor $= w = h_2 - h_1 = 1691 - 1413 = 278$ kJ/kg
Refrigerating effect, $q_0 = h_1 - h_4 = 1413 - 342 = 1071$ kJ/kg

$$\text{Ideal COP} = \frac{q_0}{w} = \frac{1071}{278} = 3.85$$

Actual COP $= 0.7 \times$ ideal COP $= 0.7 \times 3.85 = 2.697$

(i) Mass flow rate of refrigerant, $\dot{m} = \dfrac{\text{Cooling capacity of plant in kJ/s}}{\text{Refrigerating effect in kJ/kg}} = \dfrac{70.34}{1071}$

$$\dot{m} = 0.066 \text{ kg/s} \qquad \textbf{Ans.}$$

$$\text{Actual COP} = \frac{\text{Capacity of plant}}{\text{Power required}}$$

(ii) Power required $= \dfrac{70.34}{2.697} = 26.081$ kW **Ans.**

(iii) Specific volume at inlet, $v_1 = 0.62$ m³/kg [from the p–h chart]
Total actual volume = (mass of refrigerant) × (specific volume)
$= (0.066)(0.62) = 0.041$ m³/kg

$$\text{Volumetric efficiency} = \frac{\text{Actual volume}}{\text{Ideal volume}}$$

$$0.8 = \frac{0.041}{\text{Ideal volume}}$$

Ideal volume $= 0.0513$ m³/s

Ideal volume $= \dfrac{\pi}{4} D^2 L N$

$0.0513 = \dfrac{\pi}{4} \times D^3 \times 4$ [Given $D = L$ and $N = 4$ rps]

∴ $D = 0.2537$ m, $L = 0.2537$ m **Ans.**

EXAMPLE 4.35 A vapour compression refrigeration plant operates between evaporation and condensation temperatures of –10°C and 45°C, respectively. The refrigerant is dry and saturated vapour at entry to the compressor. It is discharged at 102°C from the compressor. The bore and stroke of the compressor are 80 mm each. It runs 720 rpm with a volumetric efficiency of 80%. The liquid refrigerant enters the expansion valve at 35°C. Determine:

(i) COP
(ii) Mass flow rate of refrigerant
(iii) Capacity of the plant in TR. Take specific heat of liquid refrigerant = 1.62 kJ/kg-K

You may use the properties of refrigerant tabulated as follows.

Sat. Temp.	V_g (m³/kg)	h_f (kJ/kg)	h_g (kJ/kg)	s_f (kJ/kg-K)	s_g (kJ/kg-K)
–10°C	0.233	45.4	460.7	0.183	1.762
45°C	0.046	133.0	488.6	0.485	1.587

Solution: Given: $T_1 = T_4 = -10°C = 263$ K;
$T_{sup} = T_2 = 102°C = 375$ K (superheated);
$T_3 = 45°C = 318$ K; $T_{3'} = 35°C = 308$ K (subcooled)
Bore = stroke = 80 mm; N = 720 rpm; Volumetric efficiency = 80%;
$c_{pv} = 1.0614$, $h_1 = h_{g1} = 460.7$ kJ/kg
$h_2 = h_{g2} + c_{pv}(T_{sup} - T_{sat}) = 488.6 + 1.0614(375 - 318) = 549.1$ kJ/kg
$h_3 = h_4 = h_{f3} - 1.62$(degree of undercooling)
$= 133 - 1.62(318 - 308)$
$h_3 = h_4 = 116.8$ kJ/kg

(i) Refrigerating effect, $q_L = h_1 - h_4 = 460.7 - 116.8 = 343.9$ kJ/kg
Work done, $w = h_2 - h_1 = 549.1 - 460.7 = 88.4$ kJ/kg

$$\text{Ideal COP} = \frac{q_L}{w} = \frac{343.9}{88.4} = 3.89 \quad \textbf{Ans.}$$

(ii) Theoretical volume $= \frac{\pi}{4} D^2 L N = \frac{\pi}{4} (0.08)^2 (0.08) \times \frac{720}{60}$
$= 4.83 \times 10^{-3}$ m³/s

$$\text{Volumetric efficiency} = \frac{\text{Actual volume}}{\text{Theoretical volume}}$$

Actual volume = $0.8 \times 4.83 \times 10^{-3} = 3.86 \times 10^{-3}$ m³/s
Specific volume at entry to compressor = 0.233 m³/kg

∴ $\text{Mass flow rate} = \frac{\text{Actual volume}}{\text{Specific volume}} = \dot{m} = \frac{3.86 \times 10^{-3}}{0.233} = 0.017$ kg/s
Ans.

(iii) Actual refrigerating effect = $q_L \times \dot{m}$ = 343.9 × 0.017 = 5.697 kJ/s

∴ Capacity of plant = 5.697 kJ/s or 1.62 TR **Ans.**

EXAMPLE 4.36 An NH_3 vapour compression cycle refrigerator operates between evaporation and condensation temperature of –10°C and 35°C, respectively. The refrigerant is dry and saturated at entry to the compressor and is discharged at 102°C after compression. The single-acting compressor has a bore 80 mm and stroke 80 mm. It runs at 480 rpm with a volumetric efficiency of 80%. Assuming liquid subcooling of 10°C. Calculate:
(i) COP
(ii) Mass flow rate of refrigerant
(iii) Capacity of the plant in TR

Sketch the cycle on *p–h* chart. You may use the *p–h* chart for NH_3.

Solution: Refer to *p–h* chart.
Given: T_1 = –10°C = 263 K; T_2 = 102°C = 375 K (superheated)
T_3 = 25°C = 298 K; D = 80 mm; L = 80 mm; N = 480 rpm; volumetric η = 80%
Degree of subcooling = 10°C
h_1 = 1423 kJ/kg; v_1 = 0.4175 m³/kg; h_2 = 1632 kJ/kg; h_3 = h_4 = 291 kJ/kg

(i) Refrigerating effect, $q_L = h_1 - h_4$ = 1423 – 291 = 1132 kJ/kg
Work done $w = h_2 - h_1$ = 1632 – 1423 = 209 kJ/kg

$$\text{Ideal COP} = \frac{q_L}{w} = \frac{1132}{209} = 5.42 \text{ kJ/kg} \quad \textbf{Ans.}$$

(ii) Stroke volume per min = $\frac{\pi}{4} D^2 L N = \frac{\pi}{4} (0.08)^2 (0.08)(480)$

= 0.193 m³/min

Specific volume, v_1 = 0.415 m³/kg

$$\text{Mass flow rate, } \dot{m} = \frac{\text{Stroke volume} \times \text{Volumetric efficiency}}{\text{Specific volume}}$$

$$= \frac{0.193 \times 0.8}{0.415}$$

$\dot{m}$ = 0.372 kg/min **Ans.**

(iii) Total refrigerating effect = $\dot{m} \times q_L$ = 0.372 × 1132 kJ/min

$$= \frac{421.158}{60} = 7.019 \text{ kJ/s}$$

Capacity of plant = 7.019 kJ/s = $\frac{7.019}{3.517}$ = 1.996 TR **Ans.**

EXAMPLE 4.37 A refrigeration plant of 100 TR capacity uses R22 as refrigerant. The condensing and evaporating pressures are 11.82 bar and 1.64 bar, respectively.

The refrigerant enters the compressor at dry and saturated state, whereas it leaves the condenser, subcooled by 10°C. Actual COP is 70% of its theoretical value.

Find the following:
(i) Theoretical and actual COP
(ii) Mass flow rate in kg/s
(iii) Compressor power

The properties of R22 are:

p (bar)	$T(°C)$	h_f (kJ/kg)	s_f (kJ/kg-K)	h_g (kJ/kg)	s_g (kJ/kg-K)
1.64	-30	166.1	0.8698	393.1	1.803
11.82	$+30$	236.7	1.125	414.5	1.712

c_p of vapour = 0.55 kJ/kg-K, c_p of liquid = 1.19 kJ/kg-K

Solution: Refer to Figure 4.31 for the p–h diagram.
Given: Capacity = 100 TR = 100 × 3.517 = 351.7 kJ/s
$$p_1 = p_4 = 1.64 \text{ bar}; \quad p_2 = p_3 = 11.82 \text{ bar};$$
$$T_3 = 30°C = 303 \text{ K};$$
$$c_{pv} = 0.55 \text{ kJ/kg-K}; \quad c_{PL} = 1.19 \text{ kJ/kg-K};$$
$$T_1 = T_4 = -30°C = 243 \text{ K}$$

For isentropic process 1–2; $s_1 = s_2$

$$s_{g1} = s_{g2} + c_{pv} \ln\left(\frac{T_{\text{sup}}}{T_{\text{sat}}}\right)$$

$$1.803 = 1.712 + 0.55 \ln\left(\frac{T_2}{303}\right)$$

$T_2 = 357.52 \text{ K}; \quad h_1 = h_{g1} = 393.1 \text{ kJ/kg}$
$h_2 = h_{g2} + c_{pv}(T_{\text{sup}} - T_{\text{sat}}) = 414.5 + 0.55(357.52 - 303) = 444.48 \text{ kJ/kg}$
$h_{3a} = h_4 = h_{f3} - c_{pl}(\text{degree of undercooling})$
$h_{3a} = h_4 = 236.7 - 1.19(10) = 224.8 \text{ kJ/kg}$

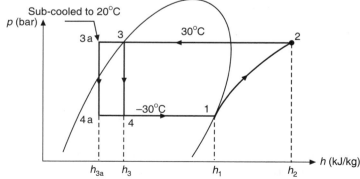

Figure 4.31 p–h diagram showing the subcooling effect (cooling liquid from point 3 to 3a)—Example 4.37.

(i) Refrigerating effect, $q_L = h_1 - h_4 = 393.3 - 224.8 = 168.5$ kJ/kg

Work done, $w = h_2 - h_1 = 444.48 - 393.3 = 51.18$ kJ/kg

$$\text{Ideal COP} = \frac{q_L}{w} = \frac{168.5}{51.18} = 3.29 \quad \textbf{Ans.}$$

$$\text{Actual COP} = 0.7 \times \text{ideal COP} = 0.7 \times 3.29 = 2.31 \quad \textbf{Ans.}$$

(ii) Mass flow rate, $\dot{m} = \dfrac{\text{Capacity of plant in kJ/s}}{\text{Refrigerating effect in kJ/kg}} = \dfrac{351.7}{168.5} = 2.08$ kg/s

Ans.

(iii) Compressor power $= \dfrac{\text{Capacity of plant in kJ/s}}{\text{Actual COP}} = \dfrac{351.7}{2.31} = 152.25$ kW

Ans.

EXAMPLE 4.38 An ice-plant produces 30 tons of ice at 0°C per day from water at 0°C. The condensation and evaporation take place at 20°C and –20°C, respectively. There is no undercooling of liquid and the vapours drawn by compressor are dry and saturated.

c_p of vapour = 1.1 kJ/kg-K. Estimate (i) the theoretical COP (ii) the rate of circulation of the refrigerant in kg/min and (iii) compressor work if the actual COP is 80% of theoretical. The properties of refrigerant are:

Sat. temp. (°C)	h_f (kJ/kg)	h_g (kJ/kg)	s_f (kJ/kg-K)
20	275.0	1462	1.043
– 20	89.6	1419	0.368

Take heat of fusion of ice as 335 kJ/kg.

Solution: Refer to Figure 4.32.

Given: Mass of ice produced = 30 tons at 0°C/day; $T_1 = T_4 = -20°C = 253$ K; $T_3 = 20°C = 293$ K; $c_{pv} = 1.1$ kJ/kg-K; Actual COP = 0.8 × ideal COP; $h_L = 335$ kJ/kg;

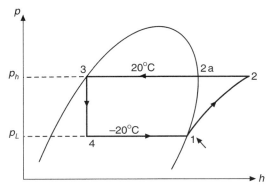

Figure 4.32 Flow diagram on p-h chart—Example 4.38.

For isentropic compression process 1–2, $s_1 = s_2$

$$s_{g1} = s_{g2} + c_{pv} \ln\left(\frac{T_{sup}}{T_{sat}}\right)$$

$$s_{f1} + \frac{h_{fg1}}{T_1} = s_{f2} + \frac{h_{fg2}}{T_{sat\,2}} + c_{pv} \ln\left(\frac{T_{sup}}{T_{sat}}\right)$$

$$0.368 + \frac{1419 - 89.6}{253} = 1.043 + \frac{1462 - 275}{293} + 1.1 \ln\left(\frac{T_2}{293}\right)$$

$T_2 = 473.66$ K; $h_1 = h_{g1} = 1419$ kJ/kg

$h_2 = h_{g2} + c_{pv}(T_{sup} - T_{sat})$
    $= 1462 + 1.1(473.66 - 293) = 1660.73$ kJ/kg

$h_3 = h_4 = h_{f3} = 275$ kJ/kg

(i) Refrigerating effect, $q_L = h_1 - h_4 = 1419 - 275 = 1144$ kJ/kg

Work done, $w = h_2 - h_1 = 1660.73 - 1419 = 241.73$ kJ/kg

$$\text{Ideal COP} = \frac{q_L}{w} = \frac{1144}{241.73} = 4.73 \quad \textbf{Ans.}$$

(ii) Heat required to be removed (actual refrigerating effect) = $mc_p\Delta T + mh_L$

$$= m(c_p \Delta T + h_L) = \frac{30 \times 1000}{24 \times 60}(4.187 \times (0) + 335)$$

$$= 6979.17 \text{ kJ/min}$$

Rate of circulation of refrigerant = Mass flow rate

$$= \frac{\text{Heat to be removed}}{\text{Specific refrigerating effect}}$$

$$= \frac{6979.17}{1144} = 6.1 \text{ kg/min} = 0.101 \text{ kg/s} \quad \textbf{Ans.}$$

(iii) Actual COP = $0.8 \times$ ideal COP = $0.8 \times 4.73 = 3.784$

$$\text{Actual COP} = \frac{\text{Actual refrigerating effect}}{\text{Compressor work}}$$

$$\text{Compressor work} = \frac{6979.17}{3.784} = 1844 \text{ kJ/min} = 30.73 \text{ kW} \quad \textbf{Ans.}$$

EXAMPLE 4.39 A cold storage of 120 TR capacity operates between the temperature limits of –30°C and +30°C. The refrigerant at the suction of compressor is dry and saturated and at the exit of condenser it is subcooled by 10°C. The actual COP is 70% of the theoretical. Find the following:

(i) Actual and theoretical COP
(ii) Mass of refrigerant circulated in kg/s

(iii) Compressor power
(iv) Piston diameter if $L/D = 1.2$, speed is 300 rpm and volumetric efficiency is 85%.
Take $c_{pv} = 0.55$ kJ/kg-K and $c_{pr} = 1.19$ kJ/kg-K
The properties of refrigerant are:

T_s (°C)	P (bar)	V_g	h_f	h_g	s_f	s_g
−30	1.6	0.136	166.2	393	0.87	1.803
+30	12	0.020	236.8	415	1.13	1.712

Solution: Refer to Figure 4.33 for p–h diagram.

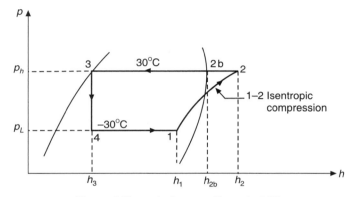

Figure 4.33 p–h diagram—Example 4.39.

Given: Cooling capacity of plant = 120 TR = 120 × 3.517 = 422.04 kW;
$T_1 = T_4 = -30°C = 243$ K; $T_3 = 30°C = 303$ K; Degree of subcool = 10°C;

Actual COP = 0.7 × ideal COP; $\dfrac{L}{D} = 1.2$; $N = 300$ rpm;

Volumetric efficiency = 0.55; $c_{pv} = 0.55$ kJ/kg-K; $c_{pl} = 1.19$ kJ/kg-K

For isentropic compression process 1–2; $s_1 = s_2$

$$s_{g1} = s_{g2} + c_{pv} \ln\left(\dfrac{T_{\text{sup}}}{T_{\text{sat}}}\right)$$

$$1.803 = 1.712 + 0.55 \ln\left(\dfrac{T_2}{303}\right)$$

$T_2 = 357.52$ K, $h_1 = h_{g1} = 393$ kJ/kg
$h_2 = h_{g2} + c_{pv}(T_{\text{sup}} - T_{\text{sat}}) = 415 + 0.55(357.52 - 303) = 444.98$ kJ/kg
$h_3 = h_4 = h_{f3} - c_{pl}(\text{degree of subcool})$
$= 236.8 - 1.19 \times 10$
$h_3 = h_4 = 224.1$ kJ/kg

(i) Specific refrigerating effect = $h_1 - h_4$ = 393 − 224.1 = 168.9 kJ/kg

Work done, $w = h_2 - h_1$ = 444.98 − 393 = 51.98 kJ/kg

$$\text{Ideal COP} = \frac{\text{Refrigerating effect}}{\text{Work done}} = \frac{168.9}{51.98} = 3.25 \quad \textbf{Ans.}$$

Actual COP = 0.7 × ideal COP = 0.7 × 3.25 = 2.275 **Ans.**

(ii) $\text{Mass of refrigerant} = \dfrac{\text{Capacity of plant in kJ/s}}{\text{Refrigerating effect in kJ/kg}} = \dfrac{422.04}{168.9} = 2.499$ kg/s **Ans.**

(iii) $\text{Compressor power} = \dfrac{\text{Capacity of plant in kJ/s}}{\text{Actual COP}} = \dfrac{422.04}{2.275} = 185.5$ kW **Ans.**

(iv) Actual volume = $m \times x_1 v_{g1}$ = 2.499 × 1 × 0.136 = 0.34 m³/s

Actual volume = Theoretical volume × Volumetric efficiency

Theoretical volume = $\dfrac{0.34}{0.85}$ = 0.39984 m³/s

Theoretical volume = $\dfrac{\pi}{4} D^2 L N$

$$0.399 = \frac{\pi}{4} D^3 \left(\frac{300}{60}\right) \quad (\because D = L)$$

D = 0.467 m, L = 0.467 m **Ans.**

EXERCISES

1. Explain the standard vapour compression cycle on p–h and T–s diagrams.
2. Explain the functions of various components of a vapour compression refrigeration system and represent the cycle on T–s and p–h diagrams. Assume that the refrigerant is in superheated condition at entry to the compressor and subcooled before expansion.
3. What do you understand by "wet compression"? Explain any one method to avoid it.
4. Discuss the working of vapour compression with the help of a schematic diagram. Also represent the cycle on T–s and p–h diagrams.
5. Discuss the relative merits of vapour compression and air refrigeration cycles.
6. Discuss the functions of various components used in a vapour compression refrigeration system.
7. Discuss the effect of variable suction and discharge pressures and subcooling on the performance of vapour compression system.
8. A liquid-to-vapour heat exchanger has negligible thermodynamic advantages. However, its use is justified. Comment.

9. "In the vapour compression refrigeration cycle, expansion is carried out using a throttling device and not an expansion engine." Comment.
10. Write short notes on the following:
    (a) Actual vapour compression cycle.
    (b) Effect of increase in condenser pressure and decrease in evaporator pressure on the performance of vapour compression cycle.
    (c) Effect of subcooling on the performance of vapour compression cycle.
11. Explain how (any one) the following methods can be used to improve the performance of a simple saturation cycle.
    (i) Subcooling of liquid refrigerant by liquid refrigerant.
    (ii) Use of liquid vapour heat exchanger.
12. Describe with T–s diagram, the ideal refrigeration cycle, theoretical and practical vapour compression cycles.

NUMERICALS

1. A Carnot refrigeration cycle absorbs heat at –20°C and rejects it at 40°C.
    (a) Calculate the coefficient of performance of this refrigeration cycle.
    (b) If the cycle is absorbing 850 kJ/min at –20°C, how much work in kJ per minute is required.
    (c) If the same cycle operates on the simple vapour compression cycle with R134a refrigerant, what are the deviations in COP and work input per kilogram of mass circulated?
2. A Carnot heat pump cycle absorbs heat at –20°C and rejects to a space maintained at 45°C.
    (a) Calculate the coefficient of performance of this heat pump cycle.
    (b) If the cycle rejects heat at the rate of 850 kJ/min at 45°C temperature, how much work in kJ/min is required?
    (c) If the above cycle operates in a simple vapour compression cycle using R134a refrigerant, what are the deviations in COP and work supplied?
3. 35°C produces a cooling effect of 8 tons with R134a refrigerant. Calculate:
    (a) Compressor work assuming isentropic compression and the refrigerant inlet to the compressor is dry saturated vapour.
    (b) If the adiabatic compression efficiency is 85%, what would be the compressor work?
    (c) If the adiabatic efficiency is 85%, and the mechanical efficiency is 90%, then what would be the compressor power?
4. A deep freezer is working between the evaporating temperature –20°C and condensing temperature. Determine the refrigerating effect per kilogram if the evaporator temperature is at –5°C and condenser temperature is 35°C. This plant operates in a saturated vapour compression cycle with R134a. If the evaporator temperature is lowered to –20°C keeping the condenser temperature constant, what would be the percentage reduction in refrigerating effect per kilogram?

5. In Bangalore, an R22 refrigeration plant produces a refrigerating effect of 5 tons at evaporating temperature of –5°C and condensing temperature of 30°C. Now the refrigeration plant is shifted to Kharagpur in West Bengal where the atmospheric temperature is about 40°C. Now evaluate the performance of this plant in respect of (a) cooling effect, (b) power requirement of the compressor, and (c) COP. Assume the simple compression cycle.

6. In Germany, a refrigerating unit with ammonia as a refrigerant operating in the simple compression cycle produces 20 tons of ice per day. Using the available water at 15°C, the evaporator temperature is –5°C. The atmospheric temperature of Germany may be about 15°C. Now the company decides to sell this unit to a party in Rajasthan (India). Considering the climatic conditions of Rajasthan, evaluate the performance of the plant, so as to decide its cooling capacity, compressor power and COP.

7. A simple saturation R134a compression system has a high absolute pressure of 0.86 MPa and a low pressure of 0.2 MPa. Find (a) the refrigerating effect per kilogram, (b) power required per kilogram, and (c) COP. If due to unavoidable reasons, the absolute pressure of the condenser increases to 1.13 MPa, what would be its effect on the above parameters?

8. A refrigeration system operates with R12 and produces 1 ton refrigerating effect at the evaporator and condenser temperatures of –5°C and 40°C, respectively. If the liquid is subcooled from 40°C to 30°C in the condenser, then calculate for the simple compression cycle and subcooled cycle the following: (a) Refrigerating effect, (b) Mass flow rate, (c) Volume of vapour handled by the compressor, (d) Power requirement, and (e) COP.
[**Ans.:** For saturated cycle, q_L = 110.8 kJ/kg, m = 0.0317 kg/s, rate of volume flow = 2.059 × 10^{-3} m^3/s, W = 0.73 kW/TR, COP = 4.8]

9. A refrigeration plant working with R134a produces 10 ton refrigerating effect and the evaporator and condenser temperatures are maintained at –5°C and 40°C, respectively. If the liquid is subcooled from 40°C to 30°C in the condenser, calculate for the simple compression cycle and for the subcooled cycle the following: (a) Refrigerating effect, (b) Mass flow rate, (c) Volume of vapour handled by the compressor, (d) Power requirement, and (e) COP.
[**Ans.:** For saturated cycle, q_L = 139 kJ/kg, m = 0.25 kg/s, v = 0.75 m^3/h, W = 7.6 kW, COP = 4.6. For subcooled cycle, q_L = 154 kJ/kg, m = 0.228 kg/s, v = 6.3 m^3/h, W = 6.8 kW, COP = 5.09].

10. An R134a refrigerant, simple vapour compression system includes a liquid suction heat exchanger in the system. The heat exchanger cools the saturated liquid coming from the condenser from 35°C to 20°C with vapour, from the evaporator at –2°C. The compressions are isentropic in both the following cases (a) and (b). (a) Calculate the coefficient of performance of the system without the heat exchanger but with the condensing temperature of 35°C and an evaporating temperature of –20°C. (b) Calculate the coefficient of performance of the system with the heat exchanger.

11. An ammonia refrigerator produces 20 tons of ice per day at 0°C. The condensation and evaporation occur at 20°C and –20°C, respectively. The temperature of vapour at the end of compression is 50°C and there is no undercooling mass flow rate of ammonia. Draw

the T–s diagram. Take the latent heat of fusion of ice = 335 kJ/kg. Specific heat of superheated vapour = 2.8 kJ/kg-K. Use the properties of ammonia as listed below.

Temp. °C	Enthalpy (kJ/kg)		Entropy (kJ/kg-K)	
	h_f	h_g	s_f	s_g
20	274.98	1461.58	1.04341	5.0919
–20	89.72	1419.05	0.3682	5.6204

[**Ans.:** mass flow rate = 0.0968 kg/s, COP (Th) = 9.04, COP (Act) = 6.329]

12. A refrigeration system uses R12 as refrigerant. The condenser temperature is 50°C and the evaporator temperature is 0°C. The refrigeration capacity is 7 tons. Assume the simple saturation cycle and determine with the help of p–h diagram:
    (i) The refrigerant flow rate
    (ii) COP
    (iii) The heat rejected in the condenser.

[**Ans.:** m = 1.02 kg/s, COP = 4.25, heat rejected = 128.5 kW]

13. An ice-plant produces 100 kg of ice per hour at 0°C from water at 25°C. Find the capacity of the plant in TR assuming latent heat of fusion of ice as 335 kJ/kg.

[**Ans.:** Capacity = 3.489 TR]

14. A refrigerating unit having capacity of 5 TR works on ammonia. The temperature limits are 40°C and 0°C. Find the power required to drive the plant and COP of the unit. Assume the refrigerant to be dry saturated at the entry to the compressor, and compression to be isentropic. There is no undercooling. What will be the effect on these values if the suction vapour is superheated by 10°C? Justify your answer by calculating power and COP.

[**Ans.:** Without superheating: Power = 3.4 kW, COP = 5.144;
With superheating: Power = 3.538 kW, COP = 4.94]

15. A food storage plant having capacity of 12 TR and using R12 as refrigerant maintains a temperature of –10°C in the food compartment. The condenser temperature is 30°C and vapour leaving the evaporator is superheated by 20°C. Find:
    (i) The theoretical COP.
    (ii) The power required if the actual COP is 75% of the theoretical COP.
    You can use the following properties of R12:

Temp. (°C)	Pressure (bar)	Liquid		Vapour		Vapour superheated by 20°C (h in kJ/kg)
		v_f	h_f	v_g	h_g	
–10	2.1912	0.7	26.9	0.0767	183.2	195.7
30	7.45	0.77	64.6	0.0235	199.6	214.3

[**Ans.:** Theoretical COP = 6.376, Actual COP = 4.7822, Power = 8.78 kW]

Chapter 5

Refrigerants

5.1 INTRODUCTION

A refrigerant is a medium of heat transfer, which absorbs heat by evaporating at low temperature and gives out heat by condensing at high temperature and pressure conditions. A refrigerant must meet certain chemical, physical and thermodynamic properties so that it can be used suitably in the vapour compression cycle. Also, a good refrigerant should have safe working properties apart from having economical aspects. There is not one refrigerant that is universally suitable for all applications. This is because of the wide differences in the conditions and requirements of the various applications.

Refrigerants can be divided into two groups: Primary and Secondary. Primary refrigerants absorb heat and generate coolness by changing their phase from liquid to vapour. Secondary refrigerants absorb heat from the bodies or space to be cooled and further transfer the same to the primary refrigerants.

The earliest primary refrigerants were water and diethyl ether; later came ammonia as early as in 1875. At about the same time, sulphur dioxide, carbon dioxide and certain hydrocarbons were also introduced as refrigerants. Carbon dioxide was disliked because of the high pressure needed and sulphur dioxide because of its odour and toxicity. The hydrocarbons are extremely inflammable due to which they did not gain much importance. Anhydrous ammonia was preferred and even today large industrial refrigeration units opt for this refrigerant.

In small refrigeration units, methyl chloride was used for a short time, but during the Second World War, it was replaced by other halocarbon compounds, commercially popular as freon, frigen, arcton, etc. Such refrigerants are fluorinated hydrocarbons derived from methane, ethane, etc. as bases. These contain fluorine and chlorine atoms along with hydrocarbons. At present these are generally referred to as chlorofluorocarbon compounds and abbreviated CFC. The term CFC covers one or more of these categories. The first CFC developed for commercial use was R12 with Dupont's registered trademarks—Arcton–12, Diaflan–12, etc.

5.2 CLASSIFICATION OF REFRIGERANTS

The classification of refrigerants is shown in Figure 5.1.

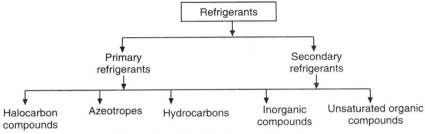

Figure 5.1 Classification of refrigerants.

5.2.1 Primary Refrigerants

Primary refrigerants are working mediums in a refrigeration system which are directly used as the carrier of heat. Primary refrigerants absorb heat and lower the temperature by changing their phase from liquid to vapour. These include halocarbon compounds, azeotropes, hydrocarbons, inorganic compounds and unsaturated organic compounds.

Halocarbon compounds

Halocarbon compounds are commercially known as freon, frigen, arcton, etc. Such refrigerants are fluorinated hydrocarbons, derived from methane, ethane, etc. as bases as discussed in Section 5.1.

Halocarbon refrigerants have a wide range of boiling points at one atmospheric pressure, hence they satisfy the requirements of domestic, commercial and industrial applications. Also, the presence of fluorine in the compounds makes them non-toxic and renders desirable physical properties. Halocarbons are ozone-unfriendly and the use of these should be totally stopped. Some of the important halocarbons are as follows:

R11–Trichloromonofluoromethane (CCl_3F)

R12–Dichlorodifluoromethane (CCl_2F_2)

R13–Monochlorotrifluoromethane ($CClF_3$)

R21–Dichloromonofluoromethane ($CHCl_2F$)

R22–Monochlorodifluoromethane $(CHClF)_2$

R40–Methylchloride (CH_3Cl)

R100–Ethylchloride (C_2H_5Cl)

R114–Tetrafluorodichloroethane ($Cl_2F_4Cl_2$)

R113–Trichlorotrifluoroethane ($C_2F_3Cl_3$)

R152–Difluoroethane ($C_2H_5F_2$)

There is also another group of refrigerants in which the chlorine atoms are replaced by bromine atoms. One such example is R13B1. The refrigerants of this group are known as *halons*.

For large industrial refrigeration units, ammonia is used while R11 with centrifugal compressors and CFCs are used for domestic refrigeration. Presently, CFCs are being replaced by new refrigerants due to ODP (Ozone Depletion Potential) problems. This is in accordance with the Montreal Protocol.

Azeotropes

A group of refrigerants whose code starts with digit 5 (for example, R502) is an azeotropic group of refrigerants. This is a mixture of different refrigerants that cannot be separated under pressure and temperature and they have fixed thermodynamic properties, e.g. R500 is the mixture of 73.8% R12 and 26.2% R152. Similarly, R502 is an azeotropic mixture of R22 and R115, R503 of R13 and R23 and R501 of R22 and R12.

Hydrocarbons

These refrigerants have the desired thermodynamics properties, but are highly inflammable. A few such hydrocarbons used as refrigerant are—R50 Methane (CH_4), R170 Ethane (C_2H_6), R290 Propane (C_3H_8), and R600 Butane (C_4H_{10}).

Inorganic compounds

These are—R717 Ammonia (NH_3), R718 Water (H_2O), R729 Air, R744 Carbon dioxide (CO_2), and R764 Sulphur dioxide (SO_2).

Unsaturated organic compounds

These are hydrocarbons with ethylene and propylene base. Examples include R1120 Trichloroethylene ($C_2H_4Cl_3$), R1130 Dichloroethylene ($C_2H_4Cl_2$), R1150 Ethylene (C_2H_4), and R1270 Propylene (C_3H_6).

5.2.2 Secondary Refrigerants

In a large refrigeration plant the secondary refrigerant is used as a cooling medium and it absorbs heat from refrigerated space or bodies to be cooled and further transfers the heat to the primary refrigerant in the evaporator. Examples of secondary refrigerants are water, brines, glycols, etc.

5.3 DESIGNATION OF REFRIGERANTS

The refrigerants are designated by the letter 'R' followed by certain numerals. For example, R12 should be really designated as R012. Three numerals of this code follow the formula—first the number of carbon atoms minus 1, second the number of hydrogen atoms plus 1, and finally, the number of fluorine atoms in that order. *Any unfilled positions are normally occupied by chlorine atoms, but occasionally by bromine* (Br) atoms. Thus, the chemical formula is $C_mH_nF_pCl_o$ in which $n + p + o = 2m + 2$.

The chemical formula for any compound derived from a hydrocarbon is denoted by

$$R - (m - 1)(n + 1)(p)$$

where
      m = number of carbon atoms
      n = number of hydrogen atoms
      p = number of fluorine atoms.

Example: R22 or R022

$m - 1 = 0$ therefore, $m = 1$ which means that the number of carbon atoms is 1

$n + 1 = 2$ therefore, $n = 1$ which means that the number of hydrogen atoms is 1

$p = 2$ It means that the number of flourine atoms is 2.

Let us write the chemical formula for R22 as $CHClF_2$.

The brominated refrigerants are denoted using an additional letter 'B' and a number following B to indicate as to how many chlorine atoms are replaced by bromine atoms, e.g. R13B1 is derived from R13 by replacing 1 chlorine atom by 1 bromine atom and its chemical formula is CF_3Br.

The inorganic refrigerants are designated according to their molecular weight added to 700, e.g. H_2O has molecular weight of 18, therefore, it is designated as R718. Ammonia (NH_3) has a molecular weight of 17, hence it is designated as R717.

Unsaturated organic compounds are designated by putting an additional number 1 before the refrigerant. For example, R120 = $C_2H_4Cl_3$, i.e. trichloroethylene, but is designated as R1120.

5.4 DESIRABLE PROPERTIES OF A GOOD REFRIGERANT

The important desirable properties of a good refrigerant are its thermodynamic, chemical and physical properties.

5.4.1 Thermodynamic Properties of Refrigerants

Latent heat of evaporation

High latent heat of evaporation (h_{fg}) per unit mass is desirable. A small quantity of refrigerant produces a large cooling effect which requires less power for its circulation. The latent heats of evaporation for common refrigerants are given in Table 5.1 for ready reference. The latent heat of evaporation for ammonia is maximum while that of refrigerant R502 is minimum among those listed in Table 5.1.

Table 5.1 Latent heat of evaporation (h_{fg}) of some common refrigerants at $-10°C$

Refrigerant	Latent heat of evaporation (h_{fg}) (kJ/kg)	Refrigerant	Latent heat of evaporation (h_{fg}) (kJ/kg)
R717	1296.4	R22	213.12
R502	153.45	R12	157.28
R11	157.28	R134a	204.85
R23	182.38		

The table shows that ammonia possesses the highest latent heat of evaporation h_{fg} compared with others. Refrigerants that have zero ODP and low GWP are harmless to the environment. Table 5.2 compares refrigerants based on these values.

Table 5.2 ODP and GWP of some common refrigerants

Group	Refrigerant	ODP	GWP	
CFCs	R11	1	1	
	R12	1	2.8–3.4	
	R12B1	3		
	R13	0.45	6.0	
	R13B1	10–13	0.8	
	R113	0.8–0.9	1.2–2.0	
	R114	0.6–0.8	3.5–4.5	
	R115	0.3–0.5	5.0–9.0	
	R500	0.74–0.87	3.38–4.87	
	R502	0.17–0.29	2.66–4.78	
HCFCs	R21	0.04		
	R22	0.04–0.06	0.2–0.35	
	R123	0.013–0.022	0.17–0.025	
	R124	0.016–0.024	0.08–0.12	
	R141b	0.07–0.11	0.08–0.11	
	R142b	0.055	0.3–0.5	
HFCs	R125	0	0.42–0.84	
	R134a	0	0.23–0.29	
	R143a	0	0.6–0.95	
	R152a	0	0.024–0.033	
	R32	0	0.14	
Mixtures: substitutes for R502				
Glide 0.4	HP 62	0	0.94	R 404A–143A/125/134A
Glide 0.4	FX 70	0	0.4	R 404A–143A/125/134A
Glide 0	AZ 50	0	0.42	143A/125
	Ammonia	0	0	

Environment protection has to be given the first priority. The CFC refrigerants have large ODP potential and their use must be stopped.

Boiling point (BP)

The refrigerants should have low boiling point temperature at atmospheric pressure. During operation, low temperature can be achieved with positive pressure of the refrigerant in evaporator coil. If the pressure inside the evaporator coil is less than the atmospheric pressure, then moisture may enter into the system through leakages. This would lead to the blockage of refrigerant flow at the expansion valve. It would not help in detecting leakage of refrigerants. Negative pressure in evaporator is undesirable. The following table shows the BP of some common refrigerants.

Refrigerant	R11	R12	R21	R22	R30	R744
BP(°C)	23.77	−29.9	8.9	−40.8	39.8	−74.6

Freezing point

Freezing temperatures should be well below the evaporator temperatures to avoid freezing of the refrigerant.

Evaporating pressure

Evaporating pressure P_L should be just above the atmospheric pressure. If too low, it would result in a large volume of suction vapour. If high, the condenser pressure and the overall pressure will be greater. A positive evaporator pressure is required in order to eliminate the possibility of the entry of air and moisture into the system. The normal boiling point of the refrigerant should be lower than the refrigeration temperature. A small difference between suction pressure and discharge pressure or a low pressure ratio P_C/P_L (the ratio of condenser pressure to evaporator pressure) is desirable for good efficiency. The pressure ratios for different refrigerants for different combinations of evaporator and condenser temperatures are shown in Table 5.3.

Table 5.3 Compression ratio p_C/p_L for different refrigerants for different combinations of evaporator and condenser temperatures

Refrigerant	−10/35°C	−10/45°C	−20/35°C	−20/45°C
R717	4.64	6.1	7.07	9.36
R22	3.57	4.5	5.24	7.04
R502	3.71	4.9	5.1	6.48
R12	3.71	4.95	5.6	7.03
R11	5.8	8.0	9.8	13.4
R134a	4.42	5.8	6.65	8.72
R123	6.5	0.9	13.0	18.15

Condensing pressure

It depends on the temperature of the condenser cooling medium which is usually water or atmospheric air. A refrigerant should have low condensing pressure to avoid robust constructions and to reduce the tendency of leakages.

Critical temperature and pressure

It is the temperature above which there will not be any phase change. The critical temperature of the refrigerant should be well above the condensing temperature for easy condensation of vapour.

Index of compression process

The work of compression per unit mass depends on the isentropic index γ ($\gamma = c_p/c_v$). The smaller the index, the smaller will be the work of compression. The compression process can follow different curves on the *p-v* diagram based on isothermal, adiabatic and polytropic processes. The isothermal compression process ($pv = c$) requires the least work but it is not a practical process. (Refer to Chapter 1.)

5.4.2 Chemical Properties of Refrigerants

The refrigerants have to satisfy certain chemical properties for safe operation of refrigeration systems. Some of the required chemical properties for a refrigerant are as follows:

Flammability: The refrigerant should not be inflammable in the presence of air or lubricating oils. CO_2 is the most inert while ethane and butane are high inflammable. Hydrocarbons such as methane, ethane, propane, butane, etc. are highly explosive and inflammable. Fluorocarbons are non-explosive and non-flammable too but have very high ODP. HFCs (R152, R143), which are alternative to the CFCs, are also flammable.

Toxicity: A refrigerant should not be toxic, poisonous or injurious. Ammonia is toxic as well as flammable, so it is not used in domestic refrigerators. R123 considered to be the replacement of R11 is found to be toxic. The freon group of refrigerants are safe from the toxic point of view. Refrigerants like methyl chloride, NH_3 and SO_2 are toxic in nature.

Action of refrigerant with water: The presence of moisture is very critical in refrigeration systems. If it is carried along with the refrigerant, it will become ice at the expansion valve or capillary tube and result into choking of the valve. This is called *moisture choking*. It is avoided by proper dehydration of the system or by placing silica gel in the pipeline before the expansion valve. Ammonia is soluble in water. Its contact with water should therefore be avoided. R134a is also very hygroscopic in nature.

Corrosiveness: The refrigerant should not have any adverse effect on the materials used in the equipment either by itself or in the presence of moisture and lubricating oils.

Leak detection: The refrigerant should possess minimum tendency to leak and should lend itself to easy methods of leak detection.

Flash point: Liquids with flash point below 21.2°C are regarded as highly inflammable.

Miscibility with oil: Since refrigerants come in contact with lubricating oils in the compressor, it is desirable that the refrigerant does not harm the properties of lubrication. This would be discussed again.

Stability: Refrigerants should be capable of withstanding the effect of pressure and temperature inside the system without decomposition. It should not react with metals or lubricants inside the system. For example, NH_3 reacts with copper and brass in presence of moisture. *Therefore,* NH_3 *as refrigerant should not be used in systems employing cuprous metals.* NH_3 is safe to work with iron and steel. Though the freon group of refrigerants do not react with steel, copper, brass, zinc and aluminium, it is corrosive with aluminium alloys having more than 2% of magnesium. The freon group should not be used in systems employing natural rubber for packings and gaskets but these refrigerants do not react with synthetic rubber.

The oil required for lubrication of the compressor is contained in the crankcase of the compressor where it is subject to contact with the refrigerant. Therefore, the refrigerant must be chemically and physically stable at the operating conditions of the oil.

5.4.3 Physical Properties of Refrigerants

Specific volume: The mass of refrigerant being pushed in every delivery stroke of the piston depends upon its density. If the specific volume is low, comparatively more mass can be pushed. It means greater cooling capacity can be achieved with a small flow rate of refrigerant. The specific volume of the refrigerant should be low in the vapour state so that the plant components can be small in size. It reduces the cost of the plant.

Viscosity: It is defined as a measure of fluid friction or as a measure of the resistance that a fluid offers to flow. Low viscosity is a desirable property to keep down the pressure losses in refrigeration piping and to obtain better heat transfer rates in evaporator and condenser.

Thermal conductivity: The refrigerant must have high thermal conductivity in liquid phase as well as vapour phase.

Dielectric strength: The refrigerant should have high electric strength, otherwise electrical short circuits are likely to occur in hermetically sealed compressors where the refrigerant vapour cools the motor winding. R12 has high dielectric strength while NH_3 has poor dielectric strength.

Handling and maintenance: The refrigerant should be safe to handle without any protective measures. The system should not pose any serious maintenance problems.

Cost and availability: The refrigerant should be available at low cost, for example, NH_3 is cheap but R12 and R22 are more expensive.

5.5 PROPERTIES OF AN IDEAL REFRIGERANT

The properties of an ideal refrigerant can be outlined as given below:

(a) It should have zero ODP and zero GWP.
(b) It should be non-toxic and non-flammable.
(c) It should be non-corrosive with most common materials.
(d) It should have high latent heat.
(e) It should have high critical pressure and temperature.
(f) It should have low condensing pressure, and the evaporating pressure should be slightly above the atmospheric pressure.
(g) It should not be miscible with lubricating oil.
(h) It should be easily available at cheap rate.
(i) The leak detection should be easy.

5.6 PROPERTIES OF IMPORTANT REFRIGERANTS

The desirable properties of refrigerants enumerated above have been exemplified by considering below certain relevant characteristics of common refrigerants.

Ammonia (NH_3) (R717)

The refrigerant NH_3 has been used for decades in industrial and larger refrigeration plants. It has no ozone depletion potential and no direct greenhouse effect. The efficiency is at least as good as that with R22, in some areas even more favourable. NH_3 has indeed very positive features, which can also be mainly exploited in large refrigeration plants.

Combining free nitrogen and hydrogen under high pressure and temperature in the presence of a catalyst produces ammonia. The process most commonly used is the Haber-Bosch method.

The nitrogen component is inert in the combustion reaction and accounts for the limited flammability of ammonia. Ammonia's lower limit of flammability and low heat of combustion substantially reduce the explosion hazards. Ammonia/air mixtures are flammable by spark ignition at concentration of 16–27% by volume in air. However, oil carried by ammonia lowers this level considerably, so that a figure of 4% by volume in air is considered the practical safe limit to prevent explosion.

Anhydrous ammonia is the most economically abundant and efficient heat transfer medium available for industrial refrigeration. Its pungent odour provides an extremely effective self-alarming and leak detecting characteristics.

Ammonia will be the best substitute for R22 due to very comparable pressure levels as shown in Figure 5.2.

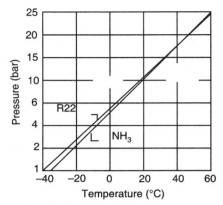

Figure 5.2 NH$_3$/R22—Comparison of pressure levels.

Some of the advantages and disadvantages of refrigerant ammonia are given below.

(i) Zero ODP and low GWP values.
(ii) Extraordinarily high enthalpy difference and as a result a relatively small circulating mass flow (approximately 13 to 15% compared to R22). The feature is favourable in large plants only.
(iii) High latent heat of vaporization (h_{fg}) per unit mass.
(iv) High COP.
(v) Good heat transfer characteristics caused by the high thermal conductivity, high latent heat, low viscosity and low liquid density compared with CFCs and HCFCs.
(vi) Low molecular mass $M_w = 17.03$ provides another advantage of low-pressure drop through compressor valves and ports. But a disadvantage with NH$_3$ is the high isentropic exponent (for ammonia $\gamma = 1.31$ and for R22, $\gamma = 1.18$ on an average, though it is a function of temperature and pressure conditions) that results in a discharge temperature which is even higher than that of R22.
(vii) Higher compression ratio and discharge temperature compared with R134a, R22, R12, R502 as shown in Figures 5.3 and 5.4, respectively.
(viii) Toxicity and unpleasant smell, for which more rigorous standards are applied to vessels, pipework, etc. make it a costlier unit for better safety.

Ammonia is corrosive to the materials containing copper. The pipelines must therefore be made of steel.

Carbon dioxide (R744)

Carbon dioxide is a safe refrigerant. Its boiling point is –78.5°C and the freezing point is –56.6°C at atmospheric pressure. It is odourless, non-toxic, non-flammable, non-explosive and non-corrosive.

At atmospheric pressure, it sublimates from solid state to liquid state producing cold at temperature of –78.5°C. It does not wet the surface in contact during sublimation. Therefore, it is used to preserve the eye-bowl (inner portion) in its state during eye operation.

R11 (CCl_3F) or Trichloromonofluoromethane

This refrigerant belongs to the CFC group. Its boiling point at atmospheric pressure is 23.8°C, which is very high, and so the systems working with this refrigerant to lower the temperature to sub-zero have to work at vacuum pressures. Due to its large specific volume, it is suitable for use with centrifugal compressors. It is used in air conditioning plants, since it is non-corrosive, non-toxic and non-flammable.

R12 (CCl_2F_2) or Dichlorodifluoromethane

This refrigerant is also from the CFC group. It is non-toxic, non-flammable, non-explosive and has low specific volume. This colourless and odourless refrigerant is safe and stable, and the system can work under positive pressures since its boiling point is –29.8°C at atmospheric pressure.

This refrigerant was earlier used in domestic refrigerators. But since its ozone depletion potential is the highest amongst the remaining refrigerants, its production and use has been totally stopped. It has been replaced by refrigerant R134a.

R22 ($CHClF_2$) or Monochlorodifluoromethane

It boils at –40.7°C at atmospheric pressure, and so it is suitable for low-temperature refrigeration. Its specific volume is very low compared with other available refrigerants for air conditioning applications and so it needs a very small compressor and is used in package type air conditioners. R22 is used for commercial air conditioning plants, frozen food plants and low temperature applications.

R500 (CCl_2F_2/CH_3CHF_2)

This refrigerant is an azeotropic mixture of 73.8% of R12 and 26.2% of R152 by weight. The mixture has thermodynamic properties entirely different from those of its constituents. It produces a higher refrigerating effect compared to R12 per kg of refrigerant. The boiling temperature of this refrigerant is –33.3°C at atmospheric pressure. Therefore, it is suitable for large capacity refrigerating plants.

Its main drawback is that it is soluble in water, and so every care has to be taken to avoid its contact with moisture.

5.7 SELECTION OF A REFRIGERANT

As stated earlier, a refrigerant is a fluid medium which absorbs heat from one area and rejects it to another, such as outdoors, usually through evaporation and condensation.

A refrigerant must satisfy many requirements, some of which do not directly relate to its ability to transfer heat. Chemical stability, safety, toxicity, cost, availability, compatibility with compressor lubricants and materials with which the equipment is constructed are also important. The environmental consequences of a refrigerant that leaks from a system must also be considered. Refrigerant selection also involves compromise between the conflicting desirable thermodynamic properties.

The pressure of evaporator corresponding to temperature of the refrigerant to be maintained is selected such that it renders positive pressures to avoid any leakages inwards. Therefore, the following six properties must be studied before choosing a right refrigerant:

- Working temperatures of the refrigerant.
- Evaporator and condenser pressures needed and the pressure ratio.
- Oil miscibility.

- High latent heat of vaporization and low specific volume.
- Toxicity, flammability, explosiveness and corrosiveness.
- Space requirements.

With all such information, a refrigerant, which is the best choice for a particular application, may be a total failure in other applications.

Irrespective of the refrigerant used, care must be taken to adapt the design of the installation to the characteristics of the refrigerant. Considering the characteristics of available refrigerants, ammonia is the best choice for large industrial units. Unfortunately, too many CFC compound installations are being built as extended commercial installations depending on their suitability. Table 5.4 shows the number of chlorofluorocarbon compounds used as refrigerants. However, R11, R22 and R502 are of most interest in industrial refrigeration.

Table 5.4 Refrigerants and their fields of applications

1.	R717 Ammonia (NH_3) Boiling point $-33.35°C$	Used with open type reciprocating, rotary and screw compressors, in cold storages, ice plants, for food preservation, etc.
2.	R718 Water (H_2O) Boiling point $100°C$	Used in the steam-ejector system only for air conditioning. Used as a secondary refrigerant.
3.	R11 (CCl_3F) Trichlorofluoromethane. Boiling point $+23.8°C$ [R123 in place of R11]	Used with centrifugal compressors in large capacity central air conditioning plants. Used as a brine at temperatures as low as $-100°C$.
4.	R12 (CCl_2F_2) Dichlorodifluoromethane Boiling point $-29.8°C$ (R134a in place of R12)	Widely used in domestic appliances. It is also used in commercial and industrial cooling installations, for example, frozen food display cabinets, cooling fountains, refrigerated trucks, railway wagons or containers. Used with all types of compressors, hermetic and open, piston, rotary, centrifugal and screw.
5.	R13 ($CClF_3$) Chlorotrifluoromethane, Boiling point $-81.4°C$	For temperatures down to $-80°C$ in cascade freezing installations using R22 or R502 in the high temperature stage.
6.	R22 Chlorodifluoromethane ($CHClF_2$). Boiling point $-40.8°C$	Used as refrigerant in domestic, commercial and industrial air conditioning; commercial and industrial refrigeration including cold storages and food processing with reciprocating and often with screw compressors.
7.	R113 (CCl_2FCClF_2) Trichlorotrifluoroethane Boiling point $+47.6°C$	Used with centrifugal compressor for cooling water or brine for commercial or industrial applications.
8.	R114 ($CClF_2CClF_2$) Dichlorotetrafluoroethane Boiling point $+3.6°C$	Used with multistage centrifugal compressors for air conditioning at high temperatures and in aircraft.
9.	R502 Azeotropic mixture of 48.8% by mass of R22 and 51.2% by mass of R115 Boiling point $-45.6°C$	It is alternative to ammonia. It is most widely used for frozen food display cabinets, also in freezing chambers and cold stores at temperatures of the order of $-35°C$.
10.	R503 Azeotropic mixture of R13 and R23 Boiling point $-88.7°C$	Used in the second stage of a cascade system with two or three stages in scientific research, test chambers, metal hardening, pharmaceutical and other processes. A temperature of about $-85°C$ is achievable using it in the lower stage of cascade.

5.8 NEW REFRIGERANTS

New refrigerants like R134a which replaces R12, and R123 which replaces R11 are already commercially available. A comparison of the properties of these refrigerants is given in Table 5.5.

Table 5.5 Comparison of physical properties of two new and two traditional refrigerants

Refrigerant properties	R12	R134a	R11	R123
Chemical formula	CCl_2F_2	CF_3-CH_2F	CCl_3F	CF_3-CHCl_2
Molecular mass	120.9	102.0	137.4	152.9
Boiling point (°C)	−29.8	−26.5	23.8	27.9
Heat of evaporation at 0°C (kJ/kg)	152.4	197.3	190.6	181.8
Critical temperature (°C)	112	101	198	184
Critical pressure (bar)	41.1	40.7	44.1	37.3
Critical sp. volume (dm^3/kg)	1.792	1.952	1.804	1.804
Flammability	NIL	NIL	NIL	NIL

R134a has physical and thermodynamic properties similar to R12, so it is the best substitute presently available. A comparison of the thermodynamic properties of the two refrigerants is made on the basis of the four main refrigerating variables:

1. Compression ratio p_C/p_L
2. Specific refrigerating effect q_L
3. Compressor discharge temperature T_2
4. Coefficient of performance COP

These four variables for the following operating conditions are investigated:

(i) Evaporating temperature, T_L, −25°C
(ii) Condensing temperature, T_C, 40°C
(iii) Suction vapour superheating, T_{sup}, 10 K
(iv) Liquid subcooling, T_{sub}, 5 K

Figures 5.3 to 5.6 show the cycle calculations in graphical form. The compression ratio p_C/p_L is higher in the case of R134a cycles, whereas the specific refrigerating effect increasingly approaches that of R12 as the evaporating temperature rises. The coefficient of performance is roughly comparable. The compressor discharge temperature under all operating conditions investigated was generally a little lower than that of R12. While the compressor discharge temperature in R134a cycles is lower, R134a can also be observed to have a volumetric refrigerating effect, on an average, up to 10% lower at low evaporating temperatures. The results obtained by compressor manufacturers have shown that disadvantages in terms of energy consumption are not likely if the compressor is adapted accordingly and the circuit is optimized.

R134a is similar to R12 in that it is compatible with all metals and metal alloys commonly used in machine and equipment manufacture. For more details, one is advised to refer to the manufacturer's specifications.

As expected, R134a, which contains no chlorine, is readily compatible with sealing materials. The recommended sealing materials for R134a are CR (chloroprene rubber), NBR (acrylonitrile butadiene rubber), etc.

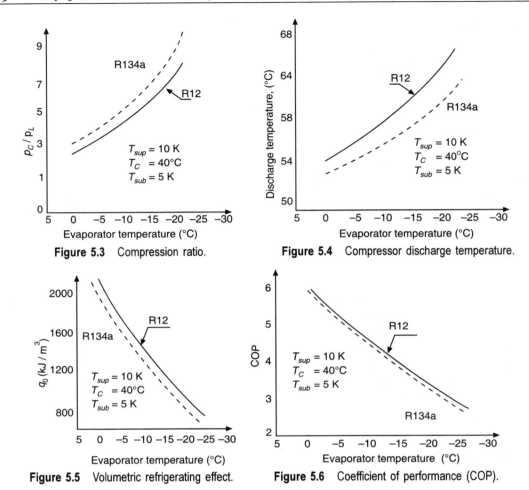

Figure 5.3 Compression ratio.

Figure 5.4 Compressor discharge temperature.

Figure 5.5 Volumetric refrigerating effect.

Figure 5.6 Coefficient of performance (COP).

R134a is almost completely immiscible with conventional mineral oil-based refrigeration oils and also with a number of synthetic lubricants such as alkylbenzenes. The synthetic compound oils are found miscible with R134a and others include polyalkylene glycols (PAGs). Polyglycols have given remarkable results in test-ring trials with open type refrigerant compressors and their extremely hygroscopic behaviour makes them difficult to handle.

As such, easter-based synthetic lubricants have been the first choice for some time now, as they are less hygroscopic in nature compared with PAGs.

R134a does not form flammable mixtures with air under normal conditions, i.e. atmospheric pressure. Flammable mixtures would form at pressures above atmospheric if the air components in the mixture exceed 60%. In leakage checks or pressure tests these refrigerants must never be used together with air or oxygen. R134a is also safe toxicologically like R12. The results indicate that the product can be used safely in domestic, commercial and industrial refrigeration.

The CFC and HCFC refrigerants will be totally phased out in the near future because of their greenhouse effect rather than the ODP. The ODP and GWP values for different refrigerants are given in Table 5.6.

The refrigerant and refrigerating systems in the future will mainly be evaluated on the basis of their energy efficiency, apart from their considerations like safety and toxicity.

Table 5.6 shows the list of refrigerants that are likely to replace various old CFC refrigerants.

Table 5.6 Traditional CFC refrigerants and their probable new substitute refrigerants

Original refrigerant	New substitute
R11	R123, R123a, R141b, R152
R114	R124a, R142b
R502	R125
R12	R134a, R152a, R22, R134a
R500	R22, R125, R134a

5.9 SECONDARY REFRIGERANTS

Secondary refrigerants are generally used in large refrigeration units. In such units, brines are chilled in the evaporator and pumped to the source of the refrigeration load. The use of brines may be advisable in order to keep coils and pipes containing a toxic refrigerant out of the occupied spaces. Secondary refrigerants eliminate long refrigerant lines and reduce the pressure drops.

In the past, only water/salt solutions were used as secondary refrigerants. Salts such as sodium chloride and calcium chloride are best used but they are very corrosive. Nowadays, glycol solutions are generally used. For special applications there are also dichloromethane (CH_2Cl_2), trichloroethylene (C_2HCl_3), alcohol solutions and acetone.

The desirable properties of secondary refrigerants are:

1. They should have low freezing point.
2. They should have good stability.
3. They should have low vapour pressure.
4. They should be non-flammable and non-toxic.
5. They should have high heat transfer coefficients.
6. They should have high specific heat. This property is advantageous in respect of the refrigerant being circulated and disadvantageous to heat transfer.

Dissolving a salt in water forms brines. The phase diagram for a brine solution is shown in Figure 5.7. The addition of salt in water decreases its freezing point. For a particular composition of salt and water, the freezing point is the lowest, however the composition is known as *eutectic composition* and the temperature as *eutectic temperature*. Figure 5.7 also shows the possible phases and mixtures existing at various concentrations and temperatures.

Consider a certain mixture of salt and water at point A. On cooling the mixture progressively to a point, say B, ice crystals begin to form. As a result the solution will become richer in salt content. Further cooling to point C results in slush which is a mixture of ice and brine. The mixture at C consists of ice at C_2 and solution at C_1. At point D the solution will have the eutectic composition corresponding to point D_1. Further cooling below point D will result into a complete solid phase.

With this discussion of phase diagram of brine solution, let us collect the information regarding the commercially used brines in industrial refrigeration installations.

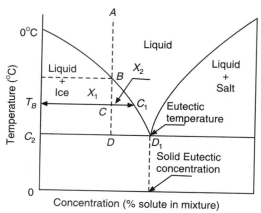

Figure 5.7 Phase diagram of a brine solution.

Table 5.7 shows the information like freezing point and solution concentration. This information is collected for –50°C of the brine solution.

Table 5.7 Comparison of brine concentrations at –50°C

Product	Freezing point (°C)	Product in 100 litres water (kg)	Concentration in solution
Calcium chloride	–51.6	62	29.9% by weight
Ethylene glycol	–51.2	134	55% by volume
Propylene glycol	–49.4	150	59% by volume

Table 5.8 shows the comparison of brine properties such as specific heat, density, etc. at 20°C while Table 5.9 indicates the comparison of viscosity of brine solutions.

Table 5.8 Comparison of brine properties at 20°C

Solution	Sp. heat (kJ/kg-K)	Density (kg/m³)	Brine quantity needed to accumulate	
			Mass (kg)	Volume (m³)
Calcium chloride	2.77	1.289	360	278
Ethylene glycol	3.28	1.070	305	286
Propylene glycol	3.38	1.042	295	283

Table 5.9 Comparison of viscosities of brine solutions

Product	Mass needed in 100 kg of water (kg)	Freezing point (°C)	Dynamic viscosity at 20°C (MPa)
Calcium chloride	62	–51.6	3.5
Ethylene glycol-I	62	–20.5	2.6
Ethylene glycol-II	134	–51.2	4.4
Propylene glycol-I	62	–18	3.9
Propylene glycol-II	150	–50	8.6

Calcium chloride is very corrosive but its intensity can be reduced by additives. It is cheap, non-toxic and has very high specific density and low viscosity. Ethylene glycol is slightly corrosive and toxic and its cost lies between the other two.

Table 5.7 shows that calcium chloride needs the smallest concentration to reach a certain low freezing point compared with ethylene glycol and propylene glycol. However, as far as the thermal energy accumulation is concerned, calcium chloride requires the highest concentration.

Advantages of secondary refrigerants

(a) Different rooms of a building can be cooled up to different temperatures by adjusting the flow rates of secondary refrigerants.
(b) These can be easily handled.
(c) Their control is easy.
(d) The size of piping required is reduced since these are in liquid form.
(e) Absolute safety in air conditioning installations as leaks of refrigerant vapours cannot reach up to the space air conditioned.

The important secondary refrigerants are water, brines, and water glycol mixture.

5.9.1 Substances Used as Secondary Refrigerants

Various substances used as secondary refrigerants include:

(a) Water if the working temperatures are above 3°C.
(b) Brines, solutions of any salt in water.
(c) Ethylene glycol and propylene glycol in solution with water.
(d) Halocarbon compounds like R11, R12, R30.

A brief description of each of the above secondary refrigerants is given below.

Water

Chilled water is used as a secondary refrigerant in air conditioning systems where the working temperatures are normally above 3°C. It is the cheapest, the least corrosive and has a high specific heat.

Brines

Brines are aqueous solutions of sodium chloride (NaCl) or calcium chloride ($CaCl_2$) in water. These are employed at temperatures below the freezing point of water, i.e. below 0°C. NaCl (common salt) brine is used in cooling the meat and fish, in ice plants and for industrial applications.

The percentage mass of $CaCl_2$ or NaCl dissolved in water varies the freezing point of water. For example, freezing point of 25% $CaCl_2$ solution is (–28.3°C) and that of 29.87% $CaCl_2$ solution is (–55°C). The freezing point of 20% NaCl solution is –16.5°C.

Ethylene glycol and propylene glycol

These mix readily with water and make colourless and odourless solutions. These have the capacity to lower the freezing temperatures. For this reason, these are used as antifreeze mixtures for I.C. engine cooling systems. These solutions become corrosive after some use, hence corrosive treatment is necessary. Joints are prone to leakages, hence must be made leakproof when these solutions are used. The freezing point of 40% ethylene glycol solution in water by mass is as low as –24.4°C.

Halocarbon compounds

Primary refrigerants like R11, R12, and R30 have low freezing points, hence can be used as secondary refrigerants for low temperatures.

5.10 TOXICITY AND SAFE HANDLING OF REFRIGERANTS

Many people consider that only ammonia is toxic in nature and not other refrigerants. Even if it is true, refrigerants require proper handling.

All CFCs and new chlorine-free refrigerants (for example, R134a, R123, etc.) have the following properties:

(i) They are non-flammable and do not form explosive mixtures with air. Partly halogenated, i.e. hydrogen-containing refrigerants such as R134a or R22 may form ignitable mixtures with air at relatively high pressures.

(ii) They are virtually odourless. Their faint odour can be identified only if they are present in relatively high concentrations in inhaled air.

(iii) They are not harmful to health if used properly. To ensure safe and proper use, certain precautions must be observed. These precautions are enlisted as follows.

1. If refrigerant containers are opened, the content can escape in the form of liquid or vapour. The higher the pressure in the container, the more violently this will occur.

2. Wear protective goggles to prevent the refrigerant from getting into the eyes as it may possibly cause severe low temperature burns.

3. Do not allow liquid refrigerant to fall on the skin. The refrigerant takes the heat required for evaporation from the surroundings, even if this happens to be the skin. Very low temperatures may occur as a result and cause localized frostbite.

4. Do not inhale refrigerant vapours in high concentrations.

5. All CFCs and new chlorine-free refrigerants start to evaporate or vaporize once the container is opened. The vapours mix with the surrounding air. If inhaled, they can have detrimental effects on human beings. The effect and the speed with which these set in depend on two factors—the grade of the refrigerant and the concentration of the refrigerant vapour in the inhaled air.

6. With all grades of refrigerants except R11 and R113, the main risk is that they may displace the oxygen needed for respiration.

7. Smoking is prohibited. Refrigerants may be decomposed by heat if they come in contact with a lit cigarette. The resulting substances are toxic and must not be inhaled.

8. During the transportation of refrigerants, utmost care must be taken so as to avoid mishappenings. The rules regarding safe transportation must be observed strictly.

9. Avoid contamination with water. Water can enter into the refrigeration circuit during repairs. It accelerates the catalytic action of all metal parts and makes the refrigerant and oil acidic. Acidic refrigerant and oil form salts when in contact with metal, the resulting salts then oxidize and breakdown the oil.

5.11 OIL AND REFRIGERANT RELATIONSHIP

The oil required for lubrication of the compressor is contained in the crankcase of the compressor where it comes in contact with the refrigerant. Therefore, the refrigerant must be chemically and physically stable with oil at the operating conditions. When contaminants such as air and moisture are present in the system in an appreciable amount, chemical reactions often take place between these contaminants, oil and refrigerant, resulting into the decomposition of oil. A high discharge temperature greatly accelerates the decomposition of oil and formation of carbonaceous deposits on discharge valves, pistons and in the compressor head and discharge line. This condition is aggravated by the use of poorly refined lubricating oils containing a high percentage of unsaturated hydrocarbons, the latter being very unstable chemically.

With all this information, one can say that use the oil recommended by the manufacturer of the compressor, never mix two oils with different characteristics. Further, make sure that the oil stays in the compressor. Oil causes problems in heat exchangers, where the oil film decreases the value of heat transfer coefficient. In refrigerant liquid pumps, it creates cavitation and in automatic controls it acts as a lubricant.

Oil miscibility (ability of the refrigerant to be dissolved into the oil and vice versa) is one, which characterizes the refrigerant oil relationship. As per miscibility, refrigerants may be divided into three groups:

1. Miscible
2. Immiscible
3. Partially miscible.

Refrigerants such as R134a, R123 and R11, are miscible with oil in all proportions under all conditions and do not pose problems. Oil that reaches the evaporator is returned to the compressor along with the refrigerant. Any oil droplets separated may also be returned to the compressor by gravity or by the high velocity return gas. The diameter of the suction tube is so designed that the velocity of the returning gas is sufficient enough to carry away the oil sticking to the walls of the tube.

The refrigerant such as ammonia, which is not miscible with oil at all, also does not pose any problems. In such a case, an oil separator is installed in the discharge line but nearer to the compressor and the separated oil is continuously returned to the crankcase of the compressor.

The refrigerants, which are miscible under conditions normally found in the condensers, but remain separate from oil normally found in the evaporator are called partially miscible refrigerants and they may pose some problems.

5.12 LUBRICATING OILS

Lubricating oils come into contact with, and often mix with the refrigerant. Therefore, the oil used to lubricate refrigeration compressors should be specially prepared and should meet certain requirements of lubrication of the compressors. For the selection of proper lubricating oils, the following properties are considered:

1. Chemical stability
2. Specific gravity
3. Pour and/or floc point

4. Flash point
5. Aniline point
6. Viscosity
7. Viscosity index
8. Neutralization number

Chemical stability

This property in fact has more weightage as far as the hermetic compressors are concerned, in which changing the oil is not practical. The lubricating oil charged once remains in these units throughout the life of the unit (ten or more years). The hermetic compressors working in refrigeration units having air-cooled condensers normally encounter a high discharge temperature. The chemical stability of oil at such a temperature is an important factor. The oil should be able to resist its decomposition.

Specific gravity

The specific gravity helps in determining the type of oil. The alkyl benzene oils, for example, are lighter and the glycol based oils are heavier than mineral oils.

A mineral oil with increased paraffin content will have a lower specific gravity than a napthene based oil.

Pour and/or floc point

The pour point of oil is the lowest temperature at which the oil can flow or pour. This property of oil is dependent on the content of wax but not on the viscosity of oil. The higher content of wax will lead to a higher point. This is especially an important consideration in selecting an oil for low temperature systems. Pour point is of special interest for oils used in R717 installations, since an oil with a low pour point is easier to drain from the low pressure side of the installation. In R717 units with an evaporating temperature lower than $-40°C$, it is recommended that an additional oil separator be installed in the discharge line between the compressor and the condenser. By this way, the oil transfer to the low pressure side of the unit is reduced. The floc point of the oil is the temperature at which wax will start to separate from R12 mixed with 10% oil. When the mixture is cooled to a certain temperature, it becomes foggy with wax particles separated out. The floc point is significant where oil and refrigerant are miscible, for example, CFCs and new refrigerants like R134a and R123.

A low floc point indicates that the oil contains low wax and is suitable for units operating especially below $-18°C$. Separation of wax may cause plugging of the expansion valve of the circuit.

Flash point

The temperature at which the oil vapour from an open heated vessel is ignited by a flame is the flash point. The flash point is useful in determining the applicability of the oil at high temperatures. An oil with a high flash point is possibly better for the separation of oil from the discharge gas in the oil separator.

Aniline point

This is the temperature at which the oil combines with pure aniline to form a homogeneous mixture. The aniline point is an indication of the quantity of unsaturated hydrocarbons which are present in the oil. The aniline point is important in assessing the likelihood of the oil attacking the rubber materials, which may come in contact with the oil. Since most of the refrigeration compressor oils have comparatively low aniline point, they cause neoprene rubber to swell without dissolving the rubber or otherwise affecting the scaling ability. This might require replacement of rubber gaskets, such as O-rings, during dismantling.

Viscosity

It is defined as a measure of fluid friction or as a measure of the resistance that a fluid offers to flow. The oil with low viscosity flows more readily than one having high viscosity. To provide adequate lubrication for the compressor, the viscosity of the oil should remain within certain limits. If the viscosity of oil is too low, the oil will not have sufficient thickness to form a protective layer between the rubbing surfaces. However, if the viscosity of the oil is too high, the oil will not have sufficient fluidity to penetrate between the rubbing surfaces.

The choice of the viscosity depends strongly on the design of the compressor and on the refrigerant used. Oil with the lowest possible viscosity must be chosen, provided that it enables a satisfactory tightness between the rubbing surfaces in combination with the refrigerant over the entire working temperature range. It is therefore, necessary to know the viscosity of the oil/refrigerant mixture at different working temperatures.

The viscosity of oil is usually expressed in Saybolt Seconds Universal (SSU). SSU is an index of the time in seconds required for a given quantity of oil at a particular temperature, usually 38°C, to flow by gravity, from a reservoir into a flask through a capillary of definite (specified) diameter. It is also expressed in other units like N-s/m^2 or m^2/s. Manufacturers of lubricating oils usually supply the information regarding the properties of refrigerant. It is also necessary to know the variation in viscosity with temperature while selecting the oil.

Viscosity index

This indicates the dependence of the viscosity on temperature. The larger the value, the lesser is the dependence. The viscosity index is used in monographs for comparison of oils.

Neutralization number

This is a measure of the degree of acidity of the oil, determined by titrating potassium hydroxide against the oil sample. The refrigeration oils are highly refined without any additives, therefore have a low neutralization number. This number indicates the content of acid particles in the oil, such as oxidizing products, which cause decomposition of the oil.

5.13 EFFECT OF CFC ON OZONE DEPLETION AND GLOBAL WARMING

The effects of CFC on ozone depletion and global warming are explained in the following sub-sections.

5.13.1 Ozone Depletion

There is a layer of ozone (O_3) gas in stratosphere up to 50 km from the earth's surface (see Figure 5.8). The ozone layer filters the ultraviolet radiations from sun entering the earth atmosphere and allows only the beneficial heat and light rays to reach the earth's surface.

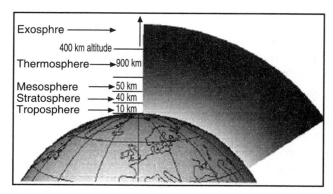

Figure 5.8 The various layers of gases in the earth's atmosphere.

Therefore, the ozone layer protects us against the harmful effects of such radiations like skin cancer.

1. Ozone is a form of oxygen. Each ozone molecule is made of three oxygen atoms.
2. Ozone, unlike oxygen, is poisonous and an increase in its concentration at ground level is not desirable.
3. In the stratosphere, ozone occurs naturally and blocks out the sun's UV-B rays, so ozone is a life-saver.

Some of the gases including CFC given out by different processes on earth reach this layer and react with ozone gas to form different compounds. This causes the depletion of ozone at particular places which are called ozone holes (see Figure 5.9).

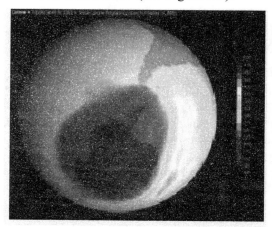

Fig 5.9 Ozone hole.

Formation and destruction of ozone

The major reasons for ozone depletion are chlorine and bromine. The chemical reactions during the formation and destruction of ozone are consolidated here in a stepwise manner.

1. Ozone in the stratosphere undergoes photodissociation by absorbing ultraviolet radiation,
$$O_3 \rightarrow O_2 + O$$
2. The free oxygen atom further reacts with another molecule of ozone resulting in the destruction of ozone,
$$O + O_3 \rightarrow O_2 + O_2$$
3. Ultraviolet radiation strikes a CFC molecule causing a chlorine atom to break away,
$$CFCl_3 \rightarrow CFCl_2 + Cl$$
4. The chlorine atom collides with an ozone molecule and scales an oxygen atom to form chlorine monoxide and leaves a molecule of ordinary oxygen,
$$Cl + O_3 \rightarrow ClO + O_2$$
$$ClO + O \rightarrow Cl + O_2$$

We find from the above that the ozone (O_3) gas is depleted to O_2 gas.

Bromine from oceans is also important and may contribute to as much as 25% of ozone reduction.

Ozone depletion potential (ODP)

ODP is the ratio of the impact of a chemical compared to the impact of a similar mass of CFC–11. Thus, ODP of CFC is 1.0.

- ODPs of other CFCs and HCFCs fall in the range of 0.01 to 1.0.
- The ODP of carbon tetrachloride is 1.2.
- HFCs have zero ODP as they do not contain chlorine.

Consequences of ozone depletion

1. Every Cl atom can destroy up to 100,000 ozone molecules.
2. 1% loss of ozone leads to 2% increase in UV-B radiations.
3. It leads to an increase in UV-B radiations at the ground level.
4. Continuous exposure to UV-B radiations affects humans, animals and plants and can lead to skin cancer, blindness, etc.
5. Increased global warming.
6. Partially dampens the greenhouse effect, resulting in increased global warming.

5.13.2 Global Warming

Global warming is another phenomenon that is a major cause of concern worldwide.

Global warming potential (GWP)

Global warming potential is a number that refers to the amount of global warming caused by a substance. GWP is the ratio of the warming caused by a substance to the warming caused by a similar mass of carbon dioxide (CO_2). Thus, the GWP of CO_2 is 1.0. The GWPs of a few refrigerants are given in Table 5.10.

Table 5.10 GWP of refrigerants

Substance	GWP per kg
CO_2	1
R12	3639
R22	560
Water	Zero
R502	5244
R134a	388

Table 5.10 enables us to assess the potential damage to the environment by the direct emission of refrigerants.

The thermal conductivity of these refrigerants in the gaseous state is very small—of the order of 0.007 W/m-K. When emitted in the atmosphere and due to low density, these refrigerants have a tendency to go to stratosphere and form a layer, which acts as an insulator. The intensity of solar radiations entering the atmosphere is very high, i.e. these have a wavelength of the order of 2–3 μm. The reflected solar radiations from the earth surface would have a tendency to escape to the space. The insulation layer formed at the stratosphere because of the said refrigerant gases will pose an added resistance to stop the reflected radiations. Hence, heat is trapped inside the atmosphere to keep the plants green. These gases are referred to as Greenhouse Gases (GHGs). The effect of these Greenhouse gases towards global warming is known as *Greenhouse* effect. This tends to increase the temperature on the earth's surface. The global warming can have frightening effects in the long run.

It should be noted that increased carbon dioxide (CO_2) released in atmosphere due to industrial processes, thermal power plants, use of vehicles, etc. is also a source of global warming. However, its effect is very small compared to that of refrigerants on per kg basis.

5.14 MONTREAL PROTOCOL

In 1974, scientists noticed that chlorofluorocarbons (CFCs) like R11 and R12 used extensively in refrigeration industry have ozone depletion potential (ODP).

CFCs are stable compounds in normal atmosphere and have long life. As explained above, when these rise to ozone layer, they decompose due to sunlight. The chlorine atom reacts with ozone and causes its depletion.

After a few preliminary discussions, in 1987 the representatives of many countries in Montreal, Canada signed an agreement regarding the phasing out of the ozone-unfriendly refrigerants. This agreement is known as *Montreal Protocol*.

This Montreal Protocol has set a timetable for phasing out of CFC gases. The salient points of the timetable are given in Table 5.11.

The subsequent conferences called for monitoring the progress of the timetable discussion also took place about the use of another group of refrigerants namely the Hydroflurocarbons (HCFCs), for example, R22.

Ozone depletion factors (ODFs) of some CFCs and HCFCs as per the Environmental Protection Agency (EPA) are given in Table 5.12.

Table 5.11 Timetable for phasing out of CFCs by developing countries

July 1996	CFCs phased out
	HCFCs frozen at 1989 levels of HCFC + 2.8% of 1989 consumption of CFCs (base level)
1 January 2004	HCFCs reduced by 35% below base levels
1 January 2005	Phase out 50% consumption
1 January 2007	Phase out 85% consumption
1 January 2010	HCFCs to be reduced by 65%
1 January 2015	HCFCs to be reduced by 90%
1 January 2020	HCFCs to be phased out, allowing for a service tail of up to 0.5% until 2030 for existing refrigeration and air conditioning equipment

Table 5.12 Ozone depletion factor for some CFCs and HCFCs as per EPA

	Refrigerant	ODF	Refrigerant	ODF
CFC	R11	1.0	HCFC R22	0.05
			HCFC R123	0.02
CFC	R12	1.0	HFC R134a	0.00
CFC	R113	0.8	HFC R152a	0.00
CFC	R114	1.0	R290 (Propane)	0.00
CFC	R502	0.3	R717 (Ammonia)	0.00

5.15 ALTERNATIVES TO CFC REFRIGERANTS

Some of the hydrochloroflurocarbons (HCFCs), hydroflurocarbons (HFCs) and hydrocarbons (HCs) have been tried out as alternatives to CFCs and are given below:

HCFC–R123 (A replacement for CFC R11)

Many properties of the refrigerant R123 are very close to the properties of R11. The R123 refrigerant is suitable as replacement for CFC, R11 which is used extensively nowadays in centrifugal compressor applications due to its high specific volume. This is close to "drop-in" refrigerant in these units. Drop-in refrigerant means charging the existing refrigeration system working on R11 with R123. The performance would be almost same as obtained with R11 under identical operating conditions. It has relatively high toxicity but less inflammability compared to R11. The use of R123 in place of R11 is only an interim solution.

HFC–R245 (A replacement for CFC R11)

It would take some time before HFC-R245 is commercialised as it is still under investigation. This refrigerant has the potential to be a long-term substitute for R11. It gives reasonable performance with the same compressor size.

5.15.1 Certain Mixtures of Refrigerants as Replacement for HCFC R22

R407

R407C is a blend of R32/R125/R134a (23/25/52% by mass). This refrigerant is extensively used in Europe and Japan where HCFC-22 is already phased out in many types of equipment, and is an excellent performance match to HCFC-22 allowing it to be used in HCFC-22 air conditioning and refrigeration systems without significant equipment changes. The major drawback is that it has approximately 5–7°C temperature glide. In most cases this deteriorates the performance of the system. The literature shows that the system with R407C gives lower cooling capacity in the range 5–10% and lower COP in the range 5–7%.

R410A

A mixture (50/50% by weight) of HFC-32 and HFC-134 is also in race for substitution of R22.

The refrigerant being considered in new air conditioning equipment from nearly every major equipment manufacturer in the world is R410A. This refrigerant offers significant performance advantages over HCFC-22. R410A is classified as an HFC refrigerant and has no undesirable ozone depleting properties. Comprised of an equal blend of HFC-32 and HFC-125, R410A is a near azeotropic refrigerant mixture that exhibits a negligible temperature glide of any 400-series refrigerant. R410A offers improved pressure-drop characteristics that allow for improved system efficiency over HCFC22. Due to its lowest temperature glide, R410A allows easy handling and near-zero change in composition in cases of major leaks. Because of its incompatibility with mineral oil, R410A systems use synthetic polyol ester (POE) oil. The operating pressures of R410A are approximately 50% higher than those of HCFC-22.

R134a

R134a ($C_2H_2F_4$) is a pure substance and an ideal refrigerant to replace HCFC-22. It has no temperature glide. However, it has much lower pressure and lower capacity and hence requires larger compressors, heat exchangers and other components. Because of its higher critical temperature, it may be used where a high condensing temperature is required.

5.15.2 Substitutes for CFC, R12 Refrigerant

R12 is the commonly used refrigerant in domestic refrigeration industry using hermetically sealed compressors for refrigerators, water coolers, etc. due to its excellent thermodynamic properties, stability and good cooling characteristics for its motor windings.

(a) **HFC-134a as replacement for R12:** Though discussed in detail in Section 5.8, the properties of HFC-134a are summarized here again.

R134a has almost the similar thermodynamic properties as R12 with some deviations:
   (i) It has higher volume flow rates.
   (ii) It is more inflammable.
   (iii) It has more tendency to leakage.
   (iv) It has higher specific heat, therefore, produces subcooled vapour in condenser and has higher refrigerating effect.
   (v) It reacts with hydrocarbon oil lubricants, therefore, the polyol ester eolin (POE) oil is needed to be used as lubricant.

(vi) This refrigerant and the oil being highly hygroscopic, the servicing of plants needs more care.

(b) **HCFC R22 as substitute to R12:** Refrigerant R22 can be used as a substitute to R12 for plants having low pressure ratios not exceeding 9. It has negligible ODP and low GWP. However, **R22 is not a "drop-in substitute"** for domestic refrigerators for the following reasons:

   (i) It has a high value of pressure ratio for the range of operating temperatures; therefore, the high discharge temperature will cause overheating of motor windings.
   (ii) Its specific volume is low at suction temperatures, thus requires a larger size of the compressor to achieve the same refrigerating effect under identical operating conditions. It requires more compressor power for the same duty.
   (iii) Increased compressor work causes the increased heat rejection in the condensers.
   (iv) Needs a large size capillary tube since it has higher discharge pressures. Alternatively, if the same capillary tube is used, the pressure after expansion will be higher requiring higher evaporator temperature, increased mass flow rate and compressor work.

(c) **Hydrocarbons propane (R290) and isobutane (R600a) as substitute to R12:** Propane hydrocarbons have zero ODP and almost zero GWP. The main drawback is their extreme inflammability and explosiveness. Propane and isobutane are being tried as alternatives to R12 refrigerant in the long run. Some of the characteristics associated with the use of these hydrocarbons are:

   (i) The density of liquid hydrocarbons is low. This improves safety in case of leaks since in a domestic refrigerator the charge of R12 is about 100 g whereas the charge of hydrocarbon will only be about 40 g.
   (ii) They need larger compressors and better safety standards due to their flammability and explosiveness.
   (iii) They need large size capillary tubes.

   However, the advantage of using these refrigerants will be that they give lower discharge temperatures, hence the compressor runs cool.

   Therefore, R290 is not a drop-in substitute refrigerant to R12.

   However, R290 has a good prospect as a substitute to R12 with smaller compressor design and with increased size of capillary tube. But, the system will be required to meet the better safety standards.

(d) **Carbon dioxide:** Carbon dioxide is non-toxic, non-poisonous and non-flammable but causes death due to suffocation when its concentration in air is higher than that recommended by the medical practitioners. It is chemically stable under all pressure and temperature conditions involved in a refrigeration system. It is compatible with all the metals and is immiscible in oil.

   It has many disadvantages. The kW required per ton of refrigeration using CO_2 as a refrigerant is nearly twice than any other commonly used refrigerant. CO_2 does not exist in liquid state at atmospheric pressure. The boiling temperature at normal pressure of 1 bar is –78.6°C. It requires a low temperature coolant in the condenser because of its low critical temperature (31°C).

5.16 KYOTO PROTOCOL AND TEWI

Some of the replacements mentioned above are already in extensive use in the USA. The HFCs, which seemed to be the solution, however, have run into rough weather due to their Global Warming Potential (GWP).

In 1997, in Kyoto, Japan, 150 countries initiated an agreement to reduce the emanation of Greenhouse Gases (GHGs).

Though carbon dioxide is the main irritant, HFCs also form a part of five other such gases. This agreement is now known as *Kyoto Protocol*. It has not yet been signed by participating countries though implementation by developed countries is already in progress.

The Protocol proposes reduction of GHGs by 5.2% below 1990 levels, by 2008 to 2012. Table 5.13 gives ODP and GWP values of some of the common refrigerants.

Table 5.13 ODPs and GWPs of some refrigerants

Refrigerant	Ozone depletion potential (ODP)	Global warming potential (GWP)
CFC-12	1.0	8100
HCFC-22	0.05	1500
HFC-134a	0	1300
R290 (propane)	0	20
R717 (ammonia)	0	< 1

The effect of this protocol is not only on refrigerants alone but also on the energy efficiency of the system as a whole. The efficiency of energy use by a system would be assessed on the basis of carbon dioxide released to produce that energy.

The refrigerants are now being analyzed along with the system by a new concept called *Total Equivalent Warming Impact* (TEWI). This concept considers global warming effect due to refrigerant emissions and release of carbon dioxide due to energy use over the lifetime of the system.

So the emphasis in the future will be on search for an alternative refrigerant which would include energy efficiency of the system along with its intrinsic GWP. Extensive research is going on to find out such an alternative.

EXERCISES

1. What is a refrigerant? Give the classification of refrigerants.
2. How are primary refrigerants classified?
3. How are primary refrigerants designated?
4. Discuss the desirable properties of a good refrigerant.
5. Enumerate the required properties of an ideal refrigerant.

6. Write a short note on the selection of a refrigerant for a refrigeration system.
7. What are secondary refrigerants? Why are these used in refrigeration and air conditioning industry? Name a few substances which are commonly used as secondary refrigerants.
8. Write short notes on:
   (a) Desirable properties of refrigerants.
   (b) Need of alternatives to CFC refrigerants.
   (c) Classification of commercial refrigerants.
   (d) Thermodynamic properties of freons.
   (e) Primary vs secondary refrigerants.
   (f) Azeotropes.
9. What are the alternatives to CFC-12 in domestic refrigerators? Compare at least three alternative refrigerants.
10. Briefly explain "Greenhouse effect and global warming".
11. What is an azeotrope? Give examples to indicate its importance.
12. Explain the ozone depletion and global warming issues. Discuss why CFC refrigerants need to be phased out. List the alternative refrigerants.
13. What happens if R22 or R290 is used as a drop-in substitute in R12 refrigerators?
14. Explain the role of refrigerant mixtures as alternative refrigerant and its classification.
15. Briefly explain the provisions of Montreal Protocol and Kyoto Protocol.
16. Write a short note on: 'Total Equivalent Warming Impact' (TEWI).
17. Discuss the suitability of carbon dioxide as a refrigerant.

Chapter 6

Multipressure Systems

6.1 INTRODUCTION

In the chapter on vapour compression systems, we learnt the working principles of the simple vapour compression system. We also understood that the parameters such as evaporating temperature, condensing temperature, suction gas superheat and subcooling of liquid at the outlet of condenser will alter the performance of a vapour compression refrigeration unit. Let us have a look on the pressure ratios invoved in the following examples:

1. A vapour compression system operating at evaporating temperature of $-5°C$ and condensing temperature of $40°C$ uses the R134a refrigerant. Now for this system, our interest is to find the pressure ratio involved. Referring to the saturated property table of R134a, we can get the evaporating pressure at $-5°C$ as 2.4 bar and the condensing pressure at $35°C$ as 10.1 bar. It has been investigated that due to pressure losses on suction side, the evaporator pressure is about 10% less than that it could be corresponding to evaporator temperature. Therefore, the actual suction pressure is about 2.16 bar. Similarly, the pressure on the discharge side of compressor is about 20% higher than that of the theoretical condensing pressure due to spring-loaded discharge valves. Therefore, the actual pressure in condenser is about 12.12 bar. Hence the actual pressure ratio is about 5.6.
2. Now, let the vapour compression system with R134a operate in the temperature range from $-20°C$ to $40°C$. The evaporating pressure at $-20°C$ is 1.3 bar and the condensing pressure at $40°C$ is 10.1 bar. The theoretical pressure ratio is about 7.76. Considering 10% pressure loss on suction side and 20% pressure loss on discharge side, the pressure ratio works out to be 10.3.

If we observe the above two examples carefully, we can see that the pressure ratio rises from 5.6 to 10.3 just because of decreased evaporating temperature from $-5°C$ to $-20°C$. Such situations frequently occur in the practical field of refrigeration for employing the refrigeration unit for varied applications.

At this stage, our interest is whether to achieve such a high pressure ratio in single-stage compression or multistage compression and why?

For higher pressure ratios, the following points are observed:

(a) As the pressure ratio increases the volumetric efficiency of a reciprocating compressor tends to zero. At higher pressure ratios, the vapour trapped in the clearance volume at the end of compression stroke is at high temperature and pressure conditions. During suction stroke, this vapour in the clearance volume expands, occupying greater space of the cylinder which otherwise could have been filled with fresh vapour. Therefore, there is a reduction in the fresh vapour entering the cylinder in every cycle. Hence the actual volume of vapour handled by the compressor decreases, and ultimately a decrease in the volumetric efficiency results.

As the volume of vapour handled by the compressor decreases, the mass flow rate in circuit decreases, and the cooling capacity of the compressor not only decreases but also tends towards zero value.

(b) The slope of entropy lines on p–h diagram decreases for the isentropics, away from the saturated vapour line. Multistaging is done to bring this isentropic line nearer to the saturated vapour line. The multistage compression is necessary to reduce the power consumption and to increase the cooling capacity at high condensing temperature (and/or low evaporator temperature applications).

(c) The temperature of the discharge gas from the compressor is higher due to overheating of the compressor components and lubricating oil.

For large capacity refrigeration plants, it is desirable to employ multistage compression only when the pressure ratio between the condenser and evaporator is greater than five. Let us look at the drawbacks of the simple vapour compression refrigeration cycle.

6.1.1 Limitations and Drawbacks of the Simple Vapour Compression Refrigeration Cycle

For a refrigeration plant required to operate at low evaporating temperature and high condenser temperature, the pressure ratio required will be very high. In such systems, the use of the simple vapour compression cycle will have the following disadvantages:

1. The specific volume of the refrigerant entering the compressor increases at low evaporator temperature. Hence, the reversible work of compression, $\int$-vdp, increases. The actual work of compression required would be much higher due to frictional heating.

2. The volumetric efficiency of the compressor decreases with an increase in pressure ratio. Consequently, the displacement volume and hence the size of the compressor required increases per kg of refrigerant circulated.

3. Referring to Figure 6.1, a vapour compression cycle (1–5–6–7) with increased pressure ratio $\dfrac{p'_c}{p_e}$ is shown compared to cycle (1–2–3–4) with pressure ratio $\dfrac{p_c}{p_e}$.

   It is observed that due to increased pressure ratio, the vapour formed after expansion has higher dryness fraction at state 7 compared to that at state 4. It means that more liquid gets evaporated while passing through expansion valve. Therefore, the flash vapour (partial evaporation of liquid refrigerant is known as flash) formed is high which does not contribute towards the refrigerating effect. As a result, the net refrigerating effect per kg of refrigerant reduces from $(h_1 - h_4)$ to $(h_1 - h_7)$.

4. For a given capacity of refrigerating plant, the mass of refrigerant to be circulated increases due to reduced refrigerating effect per kg of refrigerant. Hence, the size of evaporator and compressor required increase.
5. The COP of the system is reduced due to increase in compression work and reduction in refrigerating effect per kg of refrigerant.
6. The cost of operating the plant increases due to increase in work input.
7. It cannot meet the different loads at different evaporator temperatures as required in various industries, storage of food products, hotels, institutions, etc. without an increase in the capital and operational cost and size of the plant.

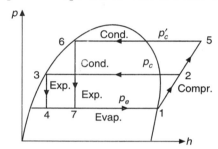

Figure 6.1 Effect of increased pressure ratio on the simple vapour compression cycle.

6.2 MULTISTAGE VAPOUR COMPRESSION SYSTEM

For large capacity refrigeration plants, it is desirable to employ multistage compression only when the pressure ratio between the condenser and evaporator is greater than five.

For small capacity refrigeration plants, multistaging is not done even for a pressure ratio of nine. Examples of these are domestic refrigerator and packaged air conditioner units.

The major operating cost of vapour compression system is the cost of energy for compressor work. Therefore, any method that leads to save the work of compression will reduce the operational cost of the system. This would be possible provided the capital and maintenance costs are not very high.

The performance, i.e. the COP of a refrigeration plant, can be increased either by increasing the refrigerating effect with fixed compressor power or by decreasing the compressor work for the same refrigeration effect.

One of the methods adopted is the use of multistage compression discussed as follows.

6.2.1 Multistage Compression with Intercooling between the Stages

Flash intercooling and water intercooling are the two methods employed for cooling at the intermediate stages. Though the water-cooled intercooler may be satisfactory for two-stage air compression, for refrigerant compression, the water is usually not cold enough. Therefore, usually in multistage compression intercooling is done either by flash gas or by the combination of flash gas and water intercooling. The work of compression can be reduced by using two or more stages of compression with intercooling in between the stages as shown in Figure 6.2 on

(p–v) diagram and in Figure 6.3 on (p–h) diagram. The isentropic process (1–a) shows the compression in a single-stage compressor between the pressure limits of p_c and p_e.

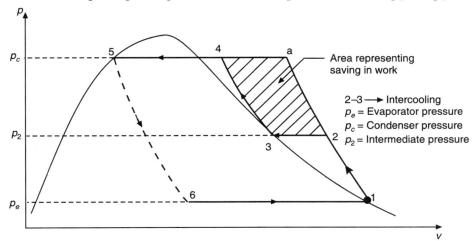

Figure 6.2 Intercooling in two-stage compression.

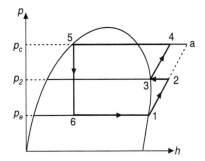

Figure 6.3 Multistage compression with intercooling.

The isentropic process (1–2) represents the compression in low pressure (LP) compressor from evaporator pressure p_e to an intermediate pressure p_2.

Compressed vapour is passed through an intercooler where it is cooled at constant pressure represented by process (2–3). The vapour cooling can be achieved either by water cooling or by refrigerant depending upon the refrigeration application.

Cooled vapour is compressed isentropically up to condenser pressure p_c represented by process (3–4) in high pressure (HP) compressor. The remainder of the cycle is same.

It can be seen that the work of compression/kg of refrigerant is reduced since the specific volume of refrigerant entering the HP compressor is reduced from v_2 to v_3 due to intercooling. The refrigerating effect remains the same during both the cycles. Due to reduced work of compression, the COP of multistage compression cycle increases.

Advantage of multistage compression cycle with intercooling

- Reduces the work of compression per kg of refrigerant.
- The coefficient of performance (COP) improves.

- Refrigerant vapour is compressed to intermediate pressure in LP compressor. So, LP cylinder will have less wall thickness. It reduces the cost of compressor.
- Heat rejected during condensation is reduced from $(h_a - h_5)$ to $(h_4 - h_5)$, so it also helps in reducing the size of condenser.
- The volumetric efficiency of the compressor improves due to small pressure ratios in each stage.
- Leakages past the piston (blow-by losses) are reduced due to the small pressure differential.
- The work of compression reduces, therefore, the operating cost is also reduced.

Disadvantages

- Increases the initial capital investment.
- The size of the plant increases due to the intercooler.

However, the above disadvantages are neutralized due to reduced energy cost over a small payback period.

6.2.2 Intermediate Pressure for Minimum Work

Let p_c and p_e be the condenser and evaporator pressures, respectively having n number of compression stages. The pressure ratio for each stage for minimum work is given by Eq. (6.1.).

$$r_p = \left(\frac{p_c}{p_e}\right)^{1/n} \tag{6.1}$$

For the two-stage compressor, $n = 2$, therefore,

$$r_p = \frac{p_2}{p_e} = \left(\frac{p_c}{p_e}\right)^{1/2}$$

i.e. intermediate pressure,

$$p_2 = \sqrt{p_c \cdot p_e} + 0.35 \tag{6.2}$$

Based on experimental results, the intermediate pressure is recommended to be as provided by the empirical relation in Eq. (6.3)

$$p_2 = \sqrt{p_e \cdot p_c \cdot \frac{T_c}{T_e}} \tag{6.3}$$

where T_c and T_e are the condenser and the evaporator temperature, respectively.

Practically, a small variation in p_2 as given by Eq. (6.3) is permitted as a small variation in p_2 will increase the work in one compressor and reduce the corresponding work in the other compressor. Therefore, the net work of compression will not be affected to a large extent.

6.3 TYPES OF MULTISTAGE VAPOUR COMPRESSION SYSTEM WITH INTERCOOLER

The intercooling of vapour refrigerant at the outlet of LP compressor in the multistage vapour compression cycle can be achieved either by water cooling or by refrigerant in a heat exchanger called intercooler.

There could be many ways of achieving the intercooling of vapour refrigerant in multistage compression; however, we have restricted our discussion to two-stage compression with intercooling for a single evaporator (i.e. single cooling load) system.

Various systems to be discussed are:

1. Two-stage compression with flash gas removal.
2. Two-stage compression with flash intercooling.
3. Two-stage compression with flash gas removal and additional gas cooler.

6.3.1 Two-stage Compression with Flash Gas Removal

In a multistage compression system, the expansion of liquid refrigerant available from the condenser outlet can be done in stages as shown in Figure 6.4. The liquid is first expanded into the flash chamber at intermediate pressure p_2. The vapour formed at this stage is removed and mixed with the hot vapour coming from the compressor-I. Therefore, the hot vapour gets cooled before it enters the second-stage compressor. The various points marked on the flow diagram (Figure 6.4) are also shown on p–h plot in Figure 6.5.

The operations involved are sequenced here:

Process 1–2: Isentropic compression in LP compressor
Process 2–3: Constant pressure cooling due to mixing of two streams
Process 3–5: Isentropic compression in HP compressor
Process 5–6: Heat rejection in the condenser
Process 6–7: Expansion of liquid to intermediate pressure p_2
Process 8–9: Expansion of liquid to evaporator pressure p_e

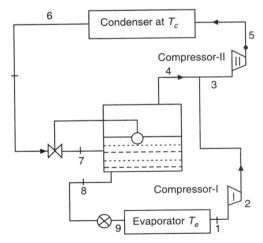

Figure 6.4 Compound compression with flash gas removal.

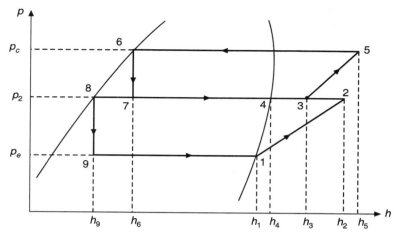

Figure 6.5 *p–h* diagram for system of Figure 6.4.

6.3.2 Two-stage Compression with Flash Intercooling

Figure 6.6 shows the arrangement for intercooling the refrigerant vapour compressed in a lower stage. This refrigerant vapour is cooled up to its saturation temperature as it is bubbled through the liquid refrigerant in the flash chamber maintained at intermediate pressure. The high pressure liquid refrigerant entering the flash chamber through the 'flow control valve' partly evaporates while cooling the vapour that comes from the lower stage compressor.

The salient points of this flow diagram are marked on the *p–h* diagram shown in Figure 6.7.

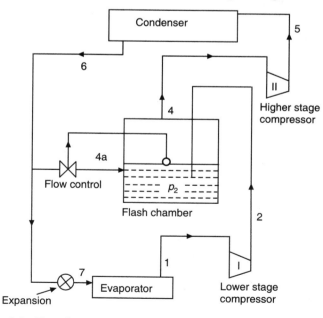

Figure 6.6 Flow diagram of a two-stage compression with flash intercooling.

The following text refers to Figure 6.6. The principle is to interrupt the compression process at a so-called inter-stage or intermediate pressure point 2, and desuperheat the refrigerant back to the saturation line at point 4. From this point the compression in the high stage restarts and terminates at point 5, at a much lower temperature than at 3, if it is a single-stage compression. There are two separate compressors—one for lower stage and another for higher stage; sometimes two-stage compression is arranged in a single compressor in which specific cylinders are allocated to different stages.

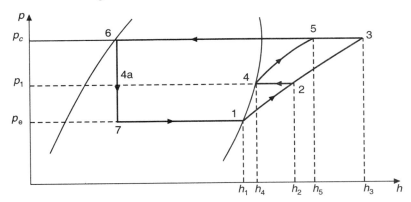

Figure 6.7 Flash Intercooling of refrigerant in two-stage compression for Figure 6.6.

EXAMPLE 6.1 An ammonia ice plant working between $-10°C$ evaporating temperature and $35°C$ condensing temperature produces 48 tonnes of ice everyday using the available water at $30°C$. If the ice manufactured is kept at $-5°C$, calculate the compressor power:

(a) In a single-stage compression
(b) In two-stage compression with flash intercooling. Assume the simple saturation vapour compression cycle.

Solution: The heat to be removed includes the following:
(i) Sensible heat to cool water from $30°C$ to $0°C$
(ii) Latent heat of fusion, 335 kJ/kg for water
(iii) Sensible heat to cool ice from $0°C$ to $-5°C$

Therefore, the refrigerating effect Q_L is calculated as follows:

Total heat removed from water to produce 48 tonnes of ice

$$= (48 \times 1000)[4.186 \times (30 - 0) + 335 + 1.95(0 + 5)] = 2229.5 \times 10^4 \text{ kJ}$$

$$\therefore \quad Q_L = \frac{2229.5 \times 10^4}{(24)(3600)} = 258 \text{ kW}$$

(a) Single-stage compression:
Refer to Figure 6.7. Properties are taken from the *p–h* chart of ammonia.

$$h_1 = 1425; \quad v_1 = 0.4182 \text{ m}^3/\text{kg}; \quad h_6 = h_7 = 320 \text{ kJ/kg}$$

Refrigerating effect per kg

$$Q_L = (h_1 - h_7) = (1425 - 320) = 1105 \text{ kJ/kg}$$

$$\text{Mass flow rate, } \dot{m} = \frac{258 \text{ kW}}{1105} = 0.233 \text{ kg/s}$$

$$= 14.00 \text{ kg/min}$$

Following the isentropic path on chart, the temperature T_3 at the end of compression is 96°C. The enthalpy at point 3, $h_3 = 1655$ kJ/kg.

Compressor power, $w = \dot{m}(h_3 - h_1) = 0.233 \times (1655 - 1425) = 53.6$ kW **Ans.**

(b) Two-stage compression with flash intercooling:

Saturated vapour pressure corresponding to temperature –10°C is 2.9 bar and for 35°C it is 13.45 bar.

$$\therefore \text{ Intermediate pressure, } p_I = \sqrt{p_e \times p_c \times \frac{T_c}{T_e}} = \sqrt{2.9 \times 13.45 \times \frac{(35+273)}{(-10+273)}}$$

$$= 6.79 \text{ bar}$$

The saturation temperature corresponding to 6.79 bar pressure is 8°C.

At saturation temperature 8°C, the following properties are noted:

$$h_4 = 1448 \text{ kJ/kg}, \quad v_4 = 0.19 \text{ m}^3/\text{kg}$$

On the p–h chart, trace the isentropic compression line from point 1 to meet the intermediate pressure line and note down the temperatures $T_2 = 45°C$ (318 K) and $h_2 = 1540$ kJ/kg. Similarly, for isentropic compression process 4–5, $T_5 = 60°C$ (333 K) and $h_5 = 1550$ kJ/kg.

The higher stage compressor must compress mass flow 14.0 kg/min. plus the rate at which the quantity of liquid that evaporates in the flash chamber. The rate of ammonia compressed in the higher stage is computed by making a heat and mass balance about the flash chamber as shown in Figure 6.8.

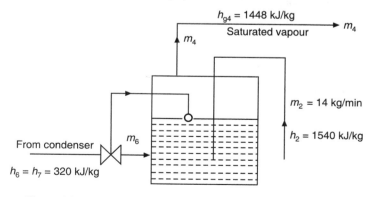

Figure 6.8 Mass balance about the flash chamber—Example 6.1.

Mass balance, $m_4 = m_6 + 14.0$
Heat balance about flash chamber,
$m_6 \times (h_7) + m_2 \times (h_2) = m_4 \times h_4$
$m_6 \times 320 + 14.0 \times 1540 = (m_6 + 14.0) \times 1448$
$m_6 = 1.14$ kg/min
$m_4 = 15.14$ kg/min $= 0.2523$ kg/s
Now the power required by the lower stage compressor in isentropic compression process 1–2 is
$w_1 = m_2 \times (h_2 - h_1) = 0.230 \times (1540 - 1524) = 26.5$ kW
Power required by the upper stage compression process 4–5,

$$w_2 = m_4(h_5 - h_4) = \frac{15.14}{60}(1550 - 1448) = 25.7 \text{ kW}$$

Total power $= w_1 + w_2 = 26.5 + 25.7 = 52.2$ kW **Ans.**

Therefore, power saving due to two-stage compression with flash intercooling is:
$$= 53.6 - 52.2 = 1.4 \text{ kW}$$

(a) If the above problem is repeated for R12 or R134a refrigerant, the power saving will be negligible. This is on account of the following properties of refrigerants R12 and R134a.

The latent heat of R12 and R134a is very small. Therefore, more liquid refrigerant would need to be evaporated in the flash chamber for the intercooling purpose. On the contrary, the higher stage compressor will be required to handle a large quantity of vapour, enhancing its power input.

(b) The slopes of isentropic lines do not change too much with respect to temperature. So the saving of specific work due to following an isentropic compression closer to the saturated vapour curve does not compensate for the increased mass flow rate in higher stage compressor. As such, flash intercooling is not beneficial for R12 and R134a multistage systems.

6.3.3 Two-stage Compression with Flash Gas Removal and Additional Gas Cooler

The expansion of liquid refrigerant from condenser pressure to the evaporator pressure is carried out in two stages. In the first step, the liquid pressure is raised to the intermediate pressure, the vapour formed is removed in the flash chamber (Figure 6.9) and in the second step only the liquid pressure is reduced to evaporator pressure from the intermediate pressure while passing through expansion valve II.

Let us have a look at the expansion process of liquid refrigerant from condenser pressure to evaporator pressure through an expansion valve. The refrigerant at the outlet of the expansion valve is a mixture of liquid plus vapour. The quantity of liquid or vapour present in the mixture depends on where does the point corresponding to the valve outlet condition lie on the two-phase region (p–h diagram). The vapour quantity in the mixture increases as this point goes away from the saturated liquid curve which is shown in Figure 6.10. The refrigerant mixture at point 10 contains more vapour quantity than it contains at state 9 per unit mixture.

In the conventional single-stage vapour compression system, the mixture from the outlet of expansion valve flows through the evaporator, picking up heat and producing cooling effect.

Now, the contribution made by the refrigerant vapour towards cooling is very small compared to liquid refrigerant. This is because of latent heat of liquid refrigerant. Now the question arises as to why the refrigerant vapour is not separated at intermediate pressure before the expansion process completes and used for intercooling in multistage compression system?

In the single-stage expansion system, i.e. without the flash chamber, the liquid could have expanded directly from point 6 to 10. But in the present system, the liquid expands in two stages, i.e. from point 6 to 7 in expansion valve-I and from point 8 to 9 in expansion valve-II. Such an expansion system results into two benefits: (i) it allows proportionately more liquid refrigerant to flow through evaporator producing more cooling effect, and (ii) the vapour separated at intermediate pressure is utilized to cool the vapour coming from the lower stage compressor. This eliminates the undesirable expansion of the vapour generated at the intermediate pressure and its recompression in the lower stage compressor. Therefore, this helps in saving the compressor power.

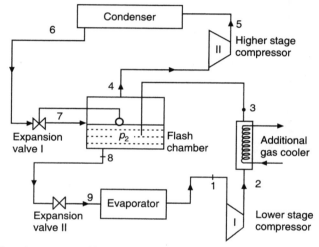

Figure 6.9 Two-stage compression system with flash gas removal and additional cooler.

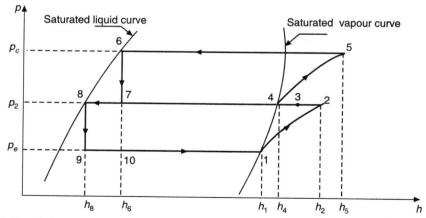

Figure 6.10 Flash gas removal with multistage compression with additional gas cooler on p–h diagram.

Such a system results in power economy and is therefore suitable for all the refrigerants.

A two-stage compression system with flash gas removal with two-stage expansion process is shown in Figure 6.9. The liquid refrigerant from the condenser at 6 first expands into a flash chamber to 7, kept at intermediate pressure p_2. The flash chamber is connected to the evaporator through another expansion valve, reducing the pressure to point 9.

EXAMPLE 6.2 One hundred ton capacity compound ammonia vapour compression system is working between $-24°C$ evaporator temperature and $36°C$ condenser temperature. The refrigerant vapour is compressed to an intermediate pressure of 5.2 bar and then cooled in a flash intercooler. The vapour is compressed isentropically in both compressors from the saturated vapour condition.

(a) Determine the mass flow rate, the compressor power and the COP if the system works in a saturated vapour compression cycle.

(b) For compound vapour compression with flash gas intercooling system, find (i) the mass flow rate in low pressure circuit, (ii) the mass flow rate in high pressure circuit, (iii) the total compressor power, and (iv) the COP.

The discharged gas from the lower stage compressor is cooled in an additional cooler up to 30°C.

Solution: Refer to Figure 6.11. For saturated vapour compression cycle 1–5a–6–10–1, the properties are taken from p–h chart

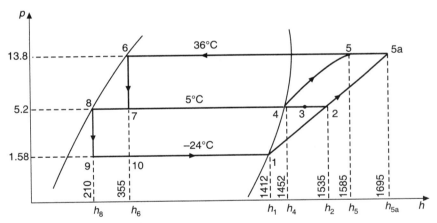

Figure 6.11(a) (i) Saturated vapour compression cycle (1–5a–6–10–1), and (ii) Compound compression with flash-gas removal (1–2–4–5–6–7–8–9–1)—Example 6.2.

Point 1: $T_{sat} = -24°C$, $p = 1.58$ bar,
$h_1 = h_{g1} = 1412$ kJ/kg, $v_1 = 0.739$ m³/kg

Point 6: $h_6 = h_{f6} = 355$ kJ/kg

Point 5a: Process 1–5a is an isentropic compression. Therefore, tracing the isentropic path on the chart, $T_{5a} = 130°C$, $h_{5a} = 1695$ kJ/kg.

Point 4: $T_{sat} = 5°C$ (273 + 5), $p = 5.2$ bar
$h_4 = h_g = 1452$ kJ/kg, $v_4 = v_g = 0.2432$ m³/kg

Process 1–2: Isentropic compression: $T_2 = 48°C$, $h_2 = 1535$ kJ/kg
Process 4–5: Isentropic compression, $T_5 = 78°C$, $h_5 = 1585$ kJ/kg

To calculate the mass flow rate in both the stages, consider the mass and energy balance about the flash chamber as shown in Figure 6.11(b).

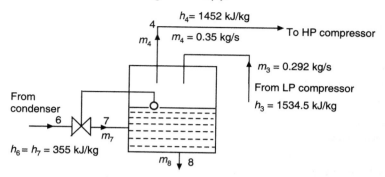

Figure 6.11(b) Mass balance about the flash chamber—Example 6.2.

Point 3: $h_3 = 1452 + 3.3(301 - 278) = 1534.5$ kJ/kg
Point 8: $h_8 = h_f$
$h_8 = h_f = -210$ kJ/kg

Mass of refrigerant to be circulated in the lower stage circuit is $m_3 = m_8$
And in higher stage circuit, it is $m_4 = m_7$
Energy balance about the flash chamber is:

$$m_3 h_3 + m_7 h_7 = m_4 h_4 + m_8 h_8$$

$$m_8 h_3 + m_7 h_7 = m_4 h_4 + m_8 h_8$$

$$m_7 = m_4$$

$$m_8 (h_3 - h_8) = m_4 (h_4 - h_7)$$

$$m_4 = \frac{m_8 (h_3 - h_8)}{(h_4 - h_7)}$$

(a) Saturated vapour compression cycle (1–5a–6–10):

$$\text{Mass flow rate} = \frac{\text{Refrigerating effect}}{\text{Refrigerating effect/kg}}$$

$$= \frac{100 \times 211}{(h_1 - h_{10})}$$

$$= \frac{100 \times 211}{(1412 - 355)} \approx 20 \text{ kg/min} = 0.33 \text{ kg/s} \quad \textbf{Ans.}$$

Compressor power, W:

$$W = m(h_{5a} - h_1)$$
$$= 0.33 \times (1695 - 1412) = 93.39 \text{ kW} \quad \textbf{Ans.}$$

$$\text{COP} = \frac{h_1 - h_{10}}{h_{5a} - h_1} = \frac{(1412 - 355)}{(1695 - 1412)} = 3.73 \quad \textbf{Ans.}$$

(b) For the compound compression system:
The rate of refrigerant flow in the low pressure (LP) circuit,

$$m_3 = m_8 = \frac{\text{Refrigerating effect}}{\text{Refrigerating effect/kg}}$$

$$= \frac{100 \times 211}{h_1 - h_9} = \frac{100 \times 211}{(1412 - 210)} = 17.6 \text{ kg/min} = 0.292 \text{ kg/s} \quad \textbf{Ans.}$$

The rate of refrigerant flow in the high pressure (HP) circuit,

$$m_4 = m_7 = m_8 \frac{(h_3 - h_8)}{(h_4 - h_7)} = \frac{0.292(1534.5 - 210)}{(1452 - 355)} = 0.35 \text{ kg/s} \quad \textbf{Ans.}$$

Compressor power W: For LP compressor, W_1

$$W_1 = m(h_2 - h_1)$$
$$= 0.33(1535 - 1412) = 40.59 \text{ kW}$$

For HP compressor, W_2
$$= m(h_5 - h_4)$$
$$= 0.35(1585 - 1452)$$
$$= 43.89 \text{ kW}$$

Net power,
$$W = W_1 + W_2$$
$$= 40.59 + 43.89 = 84.48 \text{ kW} \quad \textbf{Ans.}$$

$$\text{COP} = \frac{\text{refrigerating effect/kg}}{\text{power per kg}} = \frac{(h_1 - h_9)}{(h_2 - h_1) + (h_5 - h_4)}$$

$$= \frac{(1412 - 210)}{(1533 + 1412) + (1585 - 1452)} = 4.69 \quad \textbf{Ans.}$$

The power required for the compound system must be less than that required by the single-stage compression system.

EXAMPLE 6.3 One hundred ton refrigerating capacity compound ammonia vapour compression system is working between –24°C and 36°C, evaporating and condensing temperatures, respectively. The refrigerant vapour is compressed to an intermediate pressure of 5.2 bar and then cooled in a flash intercooler. The vapour is compressed isentropically in both the compressors. The vapour gets superheated to 10°C in the evaporator and at this temperature, it enters the low pressure compressor. In order to avoid the entry of liquid droplets, suction to the high pressure (HP) compressor is maintained at 15°C. The liquid refrigerant is subcooled to a temperature of 30°C. Assume that the discharge gas from the LP compressor is cooled by an additional water cooler to a temperature of 30°C, before it enters the flash chamber. Calculate:

(a) The mass flow rate of refrigerant in LP circuit.
(b) The mass flow rate of refrigerant in HP circuit.
(c) The compressor power kW.
(d) The COP of the system.

Solution: Refer to Figure 6.12.

Point 1: Saturated vapour, $T_{sat} = -24°C$, $p = 1.58$ bar, $h_1 = h_g = 1412$ kJ/kg
Point 1a: Superheated state, $T_{1a} = 10°C$, $h_{1a} = 1492$ kJ/kg
Point 4: Saturated vapour, $T_4 = 5°C = 278$ K, $p = 5.2$ bar

$$h_4 = h_g = 1452 \text{ kJ/kg}; \quad v_4 = v_g = 0.2432 \text{ m}^3/\text{kg}$$

Point 4a: Superheated state, $T_{4a} = 288$ K, $p = 5.2$ bar, $h_{4a} = 1455$ kJ/kg
Point 3: $h_3 = 1505$ kJ/kg, the degree of superheat is 30°C.
Point 2: Superheated condition, $T_2 = 80°C$, $h_2 = 1645$ kJ/kg
Point 5: At $p = 13.8$ bar, $T_{sat} = 36°C = 309$ K, $h_9 = 210$ kJ/kg, $h_5 = 1595$ kJ/kg

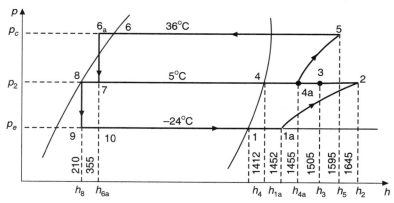

Figure 6.12 p–h diagram—Example 6.3.

Point 6a: $h_{6a} = h_f = h_7 = 355$ kJ/kg
Point 8: $h_8 = h_f = 210$ kJ/kg
Point 9: Enthalpy at point 8 = enthalpy at point 9 for isenthalpic expansion process.

(a) Rate of refrigerant in LP circuit:

$$m_3 = m_8 = \frac{\text{Net refrigerating effect}}{\text{Refrigerating effect per kg}}$$

$$= \frac{100 \times 211}{(h_{1a} - h_9)} = \frac{100 \times 211}{(1492 - 210)}$$

$$= 16.5 \text{ kg/min} = 0.275 \text{ kg/s} \qquad \textbf{Ans.}$$

(b) The rate of refrigerant flow in HP circuit:

$$m_4 = m_7 = \frac{m_8(h_3 - h_8)}{(h_4 - h_7)} = \frac{0.275 \times (1505 - 210)}{(1452 - 355)} = 0.325 \text{ kg/s} \quad \text{Ans.}$$

(c) Compressor power W:
For LP compressor, $W_1 = m(h_2 - h_{1a})$
$$= 0.275(1645 - 1492) = 42.07 \text{ kW}$$

HP compressor, $W_2 = m(h_5 - h_{4a}) = 0.325(1595 - 1455) = 45.5 \text{ kW}$
Total $W = W_1 + W_2 = 87.57 \text{ kW}$ **Ans.**

(d) $\text{COP} = \dfrac{(h_{1a} - h_9)}{(h_2 - h_{1a}) + (h_5 - h_{4a})} = \dfrac{1492 - 210}{(1645 - 1492) + (1595 - 1455)} = 4.37$ **Ans.**

EXAMPLE 6.4 A two-stage compression ammonia refrigerating system with intercooler working between the pressure limits of 1.55 bar and 14 bar is used to take a load of 50 TR. The intercooler pressure is 4.92 bar. The ammonia is cooled to 32°C in the water cooler and subcooled as liquid to 30°C. Find:
(a) The rate of ammonia circulation per minute.
(b) The power required to drive the compressors.
(c) The COP of the system.

Solution: The system is shown in Figure 6.13 on p–h diagram.

Given: $p_c = 14$ bar, $p_e = 1.55$ bar, $R_E = 50$ TR $= 50 \times 211 = 10{,}550$ kJ/min,
$T_3 = 32°C$ (after intercooler)
Subcooled liquid temperature, $T_6 = 30°C$,

From the p–h chart for ammonia refrigerant,
we get: $h_1 = 1410$ kJ/kg; $h_2 = 1560$ kJ/kg; $h_3 = 1510$ kJ/kg; $h_4 = 1665$ kJ/kg
$h_6 = h_7 = 319$ kJ/kg

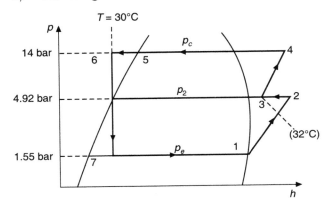

Figure 6.13 p–h diagram—Example 6.4.

(a) Rate of ammonia circulated per minute, $\dot{m}$
Refrigerating effect/kg of refrigerant,
$$q_L = h_1 - h_7 = 1410 - 319 = 1091 \text{ kJ/kg}$$
$$\therefore \quad \dot{m} = \frac{R_E}{r_E} = \frac{10{,}550}{1091} = 9.67 \text{ kg/min} \qquad \textbf{Ans.}$$

(b) Power for the compressors, P
$$P = (P)_{LP} + (P)_{HP} = m[(h_2 - h_1) + (h_4 - h_3)]$$
$$= 9.67[(1560 - 1410) + (1665 - 1510)] = 2949.4 \text{ kJ/min}$$
$$= \frac{2949.4}{60} = 49.16 \text{ kW} \qquad \textbf{Ans.}$$

(c) COP of the system:
$$\text{COP} = \frac{\text{Refrigerating effect}}{\text{Compressor power}} = \frac{10{,}550}{2949.4} = 3.58 \qquad \textbf{Ans.}$$

EXAMPLE 6.5 A compound NH_3 compression system of 175 ton capacity has two compressors, one evaporator, one flash intercooler and a liquid-to-vapour heat exchanger. The enthalpy values at various stages are as follows:

(i) Subcooled liquid NH_3 leaving HE and entering flash intercooler through a float valve = 320 kJ/kg
(ii) Saturated liquid NH_3 leaving flash intercooler and entering evaporator through an expansion valve = 197 kJ/kg
(iii) Dry saturated NH_3 vapour leaving evaporator and entering HE = 1416 kJ/kg
(iv) Superheated NH_3 vapour leaving HE. and entering LP compressor = 1433 kJ/kg
(v) Superheated NH_3 vapour leaving LP compressor and entering flash intercooler = 1593 kJ/kg
(vi) Superheated NH_3 vapour leaving flash intercooler and entering HP compressor = 1517 kJ/kg
(vii) Superheated NH_3 vapour leaving HP compressor and entering condenser = 1676 kJ/kg

Sketch the flow diagram and p–h diagram of the system. State the assumptions made, if any. Calculate:
(a) The quantity of refrigerant handled by LP and HP compressors.
(b) The power of the two compressors.
(c) The COP.

Solution: Refer to Figures 6.14(a) and (b).
Given: Refrigerating effect, $Q_L = 175$ TR $= 175 \times 211 = 36{,}925$ kJ/min; $h_9 = h_{10} = 197$ kJ/kg
$h_7 = h_8 = 320$ kJ/kg; $h_1 = 1416$ kJ/kg; $h_2 = 1433$ kJ/kg; $h_3 = 1593$ kJ/kg,
$h_5 = 1676$ kJ/kg; $h_4 = 1517$ kJ/kg

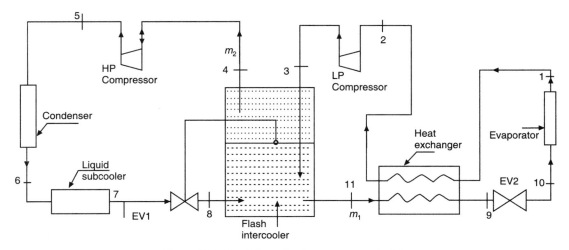

Figure 6.14(a) Schematic diagram—Example 6.5.

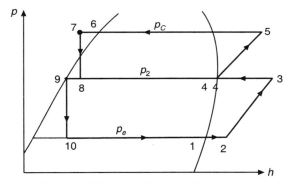

Figure 6.14(b) p–h diagram—Example 6.5.

It is assumed that there are no pressure and heat losses in the system and the processes in the compressor are isentropic.

Let m_1 be the mass flow of refrigerant through LP compressor and m_2 through HP compressor.

Applying energy balance to heat exchanger,

$$m_1(h_{11} - h_9) = m_1(h_2 - h_1)$$
$$h_{11} - 197 = 1433 - 1416$$
$$\therefore \qquad h_{11} = 215 \text{ kJ/kg}$$

(a) Quantity of refrigerant in LP and HP compressors:

Refrigerating effect/kg of refrigerant,

$$q_L = (h_1 - h_{10}) = 1416 - 197 = 1219 \text{ kJ/kg}$$

∴ Mass of refrigerant in LP compressor,

$$m_1 = \frac{Q_L}{q_L} = \frac{175 \times 211}{1219} = 30.29 \text{ kg/min} \qquad \text{Ans.}$$

By energy balance for flash intercooler, we can write

$$m_2 \cdot h_8 + m_1 \cdot h_3 = m_1 \cdot h_{11} + m_2 \cdot h_4$$

$$m_2 \times 320 + 30.29 \times 1593 = 30.29 \times 214 + m_2 \times 1517$$

$$\therefore \qquad m_2 = 34.91 \text{ kg/min} \qquad \text{Ans.}$$

(b) Total power for the two compressors,

$$W = (W)_{LP} + (W)_{HP}$$
$$= m_1(h_3 - h_2) + m_2(h_5 - h_4)$$
$$= 30.29(1593 - 1433) + 34.91(1676 - 1517)$$
$$= 10{,}395.5 \text{ kJ/min} = \frac{10{,}395.5}{60} = 173.3 \text{ kW} \qquad \text{Ans.}$$

(c) $\text{COP} = \dfrac{Q_L}{W_L} = \dfrac{175 \times 3.517}{173.26} = 3.5$ \qquad Ans.

6.4 MULTIPLE EVAPORATORS AND COMPRESSORS SYSTEMS

In the previous section we studied refrigeration systems with only one evaporator. It means the system is to produce all the cooling effect at only one temperature. In order to have the better performance of such units we have seen multiple compressors and expansion devices. However, there are many applications wherein a refrigeration plant is required to meet the various refrigeration loads at different temperatures and humidity. Examples of such systems are the air conditioning of buildings, hotels and food preservation industries. In these cases, it is necessary that each location is cooled by a separate evaporator to maintain the particular temperature and produce the required refrigeration load.

Before studying the multi-evaporator system, let us first go through the following text:

Fruits like apple and grapes are stored at −1°C for maintaining their quality. Strawberries are stored at 4°C or below. Lemons and bananas are best preserved at about 15°C, and so on.

A fruitseller wishes to preserve these fruits at the stated different temperatures. Now the question is, whether he has to purchase three refrigeration units so that each unit can be used to preserve only one type of fruit at the required temperature. If so, he would have to invest a huge amount of money. Instead, if he purchases a refrigeration unit having three evaporators working at different temperatures to meet his demand, it will have a higher operating cost. In such situations, a break point needs to be obtained considering initial investment and operating charges so as to take an advantageous decision.

Multi-evaporator systems have the advantages like less initial investment and small space requisition. One obvious disadvantage of the multi-evaporator system is that in the event of compressor breakdown all spaces served by the compressor will be without refrigeration, thereby causing the possible loss of products which otherwise should not occur.

A refrigeration system, which can meet the different refrigeration loads at different temperatures, is called a *multiple evaporator system*.

These systems may use a single or multiple or multistage compressors depending upon the application.

Some of the multiple evaporator and compressor systems are as follows:
1. Multiple evaporators at the same temperature with different refrigeration loads with a single compressor.
2. Multiple evaporators at different temperatures with a single compressor, individual expansion valves and a back pressure valve.
3. Multiple evaporators at different temperatures with a single compressor, multiple expansion valves and a back pressure valve.
4. Multiple evaporators with individual compressors and individual expansion valves.
5. Multiple evaporators with individual compressors and multiple expansion valves.
6. Multiple evaporators with compound compression, individual expansion valves with or without flash intercooling.

6.4.1 Multiple Evaporators at the Same Temperature and a Single Compressor System

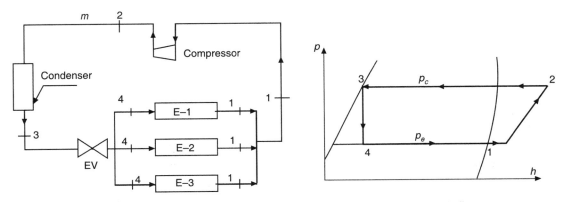

Figure 6.15(a) Schematic diagram.   **Figure 6.15(b)** *p–h* diagram.

The schematic diagram of a multiple evaporator working at the same temperature and with a single compressor system is shown in Figure 6.15(a) and its corresponding cycle is shown in *p–h* diagram in Figure 6.15(b). Such a system is needed when various food articles or other hygroscopic materials are needed to be refrigerated separately at the same temperature.

The working of such a system is similar to the simple vapour compression cycle. Let evaporators E–1, E–2 and E–3 operate at only one temperature but produce Q_{L1}, Q_{L2} and Q_{L3} amounts of refrigeration (TR), respectively. Further let m_1, m_2 and m_3 be the corresponding masses of refrigerant needed to be circulated in respective evaporators in kg/min.

Refrigeration produced/kg of refrigerant, $q_L = (h_1 - h_4)$ kJ/kg.

$$\therefore \quad m_1 = \frac{Q_{L_1} \times 211}{(h_1 - h_4)}; \quad m_2 = \frac{Q_{L_2} \times 211}{(h_1 - h_4)}; \quad m_3 = \frac{Q_{L_3} \times 211}{(h_1 - h_4)} \quad (6.4)$$

The total mass of refrigerant, m kg/min, circulated in the compressor, condenser and expansion valve will be

$$m = (m_1 + m_2 + m_3) \text{ kg/min} \quad (6.5)$$

Compressor power,

$$W = m(h_2 - h_1) \text{ kJ/min} \quad (6.6)$$

$$\text{COP} = \frac{(h_2 - h_1)}{(h_1 - h_4)} \quad (6.7)$$

EXAMPLE 6.6 A multiple evaporator R134a refrigeration system works between the temperature limits of –15°C and 30°C using a single compressor. It has three evaporators, all working at the same temperature of –15°C having the refrigeration loads of 20 TR, 10 TR and 30 TR. The refrigerant vapour leaving the evaporators is dry and there is no subcooling of refrigerant in the condenser. Determine:

(a) The refrigerating effect per kg of refrigerant.
(b) The mass of the refrigerant circulated in kg/min in each evaporator and the total mass of the refrigerant in kg/min.
(c) The compressor power.
(d) The COP of the system.
(e) The bore and stroke of the single acting compressor if the stroke is 1.2 times the bore, when the compressor runs at 300 rpm and it has a volumetric efficiency of 85%.

Saturation temperature (°C)	Specific volume, v_g (m³/kg)	Specific enthalpy (kJ/kg)		
		h_f	h_{fg}	h_g
–15°C	0.1204	179	211	390
30°C	0.0263	243	314	372.62

Solution: Refer to Figure 6.16 for the p–h diagram.

Given: $T_{se} = -15°C$, $T_{sc} = 30°C$, $Q_{L1} = 20$ TR, $Q_{L2} = 10$ TR, $Q_{L3} = 30$ TR; $L = 1.2D$, $N = 300$ rpm, $\eta_v = 85\% = 0.85$ From R134a (p–h) chart, we get $h_1 = 390$ kJ/kg; $h_2 = 418$ kJ/kg; $h_3 = h_4 = 243$ kJ/kg

(a) Refrigerating effect/kg, r_E:

$$q_L = (h_1 - h_4) = (390 - 243) = 147 \text{ kJ/kg} \quad \text{Ans.}$$

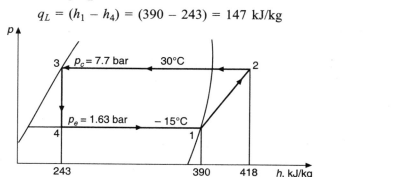

Figure 6.16 p–h diagram—Example 6.6.

(b) Mass of the refrigerant circulated in kg/min in each evaporator:
Mass of the refrigerant in E–1 evaporator,

$$m_1 = \frac{Q_{L1}}{q_L} = \frac{20 \times 211}{147} = 28.7 \text{ kg/min} \qquad \text{Ans.}$$

Mass of the refrigerant in E–2 evaporator,

$$m_2 = \frac{Q_{L2}}{q_L} = \frac{10 \times 211}{147} = 14.35 \text{ kg/min} \qquad \text{Ans.}$$

Mass of the refrigerant in E–3 evaporator,

$$m_3 = \frac{Q_{L3}}{q_L} = \frac{30 \times 211}{147}$$

$$= 43.06 \text{ kg/min} \qquad \text{Ans.}$$

Total mass of the refrigerant circulated,

$$m = m_1 + m_2 + m_3 = 28.7 + 14.35 + 43.06$$

$$= 86.11 \text{ kg/min} \qquad \text{Ans.}$$

(c) Compressor power,

$$W = m(h_2 - h_1) = \frac{86.11}{60}(418 - 390) = 40.18 \text{ kW} \qquad \text{Ans.}$$

(d) COP of the system,

$$\text{COP} = \frac{Q_{L1} + Q_{L2} + Q_{L3}}{W} = \frac{(20 + 10 + 30)\,3.517}{40.18}$$

$$= 5.25 \qquad \text{Ans.}$$

(e) Bore D and stroke L of the compressor:
From the p–h chart, $v_1 = 0.1204 \text{ m}^3/\text{kg}$
Total volume circulated,

$$V = m \times v_1 = 86.11 \times 0.1204 = 10.36 \text{ m}^3/\text{min}$$

But

$$V = \frac{\pi}{4} D^2 \times L \times N \times \frac{1}{\eta_V}$$

$$10.36 = \frac{\pi}{4} D^2 \times 1.2D \times 300 \times \frac{1}{0.85}$$

∴ Bore, $D = 0.3140 \text{ m} = 31.40 \text{ cm}$ \qquad Ans.

Stroke, $L = 1.2D = 1.2 \times 31.40 = 37.70 \text{ cm}$ \qquad Ans.

6.4.2 Multiple Evaporators at Different Temperatures with a Single Compressor, Individual Expansion Valves and a Back Pressure Valve

This system consists of one compressor and two evaporators with individual expansion valves as shown in Figure 6.17(a) and the processes are represented on the *p–h* diagram in Figure 6.17(b).

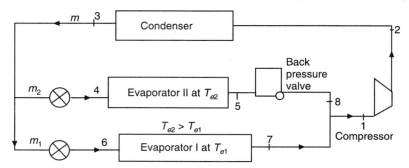

Figure 6.17(a) System with a single compressor and two evaporators with individual expansion valves and a back pressure valve system.

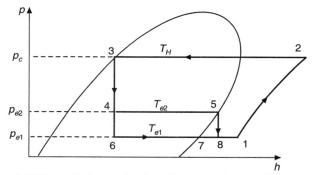

Figure 6.17(b) *p–h* diagram for the unit system shown in Figure 6.17(a).

In this system, the two evaporators are at two different temperatures and accordingly their pressures are also different. The evaporator at lower temperature will be at lower pressure in comparison with that of the other one. The pressure ratio to be maintained in the refrigerating unit is from the lower temperature evaporator T_{e1} to condenser pressure. The mass flow rate of refrigerant in each evaporator depends on the individual load.

The refrigerant mass m_2 throttled through the expansion valve flows through evaporator II. This is proportional to the load on it. The pressure of the refrigerant in evaporator II is corresponding to its saturation temperature T_{e2}, which is maintained due to the back pressure valve. The pressure of the refrigerant vapour is further reduced in back pressure valve and mixes with the vapour coming from evaporator I.

Let m_1 and m_2 be the masses of the refrigerant circulating in evaporators EI and EII, respectively. Then the mass of the refrigerant circulating in evaporators in kg/min will be

$$m_1 = \frac{Q_{L1} \times 211}{h_7 - h_6}; \quad m_2 = \frac{Q_{L2} \times 211}{h_5 - h_4} \tag{6.8}$$

All masses of refrigerant are mixed and supplied to compressor at state 1

$$\text{Total mass, } m = (m_1 + m_2) \text{ kg/min}$$

By energy balance during mixing of refrigerants, we get

$$m \cdot h_1 = m_1 \cdot h_7 + m_2 \cdot h_8$$

But

$$h_5 = h_8$$

∴

$$m \cdot h_1 = m_1 \cdot h_7 + m_2 \cdot h_5 \tag{6.9}$$

Compressor power,

$$W = m(h_2 - h_1) \text{ kJ/min} \tag{6.10}$$

$$\text{COP of the system} = \frac{\text{Total refrigerating effect}}{\text{Compressor power}}$$

$$= \frac{(Q_{L1} + Q_{L2})\,211}{m(h_2 - h_1)} \tag{6.11}$$

EXAMPLE 6.7 A R134a refrigeration unit is required to work with two evaporators at $-4°C$ and $-20°C$, having load of 4 TR and 2 TR, respectively. This unit uses one compressor and an individual expansion valve for each evaporator. The condenser is operating at 40°C temperature. The exit condition from the two evaporators is saturated vapour. There is no subcooling of liquid refrigerant. Determine:
(a) The refrigerating effect per kg in each evaporator.
(b) The mass flow rate of the refrigerant in each evaporator.
(c) The compressor power.
(d) The COP of the unit.

Solution: The saturated properties of refrigerant R134a are collected in the following table.

Temp. (°C)	Pressure (bar)	Specific volume, v_g (m³/kg)	Enthalpy (liquid) (kJ/kg)	Enthalpy (vapour) (kJ/kg)
−20	1.350	0.1538	—	385
−4	2.61	0.0769	—	395
40	11.00	0.019	258	—

Refer to Figures 6.17(a) and (b) for the solution given below.

Evaporator I: At −20°C,
(a) Specific refrigerating effect = $(h_7 - h_6) = (385 - 258) = 127.0$ kJ/kg   **Ans.**

(b) Mass flow rate, $m_1 = \dfrac{Q_L}{q_L} = \dfrac{2 \times 211}{127} = 3.32$ kg/min   **Ans.**

Evaporator II: At −4°C
(a) Refrigerating effect = $(h_5 - h_4) = (395 - 258) = 137.0$ kJ/kg   **Ans.**

(b) Mass flow rate, $m_2 = \dfrac{Q_L}{q_L} = \dfrac{(4) \times (211)}{137} = 6.16$ kg/min **Ans.**

$m_1 + m_2 = 3.32 + 6.16 = 9.48$ kg/min

Enthalpy of vapour (h_1) entering the compressor,

$$h_1 = \dfrac{m_1 h_7 + m_2 h_8}{(m_1 + m_2)} = \dfrac{(3.32 \times 385) + (6.16 \times 395)}{3.32 + 6.16} = 391.49 \text{ kJ/kg}$$

With the help of h_1 and pressure, locate point 1 on p–h plot. Draw an isentropic line from point 1 to meet constant pressure line at point 2. Then get $h_2 = 437$ kJ/kg from p–h plot.

(c) Compressor power, $W = m(h_2 - h_1)$

$= (9.48)(437 - 391) = 436$ kJ/min

$= 7.26$ kW **Ans.**

(d) $\text{COP} = \dfrac{\text{Total refrigerating effect}}{\text{Net power input}} = \dfrac{(2 + 4) \times (3.5167)}{7.26} = 2.99$ **Ans.**

EXAMPLE 6.8 An R134a unit is operating at three evaporating temperatures 6°C, –4°C and –20°C, respectively. The high and low temperature evaporators have the cooling load of 30 TR and 10 TR, respectively. The medium temperature evaporator has a load of 20 TR. Every evaporator is provided with an individual expansion valve. The vapour at the outlet of each evaporator is saturated vapour. The two high temperature evaporators have back pressure expansion valves. The vapour from the evaporators is mixed before it enters the compressor. The condensation takes place at 40°C with no subcooling effect. Determine:

(a) The refrigerating effect per kg in each evaporator.
(b) The mass flow rate of the refrigerant in kg/s in each evaporator.
(c) The compressor power.
(d) The COP of the unit.

Solution: The properties of R134a are given in the following table.

Temp. (0°C)	Pressure (bar)	Specific volume, v_g (m³/kg)	Enthalpy (liquid), h_f (kJ/kg)	Enthalpy (vapour), h_g (kJ/kg)
–20	1.3268	0.14744	173.82	386.66
–4	2.5257	0.07991	194.68	396.33
6	3.6186	0.05648	208.08	402.14
40	10.165	0.01999	256.35	419.58

Refer to Figures 6.18(a) and (b).

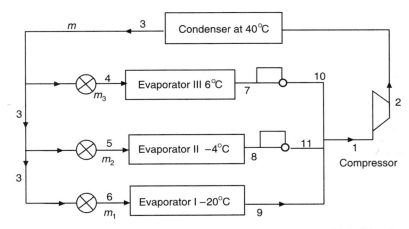

Figure 6.18(a) System with a single compressor and three evaporators with individual expansion valves and a back pressure valve—Example 6.8.

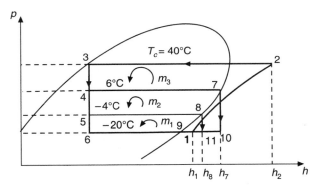

Figure 6.18(b) p–h diagram for the unit system shown in Figure 6.18(a)—Example 6.8.

Evaporator I: At temperature –20°C

(a)   Specific refrigerating effect $= (h_9 - h_6) = (386.66 - 256.35)$

$$= 130.3 \text{ kJ/kg} \quad \textbf{Ans.}$$

(b)   Mass flow rate, $m_1 = \dfrac{Q_L}{q_L} = \dfrac{(10)(211)}{130.3} = 16.19 \text{ kg/min}$

$$= 0.27 \text{ kg/s} \quad \textbf{Ans.}$$

Evaporator II: At temperature –4°C

(a)   Specific refrigerating effect $= (h_8 - h_5) = (396.33 - 256.35)$

$$= 140 \text{ kJ/kg} \quad \textbf{Ans.}$$

(b)   Mass flow rate, $m_2 = \dfrac{Q_L}{q_L} = \dfrac{(20)(211)}{140} = 30.15 \text{ kg/min} = 0.502 \text{ kg/s}$

                                                                                           Ans.

Evaporator III: At temperature 6°C

(a) Specific refrigerating effect = $(h_7 - h_4) = (402.14 - 256.35)$

$$= 145.8 \text{ kJ/kg} \quad \textbf{Ans.}$$

(b) Mass flow rate, $m_3 = \dfrac{Q_L}{q_L} = \dfrac{(30)(211)}{145.8} = 43.416 \text{ kg/min} = 0.724 \text{ kg/s}$

Ans.

(c) Total mass flow rate, $m = m_1 + m_2 + m_3 = 0.27 + 0.502 + 0.724$

$$= 1.496 \text{ kg/s}$$

The enthalpy of vapour at suction to compressor,

$$h_1 = \frac{m_1 h_9 + m_2 h_8 + m_3 h_7}{m}$$

$$= \frac{(0.27 \times 386.66) + (0.502 \times 396.33) + (0.724 \times 402.14)}{1.496}$$

$$= 397.5 \text{ kJ/kg}$$

With the help of pressure 1.3268 bar and enthalpy 397.5 kJ/kg, mark point 1 on the *p–h* chart. Then draw an isentropic line through point 1 to meet the constant pressure line, 10.165 bar at point 2 and get $h_2 = 446$ kJ/kg.

Compressor power, $W = m(h_2 - h_1) = (1.496)(446 - 397.5)$

$$= 72.536 \text{ kW} \quad \textbf{Ans.}$$

(d) $\text{COP} = \dfrac{\text{Total refrigerating effect}}{\text{Net power input}} = \dfrac{(10 + 20 + 30) \times 3.5167}{72.536} = 2.9$ **Ans.**

6.4.3 Multiple Evaporators at Different Temperatures with a Single Compressor, Multiple Expansion Valves and a Back Pressure Valve

The back pressure valve that is provided at the outlet of evaporator II helps to keep the required pressure in evaporator II. Also, it reduces the pressure of vapour at its outlet by throttling process ($h_7 = h_9$). The mass flow rate m_2 through evaporator II is enough to take its cooling load. The remaining mass flow m_1 passes through evaporator I taking care of its load.

The pressure ratio to be maintained for the unit is again based on the condenser pressure and the evaporator pressure corresponding to the lowest temperature evaporator (here evaporator I).

A back pressure valve is provided to reduce the pressure of high pressure refrigerant p_{e2} to lowest evaporator pressure p_{e1} so that the mixture of refrigerants of all evaporators at p_{e1} enters at suction of compressor at state 1.

Let m_1 and m_2 be the masses of the refrigerant flow in respective evaporators EI and EII having the refrigeration loads Q_{L1} and Q_{L2} in TR of refrigeration, respectively.

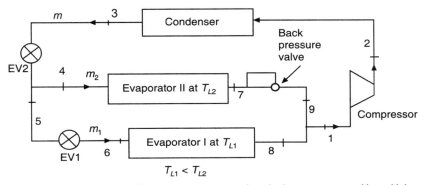

Figure 6.19(a) Refrigeration unit with two evaporators and a single compressor with multiple expansion valves and a back pressure valve.

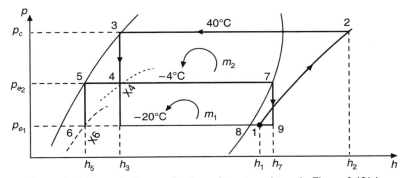

Figure 6.19(b) p–h diagram for the unit system shown in Figure 6.19(a).

(a) Mass of refrigerant circulated in each evaporator can be calculated as follows:

Evaporator I at temperature T_{L1}

Refrigerating effect/kg $= (h_8 - h_6)$

$$\therefore \quad m_1 = \frac{Q_{L1} \times 211}{(h_8 - h_6)} \text{ kg/min} \quad (6.12)$$

Evaporator II at temperature T_{L2}

Refrigerating effect per kg $= (h_7 - h_4)$

$$\therefore \quad m_2' = \frac{Q_{L2} \times 211}{(h_7 - h_4)} \text{ kg/min}$$

where m_2' represents the mass of refrigerant required to meet the load Q_{L2} at point 4. In addition, the evaporator EII is supplied all the vapour formed by mass m_1 while passing through the expansion valve 2 (EV2). Let the dryness fraction of vapour refrigerant at point 4 be x_4. The vapour associated by mass m_1 at point 4 is

$$m_2'' = \left(\frac{x_4}{1-x_4}\right) \cdot m_1$$

$$m_2 = m_2' + m_2'' = \frac{Q_{L2} \times 211}{(h_7 - h_4)} + \left(\frac{x_4}{1-x_4}\right) m_1 \qquad (6.13)$$

(b) Energy balance of mixing of refrigerants before suction to compressor:
Note that $h_7 = h_9$, due to throttling in back pressure valve.
Let the state 1 represent the suction to compressor, which is after mixing of two streams having enthalpies h_8 and h_9. By energy balance,

$$(m_1 + m_2)h_1 = m_1 \cdot h_8 + m_2 \cdot h_9 \qquad (6.14)$$

Using Eq. (6.14), the value of h_1 can be calculated.

(c) Compressor work and compressor power:
Process (1–2) is an isentropic compression process. Fix point 1 on p–h diagram at enthalpy h_1 and p_{e1}. Point 2 is obtained by intersection of $s_1 = s_2$ line and condenser pressure line p_c. Read h_2.

Compressor work/kg, $W = (h_2 - h_1)$ kJ/kg

$$\text{Compressor power, } W = (m_1 + m_2)(h_2 - h_1) \times \frac{1}{60} \text{ kW} \qquad (6.15)$$

(d) COP of the system

$$\text{COP} = \frac{\text{Refrigerating effect}}{\text{Compressor power}} = \frac{(Q_{L1} + Q_{L2}) 3.517}{W} \qquad (6.16)$$

EXAMPLE 6.9 An ammonia refrigeration unit is operating with two evaporators at –4°C and –20°C and having cooling loads of 40 TR and 20 TR, respectively. The unit uses one compressor and multiple expansion valves. The condenser pressure is 15.55 bar (saturation temperature, 40°C). The vapour leaving the evaporators is dry saturated. There is no subcooling of liquid in the condenser. Determine:

(a) The refrigerating effect per kilogram in each evaporator.
(b) The mass flow rate in each evaporator.
(c) The compressor power.
(d) The COP of the unit. Take c_p for ammonia to be 2.379 kJ/kg-K and 3.516 kJ/kg-K at saturation temperatures –20°C and 40°C, respectively.

Solution: Refer to Figures 6.19(a) and (b) for the following solution.
The saturation properties of ammonia are collected from the p–h chart.

Evaporator I: At temperature –20°C,

(a) Refrigerating effect, $\dot{q}_L = (h_8 - h_6)$ as $h_6 = h_{f5}$
$= (1408 - (-110)) = 1300$ kJ/kg **Ans.**

(b) Mass flow rate, $m_1 = \dfrac{Q_L}{q_L} = \dfrac{(20)(211)}{1300} = 3.24$ kg/min $= 0.054$ kg/s **Ans.**

Evaporator II: At temperature $-4°C$,

(a) Specific refrigerating effect, $q_L = (h_7 - h_4)$ as $h_4 = h_{f3}$

$$= (1412 - 360) = 1052 \text{ kJ/kg} \quad \textbf{Ans.}$$

$$m_2' = \left(\frac{Q_{L2} \times 211}{h_7 - h_4}\right) = \left(\frac{40 \times 211}{1052}\right)$$

$$= 8.022 \text{ kg/min} = 0.1337 \text{ kg/s}$$

(b) Mass flow rate, $m_2 = m_2' + m_1\left(\frac{x_4}{1-x_4}\right)$

$x_4 = 0.14$, obtained from the p–h chart of ammonia.

$$m_2 = 0.1337 + 0.054\left(\frac{x_4}{1-x_4}\right)$$

$$= 0.1424 \text{ kg/s} \quad \textbf{Ans.}$$

Calculate enthalpy h_1,

$$h_1 = \frac{m_2 h_9 + m_1 h_8}{(m + m_2)} = \frac{0.1424 \times 1412 + 0.054 \times 1408}{(0.054 + 0.1424)} = 1411 \text{ kJ/kg}$$

$h_2 = 1720 \text{ kJ/kg}$

(c) Compressor power W:

$$W = m(h_2 - h_1) = (0.054 + 0.1424)(1720 - 1411) = 60.7 \text{ kW} \quad \textbf{Ans.}$$

(d) $$\text{COP} = \frac{\text{Total refrigerating effect}}{\text{Net power supplied}} = \frac{(20 + 40)(3.5167)}{60.7} = 3.47 \quad \textbf{Ans.}$$

6.4.4 Multiple Evaporators with Individual Compressors and Individual Expansion Valves

The schematic arrangement with two evaporators EI, EII with individual compressors C1, C2 and individual expansion valves EV1, EV2 is shown in Figures 6.20(a) and (b). In this unit, the liquid refrigerant is expanded in the expansion valve EV2 from condenser pressure to the pressure of evaporator II (high temperature evaporator) and flows through the evaporator II to take its load and further to compressor 2. Similarly the liquid refrigerant is expanded in the expansion valve EV1 from condenser pressure to the pressure of evaporator I (high temperature evaporator) and flows through the evaporator I to take its load and further to compressor 1. At this pressure, flash gas formed is removed. This vapour does not flow through evaporator I (low temperature evaporator) but directly goes to the suction of compressor, which may help to improve the performance.

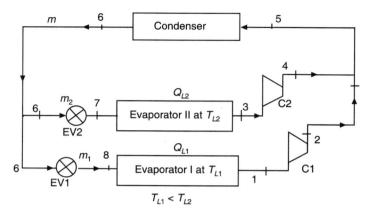

Figure 6.20(a) Refrigeration unit with multiple (two) evaporators with individual compressors and invdividual expansion valves.

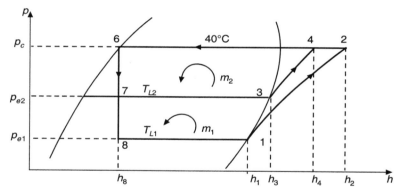

Figure 6.20(b) p–h diagram for the unit system shown in Figure 6.20(a).

Evaporator EI: At temperature, T_{L1}

q_{L1} = Refrigerating effect/kg in EII = $(h_1 - h_8)$

Mass of refrigerant through EI, $m_1 = \dfrac{Q_{L1} \times 211}{(h_1 - h_8)}$ kg/min

Evaporator EII: At temperature, T_{L2}

q_{L2} = Refrigerating effect/kg in EII = $(h_3 - h_7)$

Mass of refrigerant through EII, $m_2 = \dfrac{Q_{L2} \times 211}{(h_3 - h_7)}$ kg/min

Compressor power

Let compressor powers be W_1 and W_2 for respective compressors C1 and C2 respectively in kW. Then,

$$W_1 = \frac{m_1}{60}(h_2 - h_1) \text{ kW}$$

$$W_2 = \frac{m_2}{60}(h_4 - h_3) \text{ kW}$$

Total compressor power,

$$W = (W_1 + W_2)$$

COP of the system

$$\text{COP} = \frac{\text{Total refrigerating effect}}{\text{Compressor power}} = \frac{(Q_{L1} + Q_{L2})\,3.517}{W}$$

EXAMPLE 6.10 A R134a refrigeration system as shown in Figure 6.21(a) works with two evaporators, individual expansion valves, individual compressors and with a single condenser. Draw the cycle on p–h diagram and find:
(a) The mass of refrigerant circulated in each evaporator.
(b) The compressor power for each compressor and total power.
(c) The COP of the system.

Assume that the refrigerant leaving each evaporator is dry saturated and the liquid refrigerant leaving the condenser is subcooled to 30°C. Compression is isentropic in each compressor.

Solution: Following properties of R134a are taken from a book by Mathur and Mehta.

Temp °C	Pressure bar	h_f, kJ/kg	h_g, kJ/kg
−20	1.350	172.0	387.00
0	2.90	199.0	398.0
40	11.00	258.0	418.0

The enthalpy at point '6' i.e. $h_f = h_6 = 242$ kJ/kg is obtained by interpolation.

(a) Mass of refrigerant circulated in each evaporator:

In evaporator EI: At temperature −20°C,

$$m_1 = \frac{Q_{L1} \times 211}{h_1 - h_8} = \frac{30 \times 211}{387 - 242} = 43.65 \text{ kg/min} \qquad \textbf{Ans.}$$

In evaporator EII: At temperature 0°C,

$$m_2 = \frac{Q_{L2} \times 211}{h_3 - h_7} = \frac{25 \times 211}{398 - 242} = 33.8 \text{ kg/min} \qquad \textbf{Ans.}$$

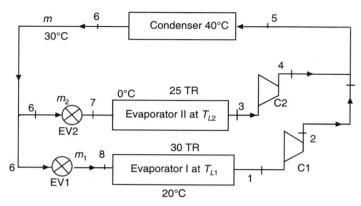

Figure 6.21(a) Refrigeration unit with two evaporators with individual compressors and individual expansion valves—Example 6.10.

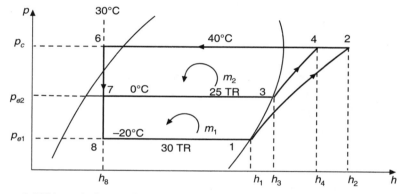

Figure 6.21(b) p–h diagram for the unit system shown in Figure 6.21(a)—Example 6.10.

(b) Compressor power for each compressor and total power:
LP compressor power, W_1

$$W_1 = \frac{m_1}{60}(h_2 - h_1) = \frac{43.65}{60}(425 - 387) = 27.6 \text{ kW} \quad \text{Ans.}$$

HP compressor power, W_2

$$W_2 = \frac{m_2}{60}(h_4 - h_3) = \frac{33.8}{60}(422 - 398) = 13.52 \text{ kW} \quad \text{Ans.}$$

Total compressor power, W

$$W = W_1 + W_2 = 27.6 + 13.52 = 41.16 \text{ kW} \quad \text{Ans.}$$

(c) COP of the system

$$\text{COP} = \frac{\text{Refrigeration load}}{\text{Compressor power, } W} = \frac{(30 + 25)\,3.517}{41.16} = 4.67 \quad \text{Ans.}$$

6.4.5 Multiple Evaporators with Individual Compressors and Multiple Expansion Valves

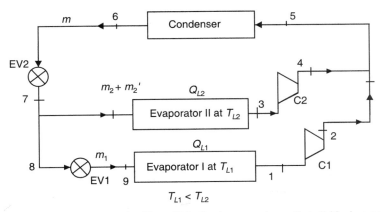

Figure 6.22(a) Refrigeration unit with multiple (two) evaporators with individual compressors and multiple expansion valves.

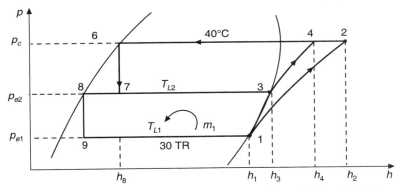

Figure 6.22(b) p–h diagram for the unit system shown in Figure 6.22(a).

The schematic arrangement of the evaporators, compressors and expansion valves is shown in Figure 6.22 (a). One should recognize the difference here. The total quantity of liquid refrigerant flows through the expansion valve EV1 and the pressure is reduced to that of evaporator pressure p_{e2}.

Liquid refrigerant flows through evaporator EII to take refrigeration load Q_{L2} in addition to all the vapours formed after leaving the EV2.

The remaining liquid refrigerant flows through EV1 and is throttled up to pressure p_{e1} of evaporator EI. Liquid refrigerant necessary to take the load Q_{L1} along with all the vapour formed flows through the evaporator EI.

Refrigerants leaving the respective evaporators are compressed in their respective compressors. It is assumed that the refrigerants leaving their respective evaporators are dry saturated at their exit.

Let m_1 and m_2 be the flow of refrigerants in evaporators EI and EII, respectively. The cooling loads are Q_{L1} and Q_{L2} in TR of refrigeration.

Evaporator EI: At temperature T_{L1}

Refrigerating effect/kg = $(h_1 - h_9)$ kJ/kg

$$\therefore \quad m_1 = \frac{Q_{L1} \times 211}{h_1 - h_9} \text{ kg/min} \tag{6.17}$$

Evaporator EII: At temperature T_{L2}

Refrigerating effect/kg = $(h_3 - h_7)$ kJ/kg

$$\therefore \quad m_2' = \frac{Q_{L2} \times 211}{h_3 - h_7} \text{ kg/min}$$

where m_2' represents the mass of refrigerant required to meet the refrigeration load Q_{L2}. In addition, evaporator EII is supplied all the vapour formed by mass m_1 while passing through the expansion valve EV2.

Dryness fraction of vapour is x_7. Therefore, 1 kg of vapour has x_7 kg of dry saturated vapour and $(1 - x_7)$ kg of liquid refrigerant. Hence, the mass of dry saturated vapour associated with m_1 kg will be

$$m_2'' = \left(\frac{x_7}{1 - x_7}\right) m_1$$

∴ Total mass flow of refrigerant through evaporator EII,

$$m_2 = m_2' + m_2'' = \frac{Q_{L2} \times 211}{h_3 - h_7} + \frac{x_7}{(1 - x_7)} \times m_1$$

Compressor power W

For compressor C1,

$$W_1 = \frac{m_1}{60}(h_2 - h_1) \text{ kW}$$

For compressor C2,

$$W_2 = \frac{m_2}{60}(h_4 - h_3) \text{ kW}$$

$$\therefore \quad W = W_1 + W_2$$

$$= \frac{1}{60}[m_1(h_2 - h_1) + m_2(h_4 - h_3)] \text{ kW}$$

Total refrigerating effect in kW,

$$Q_L = (Q_{L1} + Q_{L2}) \times 3.517 \text{ kW}$$

COP of the system

$$\text{COP} = \frac{(Q_{L1} + Q_{L2}) \times 3.517}{W}$$

EXAMPLE 6.11 A refrigeration system using R12 as refrigerant consists of three evaporators of capacities 20 TR at –5°C, 30 TR at 0°C and 10 TR at 5°C. The vapours leaving the three evaporators are dry saturated. The system is provided with individual compressors and multiple expansion valves. The condenser temperature is 40°C and the liquid leaving the condenser is saturated. Assuming isentropic compression in each compressor, find:

(a) The mass of the refrigerant flowing through each evaporator.
(b) The power required to drive the system.
(c) The COP of the system.

Solution: Refer to Figure 6.23(a) for schematic diagram and Figure 6.23(b) for the p–h diagram.

Given: $Q_{L1} = 20$ TR; $Q_{L2} = 30$ TR; $Q_{L3} = 10$ TR.

From the p–h chart for R12, we get the following properties:

$h_1 = 192$ kJ/kg, $h_2 = 212$ kJ/kg, $h_3 = 195$ kJ/kg, $h_4 = 209$ kJ/kg, $h_5 = 196$ kJ/kg
$h_6 = 207$ kJ/kg, $h_7 = h_8 = 78$ kJ/kg, $h_9 = h_{10} = 44$ kJ/kg, $h_{11} = h_{12} = 34$ kJ/kg,
$x_8 = 0.22$, $x_{10} = 0.03$.

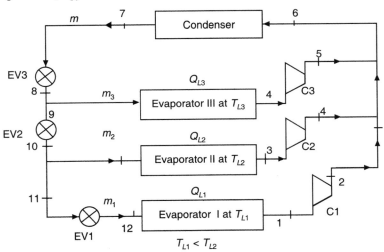

Figure 6.23(a) Refrigeration unit with multiple (three) evaporators and individual compressors and multiple expansion valves—Example 6.11.

(a) Mass of refrigerant flowing through each evaporator:
   (i) Evaporator EI: At temperature –5°C

$$m_1 = \frac{Q_{L1} \times 211}{h_1 - h_{12}} = \frac{20 \times 211}{(192 - 34)} = 26.7 \text{ kg/min} = 0.445 \text{ kg/s} \quad \textbf{Ans.}$$

   (ii) Evaporator EII: At temperature 0°C
        According to equation, we have

$$m_2 = \frac{Q_{L2} \times 211}{h_3 - h_{10}} + \left(\frac{x_{10}}{1 - x_{10}}\right) m_1$$

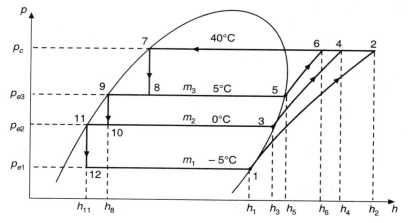

Figure 6.23(b) p–h diagram for the unit system shown in Figure 6.23(a)—Example 6.11.

$$= \frac{30 \times 211}{(195-44)} + \left(\frac{0.03}{1-0.03}\right) \times 26.71 = 42.75 \text{ kg/min} = 0.712 \text{ kg/s} \qquad \textbf{Ans.}$$

(iii) Evaporator EIII: At temperature 5°C
According to equation, we have

$$m_3 = \frac{Q_{L3} \times 211}{h_5 - h_8} + \left(\frac{x_8}{1-x_8}\right)(m_1 + m_2)$$

$$= \frac{10 \times 211}{(196-78)} + \left(\frac{0.22}{1-0.22}\right)(26.71 + 42.75)$$

$$= 37.5 \text{ kg/min} = 0.625 \text{ kg/s} \qquad \textbf{Ans.}$$

(b) Power required to drive the system, we have
Power of compressor C1, $W_1 = m_1(h_2 - h_1) = 0.445(212 - 192) = 8.9$ kW
Power of compressor C2, $W_2 = m_2(h_4 - h_3) = 0.712(209 - 195) = 9.96$ kW
Power of compressor C3, $W_3 = m_3(h_6 - h_5) = 0.625(207 - 196) = 6.87$ kW
Total compressor power, $W = 8.9 + 9.96 + 6.87 = 25.73$ kW **Ans.**

(c) COP of the system,

$$\text{COP} = \frac{(Q_{L1} + Q_{L2} + Q_{L3}) \times 3.517}{W}$$

$$= \frac{(20 + 30 + 10) \times 3.517}{25.73} = 8.195 \qquad \textbf{Ans.}$$

6.4.6 Multiple Evaporators with Compound Compression and Individual Expansion Valves

The schematic arrangement of two evaporators EI and EII with their individual expansion valves EV1 and EV2 and with compound compression is shown in Figure 6.24(a) and the corresponding cycle on the p–h diagram is shown in Figure 6.24(b).

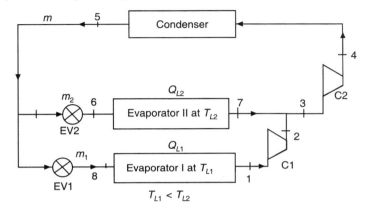

Figure 6.24(a) Refrigeration unit with multiple (two) evaporators with compound compression and individual expansion valves.

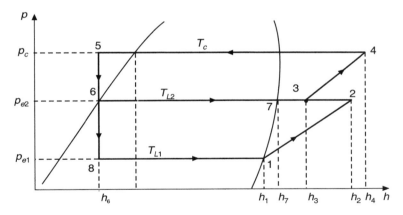

Figure 6.24(b) p–h diagram for the system of Figure 6.24(a).

Mass flow rate of refrigerant m_1 flows through evaporator EI to take the cooling load in it. The compressor C1 compresses the refrigerant m_1 circulated in evaporator EI.

The compressor C2 compresses the refrigerant circulated in evaporator EII as well as the compressed refrigerant of compressor C1 after mixing at suction.

Let Q_{L1} and Q_{L2} be the refrigeration loads in TR of refrigeration on evaporators EI and EII, respectively.

Compression is assumed to be isentropic and the refrigerant after cooling in condenser is in a saturated liquid state.

(a) Mass of refrigerant circulated:
   (i) Evaporator EI: At temperature T_{L1}
   $$m_1 = \frac{Q_{L1} \times 211}{h_1 - h_8}$$
   (ii) Evaporator EI: At temperature T_{L2}
   $$m_2 = \frac{Q_{L2} \times 211}{(h_7 - h_6)}$$

(b) Compressor power W:
   (i) Compressor C1
   $$\text{Power of compressor C1, } W_1 = m_1(h_2 - h_1)$$
   (ii) Compressor C2
   $$\text{Power of compressor C2, } W_2 = m_1(h_4 - h_3)$$
   (iii) Total compressor power in kW,
   $$W = \text{Compressor power for C1 and C2} = W_1 + W_2$$

(c) COP of the system:
$$\text{COP} = \frac{(Q_{L1} + Q_{L2}) \times 3.517}{W}$$

EXAMPLE 6.12 A typical multi-evaporator system is shown in Figure 6.25. The refrigerant is R134a. Sketch the cycle on p-h diagram. Calculate:
(a) The mass flow rate of refrigerant through each compressor.
(b) The cooling load on condenser.
(c) The overall COP.
(d) The displacement volume required for each compressor if their volumetric efficiency is 0.9.

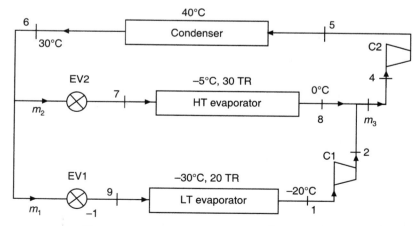

Figure 6.25 Multi-evaporator system—Example 6.12.

Given: Q_{L1} = 20 TR at –30°C and Q_{L2} = 30 TR at –5°C. The cyclic process is as shown in the p–h diagram of Figure 6.24(b). Enthalpy values from the p–h diagram are:

$$h_1 = 590 \text{ kJ/kg}; \ v_1 = 0.24 \text{ m}^3/\text{kg}; \ h_2 = 617 \text{ kJ/kg}$$

$$h_7 = 604 \text{ kJ/kg}; \ h_5 = h_6 = h_8 = 458 \text{ kJ/kg}$$

(a) Mass flow rates through compressors:
Let m_1 be the mass flow rate in the evaporator at –30°C and compressor C1,

$$m_1 = \frac{Q_{L1} \times 211}{(h_1 - h_8)} = \frac{30 \times 211}{(590 - 458)} = 47.95 \text{ kg/min} = 0.80 \text{ kg/s} \quad \textbf{Ans.}$$

Let m_2 be the mass flow rate in the evaporator at –5°C. Then,

$$m_2 = \frac{Q_{L2} \times 211}{(h_7 - h_6)} = \frac{30 \times 211}{(604 - 458)} = 43.36 \text{ kg/min} = 0.723 \text{ kg/s}$$

∴ Mass flow rate through compressor C2,

$$m_3 = m_1 + m_2 = 0.80 + 0.723 = 1.523 \text{ kg/s} \quad \textbf{Ans.}$$

(b) Cooling load on condenser:
Power for compressor C1,

$$W_1 = m_1(h_2 - h_1) = \frac{47.95}{60}(617 - 590) = 21.58 \text{ kW}$$

Vapour from compressor C1 and vapour from evaporator at –5°C are mixed before suction to compressor C2. Let the state after mixing be at point 3. By energy balance,

$$m_3 \cdot h_3 = m_1 \cdot h_2 + m_2 \cdot h_7$$

∴ $\quad 91.31 \times h_3 = 47.95 \times 617 + 43.36 \times 604$
$\quad\quad\quad h_3 = 610.8 \text{ kJ/kg}$

Mark the state point 3 at h_3 and p_{e2} at –5°C line. Draw an isentropic line (3–4) on the p–h chart up to condenser pressure p_c corresponding to saturation temperature 40°C to fix the state point 4. By reading the chart, we get h_4 = 638 kJ/kg and also specific volume, v_3 = 0.06 m³/kg.
Power for compressor C2,

$$W_2 = m_3(h_4 - h_3) = \frac{91.31}{60}(638 - 610.8) = 41.40 \text{ kW}$$

Total compressor power $W = W_1 + W_2 = 21.58 + 41.40 = 62.97 \text{ kW}$

Cooling load on condenser,

$$Q_c = m_3(h_4 - h_5) = \frac{91.31}{60}(638 - 458) = 273.93 \text{ kW} \quad \textbf{Ans.}$$

246 *Refrigeration and Air Conditioning*

Overall COP:

$$\text{COP} = \frac{(Q_{L1} + Q_{L2}) \times 3.517}{W} = \frac{(30 + 30) \, 3.517}{62.97} = 3.35 \quad \textbf{Ans.}$$

(c) Displacement volume for each compressor if $\eta_V = 0.9$ (given)
Displacement volume of LP compressor

$$= \frac{m_1 \times v_1}{\eta_V} = \frac{47.95 \times 0.24}{0.9} = 12.787 \text{ m}^3/\text{min} \quad \textbf{Ans.}$$

Displacement volume of HP compressor

$$= \frac{m_3 \times v_3}{\eta_V} = \frac{91.31 \times 0.06}{0.9} = 6.087 \text{ m}^3/\text{min} \quad \textbf{Ans.}$$

6.4.7 Multiple Evaporator System with Compound Compression, Individual Expansion Valves and Flash Intercoolers

The schematic diagram of a multi-evaporator system with compound expansion, individual expansion valves and flash intercooler is shown in Figure 6.26(a) and the corresponding cycle on the *p–h* diagram is shown in Figure 6.26(b).

In this case, flash intercooling is provided at each stage of compression. Here perfect cooling is achieved and this helps to reduce the compressor work.

The system operates with a single condenser. In this system, the refrigerant from the high pressure expansion valve EV3 is throttled to the pressure corresponding to the discharge pressure of compressor C2 and mixed with the lower stage compressed vapour of compressor C2 in the flash intercooler 2. The mixed refrigerant after cooling up to dry saturated vapour in FI2 is sent to the suction of high pressure compressor C3.

The above description is also applicable for the suction of intermediate pressure compressor C2.

Analysis of the system

Let Q_{L1}, Q_{L2} and Q_{L3} be the cooling loads on evaporators EI, EII and EIII, respectively.

(a) **Mass of refrigerant through evaporator EI and compressor C1**:

$$m_1 = \frac{Q_{L1} \times 211}{(h_1 - h_{10})} \text{ kg/min}$$

(b) **Mass of refrigerant through compressor C2**

Mass of refrigerant circulated through evaporator EII,

$$m_2' = \frac{Q_{L2} \times 211}{(h_3 - h_9)} \text{ kg/min}$$

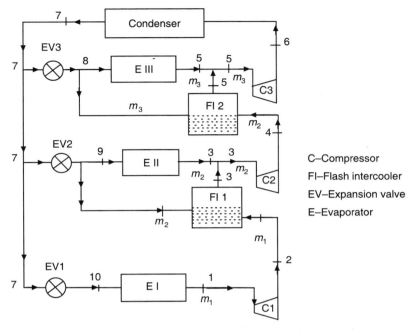

Figure 6.26(a) Multiple evaporator system with compound compression.

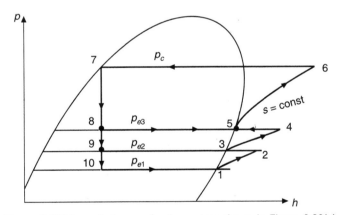

Figure 6.26(b) p–h diagram for the system shown in Figure 6.26(a).

A part of refrigerant m_2'' at point 9 from expansion valve EV2 is bled to flash intercooler 1 (FI1) for desuperheating the compressed vapour having flow rate m_1 coming from compressor C1 in superheated condition, i.e. from state 2 to state 3 shown on the p–h diagram. By energy balance,

energy gained by bled refrigerant = energy lost by compressed vapour

$$m_2''(h_3 - h_9) = m_1(h_2 - h_3)$$

∴
$$m_2'' = m_1 \left(\frac{h_2 - h_3}{h_3 - h_9}\right)$$

∴ Total mass of refrigerant circulated at suction to compressor C2:

$$m_2 = m_1 + m_2' + m_2'' = m_1 + \frac{Q_{L2} \times 211}{(h_3 - h_9)} + \left(\frac{h_2 - h_3}{h_3 - h_9}\right) m_1$$

(c) **Mass of refrigerant circulated at suction to compressor C3**

Mass of refrigerant circulated to evaporator EIII,

$$m_3' = \frac{Q_{L3} \times 211}{(h_5 - h_8)} \text{ kg/min}$$

Let m_3'' be the refrigerant bled after EVIII for desuperheating of vapour of compressor C2 in flash chamber FI2. By energy balance as explained above, we can write

$$m_3'' = \frac{(h_4 - h_5)}{(h_5 - h_8)} m_2$$

∴ Total mass of refrigerant circulated to compressor C3 will be

$$m_3 = m_2 + m_3' + m_3''$$

$$m_3 = m_2 + \frac{Q_{L3} \times 211}{(h_5 - h_8)} + \left(\frac{h_4 - h_5}{h_5 - h_8}\right) m_2$$

(d) **Compressor power, W**

$$W = W_1 + W_2 + W_3$$

or

$$W = \frac{1}{60} [m_1(h_2 - h_1) + m_2(h_4 - h_3) + m_3(h_6 - h_5)] \text{ kW}$$

(e) **Coefficient of performance of the system**

$$\text{COP} = \frac{\text{Total refrigerating load}}{\text{Total compressor power}} = \frac{(Q_{L1} + Q_{L2} + Q_{L3}) 3.517}{W}$$

EXAMPLE 6.13 A refrigeration system using R12 as refrigerant has three evaporators of capacities 30 TR at –10°C, 20 TR at 5°C and 10 TR at 10°C. The refrigerant leaving the evaporator is dry and saturated. The system is provided with compound compression, individual expansion valves and flash intercoolers. The condenser temperature is 40°C. Assuming isentropic compression in each compressor, find (a) the power required to run the system and (b) the COP of the system when the liquid refrigerant leaving the condenser is saturated.

Solution: Refer to the block diagram shown in Figure 6.26(a) and its *p–h* diagram shown in Figure 6.27.

Q_{L1} = 30 TR at –10°C, Q_{L2} = 20 TR at 5°C, Q_{L3} = 10 TR at 10°C.

Condenser temperature, $T_7 = 40°C$
From the $p–h$ diagram of R12, we get the following properties:
$h_1 = 188$ kJ/kg; $h_2 = 198$ kJ/kg; $h_3 = 194$ kJ/kg; $h_4 = 200$ kJ/kg
$h_5 = 198$ kJ/kg; $h_6 = 212$ kJ/kg; $h_7 = h_8 = h_9 = h_{10} = 78$ kJ/kg

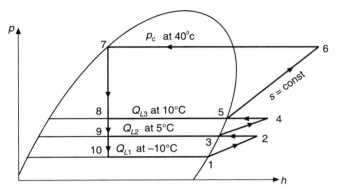

Figure 6.27 $p–h$ diagram—Example 6.13.

(a) Compressor power W:

Mass of refrigerant circulated through evaporator EI,

$$m_1 = \frac{Q_{L1} \times 211}{(h_1 - h_{10})} = \frac{30 \times 211}{188 - 78} = 57.55 \text{ kg/min} = 0.96 \text{ kg/s}$$

∴ Compressor power for compressor C1,

$$W_1 = m_1(h_2 - h_1) = \frac{57.55}{60}(198 - 188) = 9.59 \text{ kW}$$

Mass of refrigerant circulated to compressor C2 would be

$$m_2 = m_1 + \frac{Q_{L2} \times 211}{(h_3 - h_9)} + \frac{(h_2 - h_3)}{(h_3 - h_9)} m_1$$

$$= 57.55 + \frac{20 \times 211}{(194 - 78)} + \frac{(198 - 194)}{(194 - 78)} \times 57.55 = 95.91 \text{ kg/min}$$

∴ Compressor power for compressor C2,

$$W_2 = \frac{m_2}{60}(h_4 - h_3) = (200 - 194) = 9.59 \text{ kW}$$

Mass of refrigerant circulated to compressor C3 will be

$$m_3 = m_2 + \frac{Q_{L3} \times 211}{(h_5 - h_8)} + \frac{(h_4 - h_5)}{(h_5 - h_8)} m_2$$

$$= 95.91 + \frac{10 \times 211}{(198-78)} + \frac{(200-198)}{(198-78)} \times 95.91 = 115.0 \text{ kg/min}$$

Compressor power for compressor C3,

$$W_3 = \frac{m_3}{60}(h_6 - h_5) = \frac{115}{60}(212 - 198) = 26.85 \text{ kW}$$

∴ Total compressor power,

$$W = W_1 + W_2 + W_3 = 9.59 + 9.59 + 26.85 = 46.0 \text{ kW} \qquad \textbf{Ans.}$$

(b) COP of the system:

$$\text{COP} = \frac{\text{Total refrigerating effect}}{\text{Total compressor power}} = \frac{(Q_{L1} + Q_{L2} + Q_{L3})\,3.517}{W}$$

$$= \frac{(30 + 20 + 10)\,3.517}{46.0} = 4.584 \qquad \textbf{Ans.}$$

6.4.8 Refrigeration Unit with Multiple Evaporators with Compound Compression Multiple Expansion Valves and Flash Intercooler

Such a refrigeration unit with two evaporators and two compressors is illustrated in Figure 6.28(a) and the *p–h* diagram in Figure 6.28(b).

In this unit, m_1 is the mass flow rate in the low temperature evaporator I and takes care of cooling load on it. This is then compressed in compressor I to that of evaporator pressure and cooled in a flash chamber which is also kept at the pressure of evaporator II.

Let Q_{L1} and Q_{L2} be the refrigeration loads in TR on evaporators EI and EII, respectively.

(a) **Mass of refrigerant flow in evaporator EI and compressor C1**

$$m_1 = \frac{Q_{L1} \times 211}{(h_1 - h_8)} \text{ kg/min}$$

(b) **Mass of refrigerant flow at the suction to compressor C2**

Mass of refrigerant to be circulated through evaporator EII to meet the load Q_{L2} will be

$$m_2' = \frac{Q_{L2} \times 211}{(h_3 - h_6)} \text{ kg/min}$$

Mass of refrigerant required in the flash intercooler at state 6 for desuperheating the compressed refrigerant of compressor C1 to dry saturated state 3 from state 2 can be determined by the energy balance of intercooler as follows:

Energy absorbed by refrigerant from expansion valve

= Energy given by compressed vapour of compressor C1

i.e.

$$m_2''(h_3 - h_6) = m_1(h_2 - h_3)$$

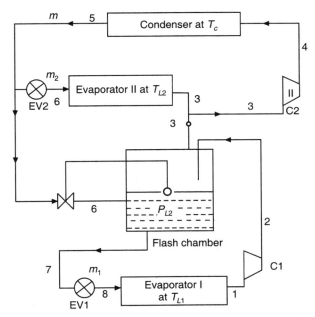

Figure 6.28(a) Refrigeration unit with multiple evaporators, with compound compression multiple expansion valves and flash intercooler.

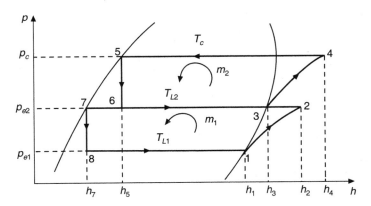

Figure 6.28(b) p–h diagram for the refrigeration unit shown in Figure 6.28(a).

$$\therefore \quad m_2'' = m_1 \frac{(h_2 - h_3)}{(h_3 - h_6)}$$

The evaporator EII is also supplied with vapour formed during expansion of mass m_1 kg/min of refrigerant while it was passing through EV2 of dryness fraction x_6.

The mass of vapour, m_2''', associated with refrigerant m_1 will be

$$m_2''' = \left(\frac{x_6}{1 - x_6}\right) m_1$$

Total mass of refrigerant at the suction to compressor C2 will be

$$m_2 = m_1 + m_2' + m_2'' + m_2'''$$

$$\therefore \quad m_2 = m_1 + \frac{Q_{L2} \times 211}{(h_3 - h_6)} + m_1 \frac{(h_2 - h_3)}{(h_3 - h_6)} + \left(\frac{x_6}{1 - x_6}\right) m_1$$

(c) **Compressor power W**

Total compressor power,

$$W = W_1 + W_2$$

$$\therefore \quad W = \frac{1}{60}[m_1(h_2 - h_1) + m_2(h_4 - h_3)] \text{ kW}$$

(d) **COP of the system**

$$\text{COP} = \frac{\text{Net refrigerating effect}}{\text{Compressor power}}$$

$$= \frac{(Q_{L1} + Q_{L2}) \, 3.517}{W}$$

EXAMPLE 6.14 An ammonia refrigeration unit is operating with two evaporators at −4°C and −20°C and have the cooling loads of 20 TR, and 40 TR respectively. The unit employs compound compression with flash intercooling at the evaporator pressure operating at higher temperature. The condenser pressure is 15.55 bar (saturation temperature 40°C). It is assumed that vapour leaving the evaporator is in dry saturated condition. Determine the following:

(a) The refrigerating effect per kilogram of each evaporator.
(b) The mass flow rate of refrigerant in each evaporator.
(c) The power required in kW.
(d) The coefficient of performance (COP).

Solution: Refer to Figures 6.28(a) and 6.28(b). (a) and (b) The specific refrigerating effect and mass flow rate of refrigerant in each evaporator are same as in Example 6.9. The required properties of ammonia have been taken from the *p–h* chart.

$h_1 = 1408$ kJ/kg, $h_2 = 1500$ kJ/kg, $T_2 = 20°C$, $h_3 = 1412$ kJ/kg and $h_4 = 1625$ kJ/kg, $h_7 = h_f = 110$ kJ/kg

The temperature of vapour at the outlet of lower stage compressor is T_2. To cool this vapour up to saturation temperature, the heat to be removed from it is

$$m_1(h_2 - h_{g2}) = (0.056)(1500 - 1412) = 4.92 \text{ kJ/s}$$

The refrigerating effect produced per kg at evaporator II is $(h_3 - h_7)$.
Mass of refrigerant evaporated, m_x

$$m_x(h_3 - h_7) = 4.92 \text{ kJ/s}, \quad h_7 = h_{f6} = 110 \text{ kJ/kg}$$

$$m_x = \frac{4.92}{(h_3 - h_7)} = \frac{4.92}{(1412 - 110)} = 3.77 \times 10^{-3} \text{ kg/s} = 0.00377 \text{ kg/s (very small)}$$

Total mass flow = $m_1 + m_2 + m_x$ = 0.056 + 0.2187 + 0.00377 = 0.278 kg/s

(c) Compressor power W

Power of lower stage compressor, $W_1 = m_1(h_2 - h_1) = 0.056(1500 - 1408) = 5.15$ kW

Power of higher stage compressor, W_2

$$W_2 = m(h_4 - h_3) = 0.278(1625 - 1412) = 59.214 \text{ kW}$$

Net power = $W_1 + W_2$ = 5.15 + 59.214 = 64.364 kW **Ans.**

(d) $\text{COP} = \dfrac{\text{Total refrigerating effect}}{\text{Net power input}} = 2.94$ **Ans.**

6.5 CASCADE SYSTEM

The evaporator temperature is usually fixed by the particular application. The higher temperature of the cycle, i.e. condensing temperature is dependent on the coolant (air or water). For the fixed condenser temperature of about 38°C, the use of single-stage vapour compression system is limited to an evaporator temperature −25°C. The main reason is the thermodynamic properties of the refrigerant. Many of the refrigerants like R11, R12, R134a, R124, R21, R22, R717 have the sub-atmospheric saturation pressures. Therefore, systems operating at sub-atmospheric pressures have many undesirable effects as stated below:

1. The pressure ratio increases abundantly, reducing the volumetric efficiency.
2. The specific volume of the refrigerant vapour will be very small requiring a large cylinder volume if reciprocating compressors are used.
3. An extremely high pressure ratio leads to overheating of the compressor.
4. Decrease in the overall performance, i.e. COP will be smaller.

To lower the intensity of the above disadvantages, compound compression with intercooling systems are preferred below −25°C. Two-stage compound compression systems are used down to about −55°C evaporator temperatures. Below this temperature, use three-stage systems. Two-stage compound compression has been discussed in the previous section.

If one needs to go below −55°C evaporator temperature, then multi-staging becomes uneconomical and the freezing point of the refrigerant also imposes limits to do so. Now the only option left is the cascade system.

A cascade system is a multi-staged application in which different refrigerants are used in the various stages. In this case, the evaporator of the higher temperature stage is the condenser for the lower temperature stage. It means it combines two or more vapour compression systems as shown schematically in Figure 6.29(a) and on p–h diagram in Figure 6.29(b).

Let T_{L2} and T_{c2} be the evaporating and condensing temperatures respectively of a high temperature cascade circuit. Similarly, let T_{L1} and T_{c1} be for the low temperature cascade circuit; but T_{L1} is much less than T_{L2}.

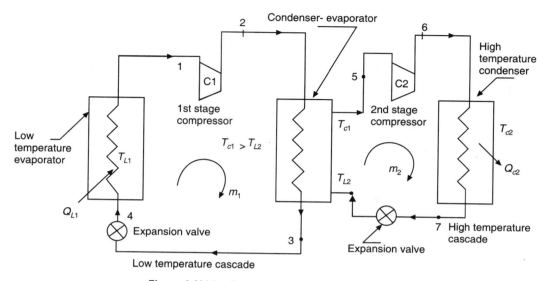

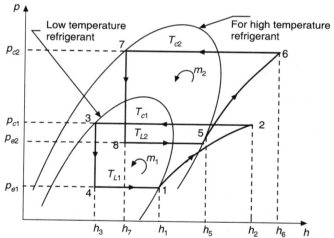

Figure 6.29(b) *p–h* diagram for the elementary cascade system shown in Figure 6.29(a).

The condenser of a low temperature cascade system discharges its heat to the evaporator of the high temperature cascade system. It means the low temperature being achieved in high temperature cascade is utilized to condense the refrigerant of low temperature cascade. Theoretically, T_{L2} and T_{c1} must be identical but for the exchange of heat between the two fluids, it practically requires a definite temperature difference. Therefore, the temperature T_{L2} is less than T_{c1} by 5 to 10°C.

Since the low temperature and high temperature portions of the system are separate, different refrigerants are so close that they produce optimum performance.

Usually high boiling point refrigerants such as R12, R14a, R22 are used for high temperature cascade. The refrigerants having high pressure, low boiling point and higher density like R13, R13B1 and R503 are used for low temperature cascade. Ammonia has been used in the high temperature cascade for the manufacture of solid carbon dioxide.

Further, such a dense refrigerant allows the compressor to have a smaller rate of displacement. With all this, the coefficient of performance will be higher. One more advantage of cascade system is that lubricating oil circulation is restricted to the respective circuit only but in compound compression oil gets circulated from one compressor to another. With two-stage cascading, a temperature of the order of –80°C is achievable but remains dependent on the refrigerants being used.

Cascade systems have the following disadvantages:

(i) Theoretically, the refrigerating capacity of the higher temperature cascade should be exactly equal to the heat being rejected by the condenser of the low temperature cascade. But practically it is very difficult to attain such a balance. This balancing is furthermore critical during pull down, wherein the refrigerating capacity of the high temperature cascade is inadequate to absorb heat from the condenser of the low temperature portion.

(ii) Practically, from the heat transfer point of view, a temperature difference between T_{L2} and T_{c1} is maintained, which reduces the efficiency of the system.

EXAMPLE 6.15 A cascade refrigeration system employs R13 on the low temperature cascade and R134a on the high temperature cascade. This cascade system produces a refrigeration effect of 10 TR at –70°C, in the low temperature evaporator with R13. The condenser of the low temperature cascade rejects heat at –20°C, while that of the high temperature cascade condenser will reject heat to water at 38°C. Assume a temperature overlap of 6°C in the condenser-evaporator (cascade condenser). Also, assume that both the cascade systems operate on saturated vapour compression cycle. Determine the following:

(a) Pressure ratio for both cascades.
(b) Mass flow rate of refrigerant for both cascades.
(c) Volume of refrigerant handled by each compressor.
(d) Compressor power for each compressor in kW.
(e) COP of each cascade and combined cascade COP.

Solution: The required saturation properties of both the refrigerants are collected from the p–h chart. Let us refer to Figures 6.29 (a) and (b) to proceed.

Low Temperature Cascade: Cycle 1–2–3–4–1

(a) Pressure ratio $= \dfrac{p_{c1}}{p_{e1}} = \dfrac{11.479}{1.8025} = 6.368$ **Ans.**

(b) Mass flow rate, $m_1 = \dfrac{Q_L}{q_L} = \dfrac{(10) \times (211)}{(h_1 - h_4)}$

$= \dfrac{(10) \times (211)}{(270 - 177)}$

$$= 22.85 \text{ kg/min} \quad \text{as } h_4 = h_{f3} = 177.24 \text{ kJ/kg}$$
$$= 0.376 \text{ kg/s} \quad \textbf{Ans.}$$

(c) Volume of refrigerant handled by compressor 1.
$$V_1 = m \cdot v_1 = (0.376)(0.08393) = 0.0316 \text{ m}^3/\text{s} = 1.896 \text{ m}^3/\text{min} \quad \textbf{Ans.}$$

(d) Compressor power, W_1: Enthalpy h_2 is noted down tracing the isentropic path.
$$W_1 = (m)(h_2 - h_1) = (0.376)(301 - 270) = 11.71 \text{ kW} \quad \textbf{Ans.}$$

(e) $\text{COP} = \dfrac{h_1 - h_4}{h_2 - h_1} = \dfrac{270 - 177}{301 - 270} = 2.99; \quad (h_4 = h_{f3})$ **Ans.**

High Temperature Cascade: Cycle 4–5–6–7–1

(a) Pressure ratio $= \dfrac{p_{c2}}{p_{e2}} = \dfrac{9.6301}{1.0164} = 9.47$ **Ans.**

(b) Mass flow rate, m_2

In order to find out the mass flow rate in the high temperature cascade cycle, one has to balance the energy in cascade condenser, i.e. energy rejected by R13 = energy absorbed by R134a

$$(m_1)(h_2 - h_3) = (m_2)(h_5 - h_8) \quad (\because h_8 = h_7 = h_{f7})$$
$$(0.376)(301 - 177) = (m_2)(382 - 253)$$
$$m_2 = 0.3615 \text{ kg/s} \quad \textbf{Ans.}$$

Enthalpy h_6 is noted down tracing the isentropic compression 5–6 on chart.
$$h_6 = 429 \text{ kJ/kg}$$

(c) The Volume of refrigerant handled by compressor 2
$$= m_2 \cdot v_5 = (0.3615)(0.18961) = 0.0685 \text{ m}^3/\text{s} = 4.1 \text{ m}^3/\text{min} \quad \textbf{Ans.}$$

(d) Compressor power, W_2
$$W_2 = (m_2)(h_6 - h_5) = (0.36)(429 - 382) = 16.95 \text{ kW} \quad \textbf{Ans.}$$

(e) $\text{COP} = \dfrac{h_5 - h_8}{h_6 - h_5} = \dfrac{382 - 253}{429 - 382} = 2.7$ **Ans.**

COP of the combined cascade system

$$= \dfrac{\text{Total refrigerating effect}}{\text{Total power input}}$$

$$= \dfrac{(10)(3.5167)}{(11.71 + 16.95)} = 1.22 \quad \textbf{Ans.}$$

EXERCISES

1. Enumerate the limitations and drawbacks of a single-stage vapour compression plant required to operate between high condenser temperature and low evaporator temperature.
2. Why is a multistage compression system used when the pressure ratios are large?
3. Write short notes on:
   (a) A refrigeration system having three evaporators at different temperatures with single compressor, multiple expansion valves and back pressure valves.
   (b) Cascade refrigeration cycle.
4. Discuss the limitations of vapour compression refrigeration system for production of low temperatures. Also, sketch and explain a cascade refrigeration system. Draw the T–s and p–h diagrams.
5. Explain the working of multiple evaporator system with compound compression and individual expansion valves with a neat sketch for its arrangement. Also, represent the cycle on the p–h diagram and explain the method of determination of COP of the system.
6. Write a few applications of low temperature refrigeration systems.
7. Explain that the use of two compressors is beneficial in NH_3 refrigeration system as compared to freon refrigeration system.
8. "Removal of flash gas and intercooling are associated with multipressure refrigeration systems." Comment.
9. Explain the working of two-stage compression with water intercooler and subcooler employed for vapour compression system.
10. Why is flash intercooler used in multistage compression?
11. What are the advantages and disadvantages of multistage compression with intercooling in between the stages?
12. "Optimum pressure ratio is not equal to the geometric mean of evaporator and condenser pressures for two-stage vapour compression system." Explain why.
13. What are the advantages of using multiple expansion valves instead of individual expansion valve in a single compressor, multiple evaporator system?
14. Why are multiple evaporator and compressor systems used in practice?
15. Why are back pressure valves used in multiple evaporator system working at different temperatures with individual expansion valves? Explain the working of such a system with the help of schematic and p–h diagrams.

NUMERICALS

1. An ammonia ice plant, working between –10°C evaporating temperature and 30°C condensing temperature, produces 60 tonnes of ice everyday using the available water at 26°C. If the ice manufactured is stored at –5°C, calculate the compressor power: (a) If a single stage compression is followed; (b) in a two-stage compression with flash intercooling. Assume saturation vapour compression cycle. [**Ans:** (a) 56 kW, (b) 54 kW]

2. An R22 refrigerant unit is required to work with two evaporators at −4°C and −20°C having load of 8 TR, and 4 TR, respectively. This unit employs a single compressor and an individual expansion valve for each evaporator. The condenser is working at 40°C temperature. The exit condition from the two evaporators is the saturated vapour. There is no subcooling of liquid refrigerant. Determine the following: (a) The refrigerating effect per kilogram in each evaporator. (b) The mass flow rate in each evaporator. (c) The compressor power. (d) The COP of the unit.

[**Ans:** Low temperature evaporator, q_L = 145 kJ/kg, m = 5.82 kg/min
High temperature evaporator, q_L = 155 kJ/kg, m = 10.8 kg/min
Compressor power = 12.5 kW, COP = 3.0]

3. A cascade refrigeration system employs R13 in the low temperature cascade and R134a in the high temperature cascade. This cascade system produces a refrigeration effect of 5 TR at −80°C in the low temperature evaporator. The condenser of low temperature rejects heat at −20°C while that of high temperature cascade condenser rejects heat to the water at 35°C. Assume a temperature overlap of 6°C in the condenser–evaporator (cascade condenser). Also assume that both cascade systems operate on saturated vapour compression cycle. Determine the following: (a) The pressure ratio for both cascades. (b) The mass flow rate in both circuits. (c) The volume flow rate of the refrigerant in each compressor. (d) The COP of each cascade and the combined cascade.

4. A cascade refrigeration system is required to produce 20 TR of refrigeration. It uses R12 in low temperature cycle with evaporator temperature of −40°C and cascade condenser temperature of −20°C. Ammonia is used in the high temperature cycle with cascade evaporator temperature of −30°C and condenser temperature of 20°C. Find the COP of the combined cycle. Assume saturated conditions of both refrigerants at the exit of evaporators and condensers and ideal cascade heat exchanger. For properties of R12 and ammonia, refer to p–h charts. [**Ans:** COP of combined cycle is 2.548]

5. A multipressure system is shown in Figure 6.30 below. R12 is used as refrigerant. Use the p–h diagram and calculate the following: (i) COP of the combined system. (ii) Displacement volume required for compressor 1 (C1) if its volumetric efficiency is 90%.

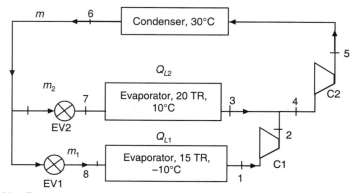

Figure 6.30 Two evaporators with compound compression and individual expansion valves.

[**Ans:** COP of the combined system = 5.946. Displacement volume of the compressor C1 is 2.2458 m³/min.]

6. A refrigeration system with R12 as refrigerant uses two evaporators of capacities 25 TR at 0°C and 30 TR at –20°C with individual compressors and individual expansion valves. The condenser is maintained at 40°C and liquid refrigerant leaving the condenser is subcooled to 30°C. The exit state of the refrigerant at each evaporator is dry saturated. Draw the arrangement and the *p–h* diagram of the system. Find the power and overall COP of the system. Refer to *p–h* chart for R12 properties.

[**Ans:** Total power = 43.16 kW, COP = 4.41]

Chapter 7

Vapour Absorption Refrigeration Systems

7.1 INTRODUCTION

The vapour absorption refrigeration system is one of the oldest methods of producing refrigerating effect. The principle of vapour absorption was first discovered by Michael Faraday in 1824 while performing a set of experiments to liquify certain gases. The first vapour absorption refrigeration machine was developed by a French scientist Ferdinand Carre in 1860. This system can be used in both domestic and large industrial refrigerating plants. The refrigerant commonly used in a vapour absorption system is ammonia.

The vapour absorption system uses heat energy, instead of mechanical energy as in vapour compression systems, in order to change the conditions of the refrigerant required for the operation of the refrigeration cycle. We have discussed in the previous chapters that the function of a compressor in a vapour compression system is to withdraw the vapour refrigerant from the evaporator. It then raises its temperature and pressure higher than the cooling agent in the condenser so that the higher pressure vapours can reject heat in the condenser. The liquid refrigerant leaving the condenser is now ready to expand to the evaporator conditions again.

In the vapour absorption system, the compressor is replaced by an absorber, a pump, a generator and a pressure reducing valve (Figure 7.1). These components in the vapour absorption system perform the same function as that of a compressor in the vapour compression system. In this system the vapour refrigerant from the evaporator is drawn into an absorption unit where it is absorbed by the weak solution of the refrigerant forming a strong solution. This strong solution is pumped to the generator where it is heated by some external source. During the heating process, the vapour refrigerant then flows into the evaporator and thus the cycle is completed.

The vapour absorption refrigeration system needs at least two fluids. One fluid acts as a refrigerant while the other as an absorber. These two ought to have certain properties which are discussed in the following subsection.

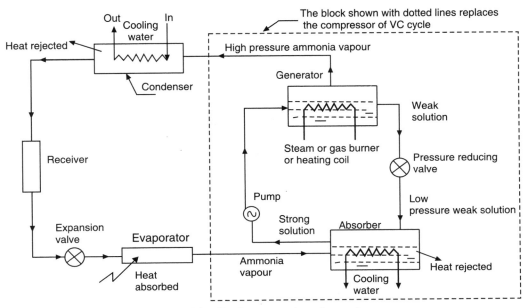

Figure 7.1 Simple vapour absorption system.

7.1.1 Refrigerant–Solvent Properties

The desirable properties of a solvent and those of a refrigerant–solvent combination are discussed here.

Desirable properties of (absorber) solvent

Some of the properties required of a solvent are:
1. The absorber should have great affinity to absorb the refrigerant.
2. Ideal absorbent should remain in liquid state under operating conditions.
3. Heat liberated during the absorption of refrigerant should be as small as possible. This reduces the heat to be rejected in the absorber.
4. It should have high boiling point.
5. The solvent should have low specific heat for better heat transfer.
6. It should have low viscosity for minimum pump work.
7. There should be a suitable pressure temperature concentration relationship to meet the actual practical conditions needed in the various components of the system.

Desirable properties of refrigerant–solvent combination

1. The desirable absorbent–refrigerant combination should have the property of high solubility at conditions in the absorber but low solubility in the generator.
2. The refrigerant should be more volatile than the absorbent for easy separation in the generator.
3. The combination of refrigerant–absorbent should be chemically stable.
4. Other properties to be considered are explosiveness, inflammability, toxicity, reaction with metals, high latent heat of vaporisation, freezing point and its availability.

5. Both should not cause corrosion.
6. The refrigerant should have high latent heat to have low mass flow rates.

Characteristics of ammonia

Ammonia is the most commonly used refrigerant in vapour absorption systems for both domestic and industrial applications. It possesses the following properties.
1. Water has large capacity to absorb ammonia vapour, e.g. 1 m^3 of water at 13°C is capable of absorbing 1000 m^3 of ammonia vapour.
2. The amount of NH_3 vapour that water can absorb increases with the increase in pressure and decreases with the increase in temperature.
3. During absorption of ammonia vapour in water, more heat is liberated. It is corrosive to metals.
4. Water can be induced to give up dissolved ammonia by heating since the boiling temperature of ammonia is −33.3°C at atmospheric pressure which is much lower than that of water at any pressure.

Some of the absorption systems are discussed in the following sections.

7.2 SIMPLE VAPOUR ABSORPTION SYSTEM

Ammonia is used as a refrigerant while water is used as an absorbent. The four components of the vapour compression cycle, as you know, are the evaporator, compressor, condenser and expansion valve. The simple vapour absorption system, as shown in Figure 7.1, consists of an absorbent, a pump, a generator and a pressure reducing valve to replace the compressor of the vapour compression system. The other components of the system are condenser, receiver, expansion valve and evaporator as in the vapour compression system.

Liquid ammonia (normally a mixture of liquid and vapour) from the expansion valve enters the evaporator. It absorbs heat from the evaporator space or it cools the secondary refrigerant in a heat exchanger. Normally these units have very large cooling capacity of the order of 80 TR and above. In such units, liquid ammonia absorbs heat from the secondary refrigerant which would be used as a medium to cool the space or products in the refrigerated space. Low pressure ammonia vapour then enters the absorber. This vapour is allowed to be mixed and absorbed in the absorber with the weak solution of aqua ammonia flowing from the generator under gravity through a pressure reducing valve. The water has the ability to absorb very large quantities of ammonia vapour and the solution thus formed is known as *aqua-ammonia*.

The absorption of ammonia vapour in water lowers the pressure in the absorber which in turn draws more ammonia vapour from the evaporator and thus raises the temperature of the solution. Some form of cooling arrangement (usually water cooling) is employed in the absorber to remove the heat of solution evolved in it. This is necessary in order to increase the absorption capacity of water because at higher temperature water absorbs less ammonia vapour. The strong solution thus formed in the absorber is pumped to the generator by a liquid pump. The pump increases the pressure of the solution up to 10 bar.

The strong solution of ammonia in the generator is heated by some external source such as gas or steam. During the heating process, the ammonia vapour is driven off the solution at high pressure leaving behind the hot weak ammonia solution in the generator. This weak ammonia solution flows back to the absorber at low pressure after passing through the pressure reducing valve. The high pressure ammonia vapour from the generator is condensed in the condenser to a high pressure liquid ammonia. This liquid ammonia is passed to the expansion valve through the receiver and then to the evaporator. This completes the simple vapour absorption cycle.

The heat required for the operation of generator can be supplied by burning kerosene or using solar energy or waste heat from process industry in the case of industrial applications.

The electrical energy required for the operation of aqua pump in this system is extremely small compared to the electrical energy needed for the compressor of a vapour compression cycle. The basic difference here is that aqua pump handles the liquid ammonia while the compressor has to work with the refrigerant vapour of high specific volume.

7.3 PRACTICAL VAPOUR ABSORPTION SYSTEM

The working principle of the simple vapour absorption system has been discussed in the previous section. In order to make the system more practical, it is fitted with an analyser, a rectifier and two heat exchangers as shown in Figure 7.2. These accessories help to improve the performance and working of the plant discussed as follows.

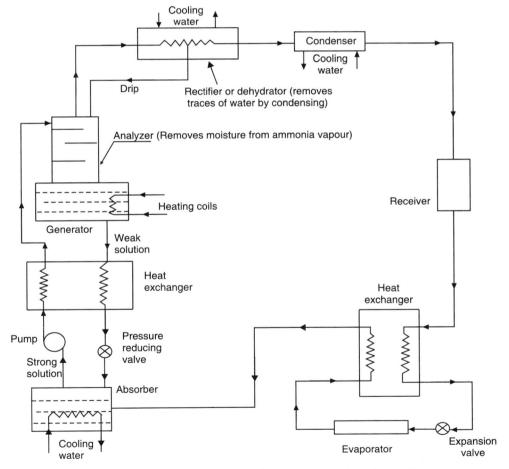

Figure 7.2 Block diagram of ammonia–water vapour absorption system.

Analyzer

The function of the analyzer is to remove moisture from the ammonia vapour leaving the generator. When ammonia is vaporized in the generator, some water is also vaporized and may flow into the condenser along with the ammonia vapours in the simple system. If these unwanted water particles are not removed before entering into the condenser, they will enter into the expansion valve where they freeze and choke the pipeline. In order to remove these unwanted water particles flowing to the condenser, an analyzer is used. The analyzer may be built as an integral part of the generator or made a separate piece of equipment. It consists of a series of trays mounted above the generator. The strong solution from the absorber and the aqua from the rectifier are introduced at the top of the analyzer and flow downward over the trays and into the generator. In this way, considerable liquid surface area is exposed to the vapour rising from the generator. The vapour is cooled and most of the water vapour condenses, so that mainly the ammonia vapour leaves the top of the analyzer. Since the aqua is heated by the vapour, less external heat is required in the generator.

Rectifier

The function of the rectifier is to condense the leftover traces of water and drain the condensate back to the analyzer, i.e. in case the water vapours are not completely removed in the analyzer, a closed type vapour cooler called rectifier (also known as dehydrator) is used. It is generally water-cooled and may be of the double pipe, shell and coil or shell and tube type. Its function is to further cool the ammonia vapour leaving the analyzer so that the water vapour still remaining in the ammonia vapour, is condensed. Thus, only dry or anhydrous ammonia vapour flows to the condenser. The condensate from the rectifier is returned to the top of the analyzer by a drip return pipe.

Heat exchangers

The heat exchanger provided between the pump and the generator cools the weak hot solution returning from the generator to the absorber using the cold strong aqua solution supplied by the aqua pump. The heat removed from the weak solution raises the temperature of the strong solution leaving the pump and going to analyzer and generator. This operation reduces the heat supplied to the generator and the amount of cooling required for the absorber. Thus, the economy of the plant increases.

The heat exchanger provided between the condenser and the evaporator may also be called liquid subcooler. In this heat exchanger, the liquid refrigerant leaving the condenser is sub-cooled by the low temperature ammonia vapour from the evaporator as shown in Figure 7.2. This subcooled liquid is now passed to the expansion valve and then to the evaporator. The functions of other components are as explained in Section 7.2.

In this system, the net refrigerating effect is the heat absorbed by the refrigerant in the evaporator. The total energy supplied to the system is the sum of the work done by the pump and the heat supplied in the generator. Therefore, the coefficient of performance of the system is given by

$$COP = \frac{\text{Heat absorbed in evaporator}}{\text{Work done by pump} + \text{Heat supplied in generator}}$$

7.4 VAPOUR ABSORPTION REFRIGERATION SYSTEM VS. VAPOUR COMPRESSION REFRIGERATION SYSTEM

Following are the advantages of the vapour absorption system over the vapour compression system:

1. In the vapour absorption system, the only moving part of the entire system is a pump which has a small motor. Thus, the operation of this system is essentially quiet and subjected to little wear. The vapour compression system of the same capacity has more wear, tear and noise due to moving parts of the compressor.
2. The vapour absorption system uses heat energy to change the condition of the refrigerant from the evaporator. The vapour compression system uses mechanical energy to change the condition of the refrigerant from the evaporator.
3. The vapour absorption systems are usually designed to use steam, either at high pressure or low pressure. The exhaust steam from furnaces and solar energy may also be used. Thus, this system can be used where the electric power is difficult to obtain or is very expensive.
4. The vapour absorption systems can operate at reduced evaporator pressure and temperature by increasing the steam pressure to the generator with little decrease in the capacity. But the capacity of a vapour compression system drops rapidly with lowered evaporator pressure.
5. The load variations do not affect the performance of a vapour absorption system. The load variations are met by controlling the quantity of aqua circulated and the quantity of steam supplied to the generator. The performance of a vapour compression system at partial loads is, however, poor.
6. In the vapour absorption system, the liquid refrigerant leaving the evaporator has no bad effect on the system except that of reducing the refrigerating effect. In the vapour compression system, it is essential to superheat the vapour refrigerant leaving the evaporator so that no liquid may enter the compressor.
7. The vapour absorption systems can be built in capacities well above 1000 TR of refrigeration. The same is not the case with the vapour compression cycle using compressors.
8. The space requirements and automatic control requirements favour the absorption system more as the desired evaporator temperature drops.

7.5 COP OF AN IDEAL VAPOUR ABSORPTION REFRIGERATION SYSTEM

The objective of the refrigerator is to remove heat (Q_L) from the cold space. To accomplish this, it needs energy input. Therefore, the COP of a refrigerator is

$$\text{COP}_R = \frac{\text{Desired effect, } Q_L}{\text{Energy input}}$$

Q_L is absorbed by the refrigerant in the evaporator.

The energy input in an absorption refrigeration system includes: (1) The heat Q_g given to the refrigerant in the generator, and (2) The heat Q_P added to the refrigerant due to pump work as shown in Figure 7.3.

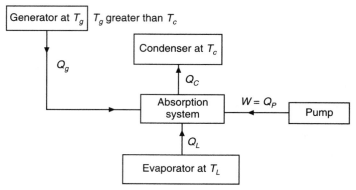

Figure 7.3 Theoretical COP of the vapour absorption system.

Let Q_C be the heat dissipated to the atmosphere or cooling water from the condenser and absorber.

According to the first law of thermodynamics,

$$Q_C = Q_g + Q_L + Q_P$$

Since the heat due to pump work Q_P is very negligible,

$$Q_C = Q_g + Q_L \tag{7.1}$$

Let

T_g be the temperature at which heat Q_g is supplied to the generator
T_C be the temperature at which heat Q_C is discharged to atmosphere or cooling water from the condenser and absorber
T_L be the temperature at which heat Q_L is absorbed in the evaporator.

Since the vapour absorption system can be considered as a perfectly reversible system, the initial entropy of the system must be equal to the entropy of the system after the change in its condition.

$\therefore$

$$\frac{Q_g}{T_g} + \frac{Q_L}{T_L} = \frac{Q_C}{T_C} \tag{7.2}$$

From Eq. (7.1), we can write

$$\frac{Q_C}{T_C} = \frac{Q_g + Q_L}{T_C}$$

or

$$\frac{Q_g}{T_g} - \frac{Q_g}{T_C} = \frac{Q_L}{T_C} - \frac{Q_L}{T_L}$$

or

$$Q_g \left(\frac{T_C - T_g}{T_g \times T_C} \right) = Q_L \left(\frac{T_L - T_C}{T_C \times T_L} \right)$$

or
$$Q_g = Q_L \left(\frac{T_L - T_C}{T_C \times T_L}\right) \left(\frac{T_g \times T_C}{T_C - T_g}\right)$$

$$= Q_L \left(\frac{T_C - T_L}{T_C \times T_L}\right) \left(\frac{T_g \times T_C}{T_g - T_C}\right)$$

$$= Q_L \left(\frac{T_C - T_L}{T_L}\right) \left(\frac{T_g}{T_g - T_C}\right) \qquad (7.3)$$

The maximum coefficient of performance (COP) of the system is given by

$$(\text{COP})_{\max} = \frac{Q_L}{Q_g}$$

$$= \left(\frac{T_L}{T_C - T_L}\right) \left(\frac{T_g - T_C}{T_g}\right) \qquad (7.4)$$

In Eq. (7.4):

1. The expression $\left(\dfrac{T_L}{T_C - T_L}\right)$ represents the COP of a Carnot refrigerator working between the temperature limits of T_L and T_C.

2. The expression $\left(\dfrac{T_g - T_C}{T_g}\right)$ represents the efficiency of a Carnot engine working between the temperature limits of T_g and T_C.

Thus, a theoretical or an ideal vapour absorption refrigeration system may be regarded as a combination of a Carnot engine and a Carnot refrigerator. The maximum COP may be written as:

$$(\text{COP})_{\max} = (\text{COP})_{\text{Carnot}} \times \eta_{\text{Carnot}}$$

EXAMPLE 7.1 In a vapour absorption refrigeration system, heat addition in the generator is at 90°C, heat rejection is at 30°C, and refrigeration takes place at the temperatures of –5°C. Find the maximum COP of the system.

Solution: Given: $T_g = 90°C = 90 + 273 = 353$ K; $T_C = 30°C = 30 + 273 = 303$ K; $T_L = -5°C = -5 + 273 = 268$ K

We know that the maximum COP of the system is

$$= \left(\frac{T_L}{T_C - T_L}\right) \left(\frac{T_g - T_C}{T_g}\right) = \left(\frac{268}{303 - 268}\right) \left(\frac{353 - 303}{353}\right) = 1.079 \qquad \textbf{Ans.}$$

268 *Refrigeration and Air Conditioning*

One question that comes to our mind is: what is significance of solving this example? Let us assume that a very large capacity ammonia vapour absorption refrigeration system is working. We know only three temperatures. With such a data under ideal conditions, the unit can produce a refrigeration effect at the rate of 1 kW at –5°C, consuming energy at the rate of 1 kW. One can also make a rough estimate that if the actual COP of the system is 50% of the maximum COP, the unit can produce cooling effect at the rate 1 kW for every 2 kW energy input. One can make a rough estimate of the COP of the system for the varying operating temperatures of T_g, T_C and T_L.

EXAMPLE 7.2 In an absorption type refrigerator, the heat is supplied to NH_3 generator by condensing steam at 1.6 bar and 80% dry. The temperature in the refrigerator is to be maintained at –5°C. Find the maximum COP possible.

If the refrigeration load is 150 TR and the actual COP is 80% of the maximum COP, find the mass of the steam required per hour. Take the temperature of the atmosphere to be 30°C.

Solution: Given: P = 1.6 bar; x = 80% = 0.8; T_L = –5°C = –5 + 273 = 268 K; Q = 150 TR; Actual COP = 80% of maximum COP; T_C = 30°C = 30 + 273 = 303 K

Maximum COP

It is interesting to ask one question here as to why is steam considered wet steam and not dry saturated or superheated one. The answer is, that steam is not generated in the boiler specifically to run the absorption refrigeration unit. Heat energy from the steam available from the process industry as a waste, or that available from the outlet of the steam turbine, is used for absorption refrigeration. The condition of steam from such a source is normally of this order. So, wet steam is assumed here.

From steam tables, we find that the saturation temperature of steam at a pressure of 1.6 bar is

$$T_g = 113.3°C = 113.3 + 273 = 386.3 \text{ K}$$

We know that maximum COP

$$= \left(\frac{T_L}{T_C - T_L}\right)\left(\frac{T_g - T_C}{T_g}\right) = \left(\frac{268}{303 - 268}\right)\left(\frac{386.3 - 303}{386.3}\right) = 1.417$$

Mass of steam required per hour:
We know that actual COP

$$= 0.8 \text{ of maximum COP} = 0.8 \times 1.417 = 1.133$$

∴ Actual heat supplied

$$= \frac{\text{Refrigeration load}}{\text{Actual COP}} = \frac{150 \times 3.5167}{1.133} = 465.5 \text{ kW}$$

Assuming that only latent heat of steam is used for heating purposes, therefore from steam tables, the latent heat of steam at 1.6 bar is

$$h_{fg} = 2220.9 \text{ kJ/kg}$$

∴ Mass of steam required per hour

$$= \frac{\text{Actual heat supplied}}{h_{fg}} = \frac{465.5}{2220.9} = 0.232 \text{ kg/s} = 838 \text{ kg/h} \qquad \text{Ans.}$$

EXAMPLE 7.3 (a) In a vapour absorption system, the heating, cooling and refrigeration temperatures are 115°C, 30°C and –10°C, respectively. Find the COP of the system.

(b) In case the heating temperature is increased to 200°C and the refrigeration temperature is reduced to –33°C with cooling temperature remaining the same, find the new COP and percentage change in COP.

Solution:
(a) **Given:**
$T_g = 115°C = 115 + 273 = 388$ K
$T_C = 30°C = 30 + 273 = 303$ K
$T_L = -10°C = -10 + 273 = 263$ K

$$\text{COP} = \left(\frac{T_L}{T_C - T_L}\right) \times \left(\frac{T_g - T_C}{T_g}\right) = \left(\frac{263}{303 - 263}\right) \times \left(\frac{388 - 303}{388}\right)$$

$$= 1.44 \qquad \text{Ans.}$$

(b) **Given:**
$T_{g1} = 200°C = 200 + 273 = 473$ K, $T_{C1} = T_C = 30°C = 303$ K
$T_{L1} = -10°C = -10 + 273 = 263$ K

$$(\text{COP})_1 = \left(\frac{T_{L1}}{T_{C1} - T_{L1}}\right) \times \left(\frac{T_{g1} - T_{C1}}{T_{g1}}\right) = \left(\frac{263}{303 - 263}\right) \times \left(\frac{473 - 303}{473}\right)$$

$$= 2.363 \qquad \text{Ans.}$$

Here, the temperature of the heat source is increased from 115°C to 200°C because of which COP had jumped from 1.44 to 2.363. It indicates that the temperature at which heat is supplied to the generator has a commendable effect on the performance of the unit.

7.6 DOMESTIC ELECTROLUX (AMMONIA–HYDROGEN) REFRIGERATOR

The 'Electrolux Company' of Luton, England first developed this *Domestic Electrolux (Ammonia–Hydrogen) Refrigerator.*

This type of refrigerator is called *three fluids absorption* system. Three fluids are, namely ammonia, hydrogen and water. Ammonia acts as a refrigerant, water as an absorbent and hydrogen gas promotes evaporation of refrigerant in the evaporator. Hydrogen does not react with ammonia and water. The main purpose of this system is to eliminate the pump so that in the absence of moving parts, the machine becomes noiseless.

270 *Refrigeration and Air Conditioning*

The principle of operation of a domestic Electrolux type refrigerator, as shown in Figure 7.4, is discussed below.

The strong ammonia solution is heated in the generator by applying heat from an external source, usually a gas burner. Due to heating process, ammonia vapours are removed from the solution and passed to the condenser. A rectifier or a water separator fitted before the condenser removes water vapour carried with the ammonia vapours so that dry ammonia vapours are supplied to the condenser. These water vapours, if not removed, will enter into the evaporator causing freezing and choking of the machine. The hot weak solution left behind in the generator flows to the absorber through the heat exchanger. This hot weak solution while passing through the exchanger is cooled. The heat removed by the weak solution is utilized in raising the temperature of the strong solution passing through the heat exchanger. In this way, the absorption is accelerated and improvement in the performance of a plant is achieved.

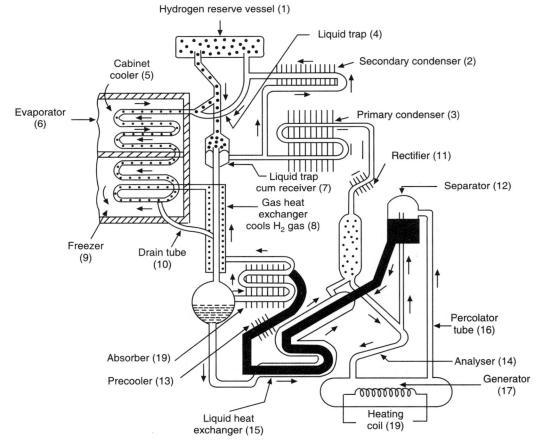

Figure 7.4 Block diagram of Electrolux refrigerator.

The ammonia vapours in the condenser are condensed by using an external cooling source. The liquid refrigerant leaving the condenser flows under gravity to the evaporator where it meets the hydrogen gas. The hydrogen gas which is being fed to the evaporator permits the liquid ammonia to evaporate at a low pressure and temperature according to Dalton's principle. During the process of evaporation, the ammonia absorbs latent heat from the refrigerated space and thus produces cooling effect.

The mixture of ammonia vapour and hydrogen is passed to the absorber where ammonia is absorbed in water while the hydrogen rises to the top and flows back to the evaporator. This completes the cycle. The coefficient of performance of this refrigerator is given by

$$COP = \frac{\text{Heat absorbed in the evaporator}}{\text{Heat supplied in the generator}}$$

Note:
1. The hydrogen gas only circulates from the absorber to the evaporator and back.
2. The whole cycle is carried out entirely by gravity flow of the refrigerant.
3. It cannot be used for industrial purposes as the COP of the system is very low.

7.7 LITHIUM BROMIDE ABSORPTION REFRIGERATION SYSTEM

In this system water works as a refrigerant and lithium bromide salt solution as an absorbent. The normal freezing point of water is 0°C. Therefore, this system is suitable for the applications where the cooling effect produced is at a temperature higher than 0°C. For example, air conditioning applications, and cold storages to store potatoes, onions, etc.

The vapour pressure of aqueous solution lithium bromide is very low. The vapour pressure is the pressure at which the liquid particles separate from the liquid surface to transform into vapour. The solution formed mixing 60% lithium bromide with 40% water at 43.5°C has a vapour pressure of the order of 6.25 mm of Hg (0.88 kPa) for which the corresponding saturation temperature for water is 6°C. It means water from the aqueous solution of lithium-bromide can evaporate at 6°C. The lithium bromide solution has strong affinity to water vapour because of its low vapour pressure. It means if water and lithium bromide solution are placed adjacent to each other in a closed evacuated system, water would evaporate. It is this principle which is utilized here.

To explain the working principle, let us consider the block diagram shown in Figure 7.5. In the evaporator, pressure of the order of 0.07 bar is maintained. Therefore, the water entering into it from the condenser evaporates. While evaporating it takes away the heat from the remaining water in it, thus producing cooling effect there. The secondary refrigerant like brine is circulated in this evaporator. It gets cooled which is supplied to the refrigerated space to take the cooling load. The water vapour formed in the evaporator then enters the absorber where it is absorbed in the strong lithium bromide salt solution coming from the generator after being cooled in the heat exchanger.

The strong solution becomes weak after absorbing water. This weak solution is pumped and heated in the heat exchanger while being supplied to the generator. Steam is circulated in the heating coil placed inside the generator. Hence, heat is supplied to the generator and so the water evaporates and this water vapour is carried to the condenser where it condenses. To remove the heat due to condensation of water vapour, cold water is circulated through the coil placed in it. The condensate enters the evaporator and the cycle of operation repeats.

In the heat exchanger, the weak solution from the absorber is heated by the strong solution flowing from the generator. It means that the hot strong liquid is cooled giving its heat to the weak solution before going into the generator. It reduces the heat requirement in the generator and the cooling load in the absorber.

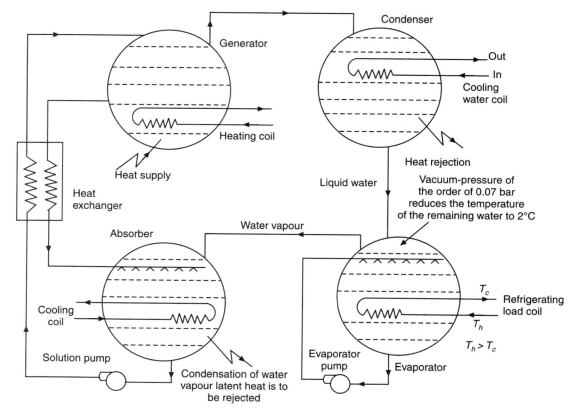

Figure 7.5 Lithium bromide absorption refrigeration system.

In the working unit, both the absorber and evaporator are placed in the common shell which operates at the low pressure (0.0078 bar) of the system. The generator and condenser are in a separate shell where the pressure of the order of 0.098 bar is maintained. This type of arrangement is shown in Figure 7.6.

The main components of the system are: (1) Generator, (2) Heat exchanger, (3) Absorber, (4) Solution pump, (5) Evaporator pump, (6) Condenser, and (7) Evaporator.

The weak Li-Br solution from the absorber is circulated through the heat exchanger by the solution pump to the generator.

In the heat exchanger, the weak solution from the absorber is heated by the strong solution flowing from the generator. It means that the weak hot liquid is cooled giving its heat to the weak solution before going into the generator. It reduces the heat requirement in the generator and cooling load in the absorber.

The water vapour refrigerant is condensed by the water circulated from cooling pond, which condenses the water vapour formed.

The condensed water vapour refrigerant from condenser at high pressure flows down from condenser. Its pressure is reduced up to evaporator pressure in the pressure reducing valve. The cooled water is sprayed in the evaporator as shown.

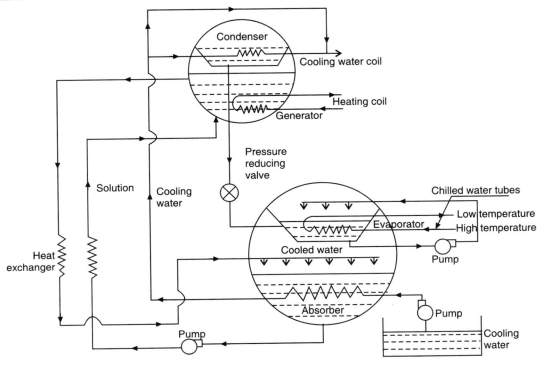

Figure 7.6 A practical lithium bromide absorption refrigeration system.

An evaporator pump sprays this cooled water where it absorbs its latent heat of vaporization from hot water circulated from air conditioned space in chilled water tubes. During the process, the hot water gives away heat and is converted into chilled water, thereby, producing the required refrigeration load for air conditioning purposes and the sprayed cool water is converted into water vapour.

The water vapour is absorbed by the strong Li-Br solution sprayed in the absorber and it is converted into weak Li-Br solution. Thus, the cycle is completed.

Advantages of the system
1. The pressure inside the system is below atmosphere, hence the system is light in weight.
2. Water as the refrigerant is non-toxic, homogeneous and it can be directly used for chilled coil for air conditioning.
3. The work energy requirements are negligible since the pressure difference between the generator and the evaporator is very small. Even the pumps can be eliminated in the system using the gravity feed by adjusting the elevation of generator, absorber and evaporator.
4. These plants have been built with up to one lakh TR capacity by Thermax Ltd. for various industries.
5. Operation and maintenance cost is very low.

Disadvantages of the system
1. Lithium bromide solution is corrosive, therefore inhibitors should be added to protect metal parts against corrosion.

2. All joints have to be made leakproof to prevent leakages since the system works under vacuum.
3. Once the system stops working, the salt solution may solidify and require replacement of the pipes, etc.

EXERCISES

1. How does an absorption refrigeration system differ from a mechanical refrigeration system?
2. How was the concept of vapour absorption system generated? Explain briefly.
3. Enumerate the desirable properties of solvent and refrigerant-solvent combination for the vapour absorption system.
4. Explain the working of a simple ammonia-water absorption refrigeration system with the help of a neat sketch.
5. Draw a schematic diagram of an actual vapour absorption refrigeration system and explain its performance.
6. Describe with a neat sketch the Li-Br and water system. What are the limitations?
7. Derive an expression for the COP of an ideal vapour absorption system in terms of heating, cooling and refrigerator temperatures.
8. Discuss the merits and demerits of the vapour absorption system compared to the vapour compression system.
9. With the aid of a neat line diagram showing all the component units, explain the working of the vapour absorption refrigeration cycle without any mechanical parts.
10. Write a short note on the Li-Br–water absorption system.

NUMERICALS

1. In an aqua ammonia vapour absorption plant, heat is supplied to the generator by condensing steam at 2 bar and 0.9 dry. The evaporator is to be maintained at 5°C. Assuming the ambient temperature as 30°C, calculate the maximum possible COP. Condensate leaves the generator at 30°C. If the actual COP is 70% of the maximum COP, calculate the mass of the steam required per day for a 20 TR plant capacity. Take the saturation temperature at 2 bar as 120°C and enthalpy of evaporation as 2200 kJ/kg.
   [**Ans:** Max COP = 2.546, Actual COP = 1.782, Steam = 71.39 kg/h]
2. The temperature of generator, condenser and evaporator of a vapour absorption system are 95°C, 30°C and –5°C, respectively. Find its maximum COP.
3. In Q. 2, if the respective temperatures of generator, condenser and evaporator are changed to 195°C, 30°C and –35°C, respectively, find the new COP and the percentage change in COP.
4. Heat to generator is supplied by steam at 2 bar, 0.9 dry in an ammonia-water absorption system. The cooling water available for condenser is at 30°C and the evaporator temperature is maintained at 5°C. (a) Find its ideal COP. (b) If the actual COP of the system is 80% of ideal COP, find the mass of steam needed in kg/h to produce a refrigeration load of 60 TR.

Chapter 8

Psychrometry

8.1 INTRODUCTION

Air conditioning means to maintain the temperature conditions that are either conducive to human comfort or are required by a product or process within a space.

Comfort air conditioning

Since the purpose of most air-conditioning systems is to provide a comfortable indoor environment, the system designer and operator should understand the factors that affect comfort.

Body heat loss

Heat is generated in the human body due to metabolism or digestion of food. This body heat is continually lost to its cooler surroundings. The factor that determines whether one feels hot or cold is the rate of loss of body heat. When the rate of heat loss is within certain limits, one feels comfortable. If the rate of heat loss is too much, one feels cold and if the rate is too low, one feels hot.

The processes by which the body loses heat to the surroundings are convection, radiation and evaporation.

In convection, the air immediately around the body receives heat from the body and becomes warm. The warm air continually moves away, by rising naturally through the cooler air around it or by being blown away, and is replaced by more air which in turn receives heat.

In radiation, the body heat is transmitted through space directly to nearby objects (e.g. walls) which are at a lower temperature than the body. However, heating sources that are warmer than the body can radiate heat towards the body, creating a feeling of warmth even at a low surrounding air temperature. This is why one feels warm in front of a fireplace even on a cold day.

The body is also cooled by evaporation: sweat on the skin, which has absorbed heat from the body, evaporates into the surrounding air taking the body heat with it.

The rate of body heat loss is affected by five conditions:
1. Air temperature
2. Air humidity
3. Air motion
4. Temperature of surrounding objects
5. Clothing

The system designer and operator can control comfort, primarily by adjusting three conditions: *temperature, humidity* and *air motion*. How are they adjusted to improve comfort?

The indoor air temperature may be lowered to increase the body heat loss in summer while in winter it may be raised to decrease the body heat loss.

In winter, humidity may be raised to decrease the body heat loss and in summer humidity may be lowered to increase the body heat loss by evaporation.

Air motion may be raised to increase the body heat loss in summer and lowered to decrease the body heat loss in winter by convection.

The occupants of buildings, of course, have some personal control over their own comfort. For instance, they can control the amount of clothing that they wear, they can use local fans to increase convection and evaporative heat loss, and they can even stay away from cold walls and windows to keep themselves warm in winter.

Indoor air quality

Another factor, air quality, refers to the degree of purity of the air. The level of air quality affects both comfort and health. Air quality is worsened by the presence of contaminants such as tobacco smoke and dust particles, biological micro-organisms and toxic gases. Cleaning devices such as filters may be used to remove particles. Adsorbent chemicals may be used to remove unwanted gases. Indoor air contaminants can be diluted in concentration by introducing substantial quantities of outdoor air into the building. This procedure is called ventilation.

The subject of Indoor Air Quality (IAQ) has been of major concern and importance in recent years. Evidence shows that there are many possible indoor air contaminants which can and have caused serious health effects on occupants. The phrases 'sick building syndrome' and 'building-related illnesses' have been coined to refer to these effects. The recommended inside design conditions both during summer and winter are given in Table 8.1(a).

Table 8.1(a) Recommended inside design conditions—summer and winter

Type of application	Summer		Winter	
	DBT (°C)	RH (%)	DBT (°C)	RH (%)
General comfort, house, hotel, office, school, apartments	23.3–24.5	45–50	23.3–24.5	35–36
Factory comfort Assembly areas, machining rooms	26.6	50–60	21.1	23.3

Industrial air conditioning

Some industrial products or processes require certain conditions, which are listed in Table 8.1(b).

Table 8.1(b) Typical samples of inside design conditions (Industrial)

Industry	Process	DBT (°C)	RH (%)	Industry	Process	DBT (°C)	RH (%)
Bakery	Dough mixer	23.8–26.6	40–50	Printing	Press room	23.8–26.6	46–48
	Fermenting	23.8–27.7	70–75		Stock room	22.7–26.6	49–51
	Crackers and biscuits	15.5–18.3	50		Storage and folding	comfort	
Textile	Cotton weaving	25.5–26.6	70–85	Manufacturing	Gear assembly	23.8–26.6	35–40
	Silk weaving	26.6	65–70		Gasket storage	37.7	50
	Rayon weaving	26.6	50–60		Honing	23.8–26.6	35–45

The conditions stated in Table 8.1(b) are only typical but may vary with applications. They may also vary as changes occur in processes, products and knowledge of the effect of temperature and humidity.

Generally, specific design conditions are required in industrial applications for one or more of the following reasons:

1. A constant temperature level may be required for a close tolerance measuring machining or grinding operations to prevent expansion or contraction of the machine parts, machined products and measuring devices. Non-hygroscopic materials such as metals, glass, plastics, etc. have a property of capturing water molecules within the microscopic surface crevices, forming an invisible, non-continuous surface film. The density of this film increases when relative humidity increases. So this film must, in many instances, be held below a critical point at which metals may etch, or the electric resistance of insulating materials is significantly decreased.

2. Control of RH is required to maintain the strength, pliability and regain of hygroscopic materials such as textiles and paper.

3. The DBT and RH control are required to regulate the rate of chemical or biochemical reactions, such as drying of varnishes or sugar coatings, preparation of synthetic fibres or chemical compounds, fermentation of yeasts, etc.

8.2 PSYCHROMETRY

This branch of science deals with the study of properties of moist air and its behaviour under different conditions. The properties of moist air include Dry-bulb Temperature (DBT), Wet-bulb Temperature (WBT), humidity, Relative Humidity (RH), etc. Such a study is important because the atmospheric air is not completely dry but is a mixture of air and water vapour. Threldkeld had proved that perfect gas relations could be used for air-conditioning calculations. The errors are less than 0.7% in calculating the humidity ratio, enthalpy and specific volume of saturated air at standard atmospheric pressure for a temperature range of –50 to 50°C. In addition to this, Gibbs–Dalton's laws for non-reactive mixtures of gases can be applied to the dry air part to obtain its properties as a single-phase substance.

The data in Tables 8.2 and 8.3 are representative of the average concentrations of many of the permanent gases and variable substances found in our atmosphere. It may be noted that in contrast with the permanent gases, the variable substances include gases as well as particulate matter.

Table 8.2 Permanent gases in air near the earth's surface

Gas	Volume or mole (%)	Parts per million (ppm)	Boiling point (K)
N_2 – Nitrogen	78.08	—	77.36
O_2 – Oxygen	20.95	—	90.18
Ar – Argon	0.93	—	87.28
Ne – Neon	0.0018	18.2	27.09
He – Helium	0.0005	5.2	4.21
Kr – Krypton	0.000114	1.14	119.83
Xe – Xenon	0.0000086	0.086	165.0
H_2 – Hydrogen	0.00005	0.5	20.3

Table 8.3 Variable substances in air near the earth's surface

Gas	Volume or mole (%)	Parts per million (ppm)	Boiling point (K)
Water vapour	0 to 4	—	373
Carbon dioxide	0.035	350	—
Methane (CH_4)	0.00017	1.7	214.3
Nitrous oxide (N_2O)	0.00003	0.3	183.7
Ozone (O_3)	0.000004	0.04	—
Particulate matter	0.000001	0.01	—
Chloroflurocarbons (CFCs)	0.00000001	0.0001	—

Of the variable substances in the atmosphere, water vapour (H_2O) is the most variable of all with concentrations ranging from 0–4% by volume. Nearly all of the water vapour in the atmosphere resides in the lower portion of the atmosphere known as troposphere. The water content in the atmosphere is often expressed as relative humidity (RH).

Dry air: The air consisting of nitrogen, oxygen, argon, etc. shown in Table 8.2 is called dry air, these being the major constituents of air. The molecular mass of dry air is 28.966 kg/kmol and characteristic gas constant R_a is 287 J/kg-K.

Moist air: It is a homogeneous mixture of dry air and water vapour. The moisture-holding capacity of the air depends upon its temperature and pressure. Air at higher temperatures can hold more moisture than the air at low temperatures. Air at high pressures holds less moisture than the air at low pressures. The molecular mass of water vapour is 18.016 kg/kmol and the characteristic gas constant R_w is 461 J/kg-K.

Saturated air: It is the condition of air that holds maximum quantity of water vapour into it at a given temperature.

Let us consider moist air represented by the state A on T–s diagram shown in Figure 8.1.

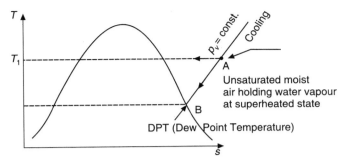

Figure 8.1 Dew point temperature.

Water vapour in the air at state A is in the superheated condition and it has a partial pressure p_v. Suppose one adds water vapour to such air till the air becomes fully saturated with water vapour (point B), the partial pressure of water vapour contained in the air would be p_s. The addition of water vapour to the air can be by spraying water in the air. At state point B, air contains the maximum amount of water vapour corresponding to its temperature. For saturated air, the maximum specific humidity (w_s or w_{max}) is given as

$$w_{max} = 0.622 \frac{p_s}{p - p_s} \tag{8.1}$$

Here, p_s = partial pressure of water vapour in a saturated air corresponding to DBT and p is the total pressure of moist air.

Dry-Bulb Temperature (DBT): It is the temperature of air measured or recorded by a thermometer. It is denoted by T or T_{db} or DBT.

Wet-Bulb Temperature (WBT): It is the temperature of air recorded by a thermometer when its bulb is covered with wet wick or cloth over which air is moving at a velocity at 2.5 to 10 m/s.

Dew Point Temperature (DPT): It is the temperature of air recorded by a thermometer when the moisture present in it starts condensing. It is denoted by T_{dp} or DPT.

Consider that a certain sample of unsaturated moist air shown by state A, in Figure 8.1, is cooled at constant pressure slowly by passing over the cooling coil. Its temperature goes on decreasing till it reaches a temperature DPT, at which the first drop of dew will be formed. It means that the water vapour in the air starts condensing. In the case of dehumidification of air, it is required to maintain the temperature of cooling coil well below DPT. During the cooling process, the partial pressure of water vapour and the specific humidity w remain constant until the vapour starts condensing.

The DPT (saturated temperature) can be found from the steam table corresponding to the partial pressure of water vapour p_v.

Humidity ratio (specific humidity): It is the ratio of mass of water vapour to the mass of dry air contained in the sample air. It is denoted by 'w'. It is normally expressed in g/kg of dry air.

$$w = \frac{\text{Mass of water vapour in air}}{\text{Mass of dry air in air}}$$

$$= \frac{m_v}{m_a} \qquad (8.2)$$

Let p_a, v_a, T, m_a, and R_a be the pressure, specific volume, DBT in K, mass and gas constant respectively for dry air.

Let p_v, v_v, T, m_v, and R_v be the pressure, specific volume, DBT in K, mass and gas constant respectively for water vapour.

The characteristic gas equation $pv = mRT$, can be applied to dry air as well as water vapour.

For dry air, $\qquad p_a v_a = m_a R_a T \qquad (8.3)$

For water vapour, $\qquad p_v v_v = m_v R_v T \qquad (8.4)$

As air and water vapour have the same volume and temperature, from Eqs. (8.3) and (8.4),

$$\frac{p_v}{p_a} = \frac{m_v R_v}{m_a R_a} \qquad (8.5)$$

$$w = \frac{m_v}{m_a} = \frac{p_v R_a}{p_a R_v} \qquad (8.6)$$

Characteristic gas constant R_a is 287 J/kg-K, R_v = 461 J/kg-K.

$$\therefore \qquad w = \frac{m_v}{m_a} = \frac{p_v \times 287}{p_a \times 461} = \frac{0.6225 \times p_v}{p_a}$$

where p_v and p_a are partial pressure of water vapour and dry air respectively.

$$\therefore \qquad w = \frac{0.6229 \times (p_v)}{(p - p_v)} \text{ kg/kg dry air} \qquad (8.7)$$

For known values of barometer pressure p and dew point temperature, the humidity ratio is determined by Eq. (8.7).

EXAMPLE 8.1 The dry-bulb temperature and dew point temperature of atmospheric air are 30°C and 14°C, respectively. If the barometer reading is 758 mm of Hg, determine the humidity ratio.

Solution: From steam tables, the partial pressure of water vapour corresponding to DP temperature 14°C is 0.015973 bar.

The atmospheric pressure is 758 mm Hg (758 × 0.0013332 = 1.0105656 bar).

$$\text{Humidity ratio, } w = \frac{0.6229 \, (p_v)}{(p - p_v)} = \frac{0.6229 \, (0.015973)}{(1.0105656 - 0.015973)}$$

$$= 0.01605 \text{ kg/kg dry air} \qquad \textbf{Ans.}$$

Absolute humidity: It is the mass of water vapour present in one cubic metre of dry air. It is expressed in terms of gram per cubic metre of dry air (g/m³ of dry air). Many a time it is expressed in terms of grains per m³ of dry air. One kg of water vapour is equal to 15,430 grains.

$$p_v V = m_v R_v T$$

where
p_v = vapour pressure in air or saturation pressure at dew point
V = volume of air, which is also of water vapour
T = dry bulb temperature
R_v = gas constant of water vapour (462 kJ/kg-K)
m_v = mass of water vapour in kg

$\therefore$ $$\text{Vapour density} = \frac{m_v}{V} = \frac{p_v}{R_v T}$$

EXAMPLE 8.2 Find the absolute humidity of the air sample which has a dew point temperature of 16°C.

Solution: From steam tables, the vapour pressure for the saturation temperature of 16°C is 0.018168 bar. Let the volume V be 1 m³.

$$R = \frac{R_0}{M_{wv}} = \frac{8314.14}{18.015} = 461.52 \text{ J/kg-K}$$

Mass of water vapour per m³ is

$$\rho = \frac{p_v}{RT} = \frac{1816.8 \times 1}{461.52 \times 289} = 0.01362$$

$$\rho = 0.01362 \text{ kg/m}^3 \qquad \textbf{Ans.}$$

This vapour density can also be directly determined using the steam tables. Take the reciprocal of the specific volume of saturated vapour (1/73.38 = 0.01362 kg/m³).

Degree of saturation (μ): It is the mass of water vapour in a sample of air to the mass of water vapour in the same air when it is saturated at the same temperature. Mathematically,

$$\mu = \frac{w_v}{w_s} \qquad (8.8)$$

where w_v and w_s are specific humidity of air and saturated air, respectively.

$$\mu = \frac{0.622 \dfrac{p_v}{p - p_v}}{0.622 \dfrac{p_s}{p - p_s}}$$

or

$$\mu = \frac{p_v}{p_s}\left(\frac{p - p_s}{p - p_v}\right)$$

where
p_s = partial pressure of water vapour when air is separated. It is obtained from steam tables corresponding to DBT (T_{db}).

p_v = partial pressure of water vapour in a moist air.
p = total pressure of moist air.

Relative humidity, $RH = \dfrac{p_v}{p_s} = 0$ when moist air is totally dry, i.e. which does not contain water vapour.

If the moist air is saturated, then $p_v = p_s$, then $RH = 1$ and $\mu = 1$. It shows that the degree of saturation varies between 0 and 1.

It is necessary to study the properties of air to control the environment in space. The study of properties of ambient air is known as *psychrometrics*. We will study these properties in subsequent sections.

Pressure: In air conditioning terms, air means a mixture of water vapour and remaining gases. So by Dalton's law of partial pressure

$$p = p_a + p_v \qquad (8.9)$$

where

p = total pressure of air
p_a = partial pressure of dry air
p_v = partial pressure of water vapour.

The partial pressure of water vapour can be found out by Carrier's equation

$$p_v = p_w - \dfrac{(p - p_w)(T_{db} - T_{wb})}{2800 - 1.3(1.8 T_{db} + 32)} \qquad (8.10)$$

where

p_w = saturation pressure of water vapour corresponding to wet bulb temperature (from steam tables)
p_v = atmospheric pressure of moist air
T_{db} = dry-bulb temperature
T_{wb} = wet-bulb temperature.

The pressure in space is usually atmospheric pressure. In some applications negative pressure is maintained to prevent objectionable particles or air from escaping to outside. In some other applications like an operating theatre a positive pressure ensures that contaminated air does not infiltrate and cause infection to the patient.

The vapour pressure and pressure difference in air conditioning are very small. So these are generally measured in mm of Hg column or water column.

Relative humidity (RH): It is the ratio of mass of water vapour in a given volume of air at any temperature and pressure to the maximum amount of mass of water vapour which the same volume of air can hold at the same temperature conditions. The air contains maximum amount of water vapour at the saturation conditions.

Let v_v and v_s be the specific volumes of water vapour in the actual and moist saturated air at temperature T and in a volume V.

$$RH = \dfrac{(V/v_v)}{(V/v_s)} = \dfrac{v_s}{v_v} \qquad (8.11)$$

Applying ideal gas equation to the state points A and B of Figure 8.2,

$$p_v v_v = p_s v_s$$

Relative humidity is therefore defined as the ratio of vapour pressure in a sample of air to vapour pressure of saturated air at the same temperature, i.e.

$$\text{RH} = \frac{\text{Vapour pressure of water vapour}}{\text{Vapour pressure of saturated air at the same temperature}}$$

$$= \frac{p_v}{p_s} \qquad (8.12)$$

Relative humidity is measured in percentage. It has great influence on evaporation of water in the air and therefore on the comfort of human beings.

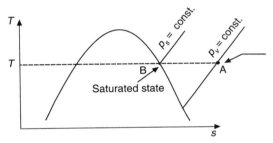

Figure 8.2 Relative humidity.

Enthalpy of air (h): Air is a homogeneous mixture of dry air and water vapour. Therefore, enthalpy of air is found taking the sum of enthalpy of dry air and enthalpy of water vapour in the moist air.

Enthalpy of air/kg of dry air = Enthalpy of dry air + enthalpy of w kg of water vapour

$$= h_a + w h_v \qquad (8.13)$$

Considering the change in enthalpy of perfect gas as a function of temperature only, the enthalpy of dry air part, above a datum of 0°C, can be found as

$$h_a = c_{pa} T_{db} = 1.005 T_{db} \text{ kJ/kg} \qquad (8.14)$$

Assuming enthalpy of saturated liquid at 0°C as zero, the enthalpy of water vapour at point A in Figure 8.3, is expressed as

$$h_v = c_{pw} T_{dp} + (h_{fg})_{dp} + c_{pv}(T_{db} - T_{dp}) \qquad (8.15)$$

where

c_{pw} = specific heat of water vapour (kJ/kg-K)
T_{db} = dry-bulb temperature
T_{dp} = dew point temperature
$(h_{fg})_{dp}$ = latent heat of vaporization at dew point temperature
c_{pv} = specific heat of water vapour (kJ/kg-K).

From reference state as 0°C,

$$h = h_a + wh_v$$

or
$$h = c_{pa}T_{db} + w[c_{pw}T_{dp} + (h_{fg})_{dp} + c_{pv}(T_{db} - T_{dp})] \tag{8.16}$$

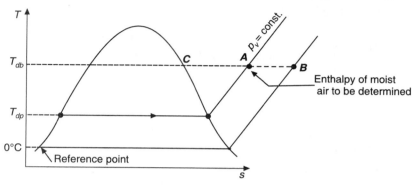

Figure 8.3 Enthalpy of air.

As the reference state of water vapour is at 0°C, $T_{dp} = 0$, instead of $(h_{fg})_{dp}$ one has to take latent heat of vaporisation at 0°C and equal to 2501 kJ/kg

$$\therefore \quad h = c_{pa}T_{db} + [w(2501 + c_{pv}(T_{db} - 0))]$$

or
$$h = 1.005 T_{db} + w(2501 + 1.88 T_{db}) \tag{8.17}$$

Equation (8.16) can also be simplified as
$$h = c_{pa}T_{db} + w[c_{pw}T_{dp} + (h_{fg})_{dp} + c_{pv}(T_{db} - T_{dp})]$$
$$= (c_{pa} + wc_{pv})T_{db} + w(h_g - c_{pv}T_{dp}) \tag{8.18}$$
$$= c_{pm}T_{db} + w(h_g - c_{pv}T_{dp}) \tag{8.19}$$

where

w = specific humidity in kg/kg of dry air
h_g = enthalpy of saturated water vapour at dew point temperature in kJ/kg
T_{db} = dry-bulb temperature in °C
T_{dp} = dew point temperature in °C
c_{pa} = specific heat of dry air = 1.005 kJ/kg-K
c_{pv} = specific heat of water vapour = 1.88 kJ/kg-K
c_{pm} = specific heat of moist air in kJ/kg of dry air-K.

Note: One of the equations from (8.17), (8.18), (8.19) can be suitably used to find the enthalpy of moist air.

Specific volume: It is the volume of air per unit mass of dry air. It is measured in m³/kg of dry air. Air flow is measured by anemometer as volume rate of flow and the heat added or cooling requires mass flow rate. So specific volume is essential to relate the two.

Thermodynamic wet-bulb temperature or temperature of adiabatic saturation: The concept of thermodynamic wet-bulb temperature can be explained with the help of equipment schematically shown in Figure 8.4.

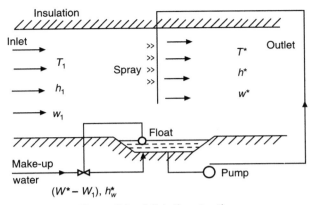

Figure 8.4 Adiabatic saturation.

This equipment consists of an insulated duct through which air flows. Water is sprayed continuously with the help of a circulating pump. The water particles evaporate and mix with air to saturate it. The latent heat required for the evaporation of water particles has to come partly from water and air. Therefore, the temperature of water decreases and after a long time reaches a temperature T^*, as same water is circulated. It means under equilibrium conditions, water will leave the chamber at the same temperature at which it enters. The temperature of water also decreases from its inlet temperature, T_1 to T^* and its humidity rises from W_1 to W^*. It means air and water have attained an equilibrium temperature T^* known as thermodynamic wet-bulb temperature. Since the system is insulated to prevent heat exchange with the surroundings, the process is called *adiabatic saturation*. As T^* can be defined for any state of moist air, there exists a temperature T^* at which the air becomes saturated due to evaporation of water into the air at exactly the same temperature T^*.

In this constant pressure process, the humidity ratio is increased from a given initial value W to the value W^*, corresponding to saturation temperature T^*. The enthalpy will also increase from a given inlet value h_1 to h^*, corresponding to the saturation temperature T^*.

The mass of water evaporated and mixed with air per unit mass of air $(W^* - W_1)$, which adds energy to the moist air amounts to $(W^* - W_1) h_w^*$. Here h_w^* denotes the specific enthalpy of the water added at the temperature T^*. If the process is adibatic the energy balance equation can be written for the constant pressure process as

$$h^* = h_1 + (W^* - W_1) h_w^* \qquad (8.20)$$

The properties W^*, h_w^* and h^* are the functions of temperature T^* only for the given constant pressure. The wet-bulb temperature (WBT) and thermodynamic wet-bulb temperature are different. Only small corrections must be applied to WBT reading to obtain T^*. These two temperatures are same for dry and water vapour mixture.

EXAMPLE 8.3 For a dry-bulb temperature of 25°C and a relative humidity of 50%, calculate the following for air when the barometric pressure is 740 mm Hg. Find without using psychrometric chart:
  (a) Partial pressure of water vapour and dry air
  (b) Dew point temperature
  (c) Specific humidity

(d) Specific volume
(e) Enthalpy.

Solution:

Given: Dry-bulb temperature, T_{db} = 25°C
Relative humidity, RH = 0.50
Atmospheric (barometric) pressure, p_b = 740 mm Hg = 740 × 133 Pa = 98,420 N/m²
$$= 98.420 \text{ kPa}$$

From steam tables, saturation pressure of water vapour corresponding to dry-bulb temperature, T_{db} = 25°C, p_s = 3.17 kPa

$$\text{Relative humidity, RH} = \frac{p_v}{p_s}$$

∴ $$0.5 = \frac{p_v}{3.17}$$

(a) Partial pressure of water vapour, p_v = 1.585 kPa **Ans.**

$$p_b = p_a + p_v$$

Partial pressure of dry air, p_a = 98.420 − 1.585 = 96.835 kPa **Ans.**

Corresponding to p_v = 1.585, the saturation temperature from steam tables = 14°C

(b) Dew point temperature, T_{dp} = 14°C **Ans.**

(c) Specific humidity, $w = 0.622 \dfrac{p_v}{p_a} = 0.622 \times \dfrac{1.585}{96.835}$

$$= 0.01018 \text{ kg/kg dry air} \quad \textbf{Ans.}$$

(d) Specific volume

$$v = v_a = \frac{R_a T}{p_a}$$

$$= \frac{287.3 \times (25 + 273)}{96.835 \times 10^3} = 0.8841 \text{ m}^3/\text{kg dry air} \quad \textbf{Ans.}$$

(e) Enthalpy of moist air

$$h = h_a + w h_v$$
$$= 1.005 T_{db} + w(2501 + 1.88 T_{db})$$
$$= 1.005 \times 25 + 0.01018(2501 + 1.88 \times 25) = 51.0636 \text{ kJ/kg dry air}$$
Ans.

EXAMPLE 8.4 A sample of moist air has a dry-bulb temperature of 43°C and a wet-bulb temperature of 29°C. Calculate the following without making use of the psychrometric chart.

(a) Partial pressure of water vapour
(b) Specific humidity
(c) Relative humidity
(d) Dew point temperature
(e) Humid specific heat
(f) Enthalpy
(g) Degree of saturation
(h) Sigma heat function.

Solution:

Given: Dry-bulb temperature, T_{db} = 43°C; Wet-bulb temperature, T_{wb} = 29°C
Assuming barometric pressure = 760 mm Hg = 1.01325 bar

(a) Partial pressure of water vapour using Carrier equation

$$p_v = p_w - \frac{(p - p_w)(T_{db} - T_{wb})(1.8)}{2800 - 1.3(1.8 T_{db} + 32)} \qquad \text{[at WBT = 29°C, } p_w = 4.013 \text{ kPa]}$$

$$= 4.013 - \frac{(101.325 - 4.013)(43 - 29)(1.8)}{2800 - 1.3(1.8 \times 43 + 32)} = 4.013 - 0.9227$$

$$= 3.090 \text{ kPa} = 23.23 \text{ mm Hg} \qquad \textbf{Ans.}$$

Corresponding to T_{db} = 43°C, saturation pressure, p_s = 8.65 kPa is obtained from steam tables.

(b) Specific humidity, $w = 0.622 \dfrac{p_v}{p_a}$

$$= 0.622 \times \frac{3.090}{(101.325 - 3.090)} = 0.01956 \text{ kg/kg dry air} \qquad \textbf{Ans.}$$

(c) Relative humidity, $RH = \dfrac{p_v}{p_s} \times 100 = \dfrac{3.090}{8.65} \times 100 = 35.72\%$ **Ans.**

(d) Dew point temperature corresponding to partial pressure, p_v = 23.23 mm Hg.

= 3.0920 kPa at 24.1°C is obtained from steam tables. **Ans.**

(e) Humid specific heat, $c_p = c_{pa} + w c_{pv}$
$$= 1.005 + 0.01956 \times 1.88 = 1.0417 \text{ kJ/kg-C} \qquad \textbf{Ans.}$$

(f) Enthalpy of moist air,
$$h = h_a + w h_v = c_{pa} \times T + w(h_{fg0} + 1.88 T)$$
$$= 1.005 \times 43 + 0.01956(2501 + 1.88 \times 43)$$
$$= 93.7157 \text{ kJ/kg dry air} \qquad \textbf{Ans.}$$

(g) Degree of saturation

$$\mu = \text{RH}\left(\frac{1 - p_s/p_b}{1 - p_v/p_b}\right)$$

$$= 0.3572 \left(\frac{1 - 8.65/101.325}{1 - 3.09/101.325}\right) = 0.3572 \times \frac{0.9146}{0.9695} = 0.3369 \qquad \textbf{Ans.}$$

(h) Sigma heat function $\Sigma = h - wh_t^*$ (h_t^* is found at 29°C)

$$= 93.7157 - 0.01959\,(121.5)$$
$$= 91.34 \text{ kJ/kg dry air} \qquad \textbf{Ans.}$$

EXAMPLE 8.5 A sample of air has dry and wet-bulb temperatures of 35°C and 25°C respectively. The barometric pressure is 760 mm Hg. Calculate without using psychrometric chart:
(a) Humidity ratio, relative humidity and enthalpy of the sample.
(b) Humidity ratio, relative humidity and enthalpy, if the air were adiabatically saturated. Only the use of steam tables is permitted.

Solution:
Given: Dry-bulb temperature, $T_{db} = 35°C$; Wet-bulb temperature, $T_{wb} = 25°C$; Barometric pressure, $p_b = 101.325$ kPa
Partial pressure of water vapour using Carrier equation,

$$p_v = p_w - \frac{(p - p_w)(T_{db} - T_{wb})(1.8)}{2800 - 1.3\,(1.8\,T_{db} + 32)}$$

$$= 3.17 - \frac{(101.325 - 3.17)(35 - 25)(1.8)}{2800 - 1.3\,(1.8 \times 35 + 32)} = 2.5168 \text{ kPa}$$

[p_w = saturation pressure of water corresponding to T_{wb} of 25°C = 3.17 kPa = 23.83 mm Hg]

Partial pressure of vapour, $p_v = 2.5168$ kPa

Now, saturation pressure corresponding to $T_{db} = 35°C$; $p_s = 5.63$ kPa

(a)   Humidity ratio, $w = 0.622 \dfrac{p_v}{p_b - p_v}$

$$= 0.622 \times \frac{2.5168}{(101.325 - 2.5168)} = 0.01584 \text{ kg/kg dry air} \qquad \textbf{Ans.}$$

Relative humidity, $\text{RH} = \dfrac{p_v}{p_s} \times 100$

$$= \frac{2.5168}{5.63} \times 100 = 44.70\% \qquad \textbf{Ans.}$$

Enthalpy of moist air, $h = h_a + wh_v$
$= c_{pa}T_{db} + w(h_{fg0} + 1.88T_{db})$
$= 1.005 \times 35 + 0.01584(2501 + 1.88 \times 35)$
$= 75.8331$ kJ/kg dry air **Ans.**

(b) Humidity ratio when air is adiabatically saturated

$$w = 0.622 \frac{p_s}{p_b - p_s}$$

$$= 0.622 \frac{5.63}{(101.325 - 5.63)} = 0.03659 \text{ kg/kg dry air}$$
Ans.

Relative humidity = 100% **Ans.**

Enthalpy of moist air, $h = c_{pa}T_{db} + w(h_{fg0} + 1.88T_{db})$
$= 1.005 \times 35 + 0.03659(2501 + 1.88 \times 35)$
$= 129.09$ kJ/kg dry air **Ans.**

EXAMPLE 8.6 Investigate the effect of humidity on the density of moist air by computing the vapour density for an air–water vapour mixture at 26°C and relative humidity of 0, 50 and 100 per cent. Also, for each case, compare the values of the degree of saturation to the values of relative humidity.

Solution:

Given: Dry-bulb temperature of air, $T_{db} = 26°C$

∴ $p_s = 3.36$ kPa (saturation pressure corresponding to 26°C)

Now, $v = v_a = \dfrac{R_a T}{p_a}$, $\quad RH = \dfrac{p_v}{p_s}$, $\quad \mu = RH \left(\dfrac{1 - p_s/p_c}{1 - p_v/p} \right)$

Case 1: When the relative humidity, RH = 0%

∴ $p_v = 0$

∴ $p_a = p_b - p_v = p_b = 1.01325 \times 10^5$ Pa

∴ $v = \dfrac{287.0(26 + 273)}{1.01325 \times 10^5} = 0.8478$ m³/kg dry air

Degree of saturation, $\mu = 0$. Vapour density, $\rho = 1.1795$ kg/m³ dry air **Ans.**

Case 2: Relative humidity, RH = 50%

$p_v = (RH)p_s = 0.5 \times 3.36 \times 10^3 = 1680$ Pa
$p_a = p_b - p_v = 1.01325 \times 10^5 - 1680 = 99{,}645$ Pa

$$v = \frac{287.3 \times (26 + 273)}{99{,}645} = 0.8621 \text{ m}^3/\text{kg dry air}$$

∴ Vapour density, $\rho = 1.161$ kg/m^3 dry air **Ans.**

Degree of saturation, $\mu = \text{RH} \left(\dfrac{1 - p_s/p}{1 - p_v/p} \right)$

or $\mu = 0.5 \left(\dfrac{1 - 3.36/101.325}{1 - 1.68/101.325} \right) = 0.4915$ **Ans.**

Case 3: Relative humidity, RH = 100%

$$p_v = (\text{RH})(p_s) = 1 \times 3.36 \times 10^3 = 3360 \text{ Pa}$$
$$p_a = p_b - p_v = 1.01325 \times 10^5 - 3360 = 97{,}965 \text{ Pa}$$

$$v = \frac{287.3 \times (26 + 273)}{97{,}965} = 0.8765 \text{ m}^3/\text{kg dry air}$$

Vapour density, $\rho = 1.1405$ kg/m^3 dry air **Ans.**

$$\mu = \text{RH} \left(\frac{1 - p_s/p}{1 - p_v/p} \right) = 1 \times \left(\frac{1 - 3.36/101.325}{1 - 3.36/101.325} \right) = 1$$

Ans.

EXAMPLE 8.7 Air at a condition of 30°C dry-bulb, 17°C wet-bulb temperature and a barometric pressure of 1050 mbar enters an equipment where it undergoes a process of adiabatic saturation, the air leaving with a moisture content of 5 g/kg higher than what it was while entering. Calculate the following.
 (a) Moisture content of the air entering the equipment.
 (b) Dry-bulb temperature and enthalpy of the air leaving the equipment.

Solution:

Given: T_{db} or DBT = 30°C, T_{wb} or WBT = 17°C

p_b = 1050 mbar = 1.050 bar = 105 kPa
p_w = saturation pressure corresponding to WBT
    = 0.01936 bar = 1.936 kPa

From Carrier equation,

$$p_v = p_w - \frac{(p_b - p_w)(T_{db} - T_{wb})(1.8)}{2800 - 1.3(1.8\, T_{db} + 32)}$$

$$= 1.936 - \frac{(105 - 1.936)(30 - 17)(1.8)}{2800 - 1.3(1.8 \times 30 + 32)} = 1.0388 \text{ kPa}$$

(a) Moisture content of air entering the equipment (specific humidity),

$$w_1 = 0.622 \frac{p_v}{p - p_v} = 0.622 \times \frac{1.0388}{(105 - 1.0388)}$$

$$= 0.006215 \text{ kg/kg dry air} \qquad \textbf{Ans.}$$

(b) $\qquad w_2 = (w_1 + 0.005) \text{ kg/kg dry air}$

$$= 0.011215 \text{ kg/kg dry air}$$

$$w_2 = 0.622 \frac{p_{v2}}{p - p_{v2}}$$

$$0.01122 = 0.622 \times \frac{p_{v2}}{105 - p_{v2}}$$

$$0.01122(105 - p_{v2}) = 0.622 \, p_{v2}$$

$$1.1781 = 0.6332 p_{v2}; \text{ Therefore, } p_{v2} = 1.8605 \text{ kPa}$$

$$p_{v2} = p_w - \frac{(p_b - p_w)(T_{db} - T_{wb})(1.8)}{2800 - 1.3(1.8 T_{db} + 32)}$$

$$1.8605 = 1.936 - \frac{(105 - 1.936)(T_{db} - 17)(1.8)}{2800 - 1.3(1.8 T_{db} + 32)}$$

$$\therefore \quad 185.5152(T_{db} - 17) = 0.755[2800 - 1.3(1.8 T_{db} + 32)]$$

$$185.5152 T_{db} - 3153.7584 = 211.4 - 0.17667 T_{db} - 3.1408$$

$$185.69187 T_{db} = 3362.0176$$

$$\therefore \quad T_{db} = 18.1052°C$$

Dry-bulb temperature of air leaving the equipment = 18.1054°C **Ans.**

Enthalpy of air leaving the equipment

$$h = c_{pa} \times T_{db} + w_2[h_{fg} + c_{pv} T_{db}]$$

$$= 1.005 \times 18.1054 + 0.01122[2501 + 1.88 \times 18.1058] = 46.6390 \text{ kJ/kg dry air}$$
Ans.

EXAMPLE 8.8 (a) Moist air is at 25°C temperature. Its dew point is measured as 20°C. The barometric pressure is 755 mm Hg. What are the values of specific and relative humidities of the air?

(b) If this air is cooled to 15°C dry-bulb temperature and 50% relative humidity, what will be the amount of total heat removed per unit mass of dry air? What will be the corresponding amount of moisture removed?

Solution: (a) $T_{db} = 25°C$; $T_{dp} = 20°C$; Barometric pressure, $p_b = 755$ mm Hg = 100.415 kPa

∴ Corresponding to T_{db}, Saturation pressure, $p_s = 3.17$ kPa;

Partial pressure of water vapour corresponding to $T_{dp} = 20°C$, $p_v = 2.34$ kPa.

Specific humidity, $w = 0.622 \dfrac{p_v}{p - p_v}$

$= 0.622 \times \dfrac{2.34}{(100.415 - 2.34)}$

$= 0.01484$ kg/kg dry air **Ans.**

Relative humidity, $RH = \dfrac{p_v}{p_s} \times 100$

$= \dfrac{2.34}{3.17} \times 100 = 73.82\%$ **Ans.**

(b) Now, enthalpy at 25°C DBT and 20°C DPT

$h_1 = h_{a1} + w_1 h_v = 1.005 T_{db} + w_1(2501 + 1.88 \times T_{db})$

$= 1.005 \times 25 + 0.01484(2501 + 1.88 \times 25)$

$= 62.9373$ kJ/kg dry air

$w_1 = 0.01484$ kg/kg dry air and $h_1 = 62.9373$ kJ/kg dry air

Now at state points 2

$T_{db} = 15°C$ and $RH = 0.5$

∴ $p_s = 1.707$ kPa

$RH = \dfrac{p_v}{p_s}$, ∴ $p_v = 0.5 \times 1.707 = 0.8535$ kPa

Specific humidity, $w_2 = 0.622 \dfrac{p_v}{p - p_s}$

$= 0.622 \times \dfrac{0.8535}{(100.45 - 0.8535)}$

$= 0.005332$ kg/kg dry air

Enthalpy of moist air, $h_2 = h_{a2} + w_2 h_{v2}$

$= 1.005 T_{db} + w_2(2501 + 1.88 \times T_{db})$

$= 1.005 \times 15 + 0.005332(2501 + 1.88 \times 15)$

$= 28.5607$ kJ/kg dry air

∴ Amount of heat removed $= h_1 - h_2 = 62.9373 - 28.5607$

$= 34.3766$ kJ/kg dry air **Ans.**

Amount of moisture removed $= w_1 - w_2 = 0.01484 - 0.005332$

$= 0.009508$ kg/kg dry air **Ans.**

EXAMPLE 8.9 (a) The temperature of air entering an adiabatic saturator is 42°C and that of the air leaving is 30°C. Compute the humidity ratio and relative humidity of the entering air.

(b) The conditions inside a room are 25°C and 50% degree of saturation. The inside surface temperature of the window glass is 10°C. Will the moisture condense from the room air upon the window glass?

Solution: (a) DBT = 42°C $\therefore p_s = 8.20$ kPa

WBT = 30°C $\therefore p_w = 4.246$ kPa

$$p_v = p_w - \frac{(p - p_w)(T_{db} - T_{wb})(1.8)}{2800 - 1.3(1.8 T_{db} + 32)}$$

$$= 4.246 - \frac{(101.325 - 4.246)(42 - 30)(1.8)}{2800 - 1.3(1.8 \times 42 + 32)}$$

$$= 3.467 \text{ kPa}$$

Humidity ratio, $w = 0.622 \dfrac{p_v}{p - p_v}$

$$= 0.622 \times \frac{3.467}{(101.325 - 3.467)} = 0.02204 \text{ kg/kg dry air} \quad \textbf{Ans.}$$

Relative humidity, $\text{RH} = \dfrac{p_v}{p_s} \times 100$

$$= \frac{3.467}{8.20} \times 100 = 42.28\% \quad \textbf{Ans.}$$

(b) Dry-bulb temperature $T_{db} = 25°C$; Relative humidity, RH = 50%
Now, saturation pressure p_s corresponding to 25°C is $p_s = 3.17$ kPa.

$$\text{RH} = \frac{p_v}{p_s}$$

$\therefore$ Partial pressure of water vapour, $p_v = 0.5 \times 3.17 = 1.585$ kPa

Now, the dew point temperature is the saturation temperature corresponding to $T_{dp} = 14°C$, $p_v = 1.585$ kPa.

Since the surface temperature of the window glass is 10°C < dew point temperature, the moisture will condense from room air upon the window glass. **Ans.**

EXAMPLE 8.10 On a particular day, the atmospheric air was found to have a dry-bulb temperature of 30°C and wet-bulb temperature of 18°C. The barometric pressure was observed to be 756 mm of Hg. Obtain the following properties, without using the psychrometric chart.

(a) Relative humidity (b) Specific humidity

(c) Dew point temperature (d) The enthalpy of air per kg of dry air
(e) Volume of moisture per kg of dry air.

Solution:

Data: DBT = 30°C and WBT = 18°C

Barometric pressure, p_b = 756 mm of Hg

From steam tables, p_w at 18°C = 0.02062 bar

Barometric pressure = $\dfrac{756}{760} \times 1.013 = 1.0077$ bar

Partial pressure, $p_v = p_w - \dfrac{(p - p_w)(T_{db} - T_{wb})}{1527.4 - 1.3\, T_{wb}}$

$= 0.02062 - \dfrac{(1.0077 - 0.02062)(30 - 18)}{1527.4 - 1.3 \times 18}$

$= 0.02062 - 0.00788 = 0.01274$ bar

Saturated vapour pressure at 30°C, $p_s = 0.04241$ bar

(a) Relative humidity, RH = $\dfrac{p_v}{p_s} = \dfrac{0.01274}{0.04241} \times 100 = 30\%$ **Ans.**

(b) Specific humidity, $w = 0.622 \dfrac{p_v}{p - p_v}$

$= 0.622 \dfrac{0.01274}{1.0077 - 0.01274} = 0.008$ kg/kg of dry air **Ans.**

(c) Dew point temperature is the saturation temperature at p_v (0.01274 bar) from the table of properties.

$$DP = 10.5°C$$ **Ans.**

(d) Enthalpy = $c_{pm} \times T_{db} + w[(h_{fg})_{dp} + c_{pw}(T_{db} - T_{dp})]$

$c_{pm} = c_{pa} + w c_{pv}$

Assume c_{pa} = 1.005 kJ/kg-K, c_{pv} = 1.88 kJ/kg-K

$c_{pm} = 1.005 + 0.008 \times 1.88$

$= 1.020$ kJ/kg-K

Therefore,

Enthalpy = $1.02 \times 30 + 0.008[2520 + 1.88(30 - 10.5)]$

$= 30.60 + 20.5 = 51.50$ kJ/kg of dry air **Ans.**

(e) Volume of mixture per kg of dry air

$$\text{Volume of dry air, } V_a = \frac{R_a T_a}{p_a}$$

$$= \frac{287 \times (273+30)}{(1.0077 - 0.01274) \times 10^5} = 0.874 \text{ m}^3\text{/kg of dry air} \quad \textbf{Ans.}$$

Volume of mixture would be the same.

EXAMPLE 8.11 Obtain all psychrometric properties of moist air at 36°C DBT and 20°C WBT without using the psychrometric chart.

Solution: p_w at WBT 20°C = 0.02337 bar

$$\text{Vapour pressure, } p_v = p_w - \frac{[p - p_w](T_{db} - T_{wb})}{1527.4 - 1.3\, T_{ab}}$$

$$= 0.02337 - \frac{[1.0133 - 0.02337](36 - 20)}{1527.4 - 1.3 \times 20}$$

$$= 0.02337 - 0.01032 = 0.01307 \text{ bar}$$

(a) Specific humidity, $w = 0.622 \dfrac{p_v}{p - p_v}$

$$= 0.622 \frac{0.01307}{1.0133 - 0.01307} = 0.00813 \text{ kg/kg of dry air} \quad \textbf{Ans.}$$

(b) Saturation pressure p_s at 36°C = 0.05862 bar

$$\text{RH} = \frac{p_v}{p_{vs}} = \frac{0.01307}{0.05862} = 22.3\% \quad \textbf{Ans.}$$

(c) Vapour density or absolute humidity $= \dfrac{p_v}{RT} = \dfrac{0.01307 \times 10^5}{462 \times (273 + 36)}$

$$= 0.00915 \text{ kg/m}^3 \quad \textbf{Ans.}$$

(d) Dew point is the saturation temperature at p_v (0.01307 bar)
From steam tables DP = 11.1°C **Ans.**

(e) Enthalpy

$$h = c_{pm} \times T_{db} + w[h_g + c_{pv}(T_{db} - T_{dp})]$$

Assume $c_{pa} = 1.005$ kJ/kg-K, $c_{pv} = 1.88$ kJ/kg-K

Specific heat of moist air, $c_{pm} = c_{pa} + w c_{pv}$

$$= 1.005 + 0.00813 \times 1.88 = 1.02 \text{ kJ/kg-K}$$

$$h = 1.02 \times 36 + 0.00812[2521 + 1.88(36 - 11.4)]$$
$$= 36.72 + 20.87 = 57.59 \text{ kJ/kg of dry air} \qquad \textbf{Ans.}$$

8.3 PSYCHROMETRIC CHART

A psychrometric chart is a graphical representation of the thermodynamic properties of moist air. These properties of moist air vary with atmospheric pressure and altitude. One such chart for atmospheric pressure of 1.01325 bar at sea level is shown in Figure 8.5. The variables shown on a complete psychrometric chart are: DBT, WBT, relative humidity, total heat, vapour pressure and the actual moisture content of the air.

As shown in Figure 8.5 the dry-bulb temperature is taken as the x-axis and the mass of water vapour per kg of dry air as the ordinate. The following illustrations will help in locating the different lines and scales on the chart.

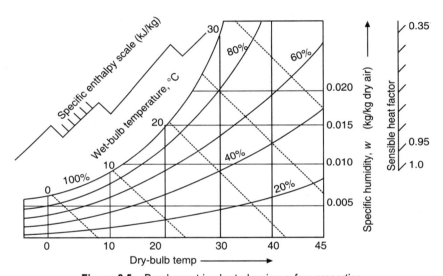

Figure 8.5 Psychrometric chart showing a few properties.

DBT lines: These dry-bulb temperature lines extend vertically upwards and there is one line for each degree of temperature.

WBT lines: The wet-bulb temperature scale is found along the 'in-step' of the chart extending from the toe to the top (Figure 8.6). These lines extend diagonally downwards to the right. There is one line for each degree of temperature.

RH lines: On the psychrometric chart, the relative humidity lines are the only curved lines on it (Figure 8.7). The various relative humidities are indicated on the lines themselves. The 100% RH line or saturation curve becomes the boundary of the chart on the left side. The region beyond this line is the supersaturated zone or fog zone.

Specific humidity lines: The scale for specific humidity is a vertical scale on the right side of psychrometric chart. The scale is in grams of moisture per kilogram dry air (Figure 8.8).

DPT lines: The scale for dew point temperature is identical to the scale of WBT lines (Figure 8.9). The DPT lines run horizontal to the right.

Specific volume lines: The specific volume lines are drawn along the sole chart and they are equally-spaced diagonal lines (Figure 8.10).

Specific enthalpy lines: The specific enthalpy scale is located along the 'in-step' of the chart (Figure 8.11). These lines are similar to WBT lines. Specific enthalpy lines indicate the total heat content. Example 8.12 explains the use of psychrometric chart.

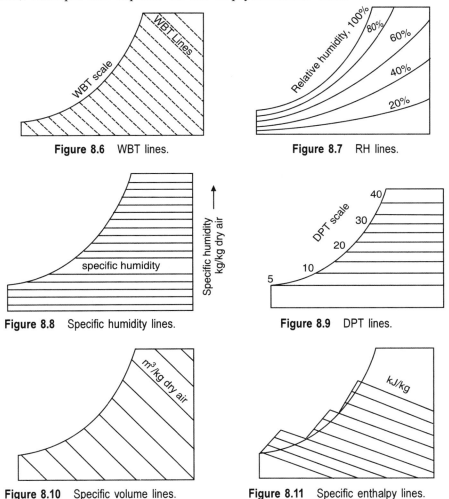

Figure 8.6 WBT lines.

Figure 8.7 RH lines.

Figure 8.8 Specific humidity lines.

Figure 8.9 DPT lines.

Figure 8.10 Specific volume lines.

Figure 8.11 Specific enthalpy lines.

EXAMPLE 8.12 The atmospheric air is at 38°C DBT and 1.01325 bar pressure. If its thermodynamic WBT is 24°C, determine:

(a) The humidity ratio (specific humidity)
(b) Specific enthalpy
(c) Dew point temperature

(d) Relative humidity

(e) Specific volume.

Solution: Locate the state point on the chart at the intersection of 38°C DBT and 24°C thermodynamic WBT lines. Then read the following values:

(a) Humidity ratio = 13 g/kg dry air
(b) Specific enthalpy = 70.8 kJ/kg
(c) Dew point temperature = 18°C
(d) Relative humidity = 32%
(e) Specific volume = 0.9 m³/kg dry air. **Ans.**

8.4 TYPICAL AIR CONDITIONING PROCESSES

Any two of the foregoing properties of air can help to locate the state of air on the psychrometric chart. The condition of air at any point on the psychrometric chart is fixed by any combination of two properties. Such combinations are unlimited to locate the point on the chart but normally DBT and DPT or WBT combination is followed.

The various air-conditioning processes can be illustrated on the psychrometric chart by marking the end conditions of the air. For all the following processes, air is considered at atmospheric pressure of 1.01325 bar.

8.4.1 Sensible Heating of Air

Sensible heat will be added to the moist air while passing it over the hot dry surface. Normally, the heating surface is steam or hot water coil, whose surface temperature is above DBT of the air. As the air comes in contact with the warm surface, the DBT of air increases and tends to approach the temperature of the heating surface.

Since no moisture is added or removed from the air the specific humidity remains constant. The process is illustrated with a schematic diagram as in Figure 8.12. The process takes place along the constant DPT line. For steady flow conditions, the required rate of heat addition is

$$Q_{1-3} = m_1(h_3 - h_1) \tag{8.21}$$

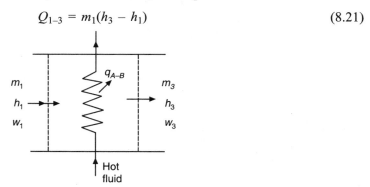

Figure 8.12 Heating of moist air by hot fluid coil.

The process takes place along the constant moisture content line, here T_{db2} or T_2 is the heater temperature. Heat added per kg of dry air can also be found as

$$Q = c_{pa}(T_{db3} - T_{db1}) + wc_{pv}(T_{db3} - T_{db1})$$
$$= (c_{pa} + wc_{pv})(T_{db3} - T_{db1})$$
$$= c_{pm}(T_{db3} - T_{db1}) \text{ kJ/kg of dry air} \qquad (8.22)$$

where c_{pm} is the specific heat of moist air.

EXAMPLE 8.13 Moist air, saturated at 10°C, flows over a heating coil at the rate of 5000 m³/h. Air leaves the coil at 40°C. Plot the process on a psychrometric chart and determine the following:

(a) WBT of air
(b) The sensible heat transferred in kW
(c) The total heat transferred in kW.

Solution: The point A is located on the saturation curve at 10°C (Figure 8.13(b)). $w_A = 7.7$ g/kg dry air and $h_A = 29.7$ kJ/kg dry air

$$v_a = 0.812 \text{ m}^3/\text{kg dry air}.$$

Locate state B. $h_B = 60.2$ kJ/kg dry air, $w_B = w_A = 7.7$ g/kg dry air

(a) WBT at state $A = 10°C$

WBT at state $B = 20.2°C$ **Ans.**

Mass flow rate of air

$$m_A = 5000/(0.812 \times 3600) = 1.71 \text{ kg/s}$$

(b) Sensible heat transfer rate = $m_A \times (h_B - h_A)$

$$= (1.71)(60.2 - 29.7)$$
$$= 52.17 \text{ kW} \qquad \textbf{Ans.}$$

(c) In this example sensible heat is also equal to total heat, as there is no change in humidity ratio. **Ans.**

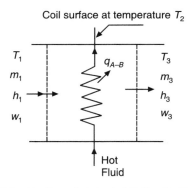

Figure 8.13(a) Heating of air by hot fluid coil—Example 8.13.

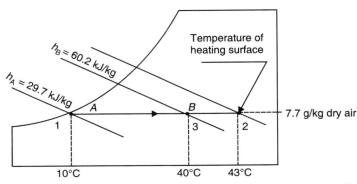

Figure 8.13(b) Heating process on psychrometric chart—Example 8.13.

Coil bypass factor: If all the air flowing over the coil comes into contact with the heating surface of the coil and remains in contact for a sufficiently for long time, then the DB temperature of the the air leaving the coil and the coil surface temperature will be same. However, practically air does not remain in contact with the heating surface for a long time and also some part of the air does not contact the surface. This means some air will bypass the coil. Therefore, the net effect is that the temperature of air at the outlet is less than the hot surface temperature. This is defined by a factor known as bypass factor. It is as indicated with respect to Figure 8.14.

The inability of coil to heat or cool the air to its temperature is indicated by a factor called Bypass Factor (BF).

$$\text{BF} = \frac{\text{Temperature drop that is not achieved}}{\text{Temperature drop that could be achieved}}$$

$$= \frac{T_2 - T_3}{T_2 - T_1} \tag{8.23}$$

where

T_1 = DBT of air entering the coil
T_3 = DBT of air leaving the coil
T_2 = mean effective temperature of coil.

The bypass factor depends upon the following.

(i) **Number of rows of coil:** Lesser is the number of rows, the higher is the BF, and greater the number of rows, the lesser is the BF. Coils are available from 2 rows to as high as 12 rows.

(ii) **Air velocity:** For higher air velocity, BF is higher. The normal air velocity range is above 160 m/min.

The bypass factors range from 0.01 to 0.15 as per the application.

EXAMPLE 8.14 Calculate the bypass factor of the heating coil defined in Example 8.13, if its mean effective temperature of the coil surface is 43°C.

Solution: Using Eq. (8.23) the coil BF is

$$\text{BF} = \frac{43 - 40}{43 - 10} = 0.09 \qquad \textbf{Ans.}$$

It means that 9% of the total air quantity passes through the heating coil without making contact with the coil surface.

8.4.2 Sensible Cooling of Moist Air

Sensible cooling of moist air can be done by passing it over a cooling coil whose surface temperature is kept below the DBT of entering air and above the DP temperature. In this process, no moisture is added or removed, and DP temperature and latent heat content of air remain the same throughout the cooling process. Therefore, the process is represented as a horizontal line from right to left, depending upon the end conditions of air [Figure 8.14(a)]. The total change in heat (enthalpy) content is equal to the change in sensible heat.

Figure 8.14(b) shows the cooling process on psychrometric chart. The air can be cooled to surface temperature of coil. But this requires contact of air with coil surface for a sufficient period. The cooling of air depends on the number of rows of coil, depth of the coil and the velocity of air approaching the coil.

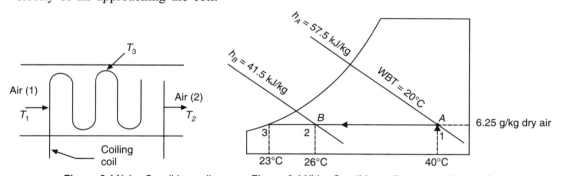

Figure 8.14(a) Sensible cooling. **Figure 8.14(b)** Sensible cooling on psychrometric chart.

The heat transfer between the air and the cooling coil indicates that the temperature difference between the air and the coil surface is large at the beginning (first few rows of the coil). At subsequent rows this difference decreases. So, the last few rows are uneconomical due to very small temperature drop achieved. Therefore, the number of rows is limited and the air is then cooled to a temperature higher than T_2.

Low velocity of air allows adequate time for cooling. But low velocity causes laminar flow and also requires a higher cross-section of coil for adequate airflow. So velocity has to be of reasonably high value. Thus, air cannot be cooled to coil temperature by low air velocity.

$$[\text{Cooling capacity of coil} = \text{Mass flow rate of air} \times \text{specific heat} \times (T_1 - T_2)] \tag{8.24}$$

It can also be calculated by using the standard values of specific heat and representative specific volume to give a formula,

$$\text{Cooling capacity} = 0.0204 \times \frac{\text{m}^3}{\text{min}} \times (T_1 - T_2) \text{ kW} \tag{8.25}$$

It can also be

$$\text{Cooling capacity} = \text{mass flow in kg/s} \times (h_1 - h_2) \text{ kW} \tag{8.26}$$

EXAMPLE 8.15 Moist air having DBT and WBT of 40°C and 20°C, respectively flows over a cooling coil at the rate of 7000 m³/h. Finally, it is cooled to 26°C DBT. Plot the process on psychrometric chart and determine:
 (a) Final WBT of air.
 (b) The total heat transferred in kW.
 If the cooling coil surface temperature is 22°C, find the bypass factor of the coil.

Solution: The process is as plotted on psychrometric chart shown in Figure 8.15(b). The point A is located on the chart at the intersection of 40°C DBT and 20°C WBT lines. Now the values: $h_A = 57.5$ kJ/kg dry air and $h_B = 41.5$ kJ/kg dry, $v_1 = 0.396$ m³/kg dry air
 (a) WBT of air at outlet = 15.5°C **Ans.**
 (b) Mass flow rate of air,

$$m_a = 7000/(0.896 \times 3600) = 2.17 \text{ kg/s}$$

Total heat transfer rate $= (m_a)(h_A - h_B)$
$= (2.17)(57.5 - 41.5) = 34.72$ kW **Ans.**

$$\text{Bypass factor} = \frac{T_2 - T_3}{T_1 - T_3} = \frac{26 - 22}{40 - 22} = 0.22 \quad \textbf{Ans.}$$

8.4.3 Cooling and Dehumidification of Moist Air

Here air is to be cooled and during cooling water vapour is to be separated from the air. Moisture separation will occur only when moist air is cooled to a temperature below its dew point temperature. Therefore, the effective surface temperature of the cooling coil kept below the initial dew point temperature of the air is called Apparatus Dew Point (ADP).

A cooling coil is shown schematically in Figure 8.15(a) and air flows uniformly across the coil. The process of cooling the air from state 1 to state 2 is as shown in Figure 8.15(b), while the coil surface temperature is kept at ADP. Although moisture separation occurs at various temperatures ranging from the initial dew point to final saturation temperature, it is assumed that

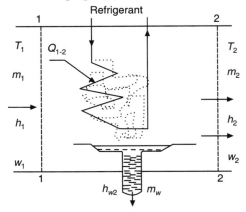

Figure 8.15(a) Schematic device for cooling and dehumidification.

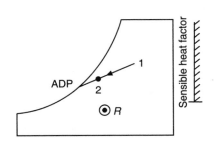

Figure 8.15(b) Cooling and dehumidification on chart.

the condensate water is cooled to the final air temperature T_2 before it drains from the system under steady state.

Cooling the air to the coil surface temperature becomes uneconomical as the size of the heat exchanger becomes very large as (ΔT) reduces. Therefore, surface area of the coil is so designed that it will have a certain bypass factor given by following formula,

$$\text{BF} = \frac{T_2 - \text{ADP}}{T_1 - \text{ADP}} \tag{8.27}$$

For the system shown in Figure 8.15(a), the energy and mass balance equations are

$$m_1 h_1 = m_2 h_2 + Q_{1-2} + m_w h_{w2}$$

$m_2 = (m_1 - m_w)$ but m_w is very small, therefore $m_2 = m_1$ is considered approximately correct.

$$m_1 h_1 = m_2 h_2 + Q_{1-2} + m_w h_{w2} \tag{8.28}$$

and

$$m_1 w_1 = m_1 w_2 + m_w \tag{8.29}$$

$$m_w = m_1 (w_1 - w_2) \tag{8.30}$$

$$Q_{1-2} = m_1 [(h_1 - h_2) - (w_1 - w_2) h_{w2}] \tag{8.31}$$

- In any cooling and dehumidification process, both sensible and latent heats need to be rejected and this is carried out by the cooling fluid circulated through the coil.
- The sum of sensible and latent heat is the total heat transferred.
- The ratio of sensible heat to the total heat transfer is known as the *Coil Sensible Heat Factor* (CSHF).

The CSHF can be obtained by drawing a line on the psychrometric chart parallel to the line marked on the protractor. (Line AB on the chart in Figure 8.16 is parallel to line ab on the protector).

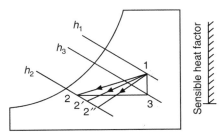

Figure 8.16 Sensible heat factor.

By knowing CSHF, we can find the sensible heat and latent heat quantities.

$$\text{CSHF} = \frac{\text{Sensible heat}}{\text{Total heat}} = \frac{Q_s}{Q_t}.$$

The enthalpy difference $(h_1 - h_2)$ represents the total heat absorbed by the coil. In the processes 1–2, 1–2′, 1–2″ the total heat absorbed is same but the proportion of sensible heat absorbed to total heat absorbed successively reduces. This is indicated by the fact that $T_2'' > T_2' > T_2$. This proportion is called the *Sensible Heat Factor* (SHF).

SHF represents the slope of the line representing the process on the psychrometric chart. So, from Figure 8.16 it can be said that,

$$\text{SHF} = \frac{h_3 - h_2}{h_1 - h_2} \tag{8.32}$$

$$\tan \theta = \frac{\Delta w}{\Delta T}$$

We see that θ is the slope of the SHF line 1–2 on the psychrometric chart which is purely a function of SHF. Thus, when a process line is to be drawn on the psychrometric chart, the following two things have to be known:

- Initial state of air
- Sensible heat factor

For the construction of SHF line on the psychrometric chart, one of the following methods can be used.

(i) In the first method, calculate $\tan \theta$. Then move vertically a certain distance Δw from the initial state, and then horizontally a distance,

$$\Delta T = (\Delta w)(\tan \theta)$$

Lastly, join the point obtained to the initial state point. However, this method is prone to a grave error since Δw is numerically small, and $\tan \theta$ tends to a value close to zero. The method is given here only to illustrate the principle involved; in practice it cannot be used.

(ii) In the second method, move vertically a certain enthalpy change Δh_L. This is proportional to the latent heat change. Then move horizontally equivalent to the sensible heat change (Δh_L) in term of enthalpy given by:

$$\Delta h_S = (\Delta h_L) \left[\frac{\text{SHF}}{1 - \text{SHF}} \right]$$

Again, join the final point to the initial point.

(iii) The third method uses a nomographic method with some charts in which a scale is provided for SHF. There is also a reference point (say point R in Figure 8.17) provided which is joined to the appropriate SHF on the scale. Then a line from the initial state point can be drawn parallel to the above line, which will give the required SHF line.

The cooling capacity of the coil can be found using Eq. (8.33).

$$\text{Cooling capacity} = \text{mass rate of flow} \times (h_1 - h_2) \text{ kW} \tag{8.33}$$

By considering a representative specific volume the cooling capacity can also be calculated by the formula,

$$\text{Capacity} = 0.02 \times \text{m}^3/\text{min} \times (h_1 - h_2) \text{ kW} \tag{8.34}$$

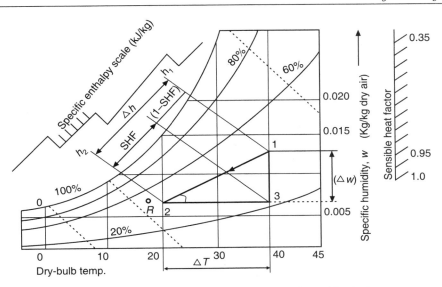

Figure 8.17 Construction of SHF line on psychrometric chart.

8.5 ADIABATIC COOLING OR COOLING WITH HUMIDIFICATION PROCESS

Adiabatic cooling is because of the water attaining the wet-bulb temperature of air due to constant evaporation. In some dry areas, the difference between dry-bulb and wet-bulb temperature is high enough to use well water as cooling medium in a coil. As the WB temperature is higher than the dew point of air, the process is sensible cooling.

Let us consider warm air blowing over the water surface as shown in Figure 8.18. The air gets cooled. For cooling, air has to lose the heat it possesses. The air physically contacts the water surface, therefore, in one way it dissipates heat to the water by conduction. Secondly, water particles separate from the water surface, evaporate and mix with the air. For the evaporation of water latent heat is to be supplied which actually comes from both air as well as water. Therefore, both air and water get cooled. The heat exchange process is only between the air and the water, so the system is said to be insulated. This process of cooling is known as *adiabatic cooling*. It is also called *evaporative cooling* or *cooling with humidification*. For this reason, water in an open lake or well is cooler than the surrounding.

The water surface can also be in the form of water spray as shown in Figure 8.18(a) to hasten evaporation and cooling.

During an adiabatic process no heat enters or leaves the system. Thus, the process line 1–2 on the psychrometric chart is along the constant enthalpy line.

The lowest possible temperature to which air can be cooled is the wet-bulb temperature (T_{wb}). Due to inefficient spray systems or uneconomical situations of providing a large number of banks of spray to get the cooling to T_{wb}, air is practically cooled to T_2 ($T_2 > T_{wb}$). The efficiency of spray is defined as

$$\text{Spray efficiency} = \frac{T_1 - T_2}{T_1 - T_{wb}} \qquad (8.35)$$

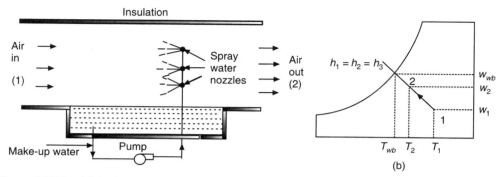

Figure 8.18(a) Adiabatic or evaporative cooling. **Figure 8.18(b)** Schematic diagram of Figure 8.18(a).

The water is recirculated with a pump. Make-up water is added to compensate for the water evaporated during the operation. The make-up water also known as *humidifier duty* in litre/h can be given by the formula,

$$\text{Humidifier duty} = \text{Make-up water}$$
$$= \text{mass rate of dry air in kg/m} \times (w_2 - w_1) \times 60 \text{ lit/h} \quad (8.36)$$

where w_1 and w_2 are specific humidities at condition 1 and 2 in kg/kg of dry air.

Humidifier duty can also be written as

$$\text{Humidifier duty} = \frac{\text{m}^3/\text{min}}{\text{Specific volume}} = (w_2 - w_1) \times 60 \quad (8.37)$$

This is the process used in air coolers/desert coolers. Here the water surface is increased by spreading water over fibre pads. The efficiency of this cooler depends upon the thickness and uniform distribution of fibres in the pad. The extent of cooling, i.e. drop in temperature, depends upon the difference between DB and WB temperature. Therefore, air-cooling is higher in dry climate where this difference is large. This process is also used in cooling towers for cooling of condenser water.

EXAMPLE 8.16 Moist air at 32°C DBT and 50% RH enters a cooling coil at 10,000 m³/h. It is desired that the air leaving the coil has a DBT of 20°C and WBT of 18°C. Determine the following:

(a) Mean effective surface temperature of the coil
(b) Bypass factor of the coil
(c) Sensible heat factor of the coil
(d) Total heat removed from air
(e) Mass of water vapour condensed.

Solution: Figure 8.19 shows the schematic solution.

State A is located at the intersection of DBT of 32°C and RH = 50%. Thus, $h_A = 70$ kJ/kg dry air and $w_A = 15.0$ g moisture/kg dry air and $v_A = 0.885$ m³/kg dry air.
State B is located on the intersection of DBT of 20°C line and WBT of 18°C line. Therefore, $h_B = 51.0$ kJ/kg dry air, $w_B = 12.0$ g moisture/kg dry air. From steam tables $h_{wb\text{-}sat}$ corresponding to temperature 17°C DPT = 71.3 kJ/kg.

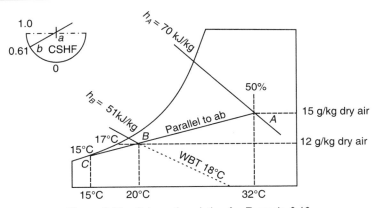

Figure 8.19 Schematic solution for Example 8.16.

(a) Extend the line *AB* to meet the saturation curve at point *C* which indicates the mean effective surface temperature of 15°C. **Ans.**

(b) The coil bypass factor $= \dfrac{T_B - T_C}{T_A - T_C} = \dfrac{20 - 15}{32 - 15} = 0.29$ **Ans.**

(c) To get the sensible heat factor of the cooling coil, draw a line parallel to the process line *AB* through the centre of the protractor given on the chart. Then read the CSHF scale on the protractor, which is 0.61 in this case. **Ans.**

(d) Total heat removed per kilogram air:
$$m_A = (10{,}000) / (0.885 \times 3600)$$
$$= 3.13 \text{ kg/s}$$
$$q_{A-B} = 3.13\,[(70 - 51) - (0.015 - 0.012) \times 71.3] = 58.8 \text{ kW} \quad \textbf{Ans.}$$

(e) Mass of water vapour condensed:
$$= m_A(w_A - w_B) = (3.13)(0.015 - 0.012) = 0.009 \text{ kg/s} \quad \textbf{Ans.}$$

8.6 HEATING AND HUMIDIFICATION

In this case both heat as well as water vapour are added to the air. To achieve this, the temperature of water to be sprayed in the air stream is kept at a temperature greater than the DBT of incoming air so that heat will be transferred to air to heat it. The unsaturated air reaches the condition of saturation and the heat of vaporisation of water is absorbed from the spray water itself so that the spray water gets cooled. The heating and humidification process is shown in Figure 8.20.

During this process, the humidity ratio, the dry-bulb temperature, the wet-bulb temperature, the dew point temperature and the enthalpy of air increase while passing through hot spray. The relative humidity may increase or decrease. The spray water is to be heated before being pumped to the spray nozzles. The air enters at state 1 and leaves at state 2, as shown in Figure 8.20.

The mass balance for water is:
$$(m_{w1} - m_{w2}) = m(w_2 - w_1) \tag{8.38}$$

where m_{w1}, m_{w2} are the mass flow rates of water entering and leaving in kg/min, respectively,

308 *Refrigeration and Air Conditioning*

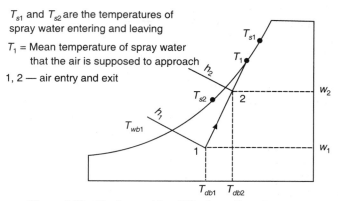

Figure 8.20 Heating and humidification with water spray.

m is the mass flow rate of air in kg/min, and w_1 and w_2 are specific humidity of air entering and leaving respectively.

The enthalpy balance,

$$(m_{w1} \times h_{w1} - m_{w2} \times h_{w2}) = m(h_2 - h_1) \tag{8.39}$$

where h_{w1} and h_{w2} are enthalpy of spray water entering and leaving, respectively.

8.6.1 Heating and Humidification by Steam Injection

Steam is normally injected into fresh outdoor air to increase humidity. High humidity air is the requirement of textile mills. The process is shown in Figure 8.21. The process can be analyzed by considering mass and energy balances. If m_v is the mass of steam supplied which has enthalpy h_v and m_a is the mass of dry air, the mass and enthalpy balances are given by

$$w_1 \times m_a + m_v = m_a w_2 \tag{8.40}$$

$$h_1 \times m_a + m_v \times h_v = m_a h_2 \tag{8.41}$$

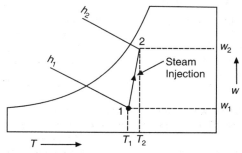

Figure 8.21 Heating and humidification by steam injection.

The dry-bulb temperature of air changes very little during the process.

EXAMPLE 8.17 Air enters a chamber at 10°C DBT and 5°C thermodynamic WBT at a rate of 100 cubic metre per minute (cmm). The barometer reads a pressure of 1.01325 bar. While passing through the chamber, the air absorbs sensible heat at the rate of 40.0 kW and picks up 45 kg/h of saturated steam at 105°C. Determine the dry and wet-bulb temperatures of the air leaving the chamber.

Solution: This is a case of simple heating and humidification of air by the addition of steam as shown in Figure 8.22(a). Specific volume of the air entering the chamber is 0.813 m³/kg dry air.

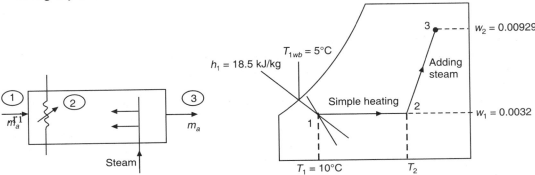

Figure 8.22(a) System for Example 8.17. **Figure 8.22(b)** Psychrometric processes for Example 8.17.

The mass flow rate of air is:
$$m_a = \frac{(\text{cmm}) \, 60}{v} = \frac{100 \times 60}{0.813} = 7380 \text{ kg dry air/h}$$

By moisture balance,
$$m_a(w_3 - w_1) = 45$$
$$w_3 = w_1 + \frac{45}{m_a} = 0.0032 + \frac{45}{7380}$$
$$= 0.00929 \text{ kg w.v./kg dry air}$$

By energy balance,
$$m_a(h_3 - h_1) = (40.0)(3600) + 45 h_v$$

where $h_v = 2683.6$ kJ/kg is the enthalpy of saturated steam at 105°C.

Thus,
$$h_3 = 18.5 + \frac{1}{7380}[144{,}000 + 45\,(2683.6)] = 54.375 \text{ kJ/kg dry air}$$

From psychrometric chart, at point 3
$$\text{DBT} = 26.5°\text{C}$$
$$\text{WBT} = 18.1°\text{C} \qquad \text{Ans.}$$

8.7 ADIABATIC MIXING OF AIR STREAMS

Mixing of two moist air streams is a common process in air-conditioning systems. For the purpose of analysis the mixing of two such streams is assumed to be an adiabatic process and is shown in Figure 8.23(a).

The pipelines carrying air streams are assumed to be perfectly insulated so that no heat enters or leaves the system [Figure 8.23(b)].

Heat balance gives

$$m_1 h_1 + m_2 h_2 = m_3 h_3 \qquad (8.42)$$

Mass balance gives

$$m_1 + m_2 = m_3 \qquad (8.43)$$

Replacing the value of m_3 from Eq. (8.43) in Eq. (8.42),

$$m_1 h_1 + m_2 h_2 = (m_1 + m_2) h_3$$

Rearranging the terms,

$$\frac{m_1}{m_2} = \frac{h_3 - h_2}{h_1 - h_3} \qquad (8.44)$$

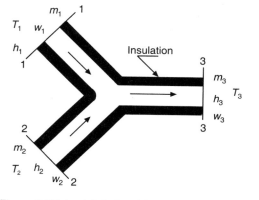
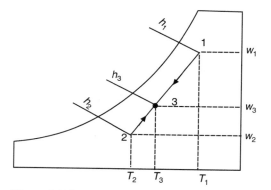

Figure 8.23(a) Adiabatic mixing of two moist streams.   **Figure 8.23(b)** Adiabatic mixing of two air streams.

Similarly, during mixing, water vapour is neither added nor removed.

So,

$$m_1 w_1 + m_2 w_2 = m_3 w_3 \qquad (8.45)$$

Substituting the value of m_3 and rearranging, we get

$$\frac{m_1}{m_2} = \frac{w_3 - w_2}{w_1 - w_3} \qquad (8.46)$$

The symmetry also gives

$$\frac{m_1}{m_2} = \frac{T_3 - T_2}{T_1 - T_3} \qquad (8.47)$$

When hot and high humid air is mixed with very cold air, the resulting mixture would contain fog and air mixture and the final condition (point 3 in Figure 8.24) on the psychrometric chart would lie to the left or above the saturation curve. The temperature of the fog is corresponding to the wet bulb temperature line passing through point 3. The fog may also result when a very fine water spray is injected into air in a greater amount than required to saturate the air. Even a small quantity of steam injected into the air stream, which may not mix properly, would also result into fog.

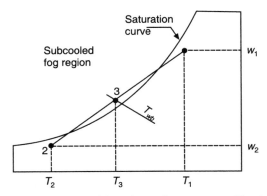

Figure 8.24 Adiabatic mixing of two air streams resulting into fog.

EXAMPLE 8.18 An air stream of 7000 m³/h at a DBT of 27°C and humidity ratio of 0.010 kg/kg dry air is adiabatically mixed with 120,000 m³/h of air having 35°C DBT and 55% RH. Find the DBT and WBT of the resulting mixture.

Solution: The process is as shown in Figure 8.25 where states 1 and 2 are marked and the specific volumes v_1 and v_2 are 0.863 and 0.9 m³/kg, respectively.

$$m_1 = 7000/(0.863 \times 3600) = 2.25 \text{ kg/s}$$

and

$$m_2 = 20{,}000/(0.9 \times 3600) = 6.17 \text{ kg/s}$$

$$\frac{\text{Line } 1{-}3}{\text{Line } 1{-}2} = \frac{m_1}{m_3} = \frac{2.25}{8.42} = 0.27$$

Consequently the length of line 1–3 = (0.27)(1–2). Accordingly, divide the line segment 1–2 to mark the point 3. The values of DBT and WBT at point 3 are read as 29°C and 20°C, respectively. **Ans.**

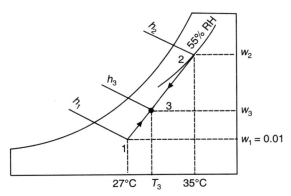

Figure 8.25 Adiabatic mixing of two air streams—Example 8.18.

8.8 AIR WASHER

Figure 8.26 shows the schematic arrangement of an air washer. Water is sprayed in the air stream. During the course of flow, air may be heated or cooled, humidified or dehumidified, etc. The temperature of water to be sprayed in the air stream is changed according to the need. Make-up water is added for any loss in the case of humidification air. Eliminator plates are placed in the path of the air to separate the water droplets.

Let T_s be the mean surface temperature of water droplets which is equal to the actual temperature of water T_w. Let T_1 be the dry-bulb temperature of air at the inlet.

(a) **Process 1–2 heating and humidification ($T_s > T_1$):** The mean surface temperature of water is greater than the dry-bulb temperature of air. Heat is supplied to water externally.

(b) **Process 1–3 humidification:** If water from pan is heated to the dry-bulb temperature of air and then sprayed, the air would only be humidified. So the only process 1–5 is humidification. This is rather impossible from the practical point of view.

(c) **Process 1–4:** The recirculated water reaches the equilibrium temperature which is equal to the thermodynamic wet-bulb temperature of air.

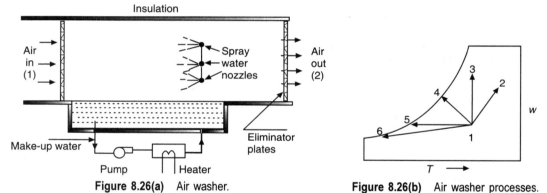

Figure 8.26(a) Air washer.   **Figure 8.26(b)** Air washer processes.

(d) **Process 1–5 sensible cooling:** If water is cooled up to the dew point temperature of air, the process 1–3 would be sensible cooling process.

(e) **Process 1–6 cooling and dehumidification:** If water is cooled to the temperature below the dew point of air, the air would not only be cooled but also be dehumidified. So the process 1–4 would be cooling and dehumidification.

Water with varied temperature can be injected into the air stream and the required effect on the air can be observed. It should be remembered here that water temperature is to be controlled. However, the heating and humidification process can give only limited heating due to very high temperature of water required for large amount of heating.

We can define the humidifying efficiency of an air washer as:

$$\eta_H = \frac{h_2 - h_1}{h_s - h_1} = \frac{w_2 - w_1}{w_s - w_1}$$

It can be shown that the bypass factor x is expressed as:

$$x = \frac{w_s - w_2}{w_s - w_1} = 1 - \frac{w_s - w_1}{w_s - w_1} = 1 - \eta_H$$

8.9 CHEMICAL DEHUMIDIFICATION OR SORBENT DEHUMIDIFICATION

Sorbents are materials that have an ability to attract and hold gases and liquids, other than water vapour—a characteristic that makes them very useful in chemical separation processes. *Desiccants are a subset of sorbents; they have a particular affinity for water.*

Many materials can attract and hold moisture but their capacity of doing so is limited (5 to 25% of their mass). In contrast, a commercial desiccant can hold up water vapour between 10 and 1100% of its dry weight.

In certain applications, the humidity requirement is very low which cannot be achieved by a cooling coil due to limitation of its temperature. In such cases air is passed through sorbents.

The solid sorbents include silica gel, zeolites, synthetic zeolites (molecular sieves), activated aluminas, carbons and synthetic polymers.

The chemical dehumidification process shown in Figure 8.27 takes place along constant enthalpy or WBT line from 1–2. Ideally the latent heat is released by condensation of moisture in the solid adsorbent from the air and picked up by air as sensible heat raising its temperature. Thus, it is an adiabatic process.

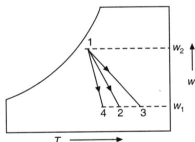

Figure 8.27 Sorbent dehumidification.

In a few cases in addition to latent heat, heat of reaction due to the chemical change in absorbent is also released. Therefore, the process is not isenthalpic but along the line 1–3. The enthalpy difference between 3 and 2, $(h_3 - h_2)$, is the additional heat of reaction per kg of dry air.

The absorption capacity of the chemical depends on its temperature. The lower the temperature, the higher would be the moisture absorption capacity and vice versa. Ideally the air should absorb all the released heat, but practically part of the heat is absorbed by the chemicals. So the chemical from the sump is cooled and then recirculated for spraying. Then the process would be 1–4 and the enthalpy difference, $(h_2 - h_4)$ or $(h_3 - h_4)$, is the heat absorbed by the chemical which is removed before recirculation.

This process, followed by sensible cooling, can give rise to some conditions unattainable by a cooling coil.

EXAMPLE 8.19 300 m³/min of moist air enters a refrigeration coil at 35°C DBT and 50% RH. The apparatus dew point (ADP) of the coil is 10°C and the bypass factor is 0.15. Determine (a) the outlet state of moist air and (b) the cooling capacity of coil in TR.

Solution: Volume flow rate = 300 m³/min

The condition of air entering the coil (35°C DBT, 50% RH) is shown with point 1 on the psychrometric chart, and the specific volume of air at state 1 is 0.895 m³/kg.

ADP = 10°C,   bypass factor = 0.15

As shown in Figure 8.28 the condition of air entering the coil is plotted on the psychrometric chart.

ADP of 10°C is plotted on the saturation curve in the chart. The bypass factor of coil is given as 0.15. Therefore,

$$BF = 0.15 = \frac{T_2 - T_3}{T_1 - T_3} = \frac{T_2 - 5}{35 - 5}$$

$$T_2 = 9.5°C$$

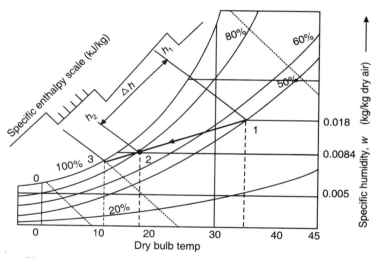

Figure 8.28 Cooling of air from state 1 to state 2—Example 8.19.

(a) Join point 1 to Apparatus Dew Point (ADP) and find point 2 having temperature 13.5°C. Read T_{wb} at point 2. The condition of air at outlet = (13.5°C DBT, 95% RH).
Ans.

(b) Cooling capacity of coil is found out by

$$\text{Mass flow rate, } m = \frac{300}{60} \times \frac{1}{0.895} = 5.586 \text{ kg/s}$$

$$\text{Capacity} = (\text{mass flow rate})(h_1 - h_2)$$
$$= 5.586(81 - 34.1)$$
$$= 262 \text{ kW} = 83 \text{ TR}$$

Ans.

EXAMPLE 8.20 500 m³/min of fresh air at 30°C DBT and 50% RH is adiabatically mixed with 1000 m³/min of recirculated air at 22°C DBT and 10°C DPT. Calculate the enthalpy, specific volume, humidity ratio and final DBT of the mixture.

Solution: From the psychrometric chart at 22°C DBT and 10°C DPT,

$h = 41.5$ kJ/kg of dry air, Specific volume $= 0.846$ m³/kg of dry air

Humidity ratio, $w = 0.0077$ kg/kg of dry air

At 30°C DBT and 50% RH,

$h = 64$ kJ/kg of dry air, Specific volume $= 0.877$ m³/kg of dry air

$w = 0.0132$ kg/kg of dry air

Recirculated air mass rate of flow $= \dfrac{1000}{0.846} = 1182$ kg/min

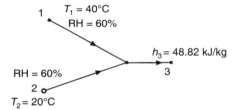

Figure 8.29(a) Mixing of two air streams—Example 8.20.

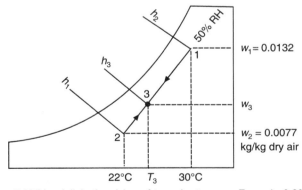

Figure 8.29(b) Adiabatic mixing of two air streams—Example 8.20.

Fresh air mass rate of flow $= \dfrac{500}{0.877} = 570.1$ kg/min

Enthalpy of mixture $= \dfrac{m_1 h_1 + m_2 h_2}{m_1 + m_2} = \dfrac{1182 \times 41.5 + 570.1 \times 64}{1182 + 570.1}$

$= 48.82$ kJ/kg of dry air **Ans.**

$$\text{Specific volume} = \frac{m_1 v_1 + m_2 v_2}{m_1 + m_2} = \frac{1182 \times 0.846 + 570.1 \times 0.877}{1182 + 570.1}$$

$$= 0.856 \text{ m}^3/\text{kg of dry air} \qquad \textbf{Ans.}$$

$$\text{Humidity ratio} = \frac{m_1 w_1 + m_2 w_2}{m_1 + m_2} = \frac{1182 \times 0.077 + 570.1 \times 0.0132}{1182 + 570.1}$$

$$= 0.00948 \text{ kg/kg of dry air} \qquad \textbf{Ans.}$$

Alternative method using the psychrometric chart. Once you find the value of enthalpy of mixture, i.e. h_3, making use of constant enthalpy line on the chart, you can draw the $h_3 = 48.82$ kJ/kg line intersecting the process line 1–2 at point 3. The remaining properties for state point 3 can be collected from the chart.

EXAMPLE 8.21 In a cooling application, moist air enters a refrigeration coil at the rate of 100 kg of air 35°C DBT and 50% RH. The apparatus dew point of the coil is 5°C and the bypass factor is 0.15. Determine (a) the outlet state of moist air and (b) the cooling capacity of coil in TR.

Solution: Mass flow rate = 100 kg of dry air/min
The condition of air entering the coil (35°C DBT, 50% RH) is shown at point 1 on the psychrometric chart, and the specific volume of air at state 1 is 0.895 m³/kg.

$$\text{ADP} = 10°\text{C}, \qquad \text{bypass factor} = 0.15$$

The condition of air entering the coil is plotted on the psychrometric chart in Figure 8.28. ADP of 5°C is plotted on the saturation curve in the chart. The bypass factor of coil is given as 0.15. Therefore,

$$\text{BF} = 0.15 = \frac{T_2 - T_3}{T_1 - T_3} = \frac{T_2 - 5}{35 - 5}$$

$$T_2 = 9.5°\text{C}$$

(a) Join point 1 to Apparatus Dew Point (ADP) and find point 2 having temperature 9.5°C. Read T_{wb} at point 2. The condition of air at outlet = (9.5°C DBT, 9.2 WBT). **Ans.**

(b) Cooling capacity of coil is found out by

$$\text{Capacity} = (\text{mass flow rate})(h_1 - h_2) = \frac{100}{60}(81 - 27.5)$$

$$= 89.1 \text{ kW} = 25.33 \text{ TR} \qquad \textbf{Ans.}$$

EXAMPLE 8.22 Room air at 20°C DBT and 60% RH is mixed with outdoor air at 40°C DBT and 40% RH in the ratio 4:1. The mixture is passed through a cooling coil whose temperature is maintained at 9°C and whose bypass factor is 0.25. Find the following.
 (a) Condition of air entering the coil.
 (b) Condition of air leaving the coil.
 (c) If 250 m³/min of air is supplied to the room, find the refrigeration load on the cooling coil.

Solution:

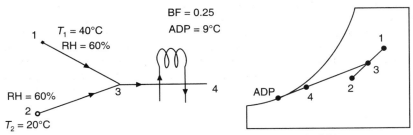

Figure 8.30 Mixing of two air streams—Example 8.22.

(a) DB temperature of mixture, $T_3 = \dfrac{1 \times 40 + 4 \times 20}{1 + 4} = 24°C$

Locate points 1 and 2, i.e. 40°C DBT and 40% RH and 20°C DBT and 60% RH, respectively.
Join 1 and 2 and find point 3 on the line 1–2, having DBT of 24°C.
The condition of air entering coil = 24°C DBT and 18.2°C WBT. **Ans.**

(b) Join 3 to point 9°C on the saturation curve, i.e. ADP.

$$BF = 0.25$$
$$= \dfrac{T_4 - ADP}{T_3 - ADP} = \dfrac{T_4 - 9}{24 - 9}$$
$$T_4 = 12.75°C$$

Locate point 4 with DBT of 12.75°C on line 3 – ADP. The condition of air leaving the coil is 12.75°C DBT, 11.6°C WBT. **Ans.**

(c) Cooling capacity $= \dfrac{\text{Flow in m}^3\text{/min}}{\text{Specific volume at } 4 \times 60}(h_3 - h_4)$

$= \dfrac{250}{0.81 \times 60}(51.5 - 33) = 93.89 \text{ kW}$ **Ans.**

EXERCISES

1. Determine the absolute humidity of the air sample having DPT of 18°C.
2. The DBT and DPT of ambient air are 40°C and 14°C, respectively. Determine the degree of saturation of air. Also determine the RH.
3. Define air conditioning.

4. Classify air conditioning as per its use.
5. What are the factors that affect human comfort and discuss their effect?
6. Define different properties of air.
7. What is fog? Show on chart how two air streams on mixing would produce fog. Why does fog occur during nights in winter and not during nights in summer?
8. Why is sling psychrometer used? And how?
9. Describe the process where SH is removed and LH is added.
10. What is the combination of processes used in winter air conditioning?
11. Discuss the processes that can be carried out by changing the water temperature of spray.
12. Write short notes on absorbent and adsorbents.
13. What is the process used for low RH requirement in a room? Describe the same.
14. What is sensible heat factor and what does it represent on psychrometric chart?

NUMERICALS

1. Moist air, saturated at 14°C, flows over a hot surface at the rate of 30,000 m³/h. Air leaves the surface at 40°C. Represent the surface on psychrometric chart and determine the following:
   (a) Final WBT of air
   (b) Sensible heat transferred in kW
   (c) Total heat transferred in kW.

2. Moist air at 32°C DBT and 65% RH enters a cooling coil at 15,000 m³/h. It is desired that the air leaving the coil should have a DBT of 22°C and WBT of 18°C. Determine the following:
   (a) Mean effective surface temperature of the coil
   (b) Sensible heat factor of the coil
   (c) Bypass factor of the coil
   (d) Total heat removed per kg of air
   (e) Mass of water vapour condensed.

3. Calculate the following properties of air at 45°C DBT and 30°C WBT.
   (a) Specific humidity
   (b) Degree of saturation
   (c) Relative humidity
   (d) Absolute humidity
   (e) Enthalpy
   (f) Dew point temperature.

4. A sling psychrometer reads 40°C DBT and 28°C WBT when the atmospheric pressure is 75 cm of Hg. Calculate the following:
   (a) Specific humidity
   (b) Relative humidity
   (c) Dew point temperature
   (d) Enthalpy
   (e) Vapour density.

   [**Ans.** : 0.0193 kg/kg of dry air, 40.7%, 24.1°C, 89.7 kJ/kg of dry air, 0.02076 kg/m^3]

Chapter 9

Cooling Load Estimation and Psychrometric Analysis

9.1 INTRODUCTION

Air conditioning means controlling the temperature, relative humidity, velocity and purity of air to be supplied to the air conditioned space. The air conditioned space may be for creating comfort conditions for human beings or it may be for creating conditions as per the needs of a specific industrial process. Such conditions have been introduced in Chapter 8.

Human body continuously generates heat at the rate varying between 100 W and 500 W depending upon the health and the activities of the individuals. Body temperature must be maintained within a narrow temperature range to avoid discomfort and danger from heat or cold stress. Therefore, heat must be dissipated in a controlled manner. Heat is neither generated uniformly throughout the body nor is it dissipated uniformly. Therefore, comfort conditions are not the same for men, women, children and old people. These also vary depending on the location and season of the year and the activities one carries out. Scientists have carried out a large number of experiments on human beings to understand the comfort conditions. As per their investigations, majority of people feel comfortable in summer at:

1. Dry-bulb temperature = $25 \pm 1°C$
2. Relative humidity = $50 \pm 5\%$
3. Air velocity = 15 to 20 m/min

For winter season the comfort conditions are:

1. Dry-bulb temperature = $22 \pm 1°C$
2. Relative humidity = 50%
3. Air velocity = 9 to 12 m/min

9.2 THERMODYNAMICS OF HUMAN BODY AND MATHEMATICAL MODEL

The rate at which body generates heat is called the *metabolic rate*. The human body is considered a heat engine with 20% thermal efficiency. This means that if the body generates heat energy of the order of 100 W due to digestion of food, heat at the rate of 80 W must be dissipated to the surroundings. If the body does not dissipate heat at this rate, it would accumulate causing discomfort. The human body works best at a particular temperature similar to that of a heat engine. The body temperature of a healthy man is 36.9°C (98.4°F) at the skin. A man with body temperature of 40.5°C (104.9°F) is considered serious and with 43.5°C (110°F) dead.

The rate of heat loss from human body to the atmosphere is through three modes—radiation, convection and evaporation of moisture. The total heat loss from the body is

$$Q = Q_r + Q_c + Q_e \tag{9.1}$$

where Q = rate of total heat loss (watt), Q_r, Q_c and Q_e are the heat loss due to radiation, convection and evaporation of the moisture respectively.

One feels comfortable only when the rate of heat generation due to metabolic rate is equal to the rate of heat loss to the surroundings. Hence, one can apply the first law of thermodynamics to a body as

$$Q_m - W = Q + Q_s \tag{9.2}$$
$$Q_m - W = Q_r + Q_c + Q_e + Q_s$$

where

Q_m = metabolic rate of heat generation, in watts and it depends on the nature of food and its consumption.
W = useful rate of work in watts
Q_s = rate of heat stored in human body in watts (for comfort, $Q_s = 0$)

(a) Convective heat loss: The heat loss from human body due to convection is given as

$$Q_c = UA(T_b - T_a) \tag{9.3}$$

where U is the overall heat transfer coefficient (W/m²-K) between the body and the surroundings. It depends on air velocity, density, thermal conductivity, specific heat of air, etc. A is the surface area of human body (m²). T_a and T_b are atmospheric and human body temperatures, respectively. In winter T_a is less than T_b, heat loss takes place at a greater pace, so one may feel cold. Therefore, one has to increase the layer of insulation by wearing woollen clothes. In summer, if T_a is greater than T_b, the only mode of heat dissipation is by evaporation of moisture. Therefore, one perspires more in summer.

(b) Radiation heat loss: The radiation heat loss from human body is given as

$$Q_r = \sigma A(T_b^4 - T_a^4) \tag{9.4}$$

where σ is the Stefan–Boltzmann constant, 5.67×10^{-8} W/m²-K⁴.

Human body gains heat from surroundings when $T_a > T_b$ and loses heat when $T_a < T_b$. The explanation as given above can also be applied here.

(c) Evaporative heat loss: Evaporative heat loss from human body is given as

$$Q_e = K_d A(p_s - p_v) h_{fg} K_c \tag{9.5}$$

where

K_d = diffusion coefficient (kg of water evaporated/m²/h/N pressure difference)
p_s = saturation pressure of water vapour corresponding to the skin temperature
p_v = vapour pressure of moisture present in the air
h_{fg} = latent heat of evaporation, kJ/kg
K_c = a factor for type of clothing.

The rate of heat dissipation from body by convection, radiation or by evaporation is affected by the quality of clothing.

9.3 EFFECTIVE TEMPERATURE

People feel comfortable only when all the three parameters, say DBT, RH and air velocity, are within certain limits as stated above. Effective temperature is a single parameter at which people feel comfortable. Effective temperature is an index of the measure of comfort.

Effective temperature (ET) is that temperature of saturated air at which people feel comfort as experienced in the actual unsaturated environment.

Effective temperature is a measure of feeling warmth or cold by the human body in response to the air temperature, RH and air motion.

Construct a constant effective temperature line say S-A on the psychrometric chart, Figure 9.1. Mark point A on the intersection of 50% RH and 25°C DBT. Draw a vertical line from x-axis corresponding to 21.7°C intersecting the saturation curve at point S. Join the points S and A indicating constant effective temperature line. This is an approximate method of obtaining a rough idea about effective temperature.

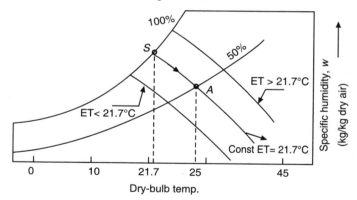

Figure 9.1 Effective temperature.

As one follows the constant effective temperature line (say from point S to A), one observes that:
- Specific humidity decreases.
- DBT increases.

But human beings feel the same comfort as ET remains the same. This is because body loses more heat in the form of latent heat, i.e. by the evaporation of sweat. An increase in DBT can be compensated by increasing the air velocity to a certain extent (say 6 m/min to 18 m/min). This can also be observed in Table 9.1.

Cooling Load Estimation and Psychrometric Analysis **323**

Table 9.1 ET is same for different DBT and RH combinations

DBT (°C)	RH (%)	% of people feeling comfort	Effective temperature (ET)	DBT (°C)	RH (%)	% of people feeling comfort	Effective temperature (ET)
21	100			25	100		
23	88			27	90		
25	76	90	21	29	80	80	25
29	64			31	70		

Based on this concept of equal comfort at different conditions, a comfort chart has been constructed by an international organisation called ASHRAE.

9.4 HUMAN COMFORT CHART

A chart is prepared showing the percentage of people feeling comfort at different effective temperatures. The collection of data regarding the percentages of people feeling so at different effective temperatures has been tabulated by ASHRAE.

The chart that gives different percentages of people (feeling comfort at different effective temperatures) is known as *comfort chart*. This chart can be prepared as follows:

1. First the prochrometric chart is redrawn taking DBT on x-axis and WBT on y-axis.
2. The 100% RH line is drawn passing through the point of intersection of DBT and WBT lines; then the other relative humidity lines are drawn as shown in Figure 9.2.
3. The effective temperature lines are drawn on this chart taking the points from Table 9.1 and similar tables published by ASHRAE.
4. The percentage of people feeling comfortable at different effective temperatures for summer and winter conditions are also shown.
5. The effective temperature lines b_1–b_1 and b_2–b_2 show that the percentage of people feeling comfort is zero in summer.
6. The maximum percentage (100%) of people feeling comfort in summer is shown by the line *BB*.
7. The maximum percentage (100%) of people feeling comfort in winter is shown by the line *AA*.

For small values of relative humidity of air, evaporation from body surface becomes rapid causing under-cooling as well as drying of the skin. On the other hand, if the relative humidity of air is high, then the evaporation from the body practically ceases and causes high discomfort due to sticky feeling on the body surface.

Generally, for the comfort conditions and softness of the skin the most desirable relative humidity range lies between 30% and 70% as shown in Figure 9.2. Skin becomes too dry when the RH is below 30% and there is a sticky feeling when RH is above 70%.

For comfortable conditions, part of the heat loss must be in the form of latent heat (evaporation loss) to maintain the softness of the skin and to avoid the shrinking of the skin surface. The heat loss by evaporation is generally 60% and by convection 40% when the man is at rest. Their proportions change with the activity of the person.

With the RH limitations, the desired comfort condition must lie between $B'B''$ for summer conditions and between $A'A''$ for winter conditions. The area $A'A''B''B'$ shown on the comfort chart is used for the year round air conditioning design.

The chart allows engineers to choose or suggest to users the most economical room conditions without jeopardising the comfort.

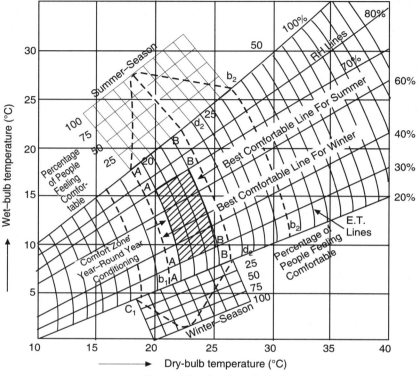

Figure 9.2 Comfort chart.

9.5 OUTSIDE DESIGN CONDITIONS

The air conditioning load is estimated to become a base for selecting the conditioning equipment. It must take into account the heat entering into the space from outdoors on a design day, as well as the heat being generated within the space. A design day is defined as:
1. A day on which the DBT and WBT are peaking simultaneously.
2. A day when there is little or no haze in the air which reduces the solar heat.
3. All of the internal loads are normal.

As far as the design of an air conditioning system is concerned, the time of peak load is important. The time of a peak load at a particular location is dependent on the outdoor conditions.

The outside ambient conditions vary from place to place and season to season. So a standard procedure to select outside design condition is given in *ASHRAE Handbook of Fundamentals*. The readers are requested to refer to the same to gather more knowledge on this topic. The outside design data is available in a tabular form. The information contained in such tables is based on the following points:

(a) Dry-bulb temperature
(b) Wet-bulb temperature
(c) Moisture content grains per kg of dry air
(d) Average daily temperature variation (average daily range)
(e) Wind velocity (summer and winter)
(f) Altitude of the place
(g) Latitude angle

It is based on the hourly recorded data of dry-bulb temperature of the last 10 years. Then it is possible to choose a dry-bulb temperature which is on 1%, 2% or 5% not exceeding basis.

The highest wet-bulb temperature coincidental with the chosen dry-bulb temperature gives us the outside design conditions of that location.

Many guide and data books like ASHRAE give outside design conditions for different places in India.

9.6 SOURCES OF HEAT LOAD

The heat gain components that contribute to the room-cooling load are classified as external or internal sources. The heat sources may be of two types, namely Sensible heat load (SH load) due to increase in room temperature and Latent heat load (LH load) due to increase in moisture content of the room air.

External sources

(a) Conduction through exterior walls, roof and glass.
(b) Conduction through interior partitions, ceilings and floors.
(c) Solar radiations through glass (SH load).
(d) Infiltration—Heat from infiltration of outside air through openings of doors (SH + LH).
(e) Ventilation—Heat from fresh outside air supplied through the system to bring freshness to the room air (SH + LH).

Internal sources

The internal load or heat generated within the space depends on the character of application. Generally, internal heat gains consist of some or all of the following items:

(a) **Occupancy**—Human body through metabolism generates and releases heat. Heat given out by people occupying the room is (SH + LH).
(b) **Lights**—Illuminants convert electrical power into light and heat. Some of the heat is radiant and is partially stored. It is in the form of SH.
(c) **Appliances**—Restaurants, hospitals, etc. have electrical, gas or steam appliances which release heat into open space. It may include SH and LH.
(d) **Equipment**—Most of the equipment give out SH loads. However, some like steriliser, coffee brewer, give out latent heat load as well.
(e) **Products brought in**—All products brought in the open space carry SH load while some like hot dishes in a restaurant bring LH load as well.
(f) **Electronic devices**—Computing machines (computers) release heat into the air-conditioned space.

(g) **Hot pipes and tanks**—Steam or hot water pipes running through the air-conditioned space or hot water tanks in the space add heat.

(h) **System heat gain**—This load is due to the air-conditioning system itself. It is of three types.

- **Duct heat gain**—Cool air in the duct absorbs heat from outside atmosphere through the duct wall (SH). It has to be reduced through insulation.
- **Duct air leakage**—Air may leak while flowing through the ducts. Such leaks should be prevented. This loss of air would result in additional load (SH + LH).
- **Blower power**—The blower or fan which circulates air through the complete system releases energy (its power) in the supply air as heat (SH).

The correct estimation of the above loads is very complex. But it would be explained in subsequent sections to a certain extent.

9.7 CONDUCTION THROUGH EXTERIOR STRUCTURES

The temperature difference between the outdoor and indoor conditions causes heat transfer through the wall or roof structure.

Heat from outside air is transferred mainly by convection to the outer surface. Then it enters the room through structure by conduction. The heat from the inside surface is transferred by convection to the room air. This is shown in Figure 9.3.

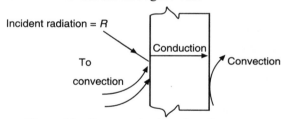

Figure 9.3 Heat transfer through wall structure.

9.7.1 Overall Heat Transfer Coefficient

Consider a plane wall shown in Figure 9.4, which is exposed to outdoor air on one side and room air on the other.

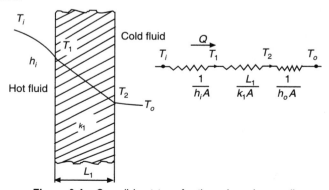

Figure 9.4 Overall heat transfer through a plane wall.

The overall heat transfer rate is calculated by the equation,
$$Q = UA\,\Delta T \qquad (9.6)$$

The overall heat transfer coefficient, $U = \dfrac{1}{(1/h_i) + (L_1/k_1) + (1/h_o)}$ \qquad (9.7)

where h_i and h_o are inside and outside heat transfer coefficients, respectively, and L_1 and k_1 are the thickness and thermal conductivity of wall. Many a time, the wall is layered with insulation. In that case the resistance to heat flow from these insulations are also accounted while finding U. Lot of research work in this area has already been carried out and the values of U are available in heat transfer data books and are used directly for calculation purposes.

9.7.2 Cooling Load Estimation by CLTD Method

The cooling loads caused by conduction heat gains through the exterior roof and wall are found from the equation
$$Q = U \times A \times CLTD_c \qquad (9.8)$$
where

Q = heat gained through roof and wall in kJ/h, kJ/s
U = overall heat transfer coefficient for roof or wall in W/m²-K,
A = area of the roof or wall in m². While finding Q for wall, area A is reduced and taken equal to the area of door and windows.
$CLTD_c$ = corrected cooling load temperature difference in °C.

The corrected cooling load temperature difference (CLTD) is not the actual temperature difference between the outdoor and the indoor air. It is a modified value that accounts for heat storage/time lag effects. The CLTD values are based on the following conditions:

1. Indoor temperature is 25.5°C.
2. Outdoor average temperature on the design day is 28.4°C.
3. Date is July 21st.
4. Location is 40°N latitude.

If the actual conditions differ from any of the above, the CLTD must be corrected as follows:
$$CLTD_c = CLTD + LM + (25.5 - T_R) + (T_a - 28.4) \qquad (9.9)$$
where

$CLTD_c$ = corrected value of CLTD (°C)
$CLTD$ = temperature available from data source
LM = correction for latitude and month from available data
T_R = room temperature (°C)
T_a = average outside temperature on a design day (°C).

9.8 HEAT GAIN THROUGH GLASS

Heat is transmitted through glass due to solar radiation. This could be direct in the form of sun rays or diffused radiation due to reflection from other objects outside. To understand the mechanism of solar radiation reaching the ordinary glass, let us study Figure 9.5.

Ordinary glass absorbs a small portion of the solar heat (5% to 6%) while about 86% of solar heat is transmitted through the glass. The net solar heat gain to the conditioned space consists of the transmitted heat plus 40% of the heat that is absorbed in the glass.

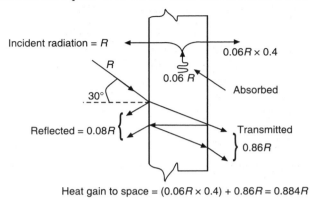

Figure 9.5 Solar radiation incident on an ordinary glass at 30° angle.

9.8.1 Factors Affecting Solar Radiation at a Place

(a) The angle of incidence of solar radiations on the glass surface depends on the latitude of the place. The peak solar radiation at latitudes beyond 23½° would be less.

(b) **Day of the year and time of the day:** The angle of incidence of solar radiations on the glass surface depends on the day of the year and time of the day. Due to relative movement of sun with respect to earth between Tropic of Cancer and Tropic of Capricorn, the radiation at a place varies with the time of the year.

(c) **Exposure of the glass:** This factor is the direction in which the glass is fixed in a space.

(d) **Altitude:** Higher altitudes cause increased solar radiation.

(e) **Haze:** Thin mist in the air absorbs solar radiations causing reduction in solar radiation.

(f) **Humidity:** Higher humidity in air reduces solar radiation at a place.

Other factors which affect the solar heat gain of the conditioned space include: (1) type of glass material, (2) colour of the glass, (3) thickness of the glass, (4) shading effect on the glass, (5) type of glass protection, (6) wind velocity. The detailed procedure to estimate the solar heat gain into conditioned space is given in ASHRAE handbooks.

9.8.2 Method of Estimation

The radiant energy from the sun passes through glass which is affected by various factors listed above. The solar heat gain can be found from the following equation:

$$Q = \text{SHGF} \times A \times \text{SC} \times \text{CLF} \qquad (9.10)$$

where

Q = solar radiation cooling load for glass (kJ/h)
A = surface area (m^2)
SHGF = maximum solar heat gain factor (W/m^2)

SC = shading coefficient

CLF = cooling/heating load factor for glass.

SHGFC is the maximum solar heat gain factor for a single clear glass for a given month, orientation and latitude. ASHRAE handbook gives all the relevant data to estimate this load.

9.9 INFILTRATION

Infiltration of air through cracks around windows or doors results in both a sensible and latent heat gain to the rooms.

9.9.1 Sensible Heat Loss Effect of Infiltration Air

Infiltration occurs when outdoor air enters through building openings, due to wind pressure. The openings of most concern to us are cracks around window sashes and door edges, and open doors. Infiltration air entering the space in summer would increase the room air temperature. Therefore, this heat gain due to infiltration air has to be removed out of the room.

The amount of sensible heat gain from infiltrating air can be determined using the equation

$$Q_s = m \times c \times TC \tag{9.11}$$

where

Q_s = sensible heat gain from outdoor air in kJ/s

m = mass flow rate of outdoor infiltration air in kg/s

c = specific heat of air in kJ/kg-K

TC = temperature change between outdoor and indoor air in °C.

If the flow rate of air is in terms of cubic metre/min, then the above Eq. (9.11) can be written as

$$Q_s = 0.0204(cmm) \times TC \tag{9.12}$$

where Q_s is in kW, cmm = air flow rate in cubic metre/min.

9.9.2 Latent Heat Loss Effect of Infiltration Air

The latent heat gain due to infiltration of outdoor air into the room is calculated using the equation

$$Q_L = 50(cmm)(w_o - w_i) \tag{9.13}$$

where

Q_L = latent heat gain due to infiltration air in kJ/s

cmm = air infiltration in cubic metre/min

w_o, w_i = specific humidity for outdoor and indoor conditions respectively in kg/kg dry air.

Finding the infiltration rate

There are two methods used to estimate the cmm of infiltration air: the crack method and the air change method.

Crack method: Infiltration depends upon wind velocity, area of window, construction of window and type of window. A reasonably accurate estimate of the rate of air infiltration per metre of crack opening can be measured or established. Energy codes list maximum permissible infiltration rates for new construction or renovation upgrading. Table 9.2 lists typical allowable infiltration rates based on a 25 metre per hour wind velocity.

The quality of installation and the maintenance of windows and doors greatly affect the resultant crack infiltration. Poorly fitted windows may have up to five times the sash leakage shown in Table 9.2.

Table 9.2 Typical infiltration rates

Component	Infiltration rate
Windows	0.011047 cmm per 0.33 m of sash crack
Residential doors	0.01415 cmm per 0.0929 m^2 of door area
Non-residential doors	0.0283 cmm per 0.0929 m^2 of door area

Air change method: This procedure for finding the infiltration rate is based on the number of air changes per hour (ACH) in a room subjected to infiltration.

One air change is defined as being equal to the room air volume.

Determination of the expected number of air changes due to infiltration is based on experience and testing. Suggested values range from 0.5 ACH to 1.5 ACH for buildings ranging from 'tight' to 'loose' construction.

Using the definition of an air change, Eq. (9.14) can be used to find the air infiltration rate in cmm.

$$\text{cmm} = \text{ACH} \times \frac{V}{60} \qquad (9.14)$$

where

cmm = air infiltration rate to room in cubic metre per minute

ACH = number of air changes per hour for room

V = room volume in m^3.

9.10 VENTILATION

Some outside air is generally admitted into a building for health and comfort reasons. The sensible and latent heat of this air is usually greater than that of the room air so it becomes part of the cooling load. The excess heat is usually removed in the cooling equipment; however, it is part of the cooling coil load and not the building load.

The sensible and latent cooling loads from ventilation air can be estimated using Eqs. (9.12) and (9.13) which are again given below:

$$Q_s = 0.0204(\text{cmm}) \times TC \qquad (9.12)$$

$$Q_L = 50(\text{cmm})(w_o - w_i) \qquad (9.13)$$

The total heat Q_t removed from the ventilation air is $Q_t = Q_s + Q_L$.

Recommended outdoor air ventilation rates for some applications are listed in Table 9.3. This table has ventilation rates similar to many state codes and standards.

The ventilation rates in Table 9.3 are often higher than the minimum listed in earlier standards. For instance, it requires 0.424 cmm per person in an office space.

There are still further changes in ventilation requirements that are being considered. For instance, the values shown in Table 9.3 do not make special allowances for the amount of indoor air pollutants being generated. Undoubtedly new standards will reflect this and other information that is being found in this rapidly developing field.

Outdoor air shall be provided at a rate no less than the greater of either (A) or (B).

A. 15 CFM (0.424 cmm) per person, times the expected occupancy rate.
B. The applicable ventilation rate from the following list, times the conditioned floor area of the space.

Table 9.3 Minimum mechanical ventilation requirement rates

Type of Use	CFM per square foot of conditioned floor area	Cmm per square foot of conditioned floor area
Auto repair workshops	1.50	0.04245
Barber shops	0.40	0.01132
Bars, cocktail lounges and casinos	1.50	0.04245
Beauty shops	0.40	0.01132
Coil-operated dry cleaning	0.30	0.00849
Commercial dry cleaning	0.45	0.01273
Hotel guest rooms (less than 500 sq ft)	30 CFM/guest room	0.848 cmm/guest room
Hotel guest rooms (500 sq ft or more)	0.15	0.004245
Retail stores	0.20	0.00566
Smoking lounges	1.50	0.04245
All others	0.15	0.004245

EXAMPLE 9.1 An auditorium seats 1000 people. The space design conditions are 25°C and 50% RH, and outdoor design conditions 35°C DBT and 24 WBT. What is the cooling load due to ventilation?

Solution: Assume ventilation air per person to be 0.424 cmm.

$$Q_s = 0.0204(\text{cmm}) \times TC$$
$$= 0.0204(0.424 \times 1000)(35 - 25) = 86.4 \text{ kW}$$
$$Q_L = 50(\text{cmm})(w_o - w_i) = 50(0.424 \times 1000)(0.014 - 0.01) = 84.8 \text{ kW}$$
$$Q_T = 86.4 + 84.8 = 171.2 \text{ kW} = 48.6 \text{ TR} \quad \textbf{Ans.}$$

9.11 OUTSIDE AIR LOAD

Both infiltration and ventilation are heat loads due to the outside air. The heat gain through infiltration air is directly in the conditioned space and ventilation air heat load is directly on the coil.

The ventilation air flow duct is joined to the return air flow duct and this in turn tends to build up pressure in the space. This positive pressure counteracts the wind pressure responsible for infiltration. Thus, it would reduce infiltration. Therefore, the infiltration and ventilation loads do not have additive effect but the higher value of the two would be the outside air load. Following three cases are discussed.

(i) When infiltration < ventilation, infiltration is completely neutralised and the outside air load is only due to ventilation.

(ii) When infiltration = ventilation, the result is same as in (i).

(iii) When infiltration > ventilation, infiltration is neutralised only to an extent due to ventilation. So the outside air load is in two parts:
In the room = infiltration air − ventilation air
On the coil = ventilation air

In some cases, ventilation quantity is increased to neutralise infiltration as infiltrating air is unfiltered air.

The outside air heat load consists of sensible as well as latent loads and which are estimated using Eqs. (9.12) and (9.13).

9.12 HEAT LOAD FROM PEOPLE

The heat gain from people is composed of two parts, sensible heat and latent heat resulting from perspiration. Some of the sensible heat may be absorbed and may lead to heat storage effect, but not the latent heat. The equations for cooling loads for sensible and latent heat gains from people are:

$$Q_s = q_s \times n \times \text{CLF} \quad (9.15)$$

$$Q_L = q_L \times n \quad (9.16)$$

where

Q_s, Q_L = sensible and latent heat gains (loads)

q_s, q_L = sensible and latent heat gains per person

n = number of people

CLF = cooling load factor for people.

The rate of heat gain from people depends on their physical activity. Table 9.4 lists values for some typical activities. The rates are suitable for a 23.3°C DB room temperature. Values vary slightly at other temperatures, as noted.

Table 9.4 Heat gain from human beings

Degree of activity	Place	Adult male	Adult female	Sensible heat (W)	Latent heat (W)
Seated at theatre	Theatre—matinee	115	97	66	31
Seated at theatre at night	Theatre—night	115	103	72	31
Seated, very light work	Offices, hotels, apartments	132	118	72	45
Light bench work	Factory	235	220	80	140
Heavy work	Factory	440	425	170	255

9.13 LIGHTING

The equation for determining the cooling load due to heat gain from lighting is

$$Q = W \times BF \times CLF \tag{9.17}$$

where

Q = cooling load from lighting (kW)

W = lighting capacity (W)

BF = ballast factor

CLF = cooling load factor for lighting.

The term W is the rated capacity of the lights in use expressed in watts. In many applications, all of the lighting is switched on at all times, but if it is not, the actual amount should be used.

The factor BF accounts for heat losses in ballasts of fluorescent lamps or other special losses. A typical value of BF is 1.25 for fluorescent lighting. For incandescent lighting, there is no extra loss and BF = 1.0.

The factor CLF accounts for storage of part of the lighting heat gain. The storage effect depends on how long the lights and cooling system are operating, as well as on the building construction, types of lighting fixtures and ventilation rate.

9.14 HEAT GAIN FROM EQUIPMENT AND APPLIANCES

The heat gain from equipment may sometimes be found directly from the manufacturer or the nameplate data with allowance for intermittent use. Some equipment produces both sensible and latent heat. Some values of heat output for typical appliances are shown in Tables 9.5(a) and (b). These values are multiplied by a usage factor depending upon the use.

Table 9.5(a) Heat gain in watts from appliances

Appliance	Size	Sensible heat gain (W)	Latent heat gain (W)	Total heat gain (W)
Restaurant, electric blender, per quart of capacity	1 to 4 qt	293	153	446
Coffee heater, per warming burner	1 to 2 burners	68	33	101
Ice maker (large)	100 kg/day	2730	0	2730
Microwave oven (heavy-duty commercial)	0.7 ft^3	2628	0	2628
Toaster (large pop-up)	10 slices	2810	2490	5300

Table 9.5(b) Heat gain from equipment working inside the conditioned space

Appliance	Size	Recommended rate of heat gain (W)
Computer Devices		
Microcomputer/word processor	16–640 Kbytes	88–528
Minicomputer	—	2200–4400
Printer (laser)	8 pages/min	293
Printer (line, high-speed)	5000 or more pages/min	735–810
Copiers/Typesetters		
Blueprint	30–67 copies/min	1145–3725
Copies (large)	—	500–1950
Miscellaneous		
Coffeemaker	10 cups sensible latent	500–450
Microwave oven	—	400
Water cooler	—	1760

The efficiency of these devices (say motors) must be taken into consideration while taking the final heat gain. For example, suppose the capacity of the motor is 300 W. If its efficiency is 95%, then the sensible load is equal to 300/0.95 = 316 W. Likewise for other devices, appropriate efficiency, usage factor, etc. need to be considered.

9.15 SYSTEM HEAT GAIN

Air-conditioning equipment is placed in the conditioned space as well as outside as per the needs and these also add up heat to it.

(a) Duct heat gain: The conditioned air flows through ducts and will gain heat from the surroundings. If the duct passes through the conditioned space, the heat gain results in useful cooling spaces and the heat results in a useful cooling effect, but for the ducts passing through unconditioned spaces, it is a loss of sensible heat that must be added to the building sensible cooling load (BSCL). The heat gain can be calculated from the heat transfer equation:

$$Q = U \times A \times TD \tag{9.18}$$

where

Q = duct heat gain (W)
U = overall coefficient of heat transfer (W/m²-K)
A = temperature difference between the air in duct and the surrounding air (°C).

It is recommended that cold air ducts passing through unconditioned areas be insulated. If there is significant heat gain to return air ducts, it should also be calculated, but it is only added to the coil sensible cooling load (CSCL) and not to the BSCL.

(b) Duct air leakage: Duct systems will leak air at joints. This leakage should be limited to 5% or less of the total cmm. If ducts are outside the conditioned space, the effect of leakage must be added to the BSCL and BLCL (building latent cooling load). If the air leaks in the conditioned space, it does useful cooling, but care should be taken that it is not distributed to the wrong location.

(c) Blower power: The power of blower, which circulates air through the system, is added to the supply air as heat. This sensible heat load is equal to the input to the blower.

9.16 ROOM COOLING LOAD

The room cooling load is the sum of each of the cooling load components (roof, walls, glass, solar, people, equipment and infiltration) in the room.

When calculating cooling loads, a prepared form is useful. A load calculation form (E-form) is shown at the end of this chapter. It can be used for individual rooms of a small building.

Room peak load: We have learned how to calculate the cooling loads, but not how to determine their peak (maximum) value. Because the air-conditioning system must be sized to handle peak loads, we must know how to find them.

The external heat gain components vary in intensity with the time of the day and the time of the year because of the changing solar radiation as the orientation of the sun changes and because of outdoor temperature changes. This results in a change in the total room cooling load. It is therefore very difficult to find the peak load accurately.

Building peak cooling load: The building cooling load is the rate at which heat is removed from all air-conditioned rooms in the building at the time the building cooling load is at its peak value.

If the peak cooling loads for each room were added, the total would be greater than the peak cooling load required for the whole building, because these peaks do not occur at the same time. Therefore, the designer must also determine the time of the year and time of the day when the building cooling load is at its peak, and then calculate it.

9.17 COOLING COIL LOAD

After the building cooling load is determined, the cooling coil load is found.

The cooling coil load is the rate at which heat must be removed by the air conditioning equipment cooling coil(s).

The cooling coil load will be greater than the building load because there are heat gains to the air conditioning system itself. These gains may include:

1. Ventilation (outside air)
2. Heat gains to ducts
3. Heat produced by the air conditioning system fans and pumps
4. Air leakage from ducts

9.18 PSYCHROMETRIC ANALYSIS OF THE AIR CONDITIONING SYSTEM

In this section, we will use the basic psychrometric processes to analyze a complete air conditioning system and will also briefly consider some more advanced concepts.

9.18.1 Determining Supply Air Conditions

The rooms in a building gain heat in summers from a number of sources. The procedures for finding these heat gains are discussed in the earlier sections of this chapter.

The rate at which heat must be extracted from a room to offset these heat gains is known as *room total cooling load* (RTCL). RTCL is composed of two parts—the room sensible cooling load (RSCL) and the room latent cooling load (RLCL).

This heat extraction or cooling effect is provided by supplying air to the room at a temperature and humidity low enough to absorb the heat gains.

These relationships are shown in Figure 9.6 and are expressed by the sensible and latent heat equations:

$$\text{RSCL} = 0.0204 \text{ cmm}_S(T_R - T_S) \tag{9.19}$$

$$\text{RLCL} = 50 \times \text{cmm}_S(w_R - w_S) \tag{9.20}$$

where

$\quad$ RSCL = room sensible cooling load (kW)
$\quad$ RLCL = room latent cooling load (kW)
$\quad$ cmm$_S$ = cmm of supply air
$\quad$ $T_R - T_S$ = temperature difference of room and supply air (°C)
$\quad$ $w_R - w_S$ = humidity ratios of room and supply air (g of wv/kg dry air).

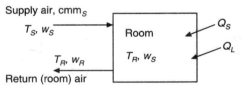

Figure 9.6 Supply air and return room air.

It is the usual practice to apply the RSCL Eq. (9.19) first to determine the supply air cmm$_S$ and T_S, and then apply the RLCL Eq. (9.20) to determine the supply air humidity ratio w_S.

1. In applying Eq. (9.19), the RSCL is known from the cooling load calculations.
2. T_R and w_R are selected in the comfort zone. This still leaves two unknowns, cmm$_S$ and T_S. One of these is chosen according to "good practice" (such as costs and job conditions), and the remaining unknown is then calculated from the equation.
3. Example (9.2) illustrates the calculation of the supply conditions.

EXAMPLE 9.2 A hair salon shop has a sensible cooling load of 16 kW and latent cooling load of 6.5 kW. The room conditions are to be maintained at 25 DBT and 50% RH. If 56 cmm of supply air is furnished, determine the required supply air DBT and WBT.

Solution:

Applying Eq. (9.19), solving it for the supply air temperature change,

$$\text{RSCL} = 0.0204 \times \text{cmm}_S \times (T_R - T_S)$$

$$T_R - T_S = \frac{\text{RSCL}}{0.0204 \times \text{cmm}_S} = \frac{16}{0.0204 \times 56} = 14°C$$

The supply air temperature is therefore

$$T_S = 25 - 14 = 11°C \qquad\qquad\textbf{Ans.}$$

The required humidity ratio of the supply air is then found from Eq. (9.20):

$$w_R - w_S = \frac{\text{RSCL}}{50 \times \text{cmm}_S} = \frac{6.5}{50 \times 56} = 0.00232 \text{ g wv/kg dry air}$$

From the psychrometric chart, $w_R = 0.01$ g wv/kg dry air, and therefore

$$w_S = 0.01 - 0.00232 = 0.00768 \text{ g wv/kg dry air}$$

Reading from the chart, WBT = 9°C. **Ans.**

9.18.2 Room Sensible Heat Factor (RSHF)

Room Sensible Heat Factor is defined as the room sensible heat to the room total heat. Mathematically, room sensible heat factor is

$$\text{RSHF} = \frac{\text{RSCL}}{\text{RTCL}} = \frac{\text{RSCL}}{\text{RSCL} + \text{RLCL}}$$

where

RSCL = room sensible cooling load
RLCL = room latent cooling load
RTCL = room total cooling load.

The supply air to the room must have the capacity to absorb simultaneously both the RSCL and the RLCL. The point S on the psychrometric chart, as shown on Figure 9.7, represents the supply air condition and the point R represents the required final condition in the room (i.e. room design condition). The line SR is called the *room sensible heat factor line* (RSHF line). The slope of this line gives the ratio of the RSCL to the RLCL. Thus, the supply air having its conditions given by any point on this line will satisfy the requirements of the room with adequate supply of such air.

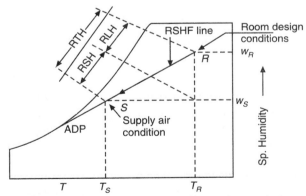

Figure 9.7 Representation of supply air condition and room design conditions.

When the supply air conditions are not known, which in fact is generally required to be found out, the room sensible heat factor line may be drawn from the calculated value of room sensible heat factor (RSHF), as explained in Example 9.3.

EXAMPLE 9.3 A shop has a sensible cooling load of 13.15 kW and a latent cooling load of 4.4 kW. The shop is maintained at 25°C DBT and 45% RH. Draw the RSHF line.

Solution: The solution is shown in Figure 9.8. The following steps are carried out.

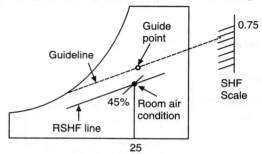

Figure 9.8 Plotting the RSHF line—Example 9.3.

1. Calculate the RSHF.

$$\text{RSHF} = \frac{\text{RSCL}}{\text{RTCL}} = \frac{13.15}{13.15 + 4.4} \approx 0.75$$

2. On the SHF scale on the psychrometric chart, locate the 0.75 slope. There is also a guide point for the SHF scale, encircled on the chart (located at 26.6°C and 50% RH.). Draw a guideline from SHF = 0.75 through the guide point.
3. Draw a line parallel to the guideline through the room condition point. This is the RSHR line because it has the RSHF slope, and passes through the room condition. (Two drafting triangles will aid in drawing an exact parallel line.)

9.19 SUMMER AIR CONDITIONING SYSTEM PROVIDED WITH VENTILATION AIR (ZERO BYPASS FACTOR)

The accumulated carbon dioxide and odours and other air contaminants of the conditioned space need to be diluted and fresh air supplied to maintain the purity of air. Accordingly, the supply air to the room comprises fresh air and recirculated room air. An amount equivalent to the fresh air is ejected from the room. The schematic diagram of the air conditioning system is shown in Figure 9.9 and the process cooling and dehumidification of air while passing over the cooling coil is shown in Figure 9.10.

The points O and R represent the outside and inside air conditions in Figure 9.10. Point 3 is the state of air after mixing the recirculated room air m_R with the ventilation air m_0. The room sensible heat factor line is drawn from the inside condition R to intersect the saturation curve at room ADP at 4. Point 4 is the supply air state for a minimum rate of supply air.

The line 3–4, therefore, represents the condition line for the apparatus and is called the *grand sensible heat factor* (GSHF) line.

It is noted that the line R–4 is the condition line for the room or the RSHF line and the GSHF line intersecting the saturation curve at coil apparatus dew point (Coil ADP). Note that, in this case, the coil ADP and the room ADP are the same.

The grand sensible heat factor is defined in the following text.

Cooling Load Estimation and Psychrometric Analysis **339**

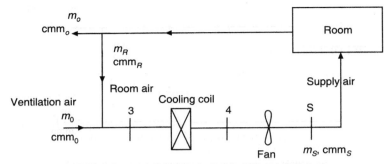

Figure 9.9 Air-conditioning system with ventilation air.

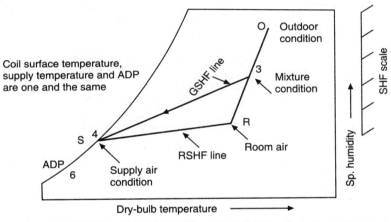

Figure 9.10 Summer air-conditioning process with ventilation air and zero bypass factor.

9.19.1 Grand Sensible Heat Factor (GSHF)

In this section a summer air-conditioning system with ventilation load and the cooling coil with bypass factor is assumed. The block diagram for this is the same as shown in Figure 9.9.

GSHF is the ratio of the total sensible cooling load (TSCL) to the grand total cooling load (GTCL), which the cooling coil or the conditioning apparatus is required to handle.

TSCL = Total sensible cooling load = RSCL + OASCL
TLCL = Total latent cooling load = RLCL + OALCL
GTCL = RSCL + RLCL + (OASCL + OALCL)

where

RSCL = room sensible cooling load
RLCL = room latent cooling load
RTCL = room total cooling load

$$\text{GSHF} = \frac{\text{TSCL}}{\text{GTCL}} = \frac{\text{TSCL}}{\text{TSCL} + \text{TLCL}} = \frac{\text{RSCL} + \text{OASCL}}{(\text{RSCL} + \text{OASCL}) + (\text{RLCL} + \text{OALCL})}$$

Outside air sensible cooling load = OASCL = $0.0204 \text{ cmm}_o(T_o - T_R)$ kW

Outside air latent cooling load, OALH = $50 \text{ cmm}_o(w_o - w_R)$ kW

Outside air total cooling load, OATCL = OASCL + OALCL

The outside air total heat can also be calculated from the following relation:

$$\text{OATH} = 0.02 \text{ cmm}_o(h_o - h_R) \text{ kW}$$

where

cmm_o = outside air or ventilation air in cubic metre per minute

T_o, T_R = dry-bulb temperature of outside air and room air respectively in °C.

w_o, w_R = specific humidity of outside air and inside air in kg/kg of dry air.

h_o, h_R = enthalpy of outside air and room air in kJ/kg of dry air.

For economical operation of the air-conditioning plant, fresh air from outside is mixed with the recirculated room air in a duct and this mixture is passed through the cooling coil where it is cooled and dehumidified. This process can be represented on the psychrometric chart shown in Figure 9.11. Point O represents the condition of the outside air, point R represents the condition of room air and point 3 represents the condition of mixture air entering the cooling coil. The mixture while passing over the coil gets cooled and dehumidified. Point 4 shows the supply air or condition of air leaving the cooling coil. When point 3 is joined with point 4, it gives a grand sensible heat factor line (GSHF line). This line, when produced up to the saturation curve, gives the apparatus dew point (ADP). Point 4, as shown, is the intersection of GSHF line and RSHF line. The GSHF line may be drawn on the psychrometric chart in a similar way as discussed for the RSHF line.

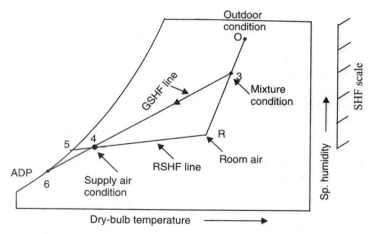

Figure 9.11 Grand sensible heat factor.

9.19.2 Winter Air Conditioning

During winters the condition of air outside the room is much different from that during summers. In winter the temperature of air is much lower (say, 5 to 10°C) and the humidity content is also very low. Therefore, in order to maintain a thermal comfort condition inside the room, supply air is to be heated so that temperature of the order 24°C is maintained in the room.

The building sensible heat losses are partially compensated by the solar heat gains and the internal heat gains such as those from occupancy, lighting, etc. Similarly, the latent heat loss due to low outside air humidity is more or less offset by the latent heat gains from occupancy. Sensible heat gains (negative loads) such as the solar heat may not be present at the time of peak load, and hence they are not counted. On the other hand, latent heat gains from occupancy, etc. are always present and should be taken into account. As a result, the design heating load for winter air conditioning is predominantly sensible.

In general, the processes in the conditioning apparatus for winter air conditioning for comfort involve heating and humidifying. Two of the typical process combinations are:

(i) Preheating the air with steam or hot water in a coil followed by adiabatic saturation and reheat with steam or hot water in a coil.

(ii) Heating and humidifying air in an air washer with pumped recirculation and external heating of water followed by reheat.

The supply conditions corresponding to point S, as shown in Figure 9.12 can be achieved by two ways.

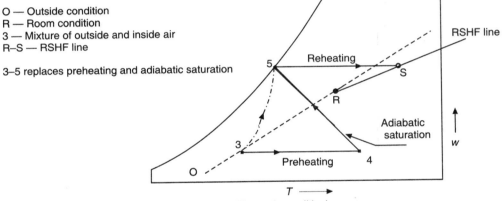

Figure 9.12 Winter air conditioning.

First method

(a) Preheating the air from point 3 and up to point 4.
(b) Adiabatic saturation
(c) Reheating the air till it reaches point S.

Second method

(a) Hot water with appropriate temperature is sprayed into the air at point 3. This helps air to heat and humidify. This is shown with the process curve 3–5.
(b) Reheating the air till it reaches point S.

The reheating process 5–S is common to both the methods. The supply air states should lie on the room sensible heat factor line. It is, therefore, determined by the RSHF and by the choice of supply air rate, which is usually known from the summer air-conditioning calculations.

9.20 EFFECTIVE ROOM SENSIBLE HEAT FACTOR (ERSHF)

ERSHF is the ratio of the effective room sensible cooling load to the effective room total cooling load.

$$\text{ERSHF} = \frac{\text{ERSCL}}{\text{ERTCL}} = \frac{\text{ERSCL}}{\text{ERSCL} + \text{ERLCL}}$$

where

ERSCL = effective room sensible cooling load = RSCL + OASCL × BPF
= RSH + 0.0204 $\text{cmm}_o(T_o - T_R)$ × BPF

ERLCL = effective room latent cooling load = RLCL + OALCL × BPF
= RLH + 50 $\text{cmm}_o(w_o - w_R)$ BPF

ERTCL = effective room total cooling load = ERSCL + ERLCL

BPF = Bypass factor

The effective room sensible heat factor line (ERSHF line) is obtained by joining the point R and point 6, i.e. ADP as shown in Figure 9.13. It is slightly different from the RSHF line. The ESHF line does not take into account the inefficiency of the cooling coil while the same is taken into consideration in ERSHF line. From point 4, draw 4–4a parallel to 3–R. Therefore, from similar triangles 6–4–4a and 6–3–R,

$$\text{BPF} = \frac{\text{Length } 4-6}{\text{Length } 3-6} = \frac{\text{Length } 4a-6}{\text{Length } R-6}$$

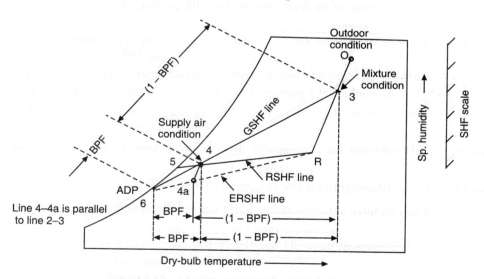

Figure 9.13 Effective room sensible heat factor.

The bypass factor is also given by

$$\text{BPF} = \frac{T_4 - \text{ADP}}{T_3 - \text{ADP}} = \frac{T_{4a} - \text{ADP}}{T_R - \text{ADP}}$$

The mass of dehumidified air is given by

$$m_d = \frac{\text{Room total load}}{h_R - h_4}$$

where
    h_R = enthalpy of air at room condition
    h_4 = enthalpy of supply air to room from the cooling coil.

EXAMPLE 9.4 In an auditorium which is to be maintained at a temperature not exceeding 24°C and a relative humidity not more than 60%, a sensible heat load of 132 kW and 84 kg/h of moisture has to be removed. Air is supplied to the auditorium at 15°C.
(a) How many kg of air per hour must be supplied?
(b) What is the dew point temperature of supply air and what is its relative humidity?

Solution:
Given, inside temperature, T_R = 24°C, RH = 60%,
Supply air to auditorium = T_S = 15°C, Room sensible cooling load (RSCL) = 132 kW

$$\text{Room latent cooling load (RLCL)} = \frac{84}{3600} \times 2444.9 \quad [h_{fg} \text{ at } 24°C = 2444.9 \text{ kJ/kg}]$$

$$= 57.04 \text{ kW}$$

$$\text{RSHF} = \frac{\text{RSCL}}{(\text{RSCL} + \text{RLCL})} = \frac{132}{(132 + 57.04)} = 0.6982$$

Draw the RSHF line as shown in Figure 9.14.

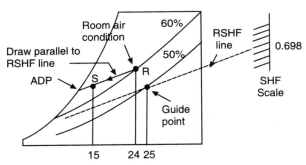

Figure 9.14 Plotting the RSHF line—Example 9.4.

Room total cooling load = RTCL = RSCL + RLCL = 189.04 kW
The supply air to the auditorium must remove this heat. Therefore, energy balance,

$$\text{RTCL} = m_a(h_R - h_S)$$

(a) Mass of air per hour that must be supplied, $m_a = \dfrac{\text{RTCL}}{h_R - h_S} = \dfrac{189.04}{52.7 - 39.7} = 14.54 \text{ kg/s}$

$$= 52.349 \times 10^3 \text{ kg/h} \qquad \textbf{Ans.}$$

(b) Supply air condition is 15°C as given in the example. It must lie on the R–ADP, so draw the vertical line DBT = 15°C, to meet the process line R–ADP at point S. Note down the RH corresponding to point S, i.e. 92%.

The point ADP indicated the DPT = 13.7°C. **Ans.**

EXAMPLE 9.5 (a) 28.5 cmm of room air at 25.5°C DBT and 50% RH is mixed with 28.5 cmm of outside air at 38°C DBT and 27°C WBT. Find the ventilation load and the condition of air after mixing.

(b) The above mixture of air is passed through an air-conditioning equipment. If the apparatus dew point (ADP) temperature of the equipment is 14.5°C, determine the heat removed by the equipment.

Solution: Collect the following properties from the psychrometric chart corresponding to the room and outside conditions.

Condition	DBT (°C)	WBT (°C)	RH (%)	w (kg wv/kg dry air)	h (kJ/kg dry air)	v (m³/kg dry air)
Room	25.5		50	0.0102	52	0.86
Outside	38	27		0.018	85	0.906
Mixture	31.59			0.014	68	

$$\text{Mass flow of room air, } m_R = \frac{28.5}{0.86} = 33.14 \text{ kg/min}$$

$$\text{Mass flow of outside air, } m_o = \frac{28.5}{0.906} = 31.46 \text{ kg/min}$$

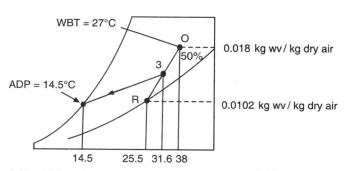

Figuer 9.15 Mixing of two streams and passing over a cooling coil—Example 9.5.

(a) Condition after mixing:

Mass balance, $m_3 = m_R + m_o = 33.14 + 31.46 = 64.6$ kg/min

$$w_1 = \frac{m_R w_R + m_o w_o}{m_3}$$

$$= \frac{33.14 \times 0.0102 + 31.46 \times 0.018}{64.6} = 0.014 \text{ kg wv/kg dry air}$$

$$T_3 = \frac{m_R T_R + m_o T_o}{m_3} = \frac{33.14 \times 25.5 + 31.46 \times 38}{64.6} = 31.59°C$$

Ventilation load = $m_o(h_o - h_R)$ = 31.46(85 − 52) = 1038.18 kJ/min

= 17.303 kW **Ans.**

(b) Corresponding to 14.5°C ADP, we get h_S = 40.6 kJ/kg dry air.

∴ Heat removed by the equipment = $m_3(h_3 - h_S)$ = $\frac{64.6}{60}$(68 − 40.6)

= 29 kW **Ans.**

Here all the air comes in contact with the surface of the coil. Therefore, the temperature of the air leaving the coil equals the surface temperature of the coil. The bypass factor is zero.

EXAMPLE 9.6 Room conditions: 26°C DBT, 19°C WBT
Outside conditions: 35°C DBT, 27°C WBT
Room heat gains:
Sensible heat: 11.1 kW
Latent heat: 3.9 kW

The conditioned air supplied to the room is 50 cmm and 25% fresh air and 75% recirculated room air. Determine the following.
(a) The DBT and WBT of supply air.
(b) The DBT and WBT of mixed air before the cooling coil.
(c) The apparatus dew point and bypass factor of the coil.
(d) The refrigeration load on the cooling coil and the moisture removed by the coil.

Solution: The following properties are collected referring to the psychrometric chart. The processes are represented as shown in Figure 9.16.

Condition	DBT (°C)	WBT (°C)	w (kg wv/kg dry air)	h (kJ/kg dry air)
Room	26	19	0.0108	54.1
Outside	35	27	0.0192	84.6
Mixture	28.25		0.0129	61.70

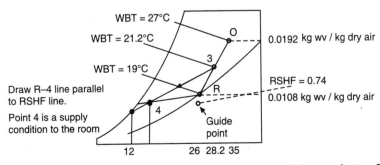

Figure 9.16 Mixing of two streams and passing over a cooling coil with bypass factor—Example 9.6.

Mixture temperature, $T_3 = 0.25 \times 35 + 0.75 \times 26$
$$= 28.25°C$$

$$\text{RSHF} = \frac{\text{RSCL}}{(\text{RSCL} + \text{RLCL})} = \frac{11.1}{(11.1 + 3.9)} = 0.74$$

Now,
$$\text{RSCL} = 0.0204 \times (\text{cmm})_S (T_R - T_S)$$
$$11.1 = 0.0204 \times 50 \times (26 - T_S)$$
$\therefore \qquad T_S = 15.12°C$

(a) Conditions of supply air are
$$\text{DBT} = 15.12°C$$
$$\text{WBT} = 14°C \qquad \textbf{Ans.}$$

(b) Condition of mixed fresh and recirculated air before the cooling coil
$$\text{DBT} = 28.25°C$$
$$\text{WBT} = 21.25°C \qquad \textbf{Ans.}$$

(c) Apparatus dew point
$$\text{Coil ADP} = 12°C$$
$$\text{Bypass factor} = \frac{T_4 - \text{ADP}}{T_3 - \text{ADP}} = \frac{15.12 - 12}{28.25 - 12} = 0.192 \qquad \textbf{Ans.}$$

(d) Refrigeration load on cooling coil
$$= m_S \times (h_3 - h_S)$$
$$= \frac{50}{0.83} \times (61.70 - 39.3) = 1349.3980 \text{ kJ/min}$$
$$= 22.49 \text{ kJ/s or kW} \qquad \textbf{Ans.}$$

Moisture removed by the coil $= m_S(w_3 - w_S)$
$$= \frac{50}{0.83}(0.0129 - 0.0093)$$
$$= 0.2168 \text{ kg/min} \qquad \textbf{Ans.}$$

EXAMPLE 9.7 500 kg of air is supplied per minute to an auditorium maintained at 21°C and 40% RH. The outside air at 5°C DBT and 60% RH is first passed over heating coils and heated until its WBT is equal to the room DBT. It is then passed through an adiabatic saturator and finally heated to 45°C before being supplied to the room. Determine.

(a) The heat added to both the heating coils.
(b) The mass of water evaporated in the air washer.

Solution: Collect the following properties from the psychrometric chart.

Condition	DBT (°C)	RH (%)	w (kg wv/kg dry air)	h (kJ/kg dry air)
Outside	5	60	0.0032	13.5
Room	21	40	0.0062	37
2	28.5			
3	13.2		0.0095	

Step I: Mark the point O on the chart for DBT = 5°C and RH = 60%
Step II: Mark the point R on the chart for DBT = 21°C and RH = 40%
Step III: Draw a line passing through point R, which is parallel to WBT line of temperature of 21°C. This will intersect the horizontal line drawn from point O, to get point 1. Point 2 is on saturation curve.
Step IV: Draw a horizontal line passing through point 2 to get point 3.

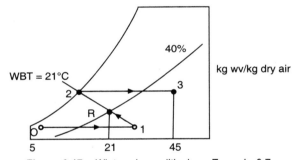

Figure 9.17 Winter air conditioning—Example 9.7.

Mass of dry air

$$m_a = \frac{500}{(1 + 0.0032)} = 498.39 \text{ kg/min}$$

$h_0 = 13.5$ kJ/kg, $h_1 = 36.8887$ kJ/kg

(a) Heat added to first coil = $m_a(h_1 - h_0)$

$\qquad = 498.39(36.8887 - 13.1285)$

$\qquad = 11841.84$ kJ/min = 197.3639 kW **Ans.**

From chart, $T_2 = 13.2$°C, $w_2 = 0.0095$ kg wv/kg dry air

$h_2 = 37.26$ kJ/kg dry air, $h_3 = 69.78$ kJ/kg dry air

Heat added to second coil

$\qquad = m_a(h_3 - h_2) = 498.39(69.7882 - 37.26)$

$\qquad = 270.1955$ kW **Ans.**

(b) Mass of water evaporated in the air washer

$$= m_a(w_2 - w_1) = 498.39(0.0095 - 0.003228) = 3.1259 \text{ kg/min} \quad \textbf{Ans.}$$

EXAMPLE 9.8 An air-conditioned space is maintained at 25°C DBT and 50% RH. The outside conditions are 40°C DBT and 25°C WBT. The space has a sensible heat gain of 24.5 kW. Conditioned air is supplied to the space as saturated air at 10°C. The equipment consists of an air washer. The air entering the air washer comprises 25% outside air. Calculate the following.
(a) Volume flow rate of air supplied to space.
(b) Latent heat gain of space.
(c) Cooling load of air washer.

Solution: The following properties of air are collected from the psychrometric chart.

Condition	DBT (°C)	WBT (°C)	RH (%)	w (kg wv/kg dry air)	h (kJ/kg dry air)
Room	25		50	0.0098	50
Outside	40	25		0.0136	76
3	28.75			0.01075	56.5

Condition of air after mixing is 25% outside air and 75% recirculated room air (Figure 9.18)

$$T_3 = 0.25 \times T_o + 0.75 \times T_R$$
$$= 0.25 \times 40 + 0.75 \times 25 = 28.75°C$$
$$w_3 = 0.25 \times w_o + 0.75 \times w_R$$
$$= 0.25 \times 0.0136 + 0.75 \times 0.0098 = 0.01075 \text{ kg wv/kg dry air}$$
$$h_3 = 0.25 \times h_o + 0.75 \times h_R = 0.25 \times 76 + 0.75 \times 50$$
$$= 56.5 \text{ kJ/kg dry air}$$

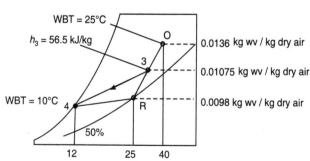

Figure 9.18 Mixing of two streams and passing over a cooling coil with bypass factor—Example 9.8.

(a) Volume flow rate of air supplied to space

$$\text{cmm} = \frac{\text{RSCL}}{0.0204 (T_R - T_S)}$$

$$= \frac{24.5}{0.0204 (25 - 10)} = 80.065 \text{ m}^3/\text{min} \quad \textbf{Ans.}$$

(b) Room latent heat gain or room latent cooling load

$$\text{RLCL} = (\text{cmm}) \times 50 \times (w_R - w_S)$$
$$= 80.065 \times 50 \times (0.0098 - 0.0076) = 8.8071 \text{ kW} \quad \textbf{Ans.}$$

(c) Cooling load of air washer

$$= m_S(h_3 - h_S)$$
$$= \frac{80.065}{0.813}(56.5 - 29.5) = 2658.98 \text{ kJ/min}$$
$$= 44.316 \text{ kW} \quad \textbf{Ans.}$$

EXAMPLE 9.9 Given for the air conditioning of a room:

Room conditions: 26.5°C DBT and 50 per cent RH
Room sensible heat gain = 26.3 kW
Room sensible heat factor = 0.82
Find the following:

(a) The room latent heat gain.
(b) The apparatus dew point.
(c) The cmm of air if it is supplied to the room at the apparatus dew point.
(d) The cmm and specific humidity of air if it is supplied to the room at 17°C.

Solution: Room sensible heat (RSH) or RSCL = 26.3 kW, RSHF = 0.82

(a) Now, room sensible cooling load

$$\text{RSHF} = \frac{\text{RSCL}}{\text{RSCL} + \text{RLCL}}$$

$$0.82 = \frac{26.3}{26.3 + \text{RLCL}}$$

$$0.82 \times \text{RLCL} + 21.566 = 26.3$$

$$\text{RLCL} = 5.773 \text{ kW} \quad \textbf{Ans.}$$

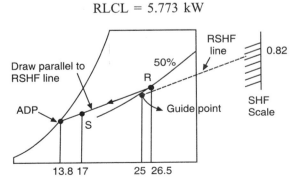

Figure 9.19 Plotting the RSHF line—Example 9.9.

(b) ADP = 13.8°C **Ans.**

(c) cmm of air supplied to the room at ADP

$$\text{cmm} = \frac{\text{RSCL}}{0.0204\,(T_R - T_{ADP})}$$

$$= \frac{26.3}{0.0204\,(26.5 - 13.8)}$$

$$= 101.51 \text{ m}^3/\text{min} \quad \textbf{Ans.}$$

(d) When air is supplied to the room at 17°C

$$\text{cmm} = \frac{\text{RSCL}}{0.0204\,(T_R - T_S)} = \frac{26.3}{0.0204\,(26.5 - 17)} = 135.70 \text{ m}^3/\text{min} \quad \textbf{Ans.}$$

Specific humidity at supply point, $w_S = 0.010$ kg wv/kg dry air **Ans.**

EXAMPLE 9.10 A hall is to be maintained at 24°C DB and 60% RH under the following conditions.

Outdoor conditions = 38°C DB and 28°C WB.
Room SH load = 46.4 kW, Room LH load = 11.6 kW
Quantity of infiltration = 1200 m³/h, ADP = 10°C
Quantity of recirculated air = 60%

If the quantity of recirculated air is mixed with conditioned air after the cooling coil, find the following.

(a) Condition of air leaving the coil.
(b) Condition of air entering the hall.
(c) The mass flow rate of air entering the cooler.
(d) The mass flow rate of total air passing through the hall.
(e) Bypass factor.
(f) The refrigeration load on the cooling coil in TR.

Solution:
The following properties of air are collected from the psychrometric chart.

Condition	DBT (°C)	WBT (°C)	RH (%)	w (kg wv/kg dry air)	h (kJ/kg dry air)
Room	24		60	0.0112	52.5
Outside	38	28		0.020	89.9

Mark points O and R on the psychrometric chart corresponding to air conditions as shown in Figure 9.20(a) and Figure 9.20(b).
Sensible heat load due to infiltration

$$= 0.0204 \times \text{cmm}\,(T_o - T_R) = 0.0204 \times \frac{1200}{60}(38 - 24)$$

$$= 5.71 \text{ kW}$$

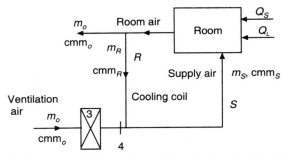

Figure 9.20(a) An air-conditioning system with ventilation air—Example 9.10.

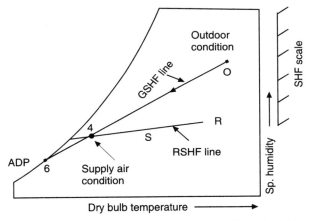

Figure 9.20(b) Grand sensible heat factor—Example 9.10.

Latent heat load due to infiltration

$$= 50 \times \text{cmm}(w_o - w_R) = 50 \times \frac{1200}{60}(0.020 - 0.0112)$$

$$= 8.8 \text{ kW}$$

Room SH load = RSH + Infiltrated SH load

$$= 46.4 + 5.72 = 52.12 \text{ kW}$$

Room LH load = RLH + Infiltrated LH load

$$= 11.6 + 8.8 = 20.4 \text{ kW}$$

Therefore, $\quad \text{RSHF} = \dfrac{\text{RSCL}}{\text{RSCL} + \text{RLCL}} = \dfrac{52.11}{52.11 + 20.4} = 0.719$

(a) Since recirculated air is mixed after the coil, air entering the coil is outside air. Join point O with ADP on the saturation curve as ADP is given. The condition of air leaving the coil would be on this line.

Draw the RSHF line as explained in earlier examples. The intersection of RSHF line with GSHF line would be the state of air leaving the coil. This air is further mixed with the recirculated room air. Therefore the mixture condition is to be found. .
Based on the psychrometric chart,
The condition of air leaving the coil = (13.5°C DB, 13°C WB) **Ans.**

(b) Air entering the hall is after mixing with the return air in proportion 40% to 60% (2 : 3)

$$T_{mixture} = \frac{2 \times 13.5 + 3 \times 24}{5} = 19.8°C$$

The condition of air entering the hall S is be on the line joining point 4 to R with $T = 19.8°C$.
The properties for point S, DBT = 19.8°C, WBT = 16.4°C WB **Ans.**

(c) It is the mass flow of air entering coil that absorbs the heat load in the room. So,

$$RTCLL = \text{mass of air} \times (h_R - h_4)$$
$$52.11 + 20.4 = \text{mass of air} \times (52.5 - 36.7)$$

Mass flow rate of air entering the cooler = $\dfrac{72.51}{15.8}$ = 4.59 kg/s **Ans.**

(d) Mass of air passing to hall has only 40% of cooled air. So,

$$\text{Mass of air to hall} = \frac{4.59}{0.4} = 11.475 \text{ kg/s} \qquad \textbf{Ans.}$$

(e) Bypass factor is given by formula

$$BF = \frac{T_4 - ADP}{T_O - ADP} = \frac{13.5 - 10}{38 - 10} = 0.125 \qquad \textbf{Ans.}$$

(f) Refrigeration load on cooling coil is given as

$$\text{Capacity of cooling coil} = \text{mass of flow} \times (h_o - h_L)$$
$$= 4.59 \times (89.9 - 36.7) = 244.188 \text{ kW}$$
$$= \frac{244.188}{3.517} = 69.43 \text{ TR} \qquad \textbf{Ans.}$$

EXAMPLE 9.11 An air conditioning system is to be designed for a restaurant with the following data:

Outside design condition = 40°C DB, 28°C WB
Inside design condition = 25°C DB, 50% RH
Solar heat gain through walls, roof, floor = 5.87 kW
Solar heat gain through glass = 5.52 kW, Occupants = 25
SH gain per person = 58 W, LH gain per person = 60 W

Internal lighting load = 15 lamps of 100 W and 10 fluorescent tubes of 80 W
SH gain from other sources = 11.60 kW, Infiltration air = 15 m³/min

If 25% fresh air and 75% recirculated air is mixed and passed through the conditioner coil, find the following.

(a) The dew point temperature of coil.
(b) The condition of supply air to the room.
(c) The amount of total air required in m³/h.
(d) The capacity of conditioning plant.

Assuming BF = 0.2, draw the schematic diagram.

Solution: Refer to Figures 9.21(a) and (b).

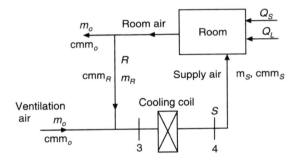

Figure 9.21(a) AC system with ventilation air—Example 9.11.

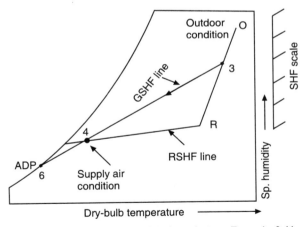

Figure 9.21(b) Grand sensible heat factor—Example 9.11.

$$\text{SH load due to occupants} = \frac{25 \times 58}{1000} = 1.45 \text{ kW}$$

$$\text{LH load due to occupants} = \frac{25 \times 60}{1000} = 1.5 \text{ kW}$$

Mark the outside condition O and the room condition R on psychrometric chart.

SH load due to infiltration air = $0.0204 \times \text{cmm}(T_o - T_R) = 0.0204 \times 15(40 - 25)$
$= 4.60$ kW

LH load due to infiltration air = $50 \times \text{cmm}(w_o - w_R) = 50 \times 15(0.0191 - 0.0098)$
$= 6.97$ kW

Room sensible heat load = sum of all SH loads
$= 5.87 + 5.53 + 1.45 + 2.46 + 4.60 + 11.6 = 31.50$ kW

Room latent heat load = sum of all LH loads
$= 1.5 + 6.97 = 8.47$ kW

$$\text{RSHF} = \frac{\text{RSCL}}{\text{RSCL} + \text{RLCL}} = \frac{31.49}{31.49 + 8.475} = 0.788$$

Draw RSHF line through point R as explained in the earlier section.
25% fresh air is mixed with 75% return air, the mixture condition is the condition of air entering the coil 3.

$$T_3 = \frac{0.25 \times T_o + 0.75 \, T_R}{1}$$

$$= \frac{0.25 \times 40 + 0.75 \times 25}{1} = 28.75°C$$

This can also be established, by finding w_3 and h_3.
$$w_3 = 0.25 w_0 + 0.75 w_R$$
$$h_3 = 0.25 h_0 + 0.75 h_R$$

Draw a line 3–ADP in such a way that the point of intersection 4 with the RSHF line is such that

$$\frac{\text{length of line 4–ADP}}{\text{length of line 3–ADP}} = 0.2$$

∴ ADP = $10.4°C$ and $T_4 = 14.2°C$

(a) The dew point temperature of coil = $10.4°C$ **Ans.**
(b) Condition 4 of supply air to room = $14.2°C$ DB, $13°$WB **Ans.**
(c) Room sensible cooling load or heat load absorbed by total air is

$$\text{RSCL} = 0.0204 \times \frac{\text{m}^3/\text{h}}{60} (T_R - T_4)$$

$$31.49 = 0.0204 \times \frac{\text{m}^3/\text{h}}{60} (25 - 14.2)$$

∴ Total air required = 8575.7 m^3/h **Ans.**

(d) Capacity of cooling coil = $0.02 \times \dfrac{m^3/h}{60}(h_3 - h_4)$

$$= 0.02 \times \dfrac{8575.6}{60}(60 - 36.5)$$

$$= 67.17 \text{ kW} = 19.1 \text{ TR} \qquad \textbf{Ans.}$$

EXAMPLE 9.12 The following data refers to summer air conditioning of a building:

Outside design condition = 43°C DB, 27°C WB
Inside design condition = 25°C DB, 50% RH
RSH = 84 MJ/h, RLH = 21 MJ/h, Bypass factor = 0.2

The room air from the room is mixed with fresh air before entering the coil in the ratio of 4 : 1 by mass. Determine:

(a) Coil ADP.
(b) Condition of air entering and leaving the coil.
(c) Fresh air cmm.
(d) Capacity of the coil in TR.

Sketch the process on psychrometric chart.

Solution: Refer to Figures 9.22(a) and (b).
Let us find the RSHF first.

$$\text{RSH} = \dfrac{84 \times 10^6}{3600 \times 1000} = 23.23 \text{ kW}$$

$$\text{RLH} = \dfrac{21 \times 10^6}{3600 \times 1000} = 5.83 \text{ kW}$$

$$\text{RSHF} = \dfrac{84}{84 + 21} = 0.80$$

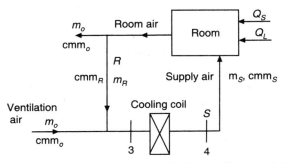

Figure 9.22(a) AC system with ventilation air—Example 9.12.

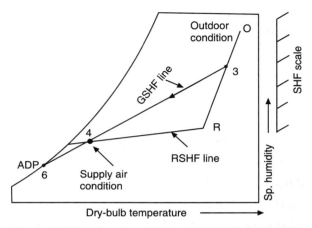

Figure 9.22(b) Grand sensible heat factor—Example 9.12.

Mark outside condition O and room condition R on the psychrometric chart. Draw the RSHF line through R as explained.

Point 3 indicates the mixture condition, at this condition the air enters the cooling coil. The condition of air entering the coil is a mixture of fresh air and recirculated air in the 1 : 4 proportion.

$$T_3 = \frac{1 \times 43 + 4 \times 25}{5} = 28.6°C$$

Condition of air entering the coil: 28.6°C DBT, 19.8°C WBT.

Draw a line 3–ADP, so that its point of intersection 4 with the RSHF line is such that

$$BF = \frac{\text{length of line } 4-\text{ADP}}{\text{length of line } 3-\text{ADP}} = 0.2$$

(a) Coil ADP = 10.6°C, T_4 = 14.2°C **Ans.**

(b) Point 4 indicates the condition of air leaving the coil: 14.2°C DBT, 12.7°C WBT **Ans.**

Supply air absorbs room total heat (RTH) while changing condition from 4 to R.

$$RTH = \frac{\text{kg/min}}{60}(h_R - h_4)$$

$$23.33 + 5.83 = \frac{\text{kg/min}}{60}(50.5 - 36)$$

Total air flow = 129.6 kg/min **Ans.**

(c) Fresh air quantity = 0.2 × 129.6 = 25.92 kg/min

Volume rate of flow = 25.92 × specific volume at O

= 25.92 × 0.919 = 23.82 cmm **Ans.**

(d) Capacity of coil = supply air kg/s × $(h_3 - h_4)$

$$= \frac{129.6}{60}(57 - 36) = 45.35 \text{ kW}$$

$$= 12.9 \text{ TR} \hspace{5cm} \textbf{Ans.}$$

EXERCISES

1. What is effective temperature? How does it account for human comfort? What is comfort equation?
2. Discuss comfort chart and assumptions made while constructing the same.
3. Discuss the factors on which solar heat through glass depends.
4. What are air films for the calculation of 'U'?
5. Write a short note on internal heat loads.
6. What is the importance of outside design conditions?
7. What is CLTD method?
8. What is system heat gain? How are they accounted for? Give examples.
9. Explain the concepts of RSHF, GSHF and ERSHF.
10. What are the difficulties in high LH load applications?
11. How is winter air conditioning carried out?

NUMERICALS

1. The room sensible and latent heat loads of an air conditioned space are 30 kW and 6 kW, respectively. The room condition is 25°C DBT and 50% RH. The outdoor condition is 40°C DBT and 50% RH. The ventilation requirement is such that on mass flow rate basis 20% of fresh air is introduced and 80% of supply air is recirculated. The bypass factor of cooling coil is 0.15. Determine:

   (a) Supply air flow
   (b) Outside air sensible heat
   (c) Outside air latent heat
   (d) Grand total heat
   (e) ERSHF.

2. For an air-conditioned room the sensible and latent heat loads are 10^5 kJ/h and 40,000 kJ/h, respectively. The make-up air is 40% of the total air supplied to the room. If:
   Ambient condition DBT = 40°C; WBT = 30°C; inside room condition—DBT 25°C and RH = 50%; and the air leaves the cooling coil at 18°C; determine the following:

   (a) RSHF
   (b) State of air entering the room
   (c) Amount of total air and make-up air
   (d) When returning air is mixed after cooling coil.

3. An air-conditioned room is maintained at 25°C DBT and 50% RH. It has sensible heat load of 12 kW and latent heat load of 8 kW when the outside conditions are 35°C DBT and 28°C WBT.

   Return air from the room is mixed with the outside air before entering the cooling coil in the ratio 4:1 and return air from the room is also mixed after the cooling coil in the same ratio 4:1. The cooling coil has the bypass factor of 0.1. The air may be reheated, if necessary before supplying to the conditioned room. Assuming the apparatus dew point to be 8°C, determine the following:
   - (i) Supply air condition to the room
   - (ii) Refrigeration load
   - (iii) Quantity of fresh air supplied.

4. The following data is for a space to be air conditioned:

   Inside design conditions: 25°C DBT, 50% RH

   Outside air conditions: 43°C DBT, 27.5°C WBT

   Room sensible heat gain, 20 kW

   Room latent heat gain, 5 kW

   Bypass factor of the cooling coil, 0.1

   The return air from the space is mixed with outside air before entering the cooling coil in the ratio 4:1 by mass. Determine the following:
   - (i) Apparatus dew point
   - (ii) Condition of air entering and leaving the cooling coil
   - (iii) Dehumidified air quantity
   - (iv) Fresh air mass flow and volume flow rate
   - (v) Total refrigeration load on the air conditioning plant.

5. An air-conditioned hall is to be maintained at 25.5°C DBT and 20°C WBT. It has a sensible heat load of 47 kW and latent heat load of 17 kW. The air is directly supplied from outside atmosphere at 38°C DBT and 27°C WBT is 1500 m^3/h into the room through ventilation and infiltration. Outside air to be conditioned is passed through the cooling coil whose apparatus dew point is 15°C. The quantity of recirculated air from the hall is 65%. This quantity is mixed with the conditioned air after the cooling coil. Determine the following:
   - (i) Condition of air after the coil and before the recirculated air mixes with it
   - (ii) Condition of air entering the hall
   - (iii) Mass of fresh air entering the cooler
   - (iv) Bypass factor of the cooling coil
   - (v) Refrigeration load on the cooling coil.

Chapter 10

Air Conditioning Systems and Equipment

10.1 INTRODUCTION

Heat always travels from a warmer to a cooler area. In summer, hot outside air continuously enters buildings that have lower temperature. To maintain the room air at a comfortable temperature, this excess heat must be continuously removed from the room. The equipment that removes this heat is called a *cooling system.*

In winter, there is continuous heat loss from room to the outdoors. If the air in the room is to be maintained at a comfortable temperature, heat must be continuously supplied to the air in the rooms. The equipment that supplies the heat required is called a *heating system.*

An air conditioning system may provide heating, cooling or both. Its size and complexity may range from a window unit for a small room to a huge system for a complex building, yet the basic principles are the same. Most heating and cooling systems have at least the following basic components:

1. A cooling source that removes heat from the fluid (air or water).
2. A heating source that adds heat to the fluid (air, water or steam).
3. Air distribution system (a network of ducts or piping) to carry the fluid to the rooms to be heated or cooled.
4. Equipment (fans or pumps) for moving the air or water.
5. Devices (e.g. radiation) for transferring heat between the fluid and the room.

Other components included are automatic controls, safety devices, valves, dampers, insulation, and sound and vibration reduction devices.

Air conditioning systems that use water as the heating or cooling fluid are called all-water or *hydronic systems*; those that use air are called all *air systems*. A system which uses both air and water is called a combination of *air and water system.*

Air conditioning systems

An air conditioning system should satisfy the need of an occupant/user at the most economical cost. The selection of the system depends upon many factors:

Customer's objectives: It could be only a relief in temperature or complete control of environment. Fresh air and air quality requirement are not given much importance in air conditioning systems of a theatre or auditorium compared to the hospital air conditioning units.

Economics: Every effort should be made in all air conditioning systems to achieve the comfort conditions with minimum energy inputs.

Occupancy: Single purpose occupancy means all occupants have the same purpose in one or more spaces. Multipurpose occupancy may need a complex system.

Thermal load: Multiplexes and supermarket buildings, these days, are provided with air conditioning systems to meet the comfort conditions. During the planning and construction of such buildings, design engineers have to find ways to minimize heat gain through external sources, especially solar radiations. The feasibility of reducing the thermal load by choosing construction options or precooling can be decisive in the design of the building.

Internal environment: Level of cleanliness, acoustics and concentration of load within the space may affect the selection of the air conditioning system.

There are a large number of variations in the types of air conditioning systems and the way they can be used to achieve the required environment in the buildings. In every installation of an air conditioning system, the engineer/contractor has to look for the best choice.

10.2 CLASSIFICATION OF AIR CONDITIONING SYSTEMS

Air conditioning systems can be classified in a number of ways. Some of these ways are as follows.

The cooling/heating fluid used

(i) **All-air systems:** These systems use only air as a heating or cooling medium.
(ii) **All-water (hydronic) systems:** These use only water for both cooling and heating purposes.
(iii) **Air-water combination systems:** Such systems use both water and air for cooling and heating purposes.

Single zone or multiple zone systems

A single zone air conditioning system can satisfy the air conditioning needs of a single zone. A multiple zone air conditioning system can satisfy the needs of air conditioning of many zones.

Unitary or central systems

Such air conditioning systems can be broadly classified into two types.

Unitary system: These systems use packaged equipment. The units consisting of fans, coils and refrigeration equipment are assembled as one unit in a factory and fitted at site. Mass production of such units with good quality is possible. Examples of the unitary systems are window air conditioners and split air conditioners.

Central or built-up systems: The components are manufactured separately and assembled and installed at site. The systems at higher capacities (> 60 TR) are suitable and more economical in initial cost as well as in running cost.

The year round air conditioning in all seasons is possible with the central systems. There are a large number of options available in these systems.

Let us now study the above common systems in more detail, beginning with the unitary system.

10.3 UNITARY SYSTEM

Such a unit is designed to be installed in or near the conditioned space. The components are contained in the unit. Unitary systems are standardized for certain applications but minor modifications are possible to suit an application. Heating components are rarely included.

10.3.1 Window Air Conditioner

The refrigeration system components—compressor, condenser, capillary tube and evaporator—are connected through copper tubes. The evaporator and condenser are at two ends such that the evaporator part is inside the room while the condenser is outside the room as shown in Figure 10.1. There is an insulated cabinet around the cooling coil with two compartments. A blower is fitted in this cabinet behind the evaporator coil which pumps air into the upper compartment. The blower pulls the room air through the cooling coil and through the filter fitted on the face of the coil. This air is then discharged back to the room through the upper compartment.

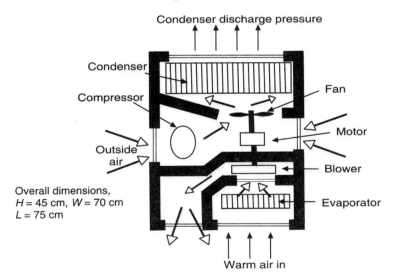

Figure 10.1 Schematic view of a typical room air conditioner.

A fan draws air from the sides and throws over the condenser coil. This helps condenser to reject heat outside the room. The control panel has three knobs. One controls the speed of the blower motor to give high cool or low cool. The second knob is of a thermostat, the bulb of which is placed at the filter to sense the temperature of the room air being sucked in by the blower. This allows the user to set the room temperature. The third knob operates a flap in the insulated cabinet to allow ventilation air supply. Refrigerant R22 is employed in this unit.

Window air conditioners are available in capacities ranging from 0.8 TR to 5 TR. One-and-half TR window air conditioner, whose height is 45 cm, width 70 cm, and depth 75 cm, is most commonly used for commercial applications. The maximum size of the window AC is limited to 5 TR due to the available capacity of the hermetic compressor. Companies such as Blue Star, Videocon, Carrier Aircon, manufacture window AC units.

In a real sense, the window AC is not an air conditioner. To a certain extent, dehumidification is possible in it while there is no provision for humidification. It is especially suitable for use in cities like Mumbai, Kolkata and Chennai where RH is more than 50%.

Limitations of window AC

(i) No humidity control though it carries out dehumidification.
(ii) Most of the window air conditioners do not provide heating for winters.
(iii) No provision for humidification in window air conditioner.
(iv) Outside temperature above 40°C can cause derating of the conditioner.

Precautions to be taken while installing a window AC

(i) It should be fitted with a small slope (3° to 5°) downwards, towards the outside which ensures that draining of condensate at the cooling coil is outside the room.
(ii) It should be ensured that the condenser is not exposed to direct solar heat to prevent undue rise in condenser pressure.
(iii) The gaps between the wall opening and the package should be blocked by insulation.

10.3.2 Split Air Conditioner

Basically it is a package unit like window AC. It splits the window air conditioner into two parts with evaporator placed inside the conditioned room while assembly of other components is placed outside the room. So it has a fan coil unit fitted inside the room and a condensing unit with an additional fan installed outside. The two units are connected by a suction line and a liquid line. In some cases, capillary is inside the condensing unit and low-pressure liquid is supplied through an insulated line to the fan coil unit.

The noise generation in a window AC is mainly due to the compressor unit which is outside in a split unit. So the split air conditioner ensures low noise level in the room. The fan coil unit has greater air throw than that of a window air conditioner. A split AC unit consumes more energy compared to window AC of the same capacity due to two reasons:

(i) There are two motors to drive two fans, one in condensing unit and another in fan coil unit.
(ii) Refrigerant flow lines are longer, so more pressure drops resulting in higher compressor power requirement.

These units are readily available in the range of 0.8 to 4 TR.

10.4 CENTRAL AIR CONDITIONING SYSTEMS

A central air conditioning system can be used for single-zone (a zone consisting of a single room or group of rooms) or multizone applications. In this section a central AC system, all-air for a single-zone application is discussed and the system is shown in Figure 10.2.

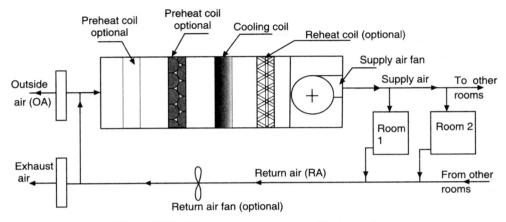

Figure 10.2 Single-zone central air conditioning system.

A single-zone air conditioning system has one thermostat that automatically controls one heating or cooling unit to maintain proper temperature in a zone comprising a single room or a group of rooms. A window air conditioner is an example of a single-zone air conditioning unit.

The system shown in Figure 10.2 is for year-round air conditioning to control both temperature and humidity. All the components shown in the figure may not be utilized in all the circumstances.

An air-handling unit (AHU) cools or heats air that is then distributed to the single zone. The supply air fan is necessary to distribute air through the ductwork to the rooms.

(i) **Cooling coil:** It cools and dehumidifies the air and provides humidity control in summer. Reheat coil is optional and is used when air temperature is to be maintained at the required level, especially in winter. In summer, it may remain idle.

(ii) **Reheating coil:** It heats the cooled air when the room heat gain is less than the maximum, thus providing humidity control in summer. The coil capacity is such that it satisfies the heating needs during winter.

(iii) **Ductwork:** It is arranged so that the system takes in some outside ventilation air (OA), the rest being return air (RA) recirculated from the rooms. The equivalent amount of outside air must then be exhausted from the building. Dampers are provided to vary the rate of ventilation air as per the requirement of fresh air in the rooms. The arrangement of dampers is shown in Figure 10.3. In some applications as in operating theatres, ventilation air can be 100%.

(iv) **Return air fan:** It takes the air from the rooms and distributes it through return air ducts back to the air conditioning unit or to the outdoors. In small systems with little or no return air ducts, the return air fan is not required because the supply fan can be used to draw in the return air.

(v) **Preheat coil:** The preheat coil may be located either in the outside air or the mixed airstream. It is required in cold climates (below freezing) to increase the temperature of air so that the chilled water cooling coils do not freeze. It is optional in milder climates and when DX (dry expansion) cooling coils are used.

(vi) **Filters:** The filters are required to clean the air.

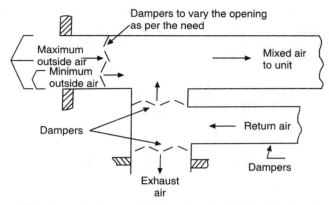

Figure 10.3 Dampers to vary the proportion of outside and return duct air.

Bypassing air around the cooling coil shown in Figure 10.4 provides another method of controlling humidity but does not give as good a humidity control in the space as with a reheat coil.

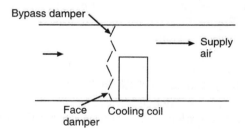

Figure 10.4 Arrangement of face and bypass dampers to provide reheat for humidity control.

A room thermostat will control the cooling coil capacity to maintain the desired room temperature. If control of room humidity is required, a room humidistat is used.

To achieve satisfactory temperature and humidity control in different zones, individual single zone units can be used for each zone. This may unacceptably increase costs and maintenance. However, there are a number of schemes that require only one air handling unit to serve a number of zones.

Four basic types of multiple-zone (all-air units and systems) systems are available:
- Reheat system
- Multizone system
- Dual duct system
- Variable air volume (VAV) system.

The reheat, multizone and dual duct systems are all constant air volume (CAV) type systems. That is, the air quantity delivered to the rooms does not vary. The temperature of this air supply is changed to maintain the appropriate room temperature. The variable air volume (VAV) varies the quantity of air delivered to the rooms.

10.5 REHEAT SYSTEM

In this system the air conditioning system is the same as with a single-zone system (air filters, cooling/heating coils, and fans as in central air conditioning). In this reheat type, a separate single duct is laid from the AHU unit to each zone or room that is to be controlled separately as shown in Figure 10.5.

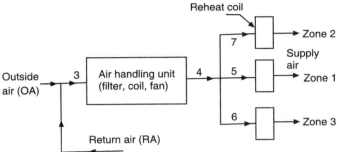

Figure 10.5 Reheat system with individual reheat coils.

A separate reheat coil is used for each zone so that one can achieve better control over both temperature and humidity. There is a wastage of energy as air is cooled in the AC system and then reheated as per the needs of each zone or room. A thermostat fitted in each room controls the temperature of the respective room or zone. The various processes of reheat system are shown on the psychrometric chart in Figure 10.6.

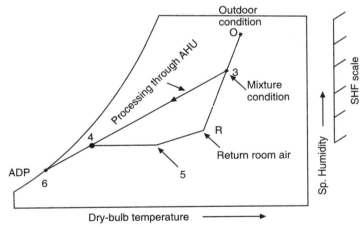

Figure 10.6 Psychrometric processes for reheat system.

10.6 MULTIZONE SYSTEM

The system uses an AHU that consists of a heating coil (hot deck) and a cooling coil (cold deck) but are placed parallely as shown in Figure 10.7. Zone dampers are installed in the unit across the hot and cold decks at the outlet of the unit. Separate ducts run for hot and cold air but are placed adjacent to each other. The hot and cold air is first mixed in a definite proportion to achieve the required temperature of an individual zone and then supplied. The duct arrangement for the multizone system is shown in Figure 10.8.

366 Refrigeration and Air Conditioning

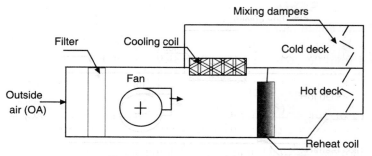

Figure 10.7 Multizone system.

The multizone system is suitable for one AHU with 12–14 zones. It is relatively inexpensive where only a few separate zones are desired and humidity conditions are not critical because one cannot control humidity accurately in this unit.

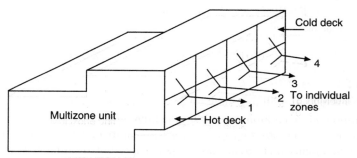

Figure 10.8 Duct arrangement for the multizone system.

10.7 DUAL DUCT SYSTEM

The system is as shown in Figure 10.9. It consists of a common filter, a common fan and individual heating and cooling coils placed in two parallel ducts. Mixing boxes are provided in the zone. The hot and cold air is mixed in the box in a definite proportion as per the needs of that zone. Placing a thermostat in each zone and sending the signal for the operation of dampers placed in the ducts leading to the box achieve this.

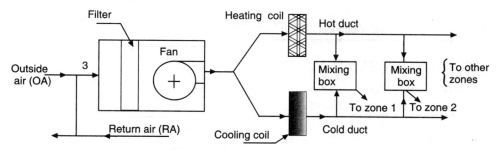

Figure 10.9 Dual duct system.

The availability of cold and warm air at all times in any proportion gives the dual duct system a great flexibility in handling many zones with widely varying loads. Dual duct systems are usually designed as high velocity air systems in order to reduce duct sizes. Fan horsepower requirements are high because large volumes of air are moved at high pressure. So the cost of the dual duct system is usually quite high.

Both the dual duct and multizone systems are inherently energy wasteful, since during part load cooling for a zone, overcooled air is reheated by mixing warm air with it—a double waste of energy.

10.8 VARIABLE AIR VOLUME (VAV) SYSTEM

In this system the air quantity supplied to each zone or room is varied to maintain the appropriate room or zone temperature.

The basic VAV system arrangement is shown in Figure 10.10. A single main duct is run from the air handling unit. Branch ducts are run from this main duct through VAV boxes to each zone. The VAV box has an adjustable damper or valve so that the air quantity delivered to the space can be varied. Room thermostats located in each zone control the dampers in their respective zone VAV boxes to maintain the desired room set-point temperature.

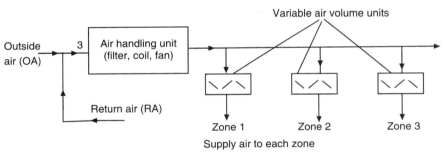

Figure 10.10 Variable air volume (VAV) system arrangement.

The psychrometric processes for a VAV system are shown in Figure 10.11 for summer cooling. The average room conditions are as indicated by point R. Zone Z1 is shown at part load when its sensible load has decreased, but its latent load has not. Its RSHF line is therefore steeper as shown. To maintain the design room DB temperature, the airflow rate to zone Z1 is throttled, and the room DB in zone Z1 is the same as R, as desired. Notice, however, that the humidity in zone Z1 is higher (point Z1) than desired.

There are many difficulties in operating the system at partial loads. At partial load of a particular zone, where cooling load continues to decrease, the reheat coil is activated. This type of VAV box can also be used to handle the problem with high latent heat loads.

At low loads, air flow rate decreases tremendously and the air circulation in the room becomes unsatisfactory, leading to uncomfortable conditions. This happens because the air supply diffusers are generally selected to give good coverage at maximum design air quantity.

The solutions to the above problems are as follows:

1. One solution is to use the reheat VAV box. When air quantity is reduced to the minimum for good air distribution, the reheat coil takes over.

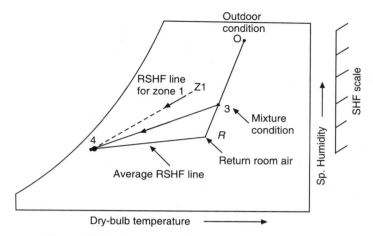

Figure 10.11 Psychrometric processes for reheat system.

2. Use variable diffusers. These diffusers have a variable sized opening. As the air flow rate decreases, the opening narrows thus resulting in better air distribution.
3. A further solution is to use fan-powered VAV boxes. This type of VAV box has a small fan. In addition to the supply air quantity, this fan draws in and recirculates some room air, thus maintaining a high total airflow rate through the diffuser.

In spite of these potential problems, VAV systems are still very popular. This is because of their significant energy saving feature when compared to other (constant air volume) multiple zone central systems. There is also another significant energy saving feature. Whenever there is part load, the air supply quantity is reduced and there is a saving of the fan power. Since a typical air conditioning system operates at part load of up to 95% of the time, this saving is considerable.

10.9 ALL-WATER SYSTEM

These are also known as *hydronic systems*. Hydronic systems distribute chilled or hot water from a central plant to each space or room. No air is distributed from the central plant. The system is schematically shown in Figure 10.12.

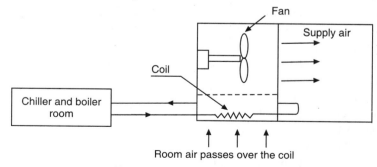

Figure 10.12 All-water system.

It is difficult to provide fresh air intake for each fan coil unit making it a room air conditioner. The fresh air supply is by opening of door or periodical opening of windows by the occupants. In case this is not acceptable, fresh air intake can be provided to the fan coil unit at a higher cost.

It takes less space and is considerably less expensive because there is no ductwork in the central air handling unit. The specific heat of water is almost four times that of air and its density is 1000 times the density of air. Therefore, for the same amount of heat dissipation very less quantity of water needs to be circulated. It requires less coil surface area. It is useful when only limited space is available. For example, installation of an AC system in existing large buildings that were not originally designed to include AC.

The all-water system has certain drawbacks too. The multiplicity of fan coil units increases the maintenance work and cost. Control of ventilation air is not precise as there is no provision for a separate ventilation arrangement. Fresh air enters only through the door and window openings. The control of humidity is also limited.

10.10 AIR-WATER SYSTEMS

A combination of air-water systems distributes both chilled and/or hot water and conditioned air from a central system to the individual rooms.

One type of air-water system uses fan-coil units as the room terminal units. Chilled or hot water is distributed to them from the central plant. Ventilation air is distributed separately from an air-handling unit to each room.

Another type of air-water system using room terminal units is called *induction unit*. It receives chilled or hot water and ventilation air from the central plant (from a central air handling unit). The central air delivered to each unit is called *primary air*. As it flows through the unit at high velocity, it induces room air (secondary air) through the unit and across the water coil. Therefore, no fans or motors are required in this type of unit, reducing maintenance greatly. The induction unit air-water system is very popular in high-rise office buildings and similar applications. Its initial cost is relatively high.

10.11 UNITARY VS. CENTRAL SYSTEMS

As already stated earlier, the classification of air conditioning systems into unitary and central systems, is not according to how the system functions, but how the equipment is arranged.

In a unitary system, the refrigeration and air conditioning components are factory selected and assembled in a package. This includes refrigeration equipment, fan, coils, filters, dampers and controls.

A central system is one where all the components are separate. The engineer has to design and install the central plant and its suitable components are based on the air-conditioning load.

Unitary equipment is usually located in or close to the space to be conditioned whereas the central equipment is usually remote from the space, and each of the components may or may not be remote from each other, depending on the desirability.

Unitary systems are generally all-air systems limited largely to the more simple types such as single-zone units with or without reheat. This is because they are factory assembled on a volume basis.

Central systems can be all-air, all-water or air-water systems and they are generally suitable for multizone units.

Unitary systems and equipment can be divided into the following three groups.
- Room units
- Unitary conditioners
- Rooftop units

These names are not standardized in the industry. For example, unitary conditioners are also called self-contained units or packaged units.

10.12 AIR CONDITIONING EQUIPMENT

A central air conditioning system has processing equipment such as air cleaner, cooling coil and heating coil, humidifiers and fans fitted in an AHU. All these items of equipment are required in different capacities to suit the requirement. Therefore, various such types of equipment are available for selection. These are described in the subsequent paragraphs.

10.13 COOLING COIL

Cooling coils may be either chilled water or evaporating refrigerant. The latter are called dry expansion (DX) coils.

Cooling coils are usually made of copper tubing with aluminium fins, but copper fins are sometimes used as shown in Figure 10.13.

Figure 10.13 Cooling coils.

The number of fins placed per centimetre is called *fin density*. The surface area provided by the fins and available for heat exchange is called *secondary area*. The inner surface area of the tubes through which water flows is called *primary area*. Fins are provided to make the heat exchangers compact. Normally the secondary area is 8 to 10 times the primary area depending upon the fin density. The coil may be constructed either with tubes in series or in parallel to reduce the water pressure drop.

When cooling coils have a number of rows, they are usually connected so that the flow of water and air are opposite to each other, called counterflow (Figure 10.14). In this way, the coldest water is cooling the coldest air, thus fewer rows may be needed to bring the air to a chosen temperature than if parallel flow were used, and therefore the chilled water temperature can be higher.

The water inlet connection should be made at the bottom of the coil and the outlet at the top, so that any entrapped air is carried through more easily. In addition, an air vent should be located at the outlet on top.

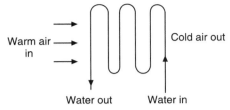

Figure 10.14 Counterflow arrangement of air and water.

10.13.1 Coil Selection

Coil selections are made from the manufacturer's tables or charts based on the required performance. The performance of a cooling coil depends on the following factors:

1. The amount of sensible and latent heat that must be transferred from the air.
2. DBT and WBT of air entering and leaving.
3. Coil construction—number and size of fins, size and spacing of tubing, number of rows.
4. Water (or refrigerant) velocity.
5. Air face velocity. The face velocity is the airflow rate in CFM divided by the projected (face) area of the coil.

Water velocities from 0.3 m/s to 2.5 m/s are used. High water velocity increases heat transfer but also results in high pressure drop and therefore requires a large pump and increased energy consumption. Velocities in the midrange of about 1–1.25 m/s are recommended.

High air velocities also result in better heat transfer and also more volume of air handled. However, if the coil is dehumidifying, the condensed water will be carried off the coil into the airstream above 125–150 cmm face velocity and eliminator baffles must be used to catch the water droplets.

The form in which manufacturers present their coil rating data varies greatly, one from another. Since using these ratings does not give much insight into how a coil performs, we will not present any rating data here.

The dehumidification effect of the cooling coil frequently results in water collecting on the coil. The water may then be carried as droplets into the moving airstream. To prevent this water from circulating into the air-conditioning ductwork, eliminators are provided downstream from the coil. This consists of vertical Z-shaped baffles that trap the droplets, which then fall into the condensate pan.

Access doors should be provided to permit maintenance. They should be located on both sides of the coils and filters. In a large-size equipment, lights should be provided inside each section.

10.14 HEATING COILS

The heating of air could be carried out in a number of ways. The heating device could be a steam coil, hot water coil, an electric heater, fuel gas furnace or a heat pump.

Steam coils

Steam generated in the boiler of a boiler room is carried through insulated pipe to the heating coils placed suitably in the conditioned space.

The steam coils consist of copper tubes connected to a common header and mounted within a metal casing. Spiral or plate fins are mounted on the tubes at spacings of 1 to 6 fins/cm depending upon the manufacturer's specifications. The coil could be one row or two rows.

The coils are available up to steam pressure of 1.2 MN/m^2. This type of coil needs a boiler and other accessories and therefore it is used only where the extent of heating justifies the expenses.

Hot water coil

Water is heated in a separate tank or in a boiler and supplied to the heating coils. The hot water coils are like single tube steam coils available in two rows. The coils may have multicircuits to reduce pressure drop and turbulators to ensure turbulent flow. Hot water coils are available up to 120°C water temperature.

Electric heaters

In open-type heaters, the heating element is exposed directly to air. These elements operate at lower temperature and their life is more. The response is also quicker.

In finned tube type, the heating element is placed in a finned tube surrounded by refractory material. It permits use of fins and cannot be damaged even by large impurities in air.

Electrical heaters have low initial cost, low installation cost, simplicity of operation and control, fast response and clean environment. Since the cost of electrical energy is very high, the use of electric heaters is limited to small-capacity units.

Duct furnaces

Gas-fired furnaces are used in air ducts. It has a burner section, a heat exchanger, a plenum and controls. This is a direct heat producing device and hence more efficient than an electric heater. Its simplicity, easy control and clean installation makes it suitable for small applications like domestic heating. Heating coils, in general, have higher face velocities than those in the cooling coil.

10.15 AIR CLEANING DEVICES (FILTERS)

Air conditioning systems that circulate air, also generally have devices that remove dust or dirt particles which result largely from industrial pollution. Occasionally, gases that have objectionable odour are also removed from the air.

Air conditioning systems should have proper filters to clean the air for the following reasons:

- Dust particles can cause serious respiratory ailments (emphysema and asthma). So, to protect human health and for comfort, dust particles have to be removed.
- To maintain cleanliness of room surfaces and furnishings.
- To protect equipment and machinery, the working of which are affected by air pollutants like dust or dirt. Some equipment will not operate properly or will wear out faster without adequate supply of clean air.
- To protect the air-conditioning machinery, for example, lint collecting on coils will increase the coil resistance to heat transfer.

10.15.1 Types of Filters/Cleaners

Air cleaners can be classified in a number of ways such as (1) types of media, (2) permanent or disposable, (3) stationary or removable, and (4) electronic air cleaners.

10.15.2 Types of Media

Viscous impingement filters

The dust particles in the air stream strike the filter media and are therefore stopped. The viscous impingement air filter has a media of coarse fibres (glass fibres and metal screens) that are coated with a viscous adhesive. Air velocities range from 90–180 mpm. This type of filter will remove larger dust particles satisfactorily but not the smaller particles. It is low in cost. Refer to Figure 10.15(a).

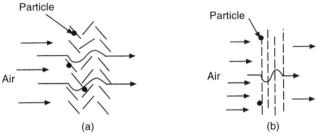

Figure 10.15 Methods of removing particles from air: (a) Impingement; (b) straining.

In straining type air filters [Figure 10.15(b)], the dust particles are larger than the space between adjacent fibres and therefore are trapped there.

Dry-type filters

These use uncoated fibre mats (glass fibres and paper) as shown in Figure 10.16. The media can be constructed of either loosely packed coarse fibres or densely packed fine fibres. By varying density, dry-type air filters are manufactured that have good efficiency only on larger particles, as with the viscous impingement type, or are also available with medium or high efficiency for removing very small particles.

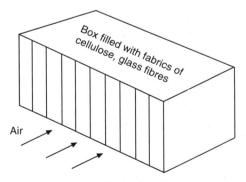

Figure 10.16 Dry filter.

HEPA (High Efficiency Particulate Air) filter

It is a very high efficiency dry-type filter for removing extremely small particles. For example, it is the only type of filter that will effectively remove viruses as small as 0.05 micron. Air face velocities through HEPA filters are very low (about 15 mpm).

The media in air filters can be arranged in the form of random fibre mats, screens or corrugated sinuous strips.

Permanent and disposable air filters

One can develop disposable type air filters so that they can be discarded. One can have permanent type too, which when saturated with dust, can be cleaned and reused. Permanent types have metal media that will withstand repeated washings, but cost more than disposable types.

Stationary and renewable air filters

Stationary air filters are manufactured in rectangular panels. The panels can be removed and either replaced or cleaned when dirty. Renewable type air filters consist of a roll mounted on a spool that moves across the airstream as shown in Figure 10.17. The fabric is wound on a take-up spool, driven by a motor. The clean fabric is continuously brought in front of the air stream by rotating the shaft. Renewable air filters are considerably more expensive than the stationary types, but maintenance costs are greatly decreased. Either fibrous materials or metal screens are used as media.

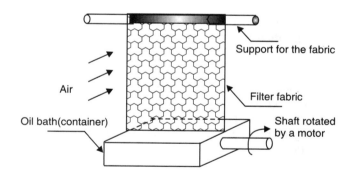

Figure 10.17 Stationary or renewable or viscous impingement filter.

10.15.3 Electronic Air Cleaners

Dust particles are given a high voltage charge by an electric grid (bank of charging plates) shown in Figure 10.18. A series of parallel plates are given the opposite electric charge. As the dust laden airstream passes between the plates, the particles are attracted to the plates. The plates may be coated with a viscous material to hold the dust. After an interval of time, the air cleaner must be removed from the service to clean the plates and remove the dirt. Electronic air cleaners are expensive, but are very efficient for removing both large and very small particles.

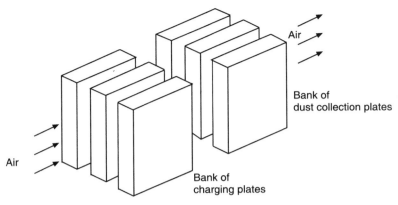

Figure 10.18 Electrostatic filters.

10.15.4 Choice of Filter

Choice of a filter depends on two factors—(1) contaminants in air and (2) performance of the filter.

Contaminants in the air

The dust particles are either solids like dust, unburnt carbon, pollens, spores, bacteria and virus or aerosols like smoke, fumes and mist. Their size, shape and concentration affect the choice of filter. The sizes of some of the contaminants are given in Table 10.1.

Table 10.1 Size of various contaminants

Tobacco smoke	0.2 micron	Bacteria	2 to 10 micron
Dust	1 to 100 micron	Pollen	10 to 150 micron
Mist	50 to 150 micron	Virus	0.05 to 2 micron

The concentration of dust particles varies from place to place. In rural areas, air may just contain mud particles while in metro cities it may contain smoke particles emitted from the vehicles.

Performance criteria

The performance of a filter is rated as per the air flow resistance and dust capacity efficiency. The filter resistance varies directly depending on the flow of air and dust content in the airflow.

The capacity of a filter is a measure of the life of the filter. The life of a filter is the period between two cleanings for cleanable filters and between replacements for disposable filters.

The efficiency of a filter is the ratio of the mass of impurities removed to the mass of total impurities present. But for fine filters, efficiency is measured by the dust spot method. The dust spot method measures the extent of blockage of light by the dust spot on the filter.

10.16 HUMIDIFIERS

These are the devices used to add moisture to the air so that the humidity increases. This becomes necessary in comfort air conditioning when the RH of atmospheric air is below 30%. The various methods of humidification are discussed in the following text.

Pan humidifier

It is the simplest humidifier. Warm air is passed over the surface of water in the pan due to which evaporation of water takes place. The vapour mixes with the flowing air. The rate of evaporation can be increased by a small heating element in the pan. This is suitable only for a small-size application due to limited capacity of humidification available.

Wetted pack humidifier

Air is passed across the water absorbent pad which is regularly wetted in a pan. The water evaporates and the air carries away the water vapour. This is also suitable for small requirements only.

Air washer or spray chamber

An air washer consists of the water sprays, arranged as shown in Figure 10.19. The baffles (not shown in the figure) are placed at the entry of air which assure uniform airflow. As air passes, the water droplets are evaporated and mixed with the air. Water falling down is collected in a pan and recirculated. An eliminator at the outlet prevents carryover of water.

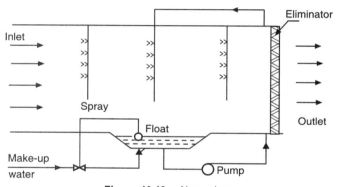

Figure 10.19 Air washer.

The air velocities through washer are 100 to 200 m/min to ensure proper functioning of the eliminator. Selection is done usually at permissible maximum velocities. A square cross-section reduces the cost of washer. Spray pressures are usually 1.5 to 2 bar. Spray density is generally 1.8 L/m^2 to 2.5 L/m^2 of face area.

Steam type humidifier

Many sections of the textile industry require RH higher than 80%. In such cases steam is introduced directly into the air for quick humidification.

10.17 FAN

Fans are necessary to distribute air through ducts to spaces that are to be conditioned. In this section, we will study the types of fans, their performance and selection, etc.

10.17.1 Types of Fans

Fans are mainly classified into two groups, namely centrifugal and axial fans.

Centrifugal fans

In a centrifugal fan, air is pulled along the fan shaft and then blown radially outwards from the shaft. The air is usually collected by a scroll casing and concentrated in one direction. Centrifugal fans may be subclassified into forward curved, radial, backward curved and backward inclined types. These differ in the shape of their impeller blades as shown in Figure 10.20. In addition, backward curved blades with a double-thickness blade are called *airfoil blades*.

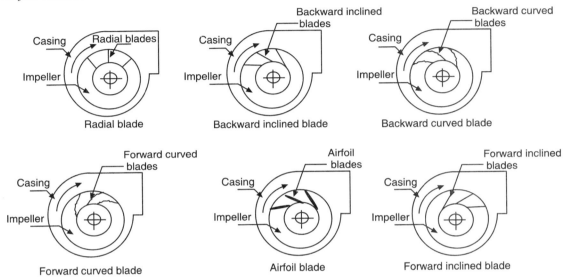

Figure 10.20 Types of centrifugal fan impeller blades.

Axial fans

In an axial flow fan, air is pulled along the fan shaft and then blown along the axis of the shaft. Propeller, tubeaxial, and vaneaxial types (Figure 10.21) are available in axial fans. The propeller fan consists of a propeller-type wheel mounted on a ring or plate. The tubeaxial fan has a vaned wheel mounted in a cylinder. The vaneaxial fan is similar to the tubeaxial types, except that it also has guide vanes behind the fan blades which improve the direction of air flow through the fan.

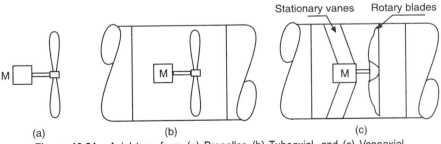

Figure 10.21 Axial type fans: (a) Propeller, (b) Tubeaxial, and (c) Vaneaxial.

10.17.2 Performance Characteristics of Fans

In general, any fluid flow through the conduit is opposed by the friction between the fluid and the conduit surface. There is a resistance to the flow of air through ducts. Therefore, a fan is required for two purposes: (1) To overcome this resistance to flow due to friction; (2) To supply the air to the conditioned space. Therefore, energy in the form of pressure must be supplied to the air. This is accomplished by the rotating fan impeller, which exerts a force on the air, resulting in both flow of the air and an increase in its pressure.

Knowledge of the fan performance is useful for correct fan selection and proper operating and troubleshooting procedures.

The volume flow rate of air (cmm) delivered and the pressure (H_t = total pressure, in mm of water gauge), created by the fan are called *performance characteristics*. Other performance characteristics of importance are efficiency (η) and brake power (BP). ME = mechanical efficiency = air power output/BP input. Fan performance is best understood when presented in the form of curves.

Forward curved blades centrifugal fan

Figure 10.22 shows the typical performance curves of a forward curved blades centrifugal fan.

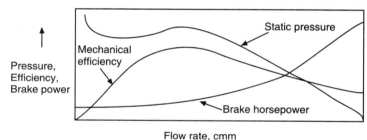

Figure 10.22 Performance curves of a forward curved blades centrifugal fan.

Some important features of this fan are follows:
1. As the flow (cubic metre per min, cmm) increases, the static pressure developed by the fan first decreases, then increases and reaches a maximum value, and then continuously decreases. The pressure developed has a slight peak in the middle range of flow, then the pressure drops off as the flow increases.
2. The BP required increases sharply with the flow.
3. Efficiency is highest in the middle ranges of flow.

Backward curved blades centrifugal fan

The performance curves of a backward curved blades centrifugal fan are shown in Figure 10.23.

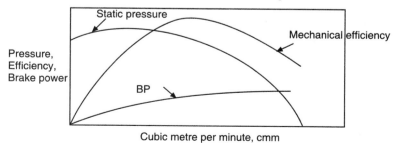

Figure 10.23 Performance curves of a backward curved blades centrifugal fan.

Some important features of this fan are follows:
1. The pressure developed has a slight peak in the middle range of flow, then the pressure drops off as the flow increases.
2. The BP increases only gradually, reaches the maximum, and then falls off.
3. Efficiency is highest in the middle ranges of flow.
4. A higher maximum efficiency can often be achieved with a backward curved blades type of fan.

10.17.3 Fan Selection

The choice of fan for a given application depends upon the pressure developed, volume flow rate and on the performance characteristics.

Centrifugal fans are very common in ducted air conditioning systems. Forward curved blades centrifugal fans are usually lower in performance. The rising BP characteristic curve could result in overloading the motor if operated at a condition beyond the selected cmm. The operating cost will often be higher due to lower efficiency.

Backward (curved or inclined) blades centrifugal fans are generally more expensive than forward curved types. These consume comparatively less power than the forward type. There is a possibility of overloading the motor if the fan is delivering more air than it was designed for. Airfoil bladed fans have the highest efficiency of any type.

Propeller fans cannot create a high pressure and are thus used where there is no ductwork. These fans are often used in packaged air conditioning units because of low cost.

Tubeaxial fans are not suitable in ducted air conditioning systems. Vaneaxial fans are suitable for ducted air-conditioning systems. They usually produce a higher noise level than centrifugal fans and therefore may require greater sound reduction treatment. Their compact physical construction is useful when the space is limited.

10.17.4 Fan Ratings

Once the proper type of fan is selected for an application, the next step is to determine the proper size to be used. Usually the fan manufacturers provide the performance characteristics of a fan for the variation in the air flow (cmm). The specifications of a fan include cmm, speed (rpm), BP, wheel diameter, etc. Based on the performance curves of a fan supplied by the manufacturers, one can get the power consumption, pressure developed and cmm of the fan.

Performance curves help the engineer to visualize changes in static pressure, BP and efficiency easily. Note that each fan curve represents the performance at a specific fan speed and air density. Fans are usually rated with air at standard conditions: a density of 1.02 kg/m^3 at 21°C. Performance curves at different air conditions may be available from the manufacturer, but if not, they may be predicted from the fan laws.

Before selecting a fan, first the duct system static pressure resistance (duct H_s) is calculated. Manufacturer's data is then used to select a fan that will produce the required cmm against the system static pressure resistance. In effect, the fan must develop a static pressure (fan H_s) and cmm equal to the system requirements. The fan may also be selected on the basis of total pressure rather than the static pressure. Either basis is satisfactory for low-velocity systems. For high-velocity systems, it is sometimes more accurate to use total pressure.

10.17.5 System Characteristics

It is quite essential to understand the system character, i.e. cmm versus pressure loss (H_f). The pressure loss due to frictional resistance in a given duct system varies as the cmm changes as follows:

$$H_{f2} = H_{f1} \left(\frac{\text{ccm}_2}{\text{ccm}_1}\right)^2 \tag{10.1}$$

Equation (10.1) can be used to find the changed pressure loss in a duct system for a changed cmm flow, if the pressure loss is known at some other flow rate.

By plotting a few of such H_f versus cmm points, a system characteristic curve can be determined. Note that the pressure loss rises sharply with cmm for any duct system, as shown in Figure 10.24.

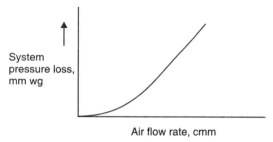

Figure 10.24 System characteristic curve.

10.17.6 Fan–system Interaction

Here we plot both the fan and system characteristic pressure versus the air flow, as shown in Figure 10.25. Then find the condition of operation of the fan and system—the intersection of the two curves gives this condition. The fan has its own performance curve and thus has its own characteristic curve.

Examining the fan and system curves is not only useful for selecting the operating condition, but aids in analysing changed conditions and in finding causes of operating difficulties. A

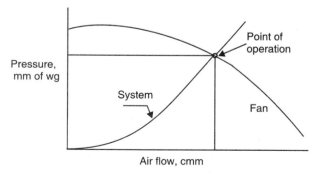

Figure 10.25 Fan and system curves plotted together.

common occurrence in air conditioning systems is that the actual system resistance for a design CFM is different from that calculated by the designer. Some reasons due to which this may happen are as follows.

1. An error in calculating pressure loss.
2. The designer adds extra resistance as a safety measure to act as a "safety factor".
3. The contractor installs the ductwork in a manner different from that planned.
4. Filters may have a greater than expected resistance due to excess dirt.
5. An occupant may readjust damper positions.

The result of this type of condition is that the duct system has a different characteristic than was planned. An examination of the fan and system curves will aid in analysing these situations.

10.17.7 Selection of Optimum Fan Conditions

Fans of different sizes or different speeds would satisfy the pressure and cmm requirements; therefore the next step in selecting fans is to decide the criteria that should be used for selecting the "best" choice.

Some of these factors will now be examined.

- Fans should be chosen for close to maximum efficiency considering the pressure cmm curve.
- Fans should not be selected to the left of the peak pressure on the fan curve. At these conditions, the system operation may be unstable, have pressure fluctuations and the system may generate excess noise.
- When using forward curved blades centrifugal fans, it should be ensured that these do not operate at significantly greater than the design cmm. If so, the motor horsepower required will increase and a larger motor may be necessary.
- Fans may have pressure curves of varying steepness. If it is expected that there would be considerable changes in system resistance, but constant cmm is required, a fan with a steep curve is desirable.

10.17.8 Fan Laws

There are a number of relationships among fan performance characteristics for a given fan operating at changed conditions, or for different size fans of similar construction; these are called *fan laws*. These relationships are useful for predicting performance if conditions are changed. We will present some of these relationships and their possible uses:

$$\text{cmm}_2 = \text{cmm}_1 \times \frac{N_2}{N_1} \tag{10.2}$$

$$H_{t2} = H_{t3} \left(\frac{N_2}{N_1}\right)^2 \text{ and } H_{s2} = H_{s1} \left(\frac{N_2}{N_1}\right)^2 \tag{10.3}$$

$$BP_2 = BP_1 \left(\frac{N_2}{N_1}\right)^3 \tag{10.4}$$

$$H_{t2} = H_{t1} \times \frac{d_1}{d_2} \text{ and } H_{s2} = H_{s1} \times \frac{d_1}{d_2} \qquad (10.5)$$

where

    cmm = volume flow rate, m³/min
    H_s = static pressure, mm of water gauge (w.g.)
    H_v = velocity pressure, mm of water gauge (mm of w.g.)
    H_t = total pressure (mm of w.g.)
    BP = brake power input
    N = speed, revolutions per min (rpm)
    d = air density (lb/ft³)
    ME = mechanical efficiency = air power output/BP input

EXERCISES

1. What are the factors that one has to keep in mind while selecting an air conditioning system?
2. Give the classification of air conditioning systems.
3. Explain what a packaged air conditioning system is.
4. Is a window air conditioner really an air conditioner or an air cooler? Explain.
5. Write a short note on split air conditioner.
6. Discuss the limitations of the window air conditioner.
7. Discuss the position of package air conditioner in the range of unitary and central systems.
8. Describe the operation of an AHU during different seasons of the year.
9. Describe a central air conditioning system.
10. Describe an all-air system.
11. Describe an all-water system and its application.
12. Explain the types of fans with neat sketches. Draw the performance curves of forward and backward curved blades centrifugal fans. What is system resistance? How does it help in fan selection? Explain fan ratings.
13. Compare the viscous impingement filter with the dry filter.
14. What are the advantages of chilled water coil?
15. Briefly describe the different types of heating and cooling devices.
16. Write a short note on the air-washer type humidifier.
17. Compare the performance of axial and centrifugal fans.
18. Compare the forward and backward curved blades centrifugal fans.
19. What are fan laws and explain their significance?
20. How are the air distribution system and fan balanced?
21. What is the effect of dirty filters on system balance?

Chapter 11

Compressors

11.1 INTRODUCTION

The compressor is one of the four essential parts of the vapour compression refrigeration system. Other parts are condenser, expansion valve and evaporator. The compressor operates in a cycle continuously, raising the refrigerant vapour pressure to the condenser pressure (corresponding to condensing temperature).

There are two principal types of compressors: positive displacement type and dynamic type.

Positive displacement compressors

These compressors increase the vapour pressure by reducing the volume through the application of work. There is a definite quantity of vapour delivered for each rotation of the crank shaft. Piston or reciprocating compressors and rotary scroll compressors are of positive displacement type.

Reciprocating compressors are further divided into the following categories: (a) single-stage compressors, (b) integral two-stage compressors, (c) open compressors, (d) hermetic compressors, and (e) semi-hermetic compressors.

Single-stage compressors: The pressure of refrigerant vapour from evaporator to condenser pressure is raised in a single stage. Small capacity reciprocating hermetic compressors having single-stage compression arrangement find widespread applications with R12, R134 and R22 refrigerants.

Integral two-stage compressors: The cylinders of the compressor are divided into low-pressure and high-pressure cylinders. These cylinders are connected through an inter-stage gas cooling system where the vapour is cooled or desuperheated. The interconnection is similar to the methods used for individual multi-stage compressor units.

Open-type compressors: In such compressors, the shaft extends through a seal in the crankcase for an external drive. The motor and compressor are enclosed in separate shells

through a drive shaft. Though shaft seals are used, some leakage of refrigerant is bound to occur. Ammonia compressors are manufactured only in the open type because of incompatibility of ammonia with motor vapour windings.

Hermetic compressors: It is a unit in which motor and compressor are enclosed in one casing with the motor shaft integral with the compressor crankshaft, the motor being cooled by the returned refrigerant vapour. There is no leakage of refrigerant and therefore these compressors are used with chloroflurocarbons (CFCs) and hydroflurocarbons (HFCs). The disadvantages of this form of construction are as follows:

- Poor cooling of motor windings
- Difficulty in repair and maintenance
- In case of motor burnout, the whole refrigeration unit gets contaminated with residues
- Electric motor repair is very complicated compared to the open type compressors
- Relatively there is more suction superheating of vapour.

Semi-hermetic compressors: The motor casing and the compressor casing are separate but bolted together. The recent development in the field of semi-hermetic compressor design seeks to eliminate some of the disadvantages of the hermetic compressor. In construction, the compressor and motor are assembled on a common crankshaft and placed inside the sealed compressor shell (casing). The electric motor stator is assembled to the outside of the housing around the motor. In this design, the motor windings make no contact with the refrigerant or any contaminants in the system.

11.2 COMPRESSION PROCESS

In the analysis of theoretical vapour compression refrigeration cycle, the compression of vapour from evaporator pressure to condenser pressure was assumed to be reversible adiabatic or isentropic. Practically, it cannot be the isentropic process. The theoretical compression processes are reversible adiabatic (isentropic) and constant temperature (isothermal), while the polytropic process is considered to be a practical one. The compression work for these processes is explained in Chapter 1 and recollected here.

The work of compression,

$$W = \oint p\, dV \tag{11.1}$$

It is also proved that $W = -\int V\, dp$

For isentropic compression ($pV^\gamma = c$): $\dfrac{T_1(s_1 - s_4)}{(T_2 - T_1)(s_1 - s_4)}$

$$W = -\int V\, dp = -\frac{\gamma(p_1 V_1)}{(\gamma - 1)} \cdot \left[\left(\frac{p_2}{p_1}\right)^{\frac{\gamma-1}{\gamma}} - 1\right] \tag{11.2}$$

For an ideal gas, $w = -c_p(T_2 - T_1)$ \hfill (11.3)

For an isothermal process ($pV = c$)

$$W = -V\int_1^2 dp = -p_1 V_1 \log \frac{p_2}{p_1} = -RT_1 \log\left(\frac{p_2}{p_1}\right)$$

$$= p_1 V_1 \log\left(\frac{V_2}{V_1}\right) = -T_2(s_2 - s_1) \qquad (11.4)$$

For polytropic compression ($pV^n = c$)

$$W = -\int V dp$$

$$= -\left(\frac{n}{n-1}\right)(p_1 V_1)\left[\left(\frac{p_2}{p_1}\right)^{(n-1)/n} - 1\right] \qquad (11.5)$$

The value of adiabatic index n for compression process may be $\geq$ or $<$ γ.

For compression, these processes are represented on p–V and T–s plots in Figures 11.1 and 11.2, respectively.

The isothermal compression process is the best as it requires minimum work. The real compression process is not isothermal, since during compression a part of the external energy is converted into internal energy. During compression, the volume decrease means that the molecules are more closely packed, so that more molecular collisions occur increasing friction as well as vibration of the molecules. In other words, the internal energy increases resulting in temperature rise. But for an isothermal process, every increase in temperature should be avoided by an infinite heat exchange with the environment. This requires an infinite intercooling between each compression step or an infinitely slow process in which the gas flow is laminar without friction between the molecules; this defines the reversible process which is impossible to realize.

The actual compression process is not an isentropic (reversible adiabatic) process following the law $pv^\gamma = c$. Such a process is not possible in practice because it needs complete avoidance of an energy exchange. In practice, heat exchange by cylinder walls, oils, etc. cannot be avoided.

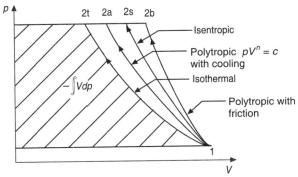

Figure 11.1 The compression process on p–V diagram.

Because of these reasons, a practical compression process is considered a polytropic process. In Figures 11.1 and 11.2, it is shown that the polytropic process follows the path 1–2a or 1–2b. Now, under what conditions does the process follow either the 1–2a path or the 1–2b path?

The compression process follows the path 1–2a when it is reversible and non-adiabatic. Non-adiabatic is in the sense that heat is removed during compression by cooling the cylinder walls. For such a process, the adiabatic index n is less than γ. Of course, the work of compression is more than that of isothermal but less than that of the isentropic process. If the value of γ for a gas is high, such as 1.4 for air and 1.3 for ammonia and when the pressure ratio is high, the compressor cylinders are cooled by water-jacketing. For CFCs and HCFCs the value of γ is not air-cooled with fins provided on them.

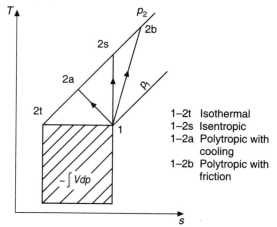

Figure 11.2 The compression process on the T–s diagram; 1–2t isothermal, 1–2a polytropic with cooling, 1–2b polytropic with friction and 1–2s isentropic.

The compression process follows the path 1–2b when it is irreversible but adiabatic. Irreversibility is due to the friction (always bound to exist) and adiabatic since there is no heat exchange. Centrifugal compressors approach such a condition.

A practical compression process in a refrigerating compressor will be accompanied by both cooling and friction (irreversible and non-adiabatic). In such cases, the compression operation may follow the path 1–2a or 1–2b depending upon the type of refrigerant, pressure ratio and cooling conditions.

Power requirement

The work required during a cycle is denoted by the area of p–V diagram shown in Figure 11.3, that is the total area (given by $-\int V\,dp$) minus the shaded area, which takes into account the effect of the clearance volume. The value of the area (or $-\int V dp$) for a polytropic process is given by Eq. (11.5).

Therefore, power required per cycle

$$= -\left(\frac{n}{n-1}\right) p_1 V_1 \left[\left(\frac{p_2}{p_1}\right)^{(n-1)/n} - 1\right] - \text{shaded area} \quad (11.6)$$

We do not have a value for n but we can use the relationship $c_p/c_v = n$ and for this, we have to apply a correction to the calculated power consumption. To do this we need to divide the theoretical power consumption p_{th} by the isentropic efficiency η_{is}. Usually η_{is} is in the range of 0.95 to 0.99 depending on the design of the compressor concerned. So, η_{is} is the ratio of the theoretical power consumption for isentropic compressor and the actual power absorbed.

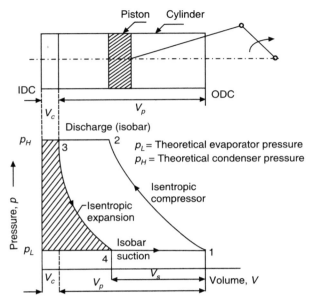

Figure 11.3 Theoretical p–V graph for the compression process.

$$\eta_{is} = \frac{m\,dh}{W_i} \quad (11.7)$$

where

m = mass flow rate of refrigerant
dh = enthalpy difference between the start and the end of the compression process
W_i = indicated power consumption or actual power added to the gas.

Here, m is obtained by dividing the total refrigeration of a plant by the specific refrigerating effect, dh can be obtained from the p–h chart and W_i from the compressor manufacturer. Generally, the manufacturers do not provide W_i but W_e (effective power consumption) is given in which the mechanical losses have already been included. So,

$$W_e = W_i / \eta_m \quad (11.8)$$

where η_m is the mechanical efficiency.

We can also use
$$\eta_e = \eta_{is} \times \eta_m = \frac{m\,dh}{W_i} \quad (11.9)$$

Note that when practical measurements of voltage and current are taken, the absorbed power can be estimated from, $W = \frac{1}{\sqrt{3}} \times VA\cos\theta$. Here V and A stand for voltage and current respectively.

11.3 INDICATOR DIAGRAMS

Before studying the practical vapour compression cycle, it is better to understand the theoretical indicator diagram shown in Figure 11.3.

During compression, the piston moves from ODC (outer dead centre) position to IDC (inner dead centre) position. It means the piston moves from extreme right (point 1) to extreme left (point 3). The volume occupied by the gas, while the piston is at IDC position is the clearance volume ($V_c = V_3$). While for suction stroke, the piston moves from IDC position to ODC position but the gas remaining in the clearance will expand to point 4. Ideally the pressure of vapour at point 4 equals atmospheric pressure. Further movement of piston towards night will reduce the pressure in the cylinder below atmospheric and the suction valve opens and remains open till the piston reaches ODC (point 1). The cylinder is filled with suction gas at constant pressure.

The piston displacement or stroke volume ($V_1 - V_3$) is equal to ($\pi D^2 L/4$), where D = diameter or bore and L is the stroke (the distance travelled by the piston from IDC to ODC). Theoretically, it is presumed that the suction valve closes at point 1 where the compression starts. The discharge valve opens at point 2 and the compressed vapour is discharged at constant pressure and the discharge valve closes at point 3 and the cycle repeats.

The work of compression W is given by

$$W = \oint p\,dV = \text{Area } 1\text{--}2\text{--}3\text{--}4\text{--}1$$

It is proved that
$$W = \oint p\,dV = \text{Area } 1\text{--}2\text{--}3\text{--}4\text{--}1$$
$$= -\oint V\,dp$$

11.3.1 Comparison of Indicator Diagrams

Let us now compare the practical indicator diagram 1–2–3–4 shown in Figure 11.4 with a theoretical one. The observed deviations are the point 4′ lying below 4 and point 2′ lying above 2. This is because in order to open the valves, the resistances to overcome are: (a) inertia of the valve ring, (b) pressure of the valve spring, (c) the valve tends to cling to the seat. Also due to throttling of vapour through suction and discharge parts, there is also volumetric loss which is not possible to show on the practical indicator diagram.

The following points are to be noted here:
1. The average evaporator pressure p_L decreases to p'_L.
2. The discharge pressure p'_3 is greater than p_3.
3. The re-expansion of vapour, remaining in the clearance will shift from point 4 to 4′ (to the right side). This reduces actual (useful) fresh vapour entry to the compressor.

The actual area of the indicator diagram 1′–2′–3′–4′–1′ is more than the ideal indicator diagram.

11.4 OVERALL VOLUMETRIC EFFICIENCY

It is the ratio of the actual or useful suction volume of gas (quantity of refrigerant vapour) divided by the piston volume of the compressor per cycle. The overall volumetric efficiency is also referred by charge coefficient.

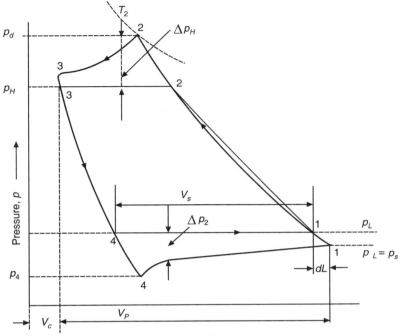

Figure 11.4 A practical indicator diagram.

The overall volumetric efficiency is affected by many factors that are discussed below:

Figure 11.3 shows the p–V diagram of the compressor operation. As the compressor delivers less than its stroke volume, we must consider volumetric efficiency.

$$\eta_v = \frac{\text{Actual volume compressed}}{\text{Stroke volume}}$$

$$= \frac{V_1 - V_4}{V_1 - V_3}$$

$$= \frac{V_1 - V_3 + V_3 - V_4}{V_1 - V_3}$$

$$= 1 - \frac{V_4 - V_3}{V_1 - V_3}$$

$$= 1 - \frac{V_3}{V_1 - V_3}\left(\frac{V_4}{V_3} - 1\right)$$

$$= 1 - \frac{V_c}{100}\left\{\left(\frac{p_H}{p_L}\right)^{1/\gamma} - 1\right\}$$

where V_c = clearance volume as percentage of stroke volume, r = compression ratio and γ = index of compression.

The volumetric efficiency depends upon the clearance volume, leakage loss, blow-by loss, etc. which are discussed in detail in the following text.

11.4.1 Clearance Volumetric Efficiency (η_{cv})

Losses due to re-expansion of gases left in the clearance volume depend on the actual space provided for clearance (V_c) as per the design. This clearance volume V_c is expressed by a factor known as *clearance factor* (c).

$$c = \frac{V_c}{V_p} \qquad (11.10)$$

Usually, c is expressed in percentage of piston displacement. The value of c for large and medium sized industrial compressors varies from 3 to 8% and 5 to 10% for small compressors. The actual size of this space determines how much gas is left in the cylinder at the beginning of the suction stroke, which re-expands before the useful gas volume can enter the cylinder. The amount of this expansion also depends on the ratio of discharge to suction pressure (p_H/p_L). Figure 11.3 illustrates the effect of the clearance volume on the clearance volumetric efficiency (η_{cv}). It is calculated in the following way using Eq. (11.1).

$$\eta_{cv} = 1 - c\left[\left(\frac{p_2}{p_1}\right)^{1/\gamma} - 1\right] \qquad (11.11)$$

Here γ is the adiabatic index whereas we know that in reality we should use n. The value of index for polytropic expansion is even lower than for polytropic compression.

11.4.2 Effect of Heat Exchange Loss

The relation for a clearance volumetric efficiency is given by Eq. (11.11). While deriving this equation, nothing is considered regarding the heat exchange that occurs between the gas and cylinder walls; even though it exists. Normally the cylinder walls of Freon compressors are air cooled. The hot gas trapped in the clearance at the end of compression stroke dissipates heat to the cylinder walls during the early part of suction stroke. But in the latter part, this trend will be reversed, i.e. heat flows from cylinder to gas because gas is cooler than cylinder. So this is a very complex phenomenon. But in total, the value of adiabatic index m for re-expansion is always less than the adiabatic index γ. Therefore, we replace γ by m in Eq. (11.11).

$$\eta_{cv} = 1 - c\left[\left(\frac{p_2}{p_1}\right)^{1/m} - 1\right] \qquad (11.12)$$

Normally, ammonia compressors are water cooled on account of the high discharge temperature. The high discharge gas temperature is due to high value of adiabatic index γ

compared to that in halocarbon compressors. As cylinder walls are being cooled by water-jacketing, there is less heat transfer from cylinder walls to the gas. This may result in higher volumetric efficiency.

11.4.3 Effect of Valve Pressure Drops

The suction valve opens at the beginning of suction stroke if the cylinder pressure is less than pressure p_1. Similarly, the discharge valve opens at the end of compression stroke only when the cylinder pressure is greater than pressure p_2 (Figure 11.4). Due to throttling effect, the cylinder pressure p_s is less than p_1 at the end of suction stroke. Actually when the compression stroke starts and when piston travels a certain distance ($dL = (V_p + V_c) - V_1$) then pressure p_1 is restored. If n is the index for this compression process to raise the pressure p'_L ($=p_s$) to p_d,

$$\eta_v = (1 + c)\left(\frac{p_s}{p_1}\right)^{1/n} - c\left(\frac{p_d}{p_1}\right)^{1/m} \qquad (11.13)$$

or

$$\eta_v = (100 + c)\left(\frac{p_s}{p_1}\right)^{1/n} - c\left(\frac{p_d}{p_1}\right)^{1/m} \qquad (11.14)$$

11.4.4 Leakage Loss

Leakage of gas past the piston rings during compression is bound to be there. Leakage of gases also takes place through suction valves, and through cylinder head and valve plate joint, etc. Considering the leakage loss, the overall volumetric efficiency η_v is written as follows:

$$\eta_v = (100 + c)\left(\frac{p_s}{p_1}\right)^{1/n} - c\lambda\left(\frac{p_d}{p_1}\right)^{1/m} \qquad (11.15)$$

where

   c = percentage clearance volume
   p_s = actual suction pressure (bar)
   p_1 = evaporator pressure (bar)
   λ = percentage of clearance gas leakage that escapes past the piston and suction valve seats. If this clearance leakage is 5%, then $\lambda = 1.05$.
   p_d = actual discharge pressure (bar)
   n, m = adiabatic index for compression and expansion processes, respectively.

In overall volumetric efficiency, the following losses are included.

(a) Leakage losses past piston and valves.
(b) Absorption of gas in oil.
(c) Losses through suction and discharge valves.
(d) Heat exchange losses.
(e) Non-tightness losses of the construction.

These losses are included in one group and included in η_{ov} (overall volumetric η). Finally, the overall volumetric efficiency is defined as

$$\eta_{ov} = \eta_{cv} \times \eta_v \qquad (11.16)$$

In practice, η_v is determined as

$$\eta_v = \frac{\eta_{ov}}{\eta_{cv}} \qquad (11.17)$$

A manufacturer determines the value of overall volumetric efficiency η_v at a particular pressure ratio and further assumes that it is proportional to T_H/T_L. The variation of η_v with the pressure ratio for refrigerants R12 and R717 (ammonia) is shown in Figure 11.5.

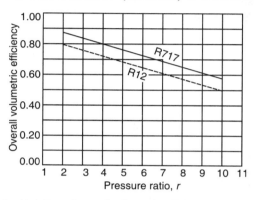

Figure 11.5 Variation of overall volumetric efficiency with pressure ratio.

In comparison with ammonia, the CFCs are highly soluble in oil. Moreover, they possess high molecular weight and poor heat transfer characteristics. Therefore, η_{ov} will be lower resulting in lower overall volumetric efficiency. This is also clear from Figure 11.5. Also high gas velocities through the valves lower the value of η_{ov}. Further, the value of η_{ov} will be still lower for R22 with its higher densities than that for R717.

It is claimed that the overall volumetric efficiency improves with a little suction superheat. If the vapour containing liquid droplets enters the cylinder, it would expand (vaporize) occupying more cylinder space.

The clearance volumetric efficiency η_{cv} will be higher for higher stroke length; provided the clearance volume V_c be kept constant. It is quite clear that the ratio V_c/V_p decreases. Comparing the losses of R22 with R717 due to heat exchange between gases and cylinder walls, we see that because of the lower discharge temperature, R22 is preferable over R717. It will be a good exercise to calculate η_{cv} for the different makes of compressor knowing the clearance volume values. In real practice, overall volumetric efficiency is calculated first with known quantities like bore, stroke and the adiabatic index. Also, η_{cv} is calculated with the help of Eq. (11.12). Then η_{ov} is derived from Eq. (11.16).

G. Lorentzen carried out some experiments to understand the influence of various losses on the overall volumetric efficiency. The experiments were conducted for R22 and R717 refrigerants for varied compression ratios and are as indicated in Figures 11.6 and 11.7, respectively. The various losses are designated as follows:

α_1 = losses due to clearance volume, i.e. η_{cv}.
β_1 = losses due to heat exchange between the gas and cylinder walls during the suction stroke.
β_2 = losses due to the lower pressure ($p'_L < p_L$) in the cylinder at the end of the suction stroke, not shown in the figure.
β_3 = tightness losses considered negligible.
β_4 = losses due to pressure and volume variation during the suction stroke.

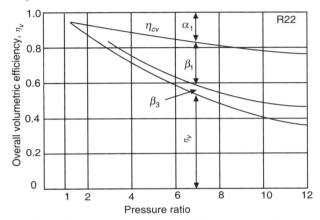

Figure 11.6 Composition of the volumetric losses for R22 compressor with varying pressure ratios.

The losses by absorption of refrigerant in the oil are not shown in Figure 11.6. The tightness losses β_3 are less important than losses due to clearance volume (η_{cv}) α_1 for both R22 and R717 compressors.

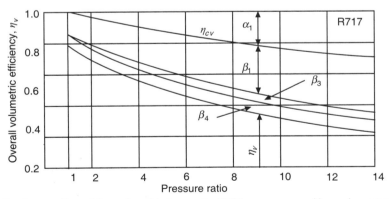

Figure 11.7 Composition of the volumetric losses for R717 compressor with varying pressure ratios.

β_1 depends on the ratio of stroke length L and bore diameter D of the cylinder for constant V_c. A large ratio of (L/D) results in an increase in the value of η_{cv}. This improves the overall volumetric efficiency η_v and the overall efficiency η_e of the compressor. G. Lorentzen also carried out experiments by changing the stroke length keeping the clearance volume V_c and the bore diametre D constant. These are as collected in Figures 11.8 and 11.9.

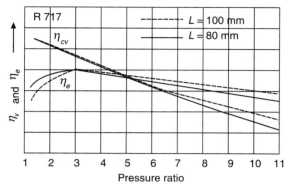

Figure 11.8 Variation of clearance volumetric efficiency and overall efficiency for different values of stroke (L) and fixed bore (D) for R717 compressor.

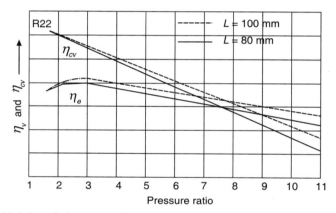

Figure 11.9 Variation of clearance volumetric efficiency and overall efficiency for different values of stroke and fixed bore for R22 compressor.

11.5 DESIGN FEATURES OF A RECIPROCATING COMPRESSOR

The important design features of a reciprocating compressor include:

(i) Cylinder bore D in m or mm.
(ii) Length of stroke L in m or mm.
(iii) Rotational speed N in rpm.

The dimensions L and D are decided based on piston speed rather than rotational speed. The piston speed U (m/s) is calculated from

$$U = \frac{2NL}{60} \qquad (11.18)$$

For a particular piston speed U, there are many combinations of N and L. Regarding the ratio L/D, it is interesting to note that the so-called square compressor ($L/D = 1$) yields the highest efficiency. This fact is shown in Figures 11.8 and 11.9 where separate diagrams for R22 and R717 compare the values of η_{cv} for $L = 100$ mm and 80 mm with a constant value of D.

Usually, L/D lies between 0.6 and 1.0. For a relatively shorter stroke (say $L/D = 0.6$) the compressor has to have a large clearance, decreasing η_{cv} and increasing the piston blow-by losses. By advanced machining operations, one can achieve fine tolerances for larger strokes and lower cylinder bores.

The piston displacement or swept volume is given by

$$V_p = \left(\frac{\pi}{4}D^2 L\right)\left(\frac{N}{60}\right) \times (n) \text{ m}^3/\text{s} \tag{11.19}$$

where n = number of cylinders.

The actual volume flow rate refrigerant vapour entering the compressor is given by

$$V_{actual} = (\eta_v)\left(\frac{\pi}{4}D^2 L\right)\left(\frac{N}{60}\right)(n) \tag{11.20}$$

where η_v is the overall volumetric efficiency.

To know the refrigeration capacity of the compressor, convert this volume flow rate V_{actual} into mass flow rate either by multiplying by density of vapour or by dividing it by the specific volume of vapour entering the compressor. The multiplication of mass flow rate (kg/s) with specific refrigerating capacity (dependent on the vapour compression refrigeration cycle) will give the total cooling capacity of the compressor.

11.6 DETERMINATION OF COMPRESSOR MOTOR POWER

Earlier in this chapter, we studied the effective power W_e of a compressor. Further, to calculate the power of an electrical driving motor, two more efficiencies must be introduced.

(i) Efficiency of driving mechanism belts or coupling η_{dr} which takes care of the driving losses.

(ii) Efficiency of electric motor η_{el} which considers losses in it.

Therefore,

Compressor motor power, $\quad W_{el} = \dfrac{W_e}{\eta_{dr}\,\eta_{el}} \tag{11.21}$

The overall COP or the performance factor of the compressor/motor combination is

$$\text{Performance factor} = \frac{Q_L\,\eta_{is}\,\eta_m\,\eta_{dr}\,\eta_{el}}{W_{is}} \tag{11.22}$$

where

Q_L = total refrigerating effect (kJ/s)
W_{is} = isentropic power $m(dh)$ (kJ/s).

11.7 PERFORMANCE OF A RECIPROCATING COMPRESSOR

The performance of a machine is an evaluation of its ability to accomplish the assigned task. Compressor performance is the result of a design compromise, which must confirm to certain physical limitations of the refrigerant, compressor and motor, while attempting to provide:

1. The higher refrigerating effect for the least power input.
2. The maximum trouble-free life expectancy.

3. The lowest cost.
4. A wide range of operating conditions.

The useful parameter of compressor performance is the capacity which may be related to the displacement and the performance factors.

Capacity is the refrigeration effect that can be accomplished by a compressor. It is the enthalpy difference between saturated liquid at condenser outlet and suction vapour multiplied by mass flow rate expressed in kJ/h (or kW).

Another term, performance factor (energy efficiency ratio) for compressor is defined as follows:

$$\text{COP or Performance factor} = \frac{\text{Cooling capacity in kW}}{\text{Power input in kW}}$$

11.7.1 Performance of Ideal Compressor

The ideal compressor is one which operates or assumes ideal conditions. For example, the compression and re-expansion processes are isentropic. Only the clearance volume influences the overall volumetric efficiency and other losses are neglected.

Effect of suction pressure

In practical usage, a refrigeration system is required to operate at different suction pressures (or evaporator pressures) and this will have more impact on mass flow rate, refrigerating capacity and power requirement if the condenser pressure is held constant. Figure 11.10 shows the variation of mass flow rate, refrigerating capacity and power for different suction pressures.

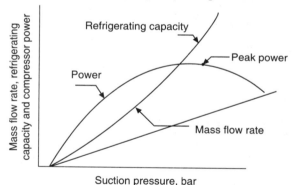

Figure 11.10 Variation of compressor power cooling capacity and mass flow rate w.r.t. suction pressure.

The mass flow rate of refrigerant of a reciprocating compressor is given by

$$m = \frac{V_p \eta_v}{v_1} \tag{11.23}$$

Equation (11.23) indicates that the mass flow rate of refrigerant is directly proportional to piston displacement and the overall volumetric efficiency and inversely proportional to specific volume of suction gas. As the suction pressure decreases, specific volume of all the refrigerants increases leading to decrease in volumetric efficiency. As a result, both mass flow as well as

refrigerating capacity decrease. This is because,
$$Q_L = m \cdot q_L \tag{11.24}$$
The power consumption of a compressor is given by
$$W = (m)(h_2 - h_1) \tag{11.25}$$
The power requirement is zero for zero mass flow rate. It increases with the increase in suction pressure and after reaching a peak value, it decreases.

At the starting of the refrigeration unit with warm evaporator, there is high load due to which the compressor faces a peak power requirement. At this state, the power requirement may be more than the motor capacity. Sometimes the motor is oversized to meet this peak power requirement during the pull-down period of the refrigeration unit.

Effect of discharge pressure

An increase in discharge pressure decreases the volumetric efficiency and thereby mass flow rate and refrigerating capacity (Figure 11.11). Let us see the performance characteristic curves of a hermetic compressor shown in Figure 11.12. These are drawn with the evaporator temperature along the *x*-axis and mass flow rate and cooling capacity on the *y*-axis while keeping the condensing temperature C_T as a parameter.

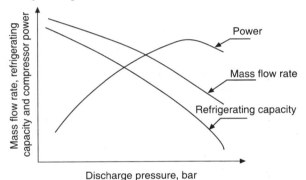

Figure 11.11 Variation of mass flow rate, refrigerating capacity and compressor power w.r.t. discharge pressure.

The mass flow rate of a refrigerant increases with the increasing evaporator temperature, while condensing temperature is held constant. The cooling capacity increases with the increasing evaporator temperature for a constant condensing temperature.

11.7.2 Actual Performance

The actual compressor performance deviates from that of ideal conditions. These deviations are in the form of losses which are associated in reality with the compressor. All these losses tend to reduce the compressor capacity and increases power consumption.

These losses are:
1. Pressure drop within the compressor due to:
    - Suction and discharge valves
    - Suction strainer
    - Suction and discharge manifolds
    - Suction and discharge mufflers

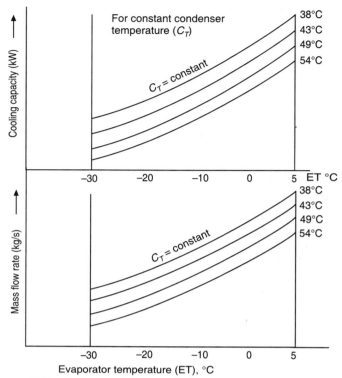

Figure 11.12 Performance characteristic curves of a hermetic compressor.

2. Heat gained by the refrigerant in compressor from the following sources:
   - Friction at the rubbing surfaces
   - Heat of compression
3. Mechanical action of valves, such as inertia and valve spring rate
4. Oil circulation
5. Gas leakage past the piston
6. Clearance volume (volumetric efficiency)
7. Deviation from isentropic compression.

11.7.3 Capacity of a Compressor

The following test conditions need to be specified for determining the compressor capacity as per the testing procedure:

(a) Evaporating temperature and pressure
(b) Condensing temperature and pressure
(c) Ambient temperature
(d) Compressor inlet or superheated return gas temperature
(e) Subcooling of liquid before entering the expansion valve

(f) Compressor cooling conditions
(g) Voltage and frequency
(h) Refrigerant.

As long as the test conditions are not defined or specified in connection with the capacity specifications, it has no real meaning.

Specifying the above conditions, the refrigerant compressor manufacturers test the compressors to know their capacity, mass flow rate and input power. The tests are carried out for constant condenser temperature varying the evaporator temperature. The results are obtained in different sets. Such performance curves are shown in Figure 11.12.

For any combination of evaporator and condenser temperature, one can get cooling capacity and mass flow rates readily from such performance curves. Such curves are readily available with the manufacturers for the range of compressors manufactured by them.

11.8 CAPACITY CONTROL OF RECIPROCATING COMPRESSOR

There are a number of ways to control the capacity of reciprocating compressors. Some of them are discussed in this section.

Speed control method

Changing or controlling the speed of a compressor can vary its capacity. This method is more efficient from the energy point of view. When an electric motor is used to drive the compressor, only two speeds are usually available so that the compressor operates either at full or 50% capacity. At lower speeds there are problems associated with lubrication and out-of-balance force. Every compressor is designed with a minimum rotational speed, so it is not desirable to reduce the speed less than this minimum speed.

Valve lifting method

Modern compressors, even the smallest ones, are multi-cylinder having 2 to 18 cylinders. To put one or more cylinders out of working, suction valves of the desired cylinders are kept open with the help of a valve lifting system. From the energy point of view, this method is relatively economical but not totally loss-free. Figure 11.13 shows the comparison of capacity control by speed control and valve lifting methods. Capacity control by suction valve lift may lead to the imbalance of crankshaft. For this reason: (a) a minimum number of cylinders must remain in operation and (b) the compressor should not operate continuously for a long period.

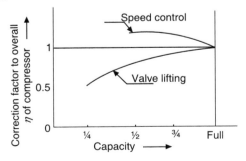

Figure 11.13 Effect of capacity control over compressor efficiency.

Bypass method

Unloading of one or more cylinders is also done by diverting the discharge gas from one or more cylinders back into the suction line as shown in Figure 11.14.

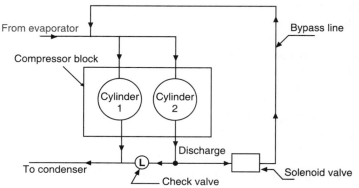

Figure 11.14 Flow diagram of cylinder bypass.

The solenoid valve shown in the circuit is operated by a pressure control which checks the suction line pressure to cut-in point and cut-out settings. As long as the suction pressure is less than the cut-in point, the solenoid valve allows the discharge from one or more cylinders to bypass to the suction line. When the suction pressure rises to the cut-out setting, the solenoid valve closes the bypass line. Then the compressor again runs with full capacity.

On-off control

A thermostat is provided on the small refrigerator and air-conditioning units. The thermostat switches the unit ON or OFF as per the setting done to meet the requirement. Any desired range of temperature can be set. It is desirable to switch the unit instead of operating at part-load since part-load efficiency is always less than full-load efficiency.

Hot gas bypass

This is shown schematically in Figure 11.15. In this method of refrigerating the capacity of compressor, the hot gas from the discharge line is bypassed to the suction line. The bypassing is through a constant pressure throttle valve. When the pressure on the suction side falls below the normal pressure (corresponding to desired evaporator temperature), then the throttle valve admits hot gas to flow to the suction side, thus keeping the suction line pressure constant. This method has several disadvantages such as there is little or no reduction of power and excessive superheating of suction gas results in overheating of compressor.

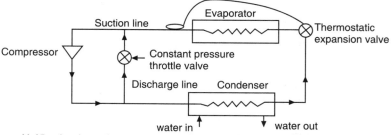

Figure 11.15 A schematic arrangement of hot gas pass from discharge line to suction.

Several modifications are introduced to Figure 11.15 in order to reduce the effect of the above mentioned disadvantages, but they cannot be eliminated completely.

11.9 ROTARY COMPRESSORS

Rotary compressors belong to the displacement group and are of three types, namely:
- Rolling piston
- Rotating vane
- Screw compressors

11.9.1 Rolling Piston Compressor

Rolling piston or fixed vane type compressors are used in domestic refrigerators and window air conditioner units in sizes up to about 2 kW. This compressor is shown in Figure 11.16.

Such a compressor employs a cylindrical steel roller mounted on the eccentric shaft with respect to the cylinder. A single vane or blade is suitably positioned in the non-rotating cylinder housing. Because of the shaft, the cylindrical roller is eccentric with the cylinder and touches the cylinder wall at the point of minimum clearance. As the shaft turns, the roller rolls around the cylinder wall in the direction of shaft rotation, always maintaining contact with the wall. The spring-supported firmly positioned blade reciprocates in the slot of cylindrical block. This reciprocating motion is caused by the eccentrically moving roller. This blade bears firmly against the roller at all times.

The displacement of this compressor can be calculated from the relation:

$$V_d = \frac{\pi}{4}.H(d_c^2 - d_r^2)$$

where
V_d = displacement per revolution (m³); H = height of cylinder block (m)
d_c = diameter of cylinder (m); d_r = diameter of roller (m).

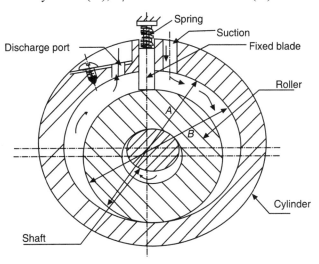

Figure 11.16 Fixed-vane, rolling-piston rotary compressor.

Suction and discharge ports are located in the cylinder wall near the blade slot but on opposite sides. The flow of vapour through both the suction and discharge ports is continuous. Also, as the suction and discharge gases are separated by the blade, the heat transfer from discharge gas to suction gas is minimum. The internal leakage from discharge side to suction side is controlled through hydrodynamic sealing. The hydrodynamic sealing depends on clearances, surface speed, surface finish and oil viscosity, close tolerances, and hence low surface finish machining is necessary to support hydrodynamic sealing and to reduce gas leakage losses.

The vibrations associated with this machine are very less, therefore, the drive motor stator and compressors are rigidly mounted in the compressor housing.

11.9.2 Rotary Vane Type Compressor

Rotary vane compressors have a low weight-to-displacement ratio. Therefore, their compact size and light weight make them suitable for transport applications. Single-stage rotary vane compressors are available in the capacity range of 2 to 37 kW for the saturated suction temperature varying from $-40°C$ to $7°C$ and for the condensing temperature up to $60°C$. Evaporating temperature up to $-60°C$ can also be achieved if two stages are employed.

A cross-sectional view of a four-bladed compressor is shown in Figure 11.17. The rotor shaft is mounted eccentrically in a steel cylinder so that the rotor nearly touches the cylinder wall on one side (but rotor and wall are separated by an oil film). Exactly opposite to this side, the clearance between the rotor and the cylinder wall is maximum. The vanes move back and forth radially in the rotor slots as they follow the contour of the cylinder wall when the rotor is turning. The vanes are held firmly against the cylinder wall by the action of centrifugal force developed by the rotor.

The suction vapour trapped between two successive blades is compressed by the reduction in volume, which results when the vanes rotate from the point of maximum rotor clearance to the point of minimum rotor clearance. The compressed vapour is discharged through the discharge ports equipped with check valve. This valve prevents the reverse flow of vapour from discharge side to the clearance space in any case. Similarly, suction ports also carry check valve to prevent the flow of compressor vapour into the suction ports.

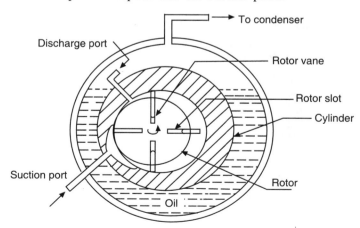

Figure 11.17 Vane type rotary compressor.

A particular compressor design results in a fixed built-in compression ratio. This ratio depends on the compressor volume ratio and the type of refrigerant. The compressor volume ratio is determined by the relationship between the volume of the cell (the space between the two successive blades) to the volume just before to opening of discharge port. The compression ratio can be calculated from the following relationship.

$$r = v_i k$$

where r = compression ratio, v_i = compression volume ratio and k = specific heat constant for the refrigerant.

The performance curves for a typical rolling piston compressor are shown in Figure 11.18. These curves are similar to the reciprocating compressor discussed earlier.

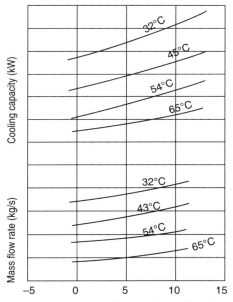

Figure 11.18 Performance curves for a typical rolling piston compressor.

11.9.3 Screw Compressors

There are two types of screw compressors, namely—single screw and twin screw. These are currently used for refrigeration and air-conditioning applications. They have the capability to operate at pressure ratios above 20:1 in a single stage and the capacity ranges from 70 to 4600 kW.

11.9.4 Single Screw Compressor

The single screw compressor consists of a single cylindrical main rotor that works with a pair of gate rotors as shown in Figure 11.19. Both the main rotor and gate rotors (star wheel) can vary widely in terms of form and mutual geometry.

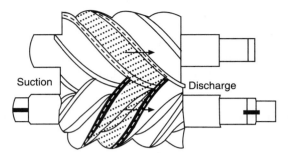

Figure 11.19 Single screw compressor.

The main rotor has six helical grooves with a cylindrical periphery and globoid root profile. Two identical star wheels each having 11 teeth are located on the opposite sides of the main rotor. The casing encloses the main rotor. It has two slots wherein teeth of the gate rotors pass. Two discharge ports are provided in the casing and are connected to a common discharge manifold.

The driver drives the main rotor shaft and the gate rotors follow by direct meshing action. In this case, the main rotor has six helical grooves while the gate rotors, each having 11 teeth, provide a speed ratio of 6:11. The geometry of the single screw compressor is such that 100% of the gas compression power is transferred directly from the main rotor to the gas.

The operation of the single screw is explained here with three different phases: suction, compression and discharge.

Suction: During rotation of the main rotor, a typical groove reaching the suction chamber gradually fills with suction gas. Further rotation of the rotor meshes with the gate rotor acting like an aspirating piston.

Compression: The main rotor engages the teeth of gate rotor located in a casing. A close clearance between the casing and the teeth of gate rotor is maintained. As the main rotor turns, the gas trapped in the space between the rotor groove and gate rotor is compressed due to the reduction in volume.

Discharge: The compressed gas in the groove space is discharged in the discharge manifold. The discharge port is located at the leading edge of the groove.

The screw rotor is normally made of cast iron and the mating gate rotors are made from plastic. The good lubricating qualities of plastic actually help in maintaining a close clearance.

Single screw compressors can operate in two ways: (i) oil-injected compressor, (ii) oil-injection free compressor

11.9.5 Oil-injected Compressor

Oil is used in single screw compressors to seal, cool, lubricate and actuate capacity control. The oil injected during the compression process absorbs the heat of compression, thereby decreasing the discharge temperature. Therefore, it helps the compressor to operate effectively at high pressure ratios. These compressors work at high pressures using common high pressure refrigerants such as R134a, R502, R114 and R717. These are commercially available in the capacity range of 15 to 1100 kW.

Oil injection requires an oil separator to remove oil from the high pressure refrigerant as shown in Figure 11.20.

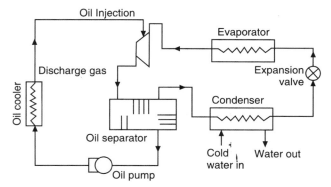

Figure 11.20 Oil injection system in a screw compressor of a refrigeration cycle.

In most of the compressors, oil can be injected without a pump because of the pressure difference between the oil reservoir (discharge pressure) and the reduced pressure in a flute during compression. This system requires a separate oil cooler and there are a number of methods for this cooling purpose.

11.9.6 Oil-injection Free Compressor

In such compressors, the fluid injected into the compression chamber is the condensate of the fluid being compressed. For refrigeration and air-conditioning applications where the pressure ratio involved is in the range of 2 to 8, the liquid refrigerant is injected into the screw compressor casing. No lubrication is required because the only power transmitted from the screw to the gate rotors is that needed to overcome small frictional losses. The job of the refrigerant injected is only to cool and seal the compressor.

This compressor has many advantages like it does not require oil separator, oil pump, refrigerant pump or external coolers.

Screw compressors are available with a secondary suction port between the primary compressor suction and the discharge port. When this port is used to admit the flashed refrigerant gas available in the economizer cycle, the useful refrigeration and compressor efficiencies improve. The theoretical economizer cycle is shown in Figure 11.21.

In this flash chamber economizer system, high pressure liquid passes through an expansion device and enters the flash chamber maintained at intermediate pressure. The gas generated from the expansion enters the compressor through the economizer port. This additional and the original charges are then compressed together to discharge conditions. The pumping capacity of the compressor at suction conditions is not affected by this additional flow through the economizer port. This total volume passes through the condenser and expansion device. As only the liquid refrigerant in saturated condition reaches the evaporator it gives a large refrigeration capacity per kilogram. In addition to this, a compressor handles more volume with less input power. This is because of flash gas removal at intermediate pressure and its introduction directly into the flute which is already at elevated pressure.

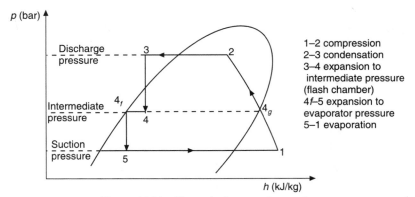

Figure 11.21 Theoretical economizer cycle.

Capacity control is achieved by controlling the speed and suction throttling as used with all positive displacement compressors, that have been discussed earlier.

Figure 11.22 shows typical efficiencies of all single screw compressor designs. These performance curves indicate that single-stage screw compressors have high isentropic and volumetric efficiencies due to the absence of suction or discharge valves and their losses and extremely small clearance volume. The curves also indicate the importance of selecting the correct volume ratio in fixed volume ratio compressors.

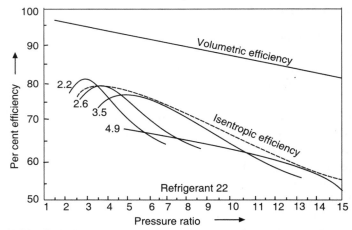

Figure 11.22 Typical compressor performance curves of screw compressors with R22.

Twin or double helical screw compressor

It consists of two mating helical grooved rotors male (lobes) and female (flutes or gullies) in a stationary housing with inlet and outlet gas ports. See Figure 11.23. The flow of gas in the rotor is mainly in an axial direction. Frequently used lobe combinations are 4 + 6, 5 + 6 and 5 + 7 (male + female). The male rotor will drive a female rotor. For instance, with a four-lobe male rotor, the driver rotates at 3600 rpm and the six-flute (gullies) female rotor follows at 2400 rpm (2/3 of rotor speed). The female rotor can be driven through synchronized timing gears. The speed of female rotor can be kept between 50% and 80% of the male rotor speed depending upon the application.

These compressors find widespread applications in industries. They can be designed to obtain a wide range of pressures from 2:1 to 20:1 and above.

As in the case of other positive displacement machines, there are three basic continuous phases of the compression cycle, namely suction, compression and discharge.

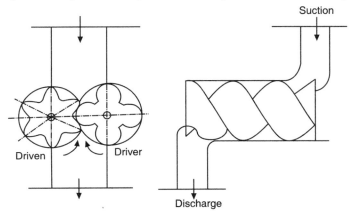

Figure 11.23 Sectional and side views of a screw compressor.

Suction: As the rotors begin to unmesh, an interlobe space between them and housing nearest to the suction end opens to the inlet port and the space is filled with gas. The suction process continues for some degrees of shaft rotation. Just prior to the point at which the interlobe space leaves the inlet port, the entire length of the interlobe space is completely filled with gas.

Compression: During further rotation, the meshing of another male to be with another female gully space occurs at the suction end and progressively compresses the gas in the direction of the discharge port. The volume occupied by the suction gas slowly decreases leading to a rise in the pressure towards the discharge port.

Discharge: At a point determined by the designed built-in volume ratio, the discharge port is uncovered and the compressed gas is discharged by further meshing of the lobe and interlobe space.

Because of the method of construction, the compression ratio or the amount by which the original suction volume reduces is fixed. We describe the compressor as having an in-built volume or pressure ratio

$$r = \frac{v_m}{v_p}$$

where v_m is the maximum interlobe space and v_p is the minimum space at the end of compression. In the ideal situation, the end compression pressure p_2 corresponds to the condensing pressure p_c of the system. Practically it is not so due to some losses.

Effect of screw compressor design on filling ratio η_v and isentropic efficiency η_{is}

In the case of reciprocating compressors, the overall volumetric efficiency has been designated as η_v. There are no pistons, valves or clearance volume in a screw compressor, therefore, the term "filling ratio" (compression volume ratio) is more suitable. In rotary compressors, the leak losses along the rotor are important. The leak losses are affected by the tolerances between rotors and rotors, and rotors and casing. By injecting sufficient amount of oil, a large part of the leakage and at the same time the heat transfer losses are reduced. Every increase of 0.01 mm tolerance may result in an increase of 1% in the volumetric losses. The tip speed of the rotor u in m/s is $u = \pi DN/60$, where D is the rotor diameter and N is the speed in rpm. The tip speed of rotor can be changed either by varying the speed N or by varying the diameter D. There is an optimum tip speed to achieve maximum efficiency for each in-built volume ratio as shown in Figure 11.24.

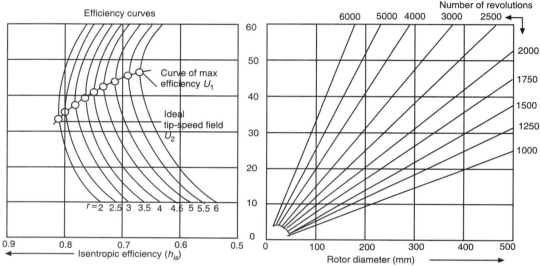

Figure 11.24 Curves relating tip speed against rotor diameter and efficiency with different values of r and rotational speed for R717 screw compressor.

Figure 11.24 shows the relationship between isentropic efficiency, tip speed, rotor speed and rotor diameter. From this, it may be concluded that small screw compressors with small rotors need to run at high speed of rotation in order to perform at optimum tip speed and highest efficiency.

When rotor speed N is increased with D constant or vice versa, the displaced gas volume per unit time increases and so the losses become relatively smaller. With this, one can conclude that the overall efficiency increases if the rotor diameter D or its speed N increases. However, this is only true for the filling ratio. As the speed increases, the friction and turbulence of the gas increase and subsequently the mechanical losses will also increase. Therefore, the isentropic efficiency η_{is} decreases.

Other parameters like the ratio of rotor diameter to its length, pressure ratio, shape of the rotor and cooling of the gas during compression process, will affect the performance of rotary compressors.

For reciprocating compressors, the volumetric efficiency is less than 0.5 at a compression ratio of 10, but for the screw compressors it is more than 0.7 at a compression ratio of 20.

One of the important design variables of the rotary compressor is the in-built volume ratio or compression ratio r.

Refrigeration plants work under variable conditions because evaporation and condensation temperatures can change while in operation. In a reciprocating compressor, section volume varies according to the varying conditions. For example, for the increase of condensing pressure the discharge valve opens later, whereas when condensing pressure decreases the discharge valve opens sooner. This is because the pressure necessary to open the discharge valve is reached more quickly. The compressor adjusts its compression ratio as per the prevailing conditions.

In a screw compressor, there is no discharge valve. The discharge takes place only when the interlobe space comes in open contact with the discharge port. The interlobe space reaching the discharge port is fixed due to the turning of the rotor. Therefore, compression ratio is only one. Suppose the condenser pressure increases, then in such a case the compressed gas reaching the discharge port may not have reached this new condenser pressure. This condition is referred to as under-compression. On the other hand, if the condenser pressure decreases, then the compressed gas reaching the discharge port might be at very high pressure resulting in over-compression. This phenomenon of under-compression and over-compression will result into poor compressor performance. The efficiency of a screw compressor can only be optimized when the in-built volume ratio corresponds to the required compression ratio. Therefore, modern screw compressors are equipped with an automatic or a manually adjustable in-built volume ratio.

Similar to the single screw compressors, the twin screw compressors can operate in two ways: (i) oil-injected compressors and (ii) oil-injection free compressors which have been explained earlier in this chapter.

Twin screw compressors are also available with a secondary suction port between the primary compressor suction and discharge ports. This port can accept a second suction load at a pressure above the primary evaporator or flash gas from a liquid subcooler vessel known as *economizer*. This economizer concept is also discussed in the preceding text.

Similar to single screw compressor performance curves, the performance curves of twin screw compressor can also be generated.

11.10 CENTRIFUGAL COMPRESSOR

Centrifugal compressor is a rotodynamic machine which continuously exchanges angular momentum between its rotating part (impeller) and steadily flowing fluid. For effective momentum exchange, the rotating speeds must be higher. Therefore, these operate with rotational speeds between 1500 and 90,000 rpm. Centrifugal compressors are used in a variety of refrigeration and air-conditioning applications with ammonia, R11 (R123), R113 (R124) and R502 (R143a) refrigerants. These refrigerants have low densities. The suction flow rates range between 0.03 and 15 m^3/s.

Figure 11.25 illustrates the basic operation of a centrifugal compressor. Essentially, it consists of four elements—inlet buckets, impeller, casing and diffuser.

Inlet buckets are attached to the rotating shaft and designed to guide the gas on the impeller in an efficient manner.

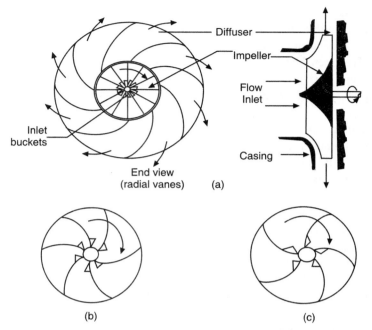

Figure 11.25 Centrifugal compressor: (a) Radial impeller vanes; (b) Backward curved vanes; (c) Forward curved vanes.

To an impeller, the inlet buckets are near the axis of rotation supply gas. The impeller has radial vanes as shown in Figure 11.25(a). It may also be constructed with backward or forward curved vanes as shown in Figures 11.25(b) and 11.25(c) with the pressure–speed characteristics as desired. But backward curved vanes are preferable for refrigerant compressors.

The casing surrounds the impeller in such a manner as to form an inlet and an outlet duct. The diffuser is placed in the radial portion of the casing.

In operation, the shaft and impeller are rotated at high speed. The air contained in its rotating passages is subjected to centrifugal force and flows radially outwards continuously. It means the air is accelerated in the impeller to impeller velocity. Thus, the action of the impeller has been to give the gas flowing through it a very high velocity of flow plus a small compression due to centrifugal force. This high velocity gas stream is then slowed down by the diffuser passage, because of its gradually increasing passage areas and the change of kinetic energy that can be converted into a further increase of pressure.

11.10.1 Work Done by Impeller

By Newton's law, it has been proved that the rate of change of angular momentum of a gas stream is equal to the torque applied to the body causing that change. Further, for an impeller of a compressor it can be shown:

$$\text{Torque} = m(v_{T2}r_2 - v_{T1}r_1) \qquad (11.26)$$

where
$\quad m$ = mass flow rate of the gas (kg/s)

v_{T1} = tangential (whirl component of) velocity at inlet (m/s)
v_{T2} = tangential (whirl component of) velocity at outlet (m/s)
r_1 = radius of impeller at inlet (m)
r_2 = radius of impeller at outlet (m).

Work done by the impeller rotating at the angular velocity ω (rad/s) is:

$$W = m\omega(v_{T2}r_2 - v_{T1}r_1)$$

But $r\omega = u$ = peripheral or tip velocity of impeller
Therefore,

$$W = m(v_{T2}u_2 - v_{T1}u_1) \tag{11.27}$$

11.10.2 Power Input

The actual work added to the gas for a mass flow rate m kg/s is given by

$$W_{actual} = \frac{m(h_2 - h_1)}{\eta_{is}}$$

Power required by the compressor (P_e), considering mechanical η_m, is

$$P_e = \frac{W_{actual}}{\eta_m} = \frac{m(h_2 - h_1)}{\eta_{is} \cdot \eta_m} \tag{11.28}$$

where h_1, h_2 are enthalpy of gas at inlet and outlet of the compressor, (kJ/kg), and η_{is} and η_m are isentropic and mechanical efficiency of the compressor, respectively.

11.10.3 Performance of Centrifugal Compressor

A centrifugal compressor does work on the gas or it adds work to the gas. The work imparted to the gas by the compressor is known as *head*. For an ideal radial flow compressor, the work added per unit mass of gas is limited by the equation,

$$h_2 - h_1 = \frac{\gamma}{\gamma - 1} p_1 V_1 \left[\left(\frac{p_2}{p_1}\right)^{\gamma/(\gamma-1)} - 1 \right] \tag{11.29}$$

For the compressor having impeller with backward curved vanes shown in Figure 11.25(b), work added or head imparted per unit mass flow rate,

$$\text{Head} = W = (v_{T2}u_2 - v_{T1}u_1)$$

But normally $v_{T1} = 0$

$$\text{Head} = W = u_2(u_2 - v_{f_2} \cdot \cot \beta_2)$$
Because $v_{T2} = u_2 - v_{f_2} \cdot \cot \beta_2$

$\therefore$
$$\text{Head} = W = u_2^2 - u_2 v_{f_2} \cot \beta_2 \tag{11.30}$$

For a compressor operating at constant speed, peripheral velocity u_2 and blade angle β_2 are constant. As the flow increases, v_{f_2} increases, the head decreases. The straight line in Figure 11.26 represents this.

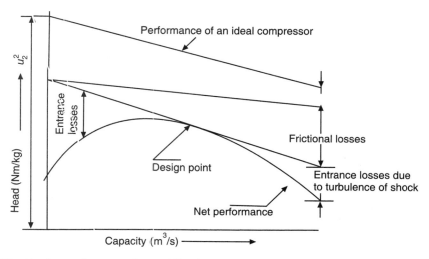

Figure 11.26 Head capacity curve of a centrifugal compressor with impeller having backward curved vanes.

The various losses have been discussed in the earlier section. Due to these losses, the characteristic curve does not remain a straight line but becomes dome-shaped. These losses are minimum for a particular flow (design point) only. Therefore, a designer or an operator is required to set the blade angle at inlet and outlet.

Surging: The head flow characteristic curves of a centrifugal compressor operating at different speeds are shown in Figure 11.27. The compressor operates in a refrigeration cycle, so accordingly the coordinates selected are refrigeration capacity and evaporating temperature.

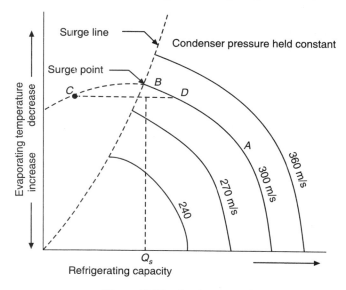

Figure 11.27 Surging effect.

It is seen that the curves for different blade tip speeds all end on a dotted line called the surge line which occurs at or just after the maximum pressure. Moving upwards on the y-coordinate indicates a decrease in the evaporating temperature or an increase of compressor work per kg mass flow rate. When the refrigeration load decreases from point A, the performance follows the path and reaches the point B. At this point B, the pressure difference between the condenser and the evaporator is just equal to the head that the compressor can develop. Further reduction of refrigeration load from point B shifts the operating condition to say point C. It means that the actual pressure difference between the condenser and the evaporator is greater than the head that a compressor can develop. Therefore, some gas flows from the compressor to the evaporator and when sufficient gas collects in the evaporator the point of operation shifts abruptly over to point D. The cycle repeats if the compressor is still in operation. This sequence of events that results in the instability of compressor is called *surging*.

Surging phenomenon causes many bad effects such as excessive stresses that may be induced in the compressor motor, shafts, etc. It follows that any one compressor is good only for a certain range of flow (or TR). Surging occurs when operating below about 30% of rated capacity.

Curve parameters

Centrifugal compressor performance curves are developed for independent variable inlet or actual volume flow versus the following parameters:

1. Polytropic head or adiabatic head
2. Discharge pressure
3. Power requirement
4. Adiabatic or polytropic efficiency
5. Pressure ratio
6. Pressure rise
7. Discharge temperature

The manufacturer can furnish any combination of these performance curves.

11.11 HERMETICALLY SEALED COMPRESSOR

The compressor and motor mounted on the same shaft are placed in a casing which is hermetically sealed. The discharge muffler is connected at the discharge of the compressor to reduce the noise, and the suction muffler is connected to the suction line. These two pipes come out of the casing and a service line is also provided for changing the refrigerant and lube oil. The suction vapour is used to cool the oil and the winding of the motor.

The power is supplied to three-phase terminals of the starting and running windings through a relay. The relay cuts off the starting winding after a start-up, leaving the running winding in circuit. The relay also has an overload protector.

As the motor is enclosed, it heats up the oil bath in the casing. In some units a special oil cooling loop is provided from the condenser. The oil must have the necessary dielectric strength for long life of the motor winding.

As the unit is sealed, its life is almost 15–20 years. There is also a reduction in noise. The use of capillary as expansion devices has been made possible only with the use of sealed unit, as the capillary has less tolerance to leakage of refrigerant.

These units are now available upto 5 TR and are found in refrigerators, water coolers, deep freezers, window/split air conditioners and even in package air conditioners.

The sealed unit has no easy access and needs to be cut open for repairs. This difficulty was overcome with the advent of a semi-sealed unit. In this unit the compressor and motor are in a casing but the casing has an opening near the head of the compressor and near the terminals of motor. So a screwed cover on the opening allows access to these critical areas in the case of repairs. Semi-sealed units are nowadays available in large capacities even up to 75–100 TR.

EXERCISES

1. Derive an expression for the volumetric efficiency of reciprocating compressor.
2. Write a short note on rotary compressor and its advantages.
3. Describe the method of capacity control in reciprocating compressors.
4. Describe the scroll compressor and its future use.
5. Compare the utility of reciprocating compressors and centrifugal compressors.
6. What are the features of a hermetically sealed unit?
7. Explain the surging phenomenon in centrifugal compressors.
8. Differentiate between the reciprocating and centrifugal compressors with neat sketches.
9. Draw the performance curves of a hermetically sealed compressor.
10. Draw the performance curves of a centrifugal compressor.
11. Explain the different types of rotary compressors.

Chapter 12

Evaporators and Condensers

12.1 INTRODUCTION

Evaporator is a component of the refrigeration unit in which the refrigerant absorbs heat from the space (object) by changing its phase from liquid to vapour. The object may be air, water or any brine solution. The size and shape of the evaporator is an important aspect while designing any refrigeration unit. Efforts are always made to reduce its size using good conducting metals, enhancing heat transfer coefficients and adapting new manufacturing technologies. In this chapter, related heat transfer theory, various types of evaporators, design concepts, etc. have been discussed.

The satisfactory operation of the refrigeration unit is dependent on the heat rejection efficiency of the condenser. Condensers of small refrigeration and air-conditioning units are air-cooled while the condensers of large capacity units are always water-cooled combined with cooling towers. The condenser is required to dissipate heat energy not only absorbed in the evaporator but also the energy inputted through compressor.

Also in this chapter, heat transfer analysis of condenser types and design concepts are included.

Heat transfer analysis of these two units is an important factor, so it is better to revise the heat transfer concepts before beginning this chapter. The three distinct modes of heat transfer are conduction, convection and radiation.

Heat exchange between two objects is possible only if the temperature gradient exists. We are interested not only in understanding the heat transfer phenomenon by these three modes but also in the rate at which the heat transfer takes place. The rate of heat transfer is an important factor in rating evaporators and condensers.

12.2 CONDUCTION

When a temperature gradient exists in a body, energy transfer occurs from the high temperature region to the low temperature region by the way of conduction. The rate of energy transfer per unit area by conduction is proportional to the normal temperature gradient as stated by Fourier's law. Mathematically,

$$\frac{Q}{A} \propto \frac{\partial T}{\partial x}$$

or

$$Q = -kA \frac{\partial T}{\partial x} \tag{12.1}$$

where Q = heat transfer rate (watts), A = area normal to the direction of heat flow (m²), k = the constant called thermal conductivity of the material (W/m-K) and $\partial T/\partial x$ is the temperature gradient in the direction of heat flow. The negative sign indicates that heat flows from a high temperature region to a low temperature region. The thermal conductivity becomes a physical property of the material and it can be determined experimentally. In the literature of heat transfer, one can find graphs for the variation of conductivity k with temperature for different metals. It is observed that the order of decreasing thermal conductivity is as follows: metals; non-metals and insulating materials; liquids and gases.

If the temperature of an object does not vary with time, then it is treated as steady state condition and in such a case, the conduction problem will be a simpler one.

On the contrary, if the temperature of the object changes with time (i.e. unsteady state) or if there are sources or sinks within the object or if the conduction of heat is in more than one direction then the problem becomes more complex. In such situations, the reader is required to refer to good books on heat transfer.

12.2.1 Steady State Heat Conduction through a Slab

In steady state conduction, the rate of thermal energy flow does not change during the process, due to which the temperature remains constant. This is the simplest form of heat transfer, which is assumed when considering heat transfer by conduction through the thickness of building insulation, tanks, in plate freezers or in plate heat exchangers.

Let us consider a plane wall of thickness Δx and thermal conductivity k, and a heat conduction Q_x normal to the wall surface ($A = z \times y$) as shown in Figure 12.1. Therefore,

$$Q_x = -k \cdot A \cdot (T_1 - T_2)/\Delta x \tag{12.2}$$

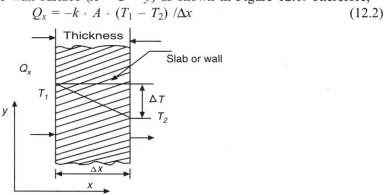

Figure 12.1 Heat conduction through a slab.

12.2.2 Steady State Heat Flow through a Cylindrical Wall

In shell and tube heat exchangers, exchange between the two fluids takes place through the curved walls of the tube.

Let us consider a cylindrical surface at a distance from the axis of a tube having inner and outer radii r_i and r_o, respectively, as shown in Figure 12.2, and let L be the length of the tube.

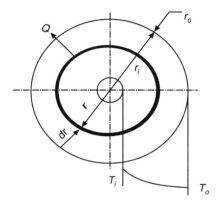

Figure 12.2 Heat conduction through a hollow tube.

Fourier's equation for radial heat conduction through the tube at any radius r is

$$Q_r = -k(2\pi rL)\frac{dT}{dr} \quad (12.3)$$

or

$$\frac{Q}{L} = 2\pi k(T_i - T_o)/\log\left(\frac{r_o}{r_i}\right) \quad (12.4)$$

12.2.3 Steady State Conduction through a Composite Wall

Many a time cold storage tanks are layered with different insulating materials to reduce the rate of heat leakage. Such problems come under this topic as discussed here. Consider a cold storage tank that is made from two materials having thickness l_1 and l_2 and having thermal conductivity k_1 and k_2, respectively, as in Figure 12.3. The thermal resistances are $R_1 = l_1/k_1A$ and $R_2 = l_2/k_2A$ for the two layers which are in series as shown in Figure 12.3.

Heat being conducted through the walls under steady state is as follows.

If the heat transfer due to convection on both sides of the wall is considered, then the resistance R_i and R_o are to be added due to convection coefficients h_i and h_o respectively to the heat transfer taking place between temperatures T_i and T_o. Therefore, Eq. (12.2) becomes

$$Q = \frac{(T_i - T_o)}{(R_i + R_1 + R_2 + R_o)} \quad (12.5)$$

where $R_i = \dfrac{1}{h_i A}$ and $R_o = \dfrac{1}{h_o A}$, $R_1 = \dfrac{l_1}{k_1 A}$, $R_2 = \dfrac{l_2}{k_2 A}$

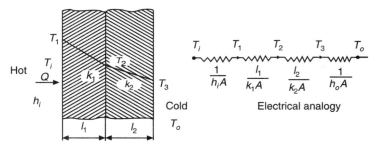

Figure 12.3 Steady heat conduction through a composite wall.

The convection heat transfer coefficient h is explained in the latter part of the chapter.

12.2.4 Steady State Conduction through a Composite Cylinder

The refrigerant lines in the open atmosphere are provided with insulation layers. For the known values of thermal conductivity, the temperature and surface area are required to calculate the heat in-leak into the system. Sometimes the problem may involve the calculation of the insulation thickness for the required rate of heat in-leak.

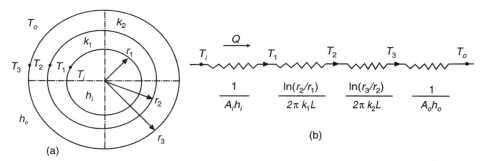

Figure 12.4 Heat flow through multiple cylindrical sections and its electrical anologue.

Let us consider a pipe with radius r_1 and negligible thickness carrying a fluid in it. Two insulation layers with thermal conductivity k_1 and k_2 are provided as shown in Figure 12.4.

$$\frac{Q}{L} = \frac{2\pi (T_i - T_o)}{(R_i + R_1 + R_2 + R_o)} \tag{12.6}$$

where $R_i = \dfrac{1}{A_i h_i}$, $R_1 = \dfrac{\ln(r_2/r_1)}{2\pi k_1 L}$, $R_2 = \dfrac{\ln(r_3/r_2)}{2\pi k_2 L}$ and $R_o = \dfrac{1}{A_o h_o}$

A_i and A_o are the surface areas for the inner and outer diameters of the composite unit. The insulation thickness provided on a pipe does not always reduce the rate of heat flow but it is dependent on the critical radius given by

$$r_o = \frac{k}{h_o} \tag{12.7}$$

where k is the thermal conductivity and h_o is the heat transfer coefficient. If the outer radius is less than the value given by this equation, the heat transfer will increase by adding more insulation. For outer radii greater than the critical value, an increase in insulation thickness will cause a decrease in heat transfer.

12.2.5 Overall Heat Transfer Coefficient

Plane wall

Consider a plane wall as shown in Figure 12.5, which is exposed to a hot fluid on one side and a cooler fluid on the other side.

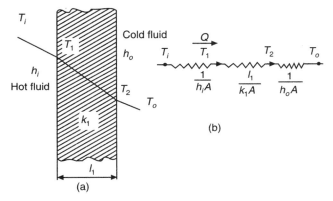

Figure 12.5 Overall heat transfer through a plane wall.

The overall heat transfer rate is calculated by the following equation,

$$Q = \frac{T_i - T_o}{\left(\dfrac{1}{h_i A} + \dfrac{l_1}{k_1 A} + \dfrac{1}{h_o A}\right)} \qquad (12.8)$$

As in earlier instances, $\dfrac{1}{hA}$ is used to represent the convection resistance. The overall heat transfer by the combination of conduction and convection is frequently expressed in terms of an overall heat transfer coefficient U, defined by the equation

$$Q = U \cdot A \cdot \Delta T_{\text{overall}} \qquad (12.9)$$

Comparing Eqs. (12.8) and (12.9),

$$U = \frac{1}{(1/h_i) + (l_1/k_1) + (1/h_o)} \qquad (12.10)$$

Consider a shell-and-tube type unit wherein the inner and outer surfaces of the tube are exposed to convection environment as shown in Figure 12.6. Let T_A and T_B be the two average fluid temperatures. Here the surface area exposed to convection heat transfer inside and outside the tube are different based on the inner and outer radii of the tube.

The overall heat transfer rate is

$$Q = \frac{T_A - T_B}{R_i + R + R_o} \qquad (12.11)$$

where R_i, R and R_o are as shown in Figure 12.6(b). The terms A_i and A_o are the inside and outside surface areas of the inner tube. The overall heat transfer may be based on either A_i or A_o of the tube. Accordingly,

$$U_i = \frac{1}{R_i + \dfrac{1}{h_i} + \dfrac{A_i \ln(r_o/r_i)}{2\pi k L} + \dfrac{A_i}{A_o}\dfrac{1}{h_o} + \dfrac{R_o}{A_o} \times A_i} \qquad (12.12)$$

$$U_o = \frac{1}{R_o + \dfrac{A_o}{A_i} \times \dfrac{1}{h_i} + \dfrac{A_o \ln(r_o/r_i)}{2\pi k L} + \dfrac{1}{h_o} + \dfrac{R_i A_o}{A_i}} \qquad (12.13)$$

Here R_i and R_o are the fouling resistance on the inside and outside of tubing, respectively. It is expressed in m²-K/W.

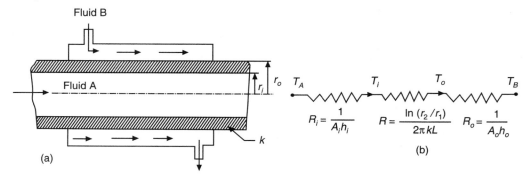

Figure 12.6 Resistance analogy for hollow cylinder with convection.

During the operation of heat exchangers, the process of deposits like rust, boiler scale, silt or any other thing getting accumulated on the surfaces, is called *fouling*. It increases the thermal resistance to heat flow.

12.3 CONVECTION HEAT TRANSFER

When a fluid comes in contact with a solid surface, which is at a different temperature, then the heat flows from the fluid to the surface or from the surface to the fluid depending upon the temperature gradient.

Newton introduced the concept of heat transfer coefficient, h (W/m²-K). He recommended the following equation to evaluate the heat transfer rate by convection.

$$Q = hA(T_s - T_w) \qquad (12.14)$$

where T_s = surface temperature, T_w = free stream temperature of fluid and A = surface area. Equation (12.14) is known as law of heating or cooling.

There are two types of thermal convection:

Natural convection: Heat transfer involving motion in a fluid caused by the difference in density and the action of gravity is called *natural* or *free convection*. Heat transfer coefficients for natural convection are generally much lower than those for forced convection. Natural convection is important in a variety of applications, such as evaporator and condenser of domestic refrigerator, gravity coils used in high humidity cold storage rooms and roof-mounted refrigerant condensers and cooling panels for air conditioning.

Forced convection: In this type of convection, mass movement of the fluid is caused by the action of external units like pump, blower or fan; and temperature gradient between the fluid and metal surface is also required for heat transfer. Forced convection is important in variety of applications like forced air coolers and heaters, forced air- or water-cooled condensers and liquid suction heat exchangers.

From fluid mechanics, we know that there are two kinds of flows, namely laminar and turbulent. Laminar flow exists at low fluid velocity while turbulent flow is at high velocity.

The correct value of heat transfer coefficient, h (W/m²-K) depends on the following characteristics of fluid flow:

1. Type of motion of the fluid
2. Fluid temperature, T (K)
3. Fluid pressure, p (N/m²)
4. Fluid velocity, v (m/s)
5. Fluid density, ρ (kg/m³)
6. Fluid thermal conductivity, k (W/m-K)
7. Thermal diffusivity, α (m²/s)
8. Specific heat of fluid, c_p (J/kg-K)
9. Fluid viscosity, μ (N-s/m²)
10. Diameter of the tube, D (m)
11. State and roughness of the surface of the objects involved in the heat transfer.

Scientists have grouped these variables into different dimensionless numbers so as to make the convection analysis easier. The natural convection involves numbers such as Reynolds number, Prandtl number, Nusselt number and Grashoff number.

For forced convection heat transfer, it is shown that Nusselt number is a function of Reynolds number and Prandtl number. However, for natural convection, Nusselt number is a function of Grashoff number and Prandtl number.

Now let us understand these numbers.

Prandtl number (Pr)

$$\Pr = \frac{v}{\alpha} = \frac{\mu/\rho}{k/\rho c_p} = \frac{c_p \mu}{k} \qquad (12.15)$$

where α = thermal diffusivity (m²/s).

The kinematic viscosity v tells us about the rate at which the momentum diffuses through the fluid because of molecular motion. The thermal diffusivity α tells us about the heat that diffuses in the fluid. Thus, the ratio of these two quantities should express the relative magnitudes of diffusion of momentum and heat in the fluid. Prandtl number connects the velocity field and temperature field.

Grashoff number (Gr)

$$\text{Gr} = \frac{g\beta \cdot \Delta T \cdot L^3}{\nu^2} \quad (12.16)$$

where
- g = gravitational acceleration (m/s²)
- β = volume coefficient of expansion = $1/T$ (K⁻¹)
- $\Delta T = (T_w - T_\infty)$ = temperature difference between the wall and free stream (K)
- L = distance from the leading edge (m)
- ν = kinematic viscosity (m²/s).

Grashoff number may be interpreted as the ratio of the buoyancy forces to the viscous forces in the free convection flow system. Its role is similar to that of Reynolds number in forced convection system. For air in free convection on a vertical flat plate, the critical Grashoff number has been observed to be about 4×10^8 by Eckert and Soehngen.

Nusselt number (Nu)

$$\text{Nu} = \frac{hd}{k} \quad (12.17)$$

where
- h = heat transfer coefficient (W/m²-K)
- d = tube diameter (m)
- k = thermal conductivity of fluid (W/m-K).

Nusselt number is the ratio of real heat flow determined by h to the heat flow by conduction through a thickness d.

There are a number of empirical correlations which can be found in books on heat transfer. Most of these are formulated based on experimental results for the specific conditions. The general correlation for turbulent flow inside tubes is

$$\frac{hd}{k} = c \cdot \left(\frac{vd}{\nu}\right)^m \left(\frac{\mu c_p}{k}\right)^n \quad (12.18)$$

where c, m and n are the constants whose values are dependent on the size of tube, type of flow, type of fluid, etc. The properties of the fluid like kinematic viscosity ν and c_p, etc. are either based on bulk temperature T_b or mean film temperature T_f.

The bulk temperature is one which represents the total energy of the flow at the particular location. Also sometimes referred to as 'mixing cup' temperature, it is the temperature the fluid would attain if placed in a mixing chamber and allowed to come to an equilibrium condition. The bulk temperature is preferably used in the convection heat transfer of the fluid flowing through tubes.

The fluid particles flowing over the surface experience the property changes. When there is appreciable variation between the wall and the free stream conditions, the properties of the fluid are evaluated at the so-called film temperature T_f, where T_f is the arithmetic mean value between the wall surface and free-stream temperature ($T_f = (T_s + T_\infty)/2$).

12.4 EVAPORATOR

It is a heat exchanger in which one of the fluids (refrigerant) remains at one temperature while the temperature of the other fluid decreases as it flows through the exchanger. This is illustrated in Figure 12.7.

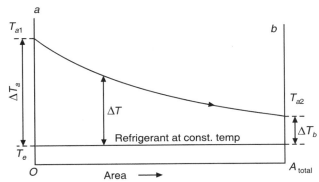

Figure 12.7 Temperature distribution in a single-pass evaporator.

An evaporator is the most important component of the refrigeration unit as it transmits 'cold' directly to the material we want to cool. The success of a refrigeration unit is dependent on efficient working of the evaporator.

There are several different ways of classifying evaporators. They may be classified as 'forced convection' or 'natural convection' depending upon whether a fan or a pump forces the fluid being cooled or whether the fluid flows naturally due to density differences of the warm and chilled fluids.

In some evaporators, refrigerant flows inside the tubes while the fluid being chilled will flow outside the tubes.

In other evaporators, the fluid being chilled flows through tubes which are immersed in the liquid refrigerant; a still another variety based on the operation is like the 'dry' and 'flooded' evaporator. These are quite interesting varieties of evaporators and we will discuss them in the following section.

12.4.1 Dry and Flooded Evaporators

A dry expansion or dry evaporator is one in which the gases in the final part of the coils are dry or (almost) free of liquid drops. It means that the amount of liquid refrigerant fed into the evaporator is limited in such a way that it evaporates completely by the time it reaches the end evaporator. Therefore, only the refrigerant vapour enters the suction line. A thermostatic expansion valve or a capillary tube is used to control the refrigerant to the evaporator which is shown in Figure 12.8. The refrigerant will be allowed to superheat to about 5 K at the end of the evaporator. This ensures entry of dry vapour to the suction of the compressor.

On the other hand, in the so-called flooded or wet evaporator, the refrigerant vapour at the outlet of the evaporator is saturated with a mist of fine liquid drops. The flooded evaporator

(Figure 12.9) is fed with an overdose of liquid, from which only a part (20% or 25%) is evaporated when the refrigerant leaves the coils that are meant to take the evaporator heat load. The other liquid refrigerant serves to keep the inner surface of the tubes wet increasing the internal heat transfer. One of the disadvantages of this evaporator is that liquid droplets may enter the suction of compressors.

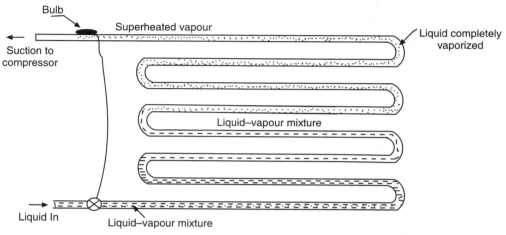

Figure 12.8 Dry expansion evaporator.

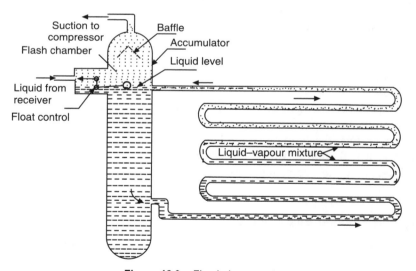

Figure 12.9 Flooded evaporator.

The dry evaporators are comparatively less efficient than the flooded evaporators. But dry evaporators are much simpler in design with less initial cost, and they require much smaller refrigerant charge and have very few oil return problems. So due to these reasons they are popular. Still there are a large number of evaporators that need to be discussed but let us consider the following points based on which the selection and design of an evaporator is made.

- Working of the evaporator whether the dry or flooded evaporator
- Heat load on the evaporator
- LMTD of the evaporator
- Value of overall heat transfer coefficient
- Extended surface
- Internal pressure drop
- Frosting and defrosting of coolers (see Section 12.5)
- Selection from manufacturer's data sheets (see Section 12.6)
- Liquid coolers (see Section 12.7)

Heat load on the evaporator

The rate at which heat flows from the refrigerated space or product to the refrigerant is called the *load of the evaporator*. The cooling capacity of the evaporator must be sufficient to bear the load on it. The cooling capacity of the evaporator is given by the equation

$$Q = U \cdot A \cdot \Delta T_m \tag{12.19}$$

where U = overall heat transfer coefficient of the heat exchanger (evaporator) (W/m²-K), A = surface area of the evaporator available for heat transfer (m²), and ΔT_m = log mean temperature difference.

The concept of overall heat transfer coefficient U is explained in Section 12.2. Also, the readers should refer to specialized heat transfer books to get the correct correlation to determine the heat transfer coefficient for boiling of the refrigerant.

As one can see, there is a wide variation in the rate of heat transfer during boiling and it is very difficult to find the heat transfer coefficient (h) accurately as the correct behavioural pattern of the refrigerant during boiling is not known. There are a large number of correlations for the estimation of heat transfer coefficient during the boiling process. We need to choose the right one for the appropriate conditions.

LMTD (ΔT_m)

With reference to Figure 12.7 the ΔT_m can be calculated as follows, i.e.

$$\Delta T_m = \frac{(T_{a1} - T_e) - (T_{a2} - T_e)}{\ln \dfrac{T_{a1} - T_e}{T_{a2} - T_e}} \tag{12.20}$$

where
T_{a1} = air inlet temperature
T_{a2} = air outlet temperature
T_e = evaporation temperature.

T_e is assumed to be constant but practically it does not remain constant on account of pressure drop in the evaporator coil.

The temperature of the air decreases progressively as the air passes through the cooling coil. The drop in temperature is the greatest across the first row of the coil and diminishes as the air passes across each succeeding row as shown in Figure 12.10.

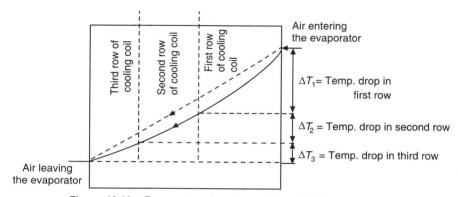

Figure 12.10 Temperature drop in a cooling coil of an evaporator.

Many a time in industry the arithmetic mean temperature difference (ΔT_{Ar}) is calculated as follows:

$$\Delta T_{Ar} = \frac{(T_{a1} - T_e) - (T_{a2} - T_e)}{2} \tag{12.21}$$

Therefore, care must be taken while selecting the evaporator as it is based on ΔT_m and not on ΔT_{Av}. The temperature difference between the outlet air temperature (T_{a2}) and evaporation temperature (T_e) must not be chosen outside the range 2–4°C for cold stores, used as storage rooms. Larger differences give a smaller air volume, meaning smaller fans but the goods stored lose more weight and cause a lower evaporation temperature, resulting in more power consumption.

Extended surface (Fin)

The evaporators that cool air by forced convection, and also many liquid chillers, are provided with fins which are actually bonded to the evaporators. The reason for the use of fins is to increase the area of heat transfer. The outer surface area of the tubes through which the refrigerant flows is known as primary area. The area available for heat transfer due to fins is known as secondary area. Normally the secondary area is 10 to 15 times the primary area. The type of fin, i.e. whether circular, triangular, lateral and longitudinal along with their efficiency should be studied at the time of evaporator design.

Pressure drop

The friction between the flowing fluid and the tube surface is unavoidable but its intensity may be reduced by smooth surfaces. The pressure of the refrigerant will drop as it passes through the evaporator coil. With pressure drop, the velocity increases thereby increasing the heat transfer coefficient which is desirable. Excessive pressure drops, however, are not economical because of the adverse effect on the coefficient of performance. In the evaporators of air-conditioning units, the pressure drop is limited to the order of 0.1–0.2 bar. One of the reasons for excessive pressure drop is the overloading of the unit. Because of excessive load, the refrigerant velocity would increase beyond the desired range and the pressure drop will be excessive. The value of pressure drop is determined by the length of the tube per circuit and the construction of the cooler.

To limit the pressure drop in the evaporator, it is required to acquire the knowledge of 'evaporators circuiting' which is not discussed here.

One of the widely used methods to supply refrigerant to the evaporator is with the refrigerant distributor and suction header as shown in Figure 12.11. The evaporator with the refrigerant distributor and the suction header is particularly effective when circuit loading is heavy as in the air conditioning coil where the temperature difference between the refrigerant and the air is large. The refrigerant through the evaporator coil and the airflow over it (Figure 12.11) are in counterflow pattern, thus giving the highest ΔT_m and subsequently the heat transfer.

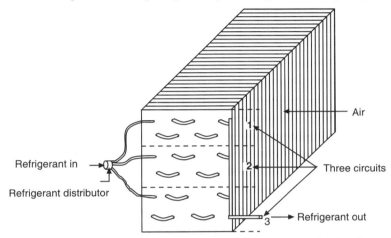

Figure 12.11 Evaporator with refrigerant distributor and suction header.

The degree of superheat is always rather limited to about 6 K. It is also required to confirm from the evaporator manufacturer that circuits are well-balanced with proper use of the refrigerant distributor. It may happen that an oversized distributor will distribute the refrigerant unevenly leading to a large subcooling.

12.5 FROSTING AND DEFROSTING OF COOLERS

If the operating temperature of the evaporator surface is below 0°C (273 K) and air is contacting the evaporator, then water vapour freezes on the coil. So one finds hoar-frost or even an ice layer. Frost formation is undesirable due to two reasons:

(i) Thick layer of frost acts like an insulator, which will cause h_o to decrease.
(ii) For forced convection coils, the effective air passage through the coils and fins will decrease. Due to this, air velocity increases, pressure drop increases (pressure drop increases as the square of the velocity) and volume flow decreases.

Because of these two reasons, the power consumption of the plant will increase making defrosting a regular requirement.

12.5.1 Methods of Defrost

The most common defrost methods are condensing unit or system off-time, electric heat and air defrost.

Condensing unit off-time

In this method, the refrigeration unit is put off and allowed to remain off until the evaporator reaches a temperature that permits defrosting and gives ample time for condensate drainage. The heat required to melt the ice is taken from the air circulated in the fixture, which is quite slow. It is usually employed in fixtures maintaining temperatures of 1°C or above. Defrost may be controlled by: (a) suction pressure control, (b) time clock initiation and termination, (c) time clock initiation and suction pressure termination, and (d) time clock initiation and temperature termination.

(a) Suction pressure control: This control is adjusted for a cut-in pressure high enough to allow defrosting during the off cycle. It is used in those fixtures that maintain temperatures from 2° to 6°C. It provides some cooling effect during the defrost period, because air is circulated over melting ice.

(b) Time clock initiation and termination: A timer initiates and terminates the defrost cycle from clock movement. The length of the defrost cycle must be determined and the clock mechanism set accordingly.

(c) Time clock initiation and suction pressure termination: In this method, defrost cycle is initiated from clock movement and is terminated by suction pressure. The duration of defrost cycle is automatically adjusted to the condition of the evaporator, insofar as frost and ice are concerned. If at all the suction pressure does not rise then in that case the timer puts off the defrost cycle as it contains such an arrangement.

(d) Time clock initiation and temperature termination: In this method, the defrost cycle is initiated by time initiation but is terminated by temperature. A temperature sensor is located on a tube of the evaporator, which always has liquid present, or in the air stream leaving the evaporator. The timer also has a fail-safe setting in its circuit to terminate the defrost cycle after a set time regardless of the temperature.

12.5.2 Latent Heat Defrost

It is also called hot gas defrost and uses heat in the discharge gas from the compressor to defrost the evaporator. It means that hot gases at the outlet of the compressor discharge are diverted and circulated through the evaporator coil. A typical hot gas defrost arrangement is shown in Figure 12.12. This is a very effective energy recovery method used in large ammonia installations. The available heat of the plant is utilized.

Some part of this energy gain will, however, be lost under the following conditions:

  (i) If the compressor is new to producing hot gas.
  (ii) If the condenser pressure is less or if in winter, the temperature of the discharge gas is not sufficiently high to cause defrosting.

12.5.3 Electric Defrost

In this method, heat is supplied to the evaporator by an external electric heater. This method requires a longer defrost period, 1.5 times the latent heat defrost method. The heater element may be in direct contact with the evaporator coil or it may be located between the evaporator and the fan. This heater will have a temperature-limiting device to switch off when the required temperature is sensed.

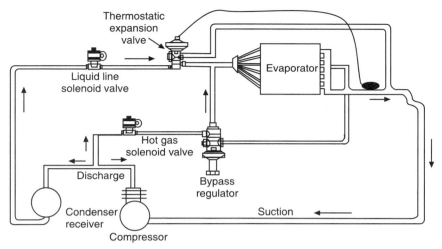

Figure 12.12 Typical schematic diagram of 'hot gas' bypass from compressor discharge to evaporator inlet.

Depending upon the application, the defrosting method of the evaporator is decided. Moreover, defrosting of frozen food products like eggs, fishes and meat is carried out in different ways. Further, the above discussed methods indicate the basic idea of defrosting.

The task of determining the period and frequency of optimum defrosting is very difficult. Frequent defrosting will result into more energy wastage as indicated by the experiments being conducted in this respect. The optimum frequency obtained after conducting experiments ranges between 6–8 defrosting cycles per day. Defrosting periods of 15-20 minutes with hot gases were determined, while for electrical defrosting, this figure was about 30 minutes.

12.6 SELECTION FROM MANUFACTURER'S DATA SHEETS

The performance of an evaporator is dependent on a large number of variables which are difficult to consider simultaneously. A practical approach is to evaluate the performance of an evaporator at different TDs (difference in temperature between the temperature of air entering the evaporator and evaporation temperature of refrigerant corresponding to the pressure at the evaporator outlet). While conducting the test, the relative humidity (RH) of the air entering the coil is normally kept at 85% and the degree of superheat of the refrigerant at the outlet is about 6 K (as specified in the ISI test code). When a manufacturer's test data varies from the test conditions, then the corrections are to be applied.

The test results of the cooler are at a particular TD and the practical employment may be at a different TD. Again, the necessary corrections need to be applied to meet the field requirements. For a decreasing TD, the heat transfer coefficient h_i on the inside of the cooler coil decreases, so the cooling capacity decreases more than proportionally with TD. The U-value also decreases. For the decreased capacity, the internal pressure drop will decrease. The net effect is that the difference between the air inlet and outlet temperatures and the evaporation temperature decreases. This leads to a higher SHF (sensible heat factor), and less latent heat leads to more sensible heat. Therefore, it is also necessary that the manufacturer must specify which SHF has been used and should give correction factors for other SHF values.

The manufacturer's data sheet must also contain the other information, such as:

(i) Cooling capacity value indicated is gross or net.
(ii) The evaporator testing condition (dry evaporation or flooded evaporation).
(iii) The refrigerant used.

12.7 LIQUID COOLERS (CHILLERS)

The liquid coolers are essentially the recuperative type of heat exchangers in which the liquid refrigerant and the liquid to be cooled exist on either side of the tube. The liquids being cooled are water, brine solutions, milk, beer and other products with low viscosity. The liquid refrigerant is evaporated producing cold and is transferred to these fluids. The types of liquid coolers are:

- Direct expansion type
- Flooded shell-and-tube coolers
- Baudelot cooler
- Shell-and-coil types.

12.7.1 Direct Expansion Type

Figure 12.13 shows a typical shell-and-tube cooler. The refrigerant evaporates inside the tubes while the liquid being cooled occupies the space of the shell. The liquid on the shell side flows at right angles to the tubes. The baffles will help to increase the velocity thereby increasing the heat transfer coefficient and subsequently the rate of heat transfer. These are used with positive displacement compressors like reciprocating, rotary and screw compressors.

The uniform distribution of refrigerant in all the tubes is difficult to achieve, therefore a spray distributor is a must for this purpose. The number refrigerant passes is another important item in the performance of a direct expansion cooler. A single-pass cooler must evaporate all the liquid refrigerant, before it reaches the end of the tubes, for which one has to provide long tubes. A multi-pass cooler of perhaps a smaller size but uniform distribution is difficult to achieve after first pass.

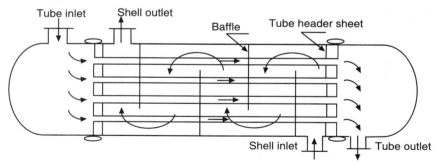

Figure 12.13 Shell-and-tube heat exchanger.

Another type of heat exchanger, say, a tube-in-tube (double pipe) cooler consists of one or more pairs of coaxial tubes. The refrigerant flows through the shell while the liquid is cooled through the tubes. The liquid side surface can be cleaned if access is provided through the header. The compact heat exchangers are also used but its construction features do not allow any cleaning.

12.7.2 Flooded Shell-and-Tube Coolers

In this type, the liquid being cooled flows through the tubes, which are immersed in the liquid refrigerant as shown in Figure 12.14. These are normally used with rotary screw or centrifugal compressors to cool water or brine.

The liquid refrigerant is fed from the bottom side while the refrigerant vapour separated at the upper side is sucked by the compressor. The velocity of the liquid being cooled is kept between 1 and 3 m/s inside the tubes. These are generally suitable only in the horizontal orientation.

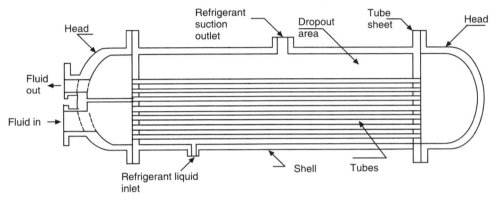

Figure 12.14 Flooded shell-and-tube cooler.

12.7.3 Baudelot Cooler

This cooler is used to cool liquids to a temperature very close to their freezing points. Such coolers are useful in milk chilling plants. In the evaporator, a number of columns of horizontal tubes or vertical plates are arranged over which the liquid being cooled is circulated by gravity. The liquid gathers in a collector tray at the bottom of the coil, from which it may be recirculated over the cooler or pumped to its destination in the industrial process.

12.7.4 Shell-and-Coil Cooler

It consists of a simple tank, filled with the liquid to be cooled while the refrigerant coil (tubing) is immersed in it (Figure 12.15). These units are suitable to cool drinking water, and also for use in bakeries and photographic laboratories. The main advantage lies in their easy access for cleaning purposes.

Heat transfer for liquid coolers can be calculated by the equation $Q = U \cdot A \cdot \Delta T_m$. The surface area A is evaluated if the geometry of the cooler is known. To determine U, readers may refer to a good book on heat transfer.

Pressure drop (Δp) is normally minimum in Baudelot and shell and coil heat exchanger but it is considerable in direct expansion and flooded coolers. Turbulent flow is maintained in both the direct expansion and flooded coolers to achieve high heat transfer coefficients, but these coolers experience a higher drop in pressure. The following equation is used to determine the pressure drop for the changed fluid flows.

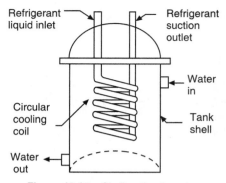

Figure 12.15 Shell-and-coil cooler.

$$\text{New } \Delta p = \text{original } \Delta p \left[\frac{\text{New flow}}{\text{original flow}} \right]^{1.8} \qquad (12.22)$$

The pressure vessels (heat exchanger) have to be tested as per the ISI test conditions to observe the safety measures.

12.8 CONDENSERS

Like the evaporator, the condenser is also a heat exchanger which rejects the heat of a refrigeration cycle. The heat rejected by this component consists of heat absorbed by the evaporator plus the energy input to the compressor. The compressor discharges the high pressure and high temperature gas to the condenser wherein the heat flows from the refrigerant gas through the condenser tube wall to the cooling medium which is water or air.

The temperature of the refrigerant during condensation remains constant while the temperature of the cooling medium increases from inlet to the outlet. This is represented by the flow diagram shown in Figure 12.16.

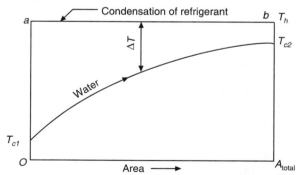

Figure 12.16 Temperature distribution in a single-pass condenser.

The condensers are classified into three groups:
- Water-cooled condenser
- Air-cooled condenser
- Evaporative (air–water cooled) condenser

12.8.1 Water-cooled Condensers

The heat rejection rate (Q_h) is determined from the known values of evaporator load (Q_L) and the energy input W_c to the compressor, i.e

$$Q_h = Q_L + W_c \qquad (12.23)$$

The values of Q_L, Q_h and W_c are in watts or kW.

The condenser load (heat rejection rate) Q_h depends on the operating conditions of the system. Work of compression varies with the compression ratio, type of compressor, type of refrigerant, etc. When all such details are not available, then the condenser load is determined as follows:

$$\text{Condenser load} = (\text{Evaporator capacity})(\text{Heat rejection factor}) \qquad (12.24)$$

The heat rejection factor is obtained from the curves given in Figure 12.17 for open compressors and from Figure 12.18 for hermetic compressors.

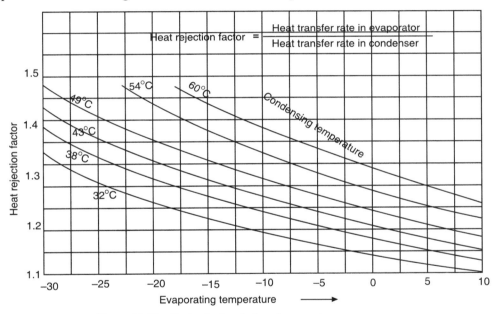

Figure 12.17 Heat rejection factors for open-type compressors.

EXAMPLE 12.1 Determine the heat rejection rate in the condenser of a refrigeration unit. The unit employs an open-type compressor having 8.8 kW capacity when operating at $-12°C$ suction temperature and $43°C$ condensing temperature.

Solution: The heat rejection factor = 1.28 as obtained from the curves given in Figure 12.17 for suction temperature $-12°C$ and condensing temperature $43°C$.

Heat rejection rate in condenser = $(8.8)(1.28) = 11.26$ kW **Ans.**

EXAMPLE 12.2 Calculate the condenser load for an evaporator having cooling capacity of 1.2 kW at $-23°C$ suction temperature and $38°C$ condensing temperature.

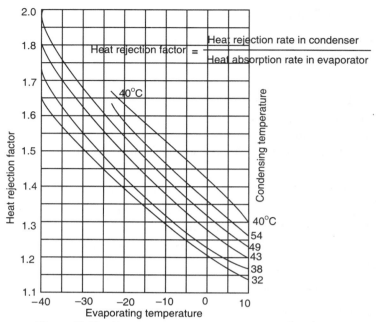

Figure 12.18 Heat rejection factors for hermetic compressors.

Solution: Heat rejection factor of 1.46 is obtained from the curves given in Figure 12.18 at −23°C suction temperature and 38°C condensing temperature.

The condenser load = (1.2)(1.46) = 1.75 kW **Ans.**

The volumetric flow rate of water to be circulated in the condenser may be found from the equation

$$V_w = 1000 \times Q_h / [\rho \cdot c_p(T_2 - T_1)] \qquad (12.25)$$

where V_w = volume flow rate of water in lit/s or kg/s, ρ = density of water (kg/m³), Q_h = heat rejection rate and T_1 and T_2 are the temperatures of water entering and leaving the condenser respectively, and c_p = specific heat of water J/kg-K.

Heat transfer

The heat rejection process in the condenser occurs in three stages: (i) First the cooling of superheat vapour to the saturated vapour condition takes place. (ii) Then, the condensation of refrigerant vapour to the saturated liquid state occurs. (iii) Finally, there is subcooling of the liquid refrigerant. In all the three processes, the heat transfer rates are different, but the heat transfer rate is very high during condensation compared to that in the other two processes. The average heat transfer rate obtained from Eq. (12.26) also gives satisfactory result.

$$Q = U \cdot A \cdot \Delta T_m \qquad (12.26)$$

The overall heat transfer coefficient based on outside or inside surface may be obtained referring to Section 12.2. The heat transfer coefficients on water side and on refrigerant side need to be calculated.

The basic correlations to calculate the heat transfer coefficient h_o for refrigerant vapour condensing on the outside of a circular tube are given in any good book on heat transfer.

Water pressure drop

The water being a liquid with large density causes a high pressure drop which is an important consideration for water-cooled condensers. Usually, the cooling tower has the accompanying water-cooled condensers. The pressure drop through the condenser should be lower than the available pressure provided by the pumps.

Therefore, to determine the pressure drop correctly one is required to refer to the Darcy-Weisbach equation. Predicting pressure drop for shell-and-coil condensers is more difficult than it is for shell-and-tube condensers because of their complex construction.

Water-cooled condenser types

The most common types of water-cooled refrigerant condensers are:

1. Shell-and-tube type
2. Shell and coil
3. Tube in tube
4. Brazed plate

The selection of a particular type condenser for applications depends on the following factors:

(i) Cooling load (condenser load)
(ii) The refrigerant used
(iii) The quality and temperature of available water
(iv) Location and space allotment
(v) The required operating pressure on refrigerant and on water side
(vi) Maintenance cost considerations

Shell-and-tube type condensers: These are similar in construction to that shown in Figure 12.13. The water is circulated through the tubes while the refrigerant condensers on the outside of the tubes (shell side). These are commercially available in sizes from 3.5 to 35000 kW. The inside tube surface is modified many a time to longitudinal or spiral grooves and ridges, internal fins or any other feature to promote turbulence and augment heat transfer. With such modifications, fouling, however, becomes relatively more.

Shell-and-coil condensers: These are as shown in Figure 12.15. The cooling water is circuited through the coils while the refrigerant will condense on the outside of the coil in the shell. These are available in sizes from 2 to 50 kW. Because of the nature of construction, the tubes are neither replaceable nor mechanically clearable.

Tube-in-tube condensers: The arrangement of tubes is similar to the one shown in Figure 12.15. The refrigerant vapour may condense either in the annular space or inside the tubes as per the connection of water flow and refrigerant. These are built in sizes from 1 to 180 kW. The tube-in-tube condenser design differs from that of the other two types of condensers. In this the refrigerant condensation is possible inside the tubes as well. Therefore, condensing coefficients are more difficult to predict for inside tubes. Normally, water flow is maintained through tubes for easy design and maintenance problems.

Brazed plate condensers: The plates are brazed in such a way that they form separate channels. These are built in sizes from 1.5 to 350 kW. The plates may be of stainless steel or other metals compatible with the refrigerant having fine smooth surfaces which help in keeping scaling to a low level.

Non-condensable gases

In the case of the refrigeration unit, the components assembled for the first time contain air and water vapour. Even as these are driven out by a good evacuation process, they still remain in the residual form. Air also enters the system through leaks, if any, in the evaporator since its pressure is below the atmospheric pressure. Water vapour may enter the system along with lubricants of hygroscopic nature. For example, the synthetic oils usable with R134a and R123 are hygroscopic in nature. When the entry of non-condensable gases in the system is not fully eliminated, a purge system, which automatically expels these gases, is to be provided. The ill effects of non-condensable gases are as follows:

 (i) Their presence increases the condenser pressure due to which the power consumption increases and the system capacity decreases.
 (ii) These gases form a resistance film over some of the condensing surface, thus lowering the heat transfer coefficient. It was found that for 2.89% of air by volume in a steam chest the heat transfer coefficient dropped from 11.4 to about 3.5 kW/m^2-K.
 (iii) If a purge system is provided, the loss of refrigerant occurs. This loss associated with the purge system is as shown in Figure 12.19.

Pressure vessels must be constructed and tested under the rules of national, state and local codes. The introduction of the current ASME boiler and pressure vessel code, section-VIII, gives guidance on rules and exemptions.

The ARI standard 450-87, the standard for the water-cooled refrigerant condensers of remote type, covers the industry criteria for standard equipment, standard safety provisions, marking fouling factors and recommended testing methods.

ASHRAE standard 22-92, methods of testing for rating water-cooled refrigerant condensers, covers the recommended testing methods.

The fouling of a condenser surface is unavoidable as the cooling water being circulated through, contains mineral solids which precipitate out of water and adhere to the surface of the condenser. This scaling adds extra resistance to heat flow through the tube walls. Therefore, both condenser temperature and pressure increase. Compressors working at this high pressure require more power. It means fouling or scaling deteriorates the performance of the condenser. Therefore, the condenser surface should be cleaned regularly keeping in mind the economic considerations.

The rate of tube fouling is affected by the following factors:

 (i) The state of purification of water used in respect of its content of mineral solid particles
 (ii) The condensing temperature
 (iii) The frequency of tube cleaning.

Many of the manufacturers of water-cooled condensers quote the condenser rating based on clean tubes. The scale formation rates for different qualities of water are available in standard heat transfer books (or may be obtained from the condenser books).

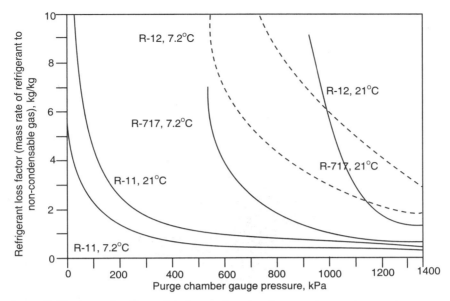

Figure 12.19 Loss of refrigerant during purging at various gas temperatures and pressures.

Rating and selection of water-cooled condenser

The heat rejection rate from the condensing refrigerant in water-cooled condensers is dependent on three parameters: condensing temperature, rise of water temperature and the fouling factor. Therefore, the rating is done based on these three parameters.

To select a suitable condenser from a manufacturer's catalogue, it is necessary to have the knowledge of following quantities:

- The heat rejection rate in kW
- Evaporator temperature
- Condensing temperature
- Water temperature entering the condenser
- Water temperature at the outlet of condenser
- Fouling factor or type of water

The water flow rate in each tube should not be less than 0.032 L/s so as to maintain turbulent flow. Also as per ARI standards, the water velocity should not exceed 2.6 m/s per tube for the rated condensers.

12.8.2 Air-cooled Condensers

The refrigerant gas enters the condenser in a superheated state. The process of condensation takes place in three stages—desuperheating of the gas to the saturated vapour state, the actual condensation of dry saturated vapour to the state of saturated liquid state and the subcooling of liquid.

All this heat is dissipated to the cooling medium, which is air in this case. Air circulation may be due to natural convection or by forced convection using a fan or blower.

As far as utilization of condenser area is concerned, 85% of the area is for the condensation process and the remaining 15% approximately is shared by desuperating and subcooling. This approximate pattern of area utilization is true for all refrigerant condensers. The most commonly followed temperature pattern is shown in Figure 12.20.

Natural convection condensers are useful only in small refrigeration units, say for example, a domestic refrigerator wherein the condenser is either a plate surface or finned tubing.

Forced convection air-cooled condensers are of two types—chassis mounted and remotely located.

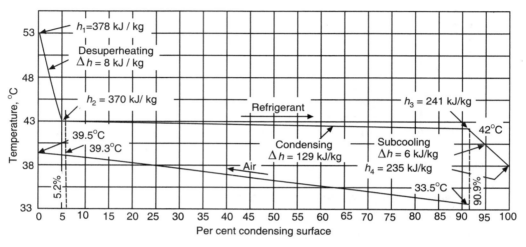

Figure 12.20 Temperature and enthalpy changes in air-cooled condensers.

A compressor, compressor driver and condenser—all are mounted on a common chassis, which forms a package type air-cooled condenser. One such example is the window air-conditioner unit. In other types, i.e. remote located, the condenser is placed away from the compressor so that it gets maximum benefit of the fresh outdoor air.

Similar to the water-cooled condenser, an air-cooled condenser also requires a definite quantity of air circulated over to carry away the heat. Usually, in air circulated by a fan the correct velocity of air reaching the condenser is very difficult to predict as the air is not channelized. In such cases the velocity of air passing through a condenser is a function of face area of the condenser and volume of air delivered by the fan. The air velocity is given by

$$\text{Air velocity (m/s)} = \frac{\text{Volume of air (m}^3\text{/s)}}{\text{Face area of condenser (m}^2\text{)}} \qquad (12.27)$$

The optimum velocity of air for a particular condenser is best found by experimentation. Therefore the manufacturers provide air-cooled condensers along with a built-in fan arrangement.

Heat transfer and pressure drop

The heat to be dissipated to the air by the condensing refrigerant is given by Eq. (12.26). The overall heat transfer coefficient U based on outside surface is given by Eq. (12.28).

Normally, all air-cooled condensers are provided with fins to enhance the surface area of heat transfer.

$$U_o = \frac{1}{\left(\dfrac{1}{\eta h_a} + \dfrac{B}{h_r}\right)} \qquad (12.28)$$

where η = fin effectiveness, h_a = heat transfer coefficient on the air side, h_r = heat transfer coefficient on the refrigerant side and B is the surface ratio ranging from 1.03 to 1.15 for bare-pipe coils and 10 to 30 for finned coils.

The values of h_a and h_r are very difficult to predict but for approximate analysis, the correlations given in the literature can be used.

Rating and selection of air-cooled condensers

Air-cooled condensers are rated in terms of total heat rejection (THR). This value of THR is the difference between the enthalpy of refrigerant entering the condenser and the enthalpy of liquid refrigerant at the outlet of the condenser.

A condenser is also rated in terms of net refrigeration effect (NRE). For open compressors, the THR is NRE plus power input to the compressor. For hermetic compressors, THR is the sum of NRE and power input to the compressor minus the heat losses from the compressor surfaces and discharge line.

THR is also equal to $UA\Delta T_m$. For constant values of U and A, THR is dependent only on the value of ΔT_m. U and A are decided at the time of manufacture. Since most of the condensers are built with fans or blowers, the quantity of air flowing over the condenser is also fixed. Therefore, the average temperature of air passing over the condenser depends only on the dry-bulb temperature (DBT) of the entering air. Therefore, the capacity of the condenser is directly dependent on TD (temperature difference between the entering air and the condensing refrigerant). Hence, we can conclude that the condenser rating is based only on TD.

Based on this TD, referring to the manufacturer's catalogue of condensers, the selection is made. For example, for a condensing temperature of 38°C, if the DBT is 28°C, the condenser can be selected for a TD of (38 – 28) 10°C, whereas if the DBT is 25°C, the condenser should be selected for a 13°C TD.

Control of air-cooled condensers

For proper function of a condenser, its pressure and temperature must be kept within certain limits. As discussed earlier, the increase in condensing temperature is following its increased pressure, for which the compressor power requirement will also increase. On the other hand, low condensing pressure hinders the flow of subcooled liquid refrigerant through the liquid feed device. Because of this, there will be starving of the evaporator, leading to the system trip-out at low evaporator pressure. This is possible especially in winter seasons, when the atmospheric ambient temperature is low. To avoid such low pressure failures, there must be a device that ensures flow of liquid refrigerant to evaporator.

To prevent excessive low pressure during winter operation, two control methods—refrigerant side control and air-side control—should be adopted.

The control on the refrigerant side is achieved by the following:

1. The active surface area of the condenser is modulated by flooding it with more liquid refrigerant by the storage of refrigerant. It requires a receiver and a large amount of refrigerant. The liquid refrigerant quantity in the condenser is controlled with the help of a valve arrangement. In this way, we can control the required temperature and pressure in the condenser.
2. The condenser surface is designed and constructed in two equal and parallel sections, each accommodating half of the load during normal summer operation.

The air-side may be controlled by one of these three methods: (i) cycling of fans, (ii) modulating dampers and (iii) fan speed control.

The fan cycling method is possible only if multiple fans are employed for circulating air over the condenser. In this method, one or more fans are switched off from the arranged fans as required in cold season. Less circulation of air means less condensation rate, which means high pressure and temperature, as required.

Dampers are used to control the air flow through the condenser coil—100% to zero per cent. By dampers, it is possible to control the condenser pressure as per the requirement.

12.8.3 Evaporative Condensers

This is essentially a combined effect of a condenser and a cooling tower. A diagram of a typical evaporative condenser is shown in Figure 12.21.

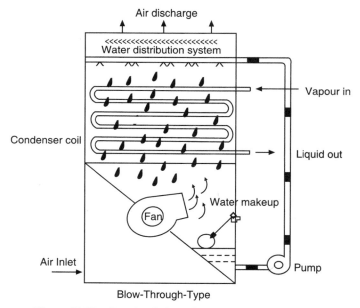

Figure 12.21 Functional view of an evaporative condenser.

An evaporative condenser consists of a condenser coil into which a high temperature and high pressure gas enters. Water is sprayed over this coil but in the opposite direction of the flow of water and air is circulated with the help of a fan or a blower.

The heat flows from the refrigerant vapour inside the condenser tubes to its outer surface. The water wetting this outer surface gets partly evaporated, absorbing heat from the coil surface, which causes cooling and condensation of the vapour refrigerant.

A water-cooled condenser requires a cooling tower and a water treatment arrangement to purify the water circulated in it. In an evaporative condenser, the same water is recirculated, so it requires less power and no cooling tower. Compared to an air-cooled condenser, the evaporative condenser requires less surface area and volume of air for the same heat rejection rate. Condensing temperature is also kept low, thus requiring less compressor power. The operation of an air-cooled condenser is limited due to DBT of the entering air while the operation is by WBT in the case of an evaporative condenser. The WBT is always lower than DBT by about 10–14°C.

Heat transfer

In an evaporative condenser, the heat transfer from the condensing refrigerant to water film and air outside the tubes occurs through the tube wall. Practically, the heat transfer process in an evaporative condenser is very complex in nature and is not discussed here. But in reality it is an evaporative cooling process.

The variation of temperature of water and air in a counterflow evaporative condenser is shown in Figure 12.22.

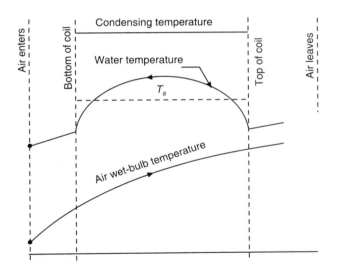

Figure 12.22 Temperature variation for an evaporative condenser.

Rating and selection of evaporative condensers

Heat rejected from the condensing refrigerant in the evaporative condenser depends upon the saturated condensing temperature and WBT of the entering air. The type of refrigerant and the superheat quality of refrigerant vapour entering the condenser affects this rate. So it requires some correction to be applied. Then selection is made referring to the manufacturer's catalogue.

EXERCISES

1. What are evaporators and condensers?
2. Explain the dry and flooded types of evaporators.
3. List the points to be considered while designing the evaporator.
4. Explain the different defrosting methods.
5. Explain liquid chillers.
6. Explain the different types of water-cooled condensers.
7. What is heat rejection factor? Discuss its significance.
8. Explain how the water-cooled condensers are rated and selected.
9. Explain how the air-cooled condensers are rated and selected.
10. Explain the working of evaporative condensers. How are they rated and selected?

Chapter 13

Expansion Devices

13.1 INTRODUCTION

An expansion device is one of the main components of the vapour compression refrigeration system. Though it is a more difficult component to understand, its basic function is extremely simple. The purpose of the expansion device is to reduce the pressure of the liquid refrigerant and to control the flow of refrigerant from high pressure side to the low pressure side of the system.

The proper control of refrigerant flow is also important for the following reasons:

1. The proper working of the evaporator depends upon the correct amount of liquid refrigerant and its flow pattern inside the evaporator tubes. Too much or too little refrigerant will provide less efficiency during the heat transfer process. To obtain the best heat transfer, the inside surface of the evaporator tubes must be completely wetted by the liquid, except in the last section of the evaporator. The last section is utilized to superheat the refrigerant.
2. The condition of the refrigerant at the outlet of evaporator must be in a superheat state, otherwise liquid droplets may enter the suction side of the compressor and cause damage to the compressor valves and bearings. This is generally known as *compressor slugging* and must be avoided.

An expansion device is one in which pressure drop occurs due to the flow through a restricted passage. Basically, there are two types of expansion devices—constant restriction type and variable restriction type.

The example of the constant restriction type expansion device is a capillary tube.

The following expansion or flow control devices fall under the category of variable restriction.

1. Hand-operated expansion device
2. Automatic expansion valve
3. Thermostatic expansion valve

4. Capillary tube
5. Low pressure float valve
6. High pressure float valve

There are some other flow control devices which are also important accessories to many systems. These are solenoid valves which are used in liquid line and gas line, and pressure regulating valves which are used to control evaporating pressure and condensing pressure.

13.2 HAND-OPERATED EXPANSION VALVE

The hand-operated expansion valve is one of the earliest types of metering devices (Figure 13.1). The flow through the valve is dependent on the pressure differential across the valve and the orifice opening. It consists of a needle nosed plunger that can be adjusted to regulate the flow through an orifice. The pressure drop across the valve is constant. Whenever the load changes, the valve must be readjusted, thus requiring the attention of an operator. When the system is shut down, the valve must be closed and opened again at the beginning of system start. This valve is seldom used but is suitable for ammonia systems.

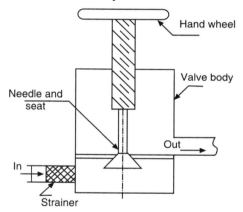

Figure 13.1 Hand expansion valve—Globe type.

13.3 AUTOMATIC EXPANSION VALVE (AEV)

The AEV is generally so termed because it opens and closes automatically without the aid of an external device. It maintains a constant refrigerant pressure in the evaporator. Therefore, its working is fully dependent on evaporator pressure but not on the load on it. The schematic diagram of AEV is shown in Figure 13.2. This type of valve consists of a diaphragm (bellow), a control spring and the basic valve needle and seat.

The control spring exerts a force to move the diaphram downwards to open the valve. But the evaporator pressure exerts a force on the diaphragm upwards to close the valve. These two forces oppose each other. The difference of these two forces will cause the particular diaphragm position. The position of the diaphragm causes the needle position to adjust and allows a definite rate of liquid refrigerant to flow to the evaporator. The spring force is constant as per the initial setting. The evaporator pressure is the only variable according to which the valve is required to function.

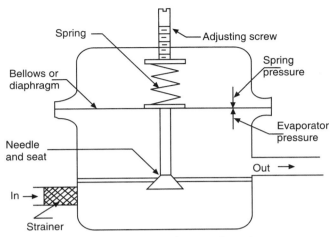

Figure 13.2 Automatic expansion valve.

The AEV is installed at the evaporator inlet as a device to control the flow of the refrigerant (Figure 13.3). The valve meters the refrigerant to maintain a constant evaporator pressure during the system operation.

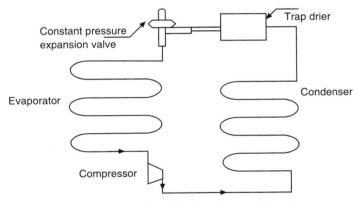

Figure 13.3 Location of automatic expansion valve.

During the off cycle, the valve closes slowly when pressure is on the evaporator side and exceeds the spring force. It remains in the closed condition till the system starts again. When the refrigeration system starts, the pressure in the evaporator starts falling and when this pressure is lower than that exerted by the control spring then the valve opens. It opens wide to meet the required flow of the system.

The tension of the control spring above the diaphragm can be increased or decreased by setting the adjusting screw. So the valve can be made to operate at a pressure lower or higher than the normal setting. If one needs to set the valve at any new evaporator pressure, then the refrigeration unit should be run for more than 24 hours. During this time the refrigerant and the oil in the system get distributed and the evaporator becomes cold enough. Valve setting can be adjusted to open at a predetermined pressure within the range of control spring.

Following are some features of AEV.

Protection against icing of evaporator: An automatic expansion valve maintains a constant low side pressure and, therefore, a constant evaporator temperature. Adjusting the valve for an evaporator temperature just above the freezing point of water will completely eliminate the possibility of frost formation regardless of the ambient temperature, heat load or the length of time the unit is operated. Therefore, AEV is the best choice for drinking water coolers, soda fountain water coolers, photo developing tanks and various industrial liquid chillers.

Motor overload protection: With AEV the low side pressure is constant and does not vary as the load fluctuates. Therefore, there are no variations in motor current requirement. During the high load condition too, AEV allows the refrigerant to flow at a fixed rate. Therefore, the motor cannot draw more current and is therefore protected.

Simplification of field service: The major problem in charging the capillary tube system lies in charging the correct quantity of refrigerant corresponding to the ambient temperature at the time of charging. An overcharged system with the use of capillary leads to operating difficulties. If an expansion valve system is overcharged, the valve will automatically adjust the flow of refrigerant to feed the evaporator properly. Also, the atmospheric temperature does not affect the working of an expansion valve equipped system.

An ideal valve for low starting torque motors: The expansion valve permits system unloading from high pressure side to low pressure side during the system off cycle. Therefore, motors require low starting torque.

The following factors affect the valve capacity:
   (i) Orifice size
   (ii) Range of needle movement
   (iii) Pressure difference across the valve
   (iv) Type of refrigerant
   (v) Condensing pressure or temperature
   (vi) Size of the bleed slot
   (vii) Evaporator temperature or pressure
   (viii) Liquid subcooling

The big disadvantage of the valves is its instability to increase flow with increased load. Actually, an increase in load will attempt to raise the suction pressure, causing the valve to move towards the closed position. A decrease in load will tend to open the valve.

13.4 THERMOSTATIC EXPANSION VALVE (TEV)

TEV is the most widely used type of metering device. The reasons are:
   (i) TEV is load-oriented. Thus, when the load is increased, the valve opens to increase the refrigerant flow.
   (ii) It maintains a constant degree of suction superheat at the evaporator outlet. This helps to keep the evaporator filled with refrigerant at all conditions of load.
   (iii) TEV protects the compressor by evaporating the liquid refrigerant before it leaves the evaporator.
   (iv) TEV can be used to feed an evaporator that has higher than average pressure drop. This is accomplished by the use of an external equalizer.

The schematic diagram of TEV is shown in Figure 13.4. Its main parts comprise (1) a needle and seat, (2) a pressure bellows or diaphragm, (3) a fluid-charged bulb mounted at the outlet of evaporator, which opens on one side of the diaphragm and (4) a spring, the tension of which is adjustable by the screw.

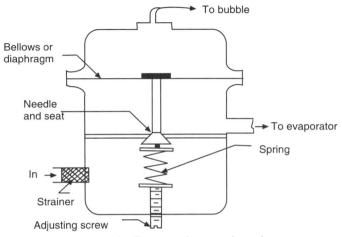

Figure 13.4 Thermostatic expansion valve.

The valve operates as a result of the balance of three pressures: (1) suction pressure, (2) spring pressure, and (3) bulb pressure. The saturated liquid-vapour mixture in the bulb exerts pressure on the top of the diaphragm. The other two pressures, suction and spring pressures, oppose it.

The actual function of the valve is to control the amount of superheat in the evaporator. With reference to Figure 13.5, assume that the evaporator pressure is 2.7 bar(g) (4°C suction temperature corresponding to 3.7 bar pressure for R134a). Usually the factory setting of the spring adjustment is to maintain 6°C of superheat. Therefore, the bulb pressure corresponding to 10°C (4 + 6 = 10°C) is 3.1 bar(g) for R134a. Hence, the spring pressure is 0.4 bar(g) (3.1 − 2.7). The spring pressure (0.4 bar(g)) plus the evaporator pressure (2.7 bar(g)) is equal to the bulb pressure (3.1 bar(g)), when the valve is in equilibrium. The valve will remain in equilibrium till a change in the degree of suction superheat unbalances the forces and causes the valve to move in one direction or the other.

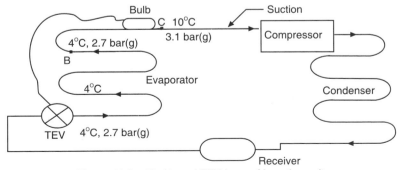

Figure 13.5 Working of TEV in a refrigeration unit.

Usually the bulb is charged with the refrigerant, which is same as that of the system. From Figure 13.5 it is clear that the refrigerant temperature from the inlet of evaporator and up to point B is 4°C (2.7 bar(g)). As the refrigerant flows from point B to C, its temperature rises to 10°C (3.1 bar(g)). For the above analysis, pressure drop in the evaporator is assumed to be zero.

To understand the working of TEV, suppose the load on the system increases. It means the degree of superheat becomes greater than 6°C, the pressure in the remote bulb exceeds the combined evaporator and spring pressures and the valve opens wide, thereby increasing the flow of liquid into the evaporator until the superheat reduces to 6°C. On the other hand, if the load on the system is decreased, then the degree of superheat falls below 6°C. At the same time, the pressure in the bulb will be less than the combined evaporator and spring pressures. The resultant force tends to close the valve, reducing the valve opening. The refrigerant flow is controlled until the degree of superheat becomes 6°C. In this way, the cycle of operation repeats.

The degree of superheat can be varied by adjustment of the screw. The spring adjustment is referred to as 'superheat adjustment'. Increasing the spring tension increases the degree of superheat and decreasing the spring tension decreases the superheat.

The automatic expansion valve maintains evaporator temperature and pressure at constant values irrespective of the load, while the TEV maintains suction superheat at a constant value as per the load variations irrespective of the evaporator pressure and temperature.

13.4.1 External Equalizer

While studying TEV, the inlet pressure of evaporator p_o was assumed constant throughout the length of the evaporator. It means pressure drop in the evaporator was totally neglected. But practically there occurs a significant pressure drop (Δp) in the evaporator due to its long length and number of bends as indicated in Figure 13.6. Therefore, the saturation temperature of refrigerant corresponding to this lower evaporator pressure ($p_{o1} = p_o - p$) at the outlet must be lower compared to evaporator inlet. Now in such a situation the refrigerant requires more superheat to raise its low temperature (T_{o1}) to the required degree of superheat, for which a greater length of evaporator coil at the outlet must be filled with vapour. This is nothing but poor utilization of evaporator surface. To improve utilization of evaporator surface for better cooling purposes, the remedy is an external equalizer connection.

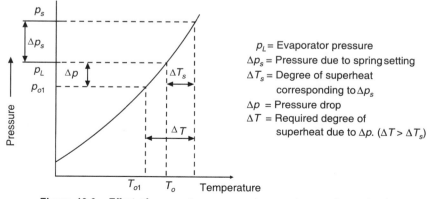

Figure 13.6 Effect of evaporator pressure drop on degree of superheat.

A TEV with an external equalizer connection is shown in Figure 13.7. A small tube connection from the suction line to the thermostatic expansion valve is called an external-equalizer connection. Such a connection transmits the pressure at the outlet of the evaporator to the inside of the diaphragm of TEV. The space between the two diaphragms (Figure 13.7) is in no way connected to the inlet of evaporator. Thus, the pressure between the two diaphragms is always equal to the pressure at the evaporator outlet irrespective of the value of pressure drop in the evaporator tubing. Now the pressure to counteract the remote bulb pressure on the diaphragm is spring pressure plus pressure at the outlet of evaporator.

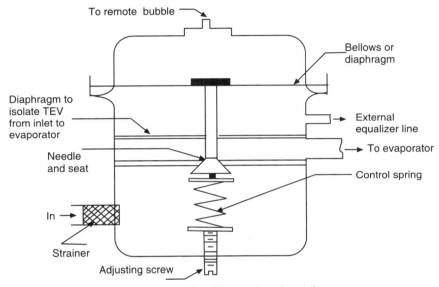

Figure 13.7 TEV with an external equalizer.

Many a time the evaporator is provided with a distributor to assure uniform refrigerant feeding to all parts of the coil. The use of TEV with an external equalizer connection is shown in Figure 13.8. The figure shows that the evaporator carries a refrigerant distributor and a suction header.

Although an external equalizer does not reduce the evaporator pressure drop in any way, it permits the full and effective use of the evaporator surface.

Following are some disadvantages of the TEV. (1) It allows to overload the compressor driver because of excessive evaporator pressure and temperatures during the periods of heavy loading. (2) It opens wide and overfeeds the evaporator when the compressor is started. Further due to this overfeeding, liquid refrigerant may enter the compressor causing damage to it.

These disadvantages are because of the fact that the TEV limits the suction superheat and not the evaporator pressure. These operating difficulties can be overcome by limiting the evaporator pressure. The TEV has an in-built pressure limiting device. When the evaporator pressure rises above the predetermined maximum, then an in-built pressure-limiting device of TEV acts to throttle the flow of liquid to the evaporator and limits its pressure. The maximum operating pressure (MOP) of TEV is limited either by mechanical means or by the gas-charged remote bulb.

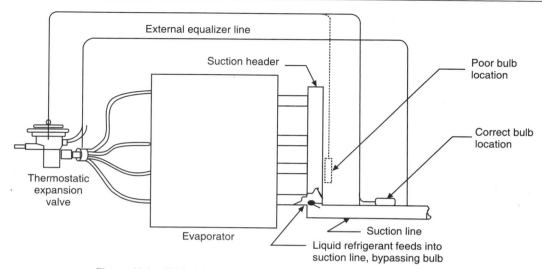

Figure 13.8 TEV with an external equalizer connected to an evaporator.

13.4.2 Gas-charged TEV

Gas-charged TEV's name itself suggests that the bulb is charged with a different gas. The bulb is also charged with a limited quantity of system refrigerant. The quantity charged in the bulb is such that it will completely vaporize at the predetermined temperature. Once the charge in the bulb is in saturated vapour state, its pressure does not increase with an increase in the temperature. The pressure exerted by the vapour of the bulb on the diaphragm is always constant and is equal to spring pressure + evaporator pressure. Again spring pressure is also constant, therefore, the working of TEV is directly dependent on the evaporator pressure. At the same time it also maintains a constant superheat condition at the suction line. Whenever the evaporator pressure exceeds the maximum operating pressure (MOP), then the TEV will be closed stopping the flow of liquid to the evaporator.

Here, we can say indirectly that the pressure of evaporator is limited by the pressure of the charge in the bulb. Any change in the spring (superheat) setting will also change the MOP. The bulb pressure is the sum of spring pressure plus evaporator pressure, and increasing the superheat (spring) setting will decrease the MOP. Similarly for decreasing the superheat, we have to increase the MOP of the valve.

Gas-charged valves must be carefully applied so that TEV should not become a pressure control valve losing the advantages of bulb control. The TEV should be located at a warm place compared to the bulb. If the diaphragm chamber becomes colder than the bulb, the small amount of charge in the element may condense and the bulb may contain only gas. In such a case the valve will throttle or close.

13.4.3 Refrigerant-charged Expansion Valves

A thermostatic expansion valve having its bulb charged with the system refrigerant is suitable for medium and high evaporator temperature applications. For a sub-zero evaporator temperature, the bulb requires charging with a different fluid. The reason is pressure–temperature variation of the system refrigerant as shown in Figure 13.9.

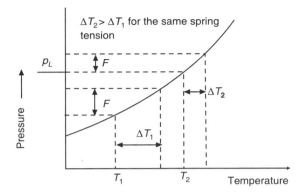

Figure 13.9 Increase in superheat at lower evaporator temperature.

We know that the degree of superheat in the suction line is adjusted by adjusting the tension of control spring of TEV. Let this tension of spring be F at evaporator temperature T_1 corresponding to a degree of superheat temperature T_1. Now let the refrigeration unit operate at a new lower evaporator temperature T_2. For the same value of spring tension F, the degree of superheat required will be increased to T_2 ($T_2 > T_1$). Now if the system refrigerant is charged in the bulb for the evaporator operating at T_2, a considerable portion of the evaporator is not effective since a large area is required for superheating the gas. For this reason, a fluid for the bulb will be so selected that its degree of superheat remains unaffected by a change in the evaporator temperature. Expansion valves whose bulbs are charged with fluids other than the system are called *cross-charged* valves. The fluid charged into the bulb other than the system refrigerant is called *power fluid*. The pressure–temperature curve of the power fluid crosses the pressure–temperature curve of the system refrigerant as shown in Figure 13.10.

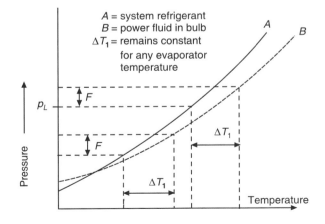

Figure 13.10 Maintaining uniform superheat by the use of cross-charged expansion valve.

The cross-charged valves during pull-down permit a large temperature change with a small amount of change in diaphragm pressure, improving the pull-down feature of the valve, thus, providing some limitation on the overload characteristics of the valve. The advantages include:
- Moderately slow pull down.
- Insensitivity to cross ambient control.
- Dampened response to suction line temperature changes, thus minimizing 'hunting'.
- Superheat characteristics that can be tailored for specific applications.

13.4.4 Behaviour of Charge in the Bulb

At this stage we know that the bulb can be charged with the system refrigerant or a power fluid or even quantity of which is decided precisely. We have also studied why one has to go for a power fluid (cross-charged TEV) and what its advantages are. But we did not discuss the exact behaviour of fluid in the bulb which is necessary to understand. For the purpose, consider the p-v diagram of the power fluid and indicate different states of the same as shown in Figure 13.11.

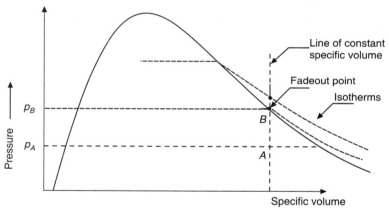

Figure 13.11 Fadeout point of power fluid.

The volume of the power fluid in the bulb, capillary passage from bulb to diaphragm and some space above the diaphragm remain constant throughout the working. The heating and cooling of the power fluid occurs at constant volume and lies on the vertical line.

At point A the pressure of the power fluid is corresponding to its temperature (mixture of liquid + vapour). As the temperature of the fluid in the bulb increases from point A to B, its pressure also increases to p_B. Point B is on the saturation curve. Further heating of the fluid increases the pressure very slightly because at point B, all the liquid is converted into vapour. Therefore, the point B is called the *fadeout point*. The pressure exerted by the fluid corresponding to point B is almost the maximum pressure. For this fadeout point (maximum pressure) the valve opening will be maximum. It means that the pressure of the power fluid does not increase significantly beyond B, the valve opens no wider and thereby limits the pressure in the evaporator to the 'maximum operating pressure'.

It is very necessary to limit the MOP of the evaporator especially during the pull-down period. Otherwise the motor of the compressor will draw more current to meet the heavy load, which is undesirable. It is necessary to guard the motor against overloading during peak load.

For this purpose, suction pressure should be limited which is possible by charging a precise amount of power fluid to the bulb. This, of course, limits the suction pressure to fadeout point.

13.4.5 Rating and Selection of TEV

The first step in the selection is to decide which type of valve is required based on the nature of application. For type selection the factors considered are:

(1) Pressure limits, (2) possible need of an external equalizer, (3) bulb charge, and (4) size of the valve inlet and outlet connections.

After deciding a particular type of valve, refer to the manufacturer's catalogue ratings. The valves are rated based on the refrigerating capacity corresponding to a condensing temperature of 38°C with zero degree subcooling.

In order to select the proper size of the valve, the following data are required.

1. The evaporator temperature
2. Refrigeration capacity, kW
3. Available pressure difference across the valve

These data help to decide the required flow rate and orifice size. The flow rate is proportional to the pressure difference across the valve. Therefore, a correct estimation of pressure difference across the valve is very necessary. Otherwise it may lead to improper selection of valve. With this correct value of pressure difference, select a valve from a manufacturer's rating table, which has a capacity equal to or slightly greater than the system capacity at the desired operating conditions.

13.5 CAPILLARY TUBE

The capillary tube is the simplest type of refrigerant flow control device used in modern refrigeration systems. However, its application is limited to a single evaporator system. The capillary tube is a tube 0.6 m to 6.0 m long with an inside diameter ranging from 0.6 mm to 2.5 mm. The capillary tube acts in the same manner as a small diameter water pipe, which holds water back, allowing a higher pressure to be built up behind the water column with only a small rate of flow. In a similar manner, the small diameter capillary tube holds back the liquid refrigerant enabling a high pressure to be built up in the condensing unit during operation of the unit, and at the same time permitting the liquid refrigerant to flow slowly into the evaporator. As the liquid refrigerant flows through the tube, the pressure drops because of friction and acceleration of the refrigerant. Some of the liquid flashes into vapour as the refrigerant flows through the tube.

Since the orifice is fixed, the rate of feed is relatively not flexible. Under conditions of constant load and constant discharge and suction pressures, the capillary tube performs satisfactorily. However, any change in the evaporator load or fluctuations in head pressure can result in underfeeding or overfeeding of the evaporator with the refrigerant.

To obtain the desired pressure drop and flow, numerous combinations of capillary bore and length are possible. A specific choice of capillary tube with its bore and length is made and installed in the refrigeration system. Once this is done, the tube cannot adjust itself to variations in discharge pressure, suction pressure and load. For balanced operation of the system, it is necessary that the compressor pumps the refrigerant at the same rate at which the capillary tube feeds the evaporator. An unbalanced condition, if any, should be of a short duration. When the system remains in an unbalanced condition for a long duration, then the two undesirable effects are as follows: Either the capillary tube overfeeds the evaporator and further entry of liquid into suction may result into slugging of the compressors. Or the capillary underfeeds the evaporator, resulting in poor efficiency, due to its starvring.

Let us consider the performance curves (suction pressure versus mass flow rate) of a reciprocating compressor superimposed on the characteristic curves of a capillary tube (mass flow rate versus suction pressure) that will give the balance points of operation. These are represented in Figure 13.12.

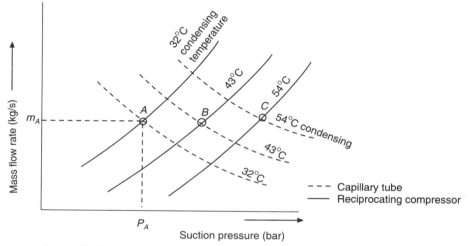

Figure 13.12 Balance points with reciprocating compressor and capillary tube.

The rate of refrigerant flow through the capillary tube depends on the pressure difference between the condenser and suction pressure. This pressure difference varies according to the changes either in condenser pressure or in suction pressure. For constant condenser pressure, the increase in suction pressure reduces the pressure difference. Therefore, the flow rate through the capillary tube decreases. Such variations for constant condenser temperature (pressure) are shown in Figure 13.12 by dotted lines. It is shown that the mass flow rate of a compressor increases with increase in suction pressure and is shown by full lines. Points A, B and C are the balance points corresponding to their fixed condensing temperatures. What does it mean? It means, for example, that the point A corresponds to a condensing temperature 32°C, the capillary tube must allow a mass flow rate of m_A so that the suction pressure is kept to p_A. Moreover, at this suction pressure the rate of heat transfer through the evaporator is such that all the refrigerant entering in it should be evaporated. Otherwise, any variation of heat transfer rate through the evaporator may result into an imbalance even though the balance point between the capillary tube and the compressor is achieved.

As stated earlier, the imbalance results in overfeeding or underfeeding of the refrigerant to the evaporator. Now to understand this clearly, study Figure 13.13, where the point A represents a balance point.

When does flooding occur? To answer this, consider the balance point A at which the capillary tube is feeding the evaporator at the rate of flow of m_A. The evaporator load is

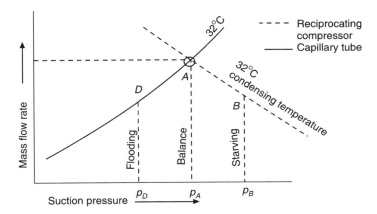

Figure 13.13 Unbalanced condition.

supposed to be just sufficient to evaporate this refrigerant rate m_A and the compressor pumping is also at the same rate. Now suppose that the load on the evaporator is reduced (by removing the stored items from the refrigerated space) but the capillary is still feeding at the previous rate. Therefore, all this liquid will not be evaporated, thus accumulating some liquid in the evaporator coil. Thus, flooding of the evaporator results. It is continued for a long time in which case the liquid may enter the suction of compressor damaging the valves and lubrication. Such working of the compressor is termed sluggish operation. Therefore, the refrigeration unit must be precisely charged so that no more accumulation of the liquid in the evaporator takes place.

For the reduced load, system balances at point D as indicated in Figure 13.13, which falls in the vapour mixture area. This indicates that the entry of refrigerant to the capillary tube is a mixture of liquid and vapour. Because of this, the flow rate through the capillary tube decreases. The refrigerating effect q_o is less than the subcooled liquid condition, even though the compressor work is unchanged.

Let us see when does the starving of the evaporator occur? The answer is simple. When the evaporator temperature (pressure) increases, then there is a rise of load on it. This high load condition is indicated by point D in Figure 13.14. In order to reduce the rate of gas flashing in the capillary, it is attached (bounded) to the suction line.

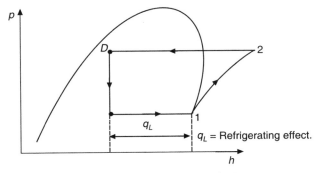

Figure 13.14 Reduction of refrigerating effect when vapour enters the capillary tubes.

At the high load condition (point B) the capillary tube feeds at the normal rate (which is insufficient to meet the high load). At suction pressure, point B, the compressor draws more refrigerant mass from the evaporator than evaporator receives from capillary. Therefore, the evaporator soon becomes short of refrigerant, i.e. starving. Such starving of evaporator may not continue for a long time, and so some corrective action must take place to strike a balance condition. The corrective condition, which normally occurs, is that liquid backs up into the condenser. Therefore, condenser surface area is reduced which raises the condenser pressure. With elevated condenser pressure, the compressor capillary is reduced and the capillary rate of feed to the evaporator is increased until the balance is restored. Another possibility for regaining a balanced condition (flow) is that the heat transfers coefficient in the starved evaporator decreases. A greater temperature difference must develop between the fluid being cooled and the refrigerant in the evaporator. This greater temperature difference occurs when the suction pressure drops back to pressure point A and restores the balanced flow.

Following are a few advantages of capillary tube.
- It is simple, non-moving parts and inexpensive.
- During off cycle, the liquid from condenser side flows to evaporator side through capillary tube, equalizing the pressure. Therefore, motor requires low starting torque.

The disadvantages are:
- Orifice of the capillary tube is fixed, thus suitable for constant load and constant discharge and suction pressure. Under fluctuating conditions, its performance is poor due to starving of evaporator and even results into sluggish compressor operation.
- Because of the small bore, it is essential that the system be kept free from dirt and foreign matter. Usually a filter is placed before the capillary tube. It is not suitable for open type systems.
- Application is limited to only small units with single evaporator.

13.5.1 Selection of Capillary Tube

A designer is required to decide the bore as the well as the length of the capillary tube for a new refrigerating unit. The capillary tube selected must give a balanced flow for the evaporator and compressor operation; depending upon the condenser pressure, suction pressure and refrigerating capacity.

13.6 FLOAT VALVES

There are two types of float valves used to control the refrigerant flow to the flooded evaporators. They are—low side float valve, and high side float valve.

These valves have a main component called a *float*. It is hollow spherical in shape and is made of metal or plastic. It senses the liquid level and acts to open or close the valve to admit either more or less refrigerant into the evaporator. These valves may be in operation continuously so as to admit the liquid refrigerant in accordance with the refrigerant evaporation. Refrigerant evaporation will be as per the load variation on the evaporator. The liquid level in the flooded evaporator is so maintained that the inside surface of the evaporator coil is always wetted with liquid. As soon as the liquid refrigerant becomes vapour, it will be drawn by the compressor. Therefore, evaporator capacity is better utilized and performance is also better. There are many reasons for this which were discussed in the earlier part of this chapter. Let us study the two float valves.

13.6.1 Low Pressure Float Valve

This valve acts to maintain a constant level of liquid in the evaporator. To ensure that it controls the refrigerant flow to the evaporator, refrigerant flow is in accordance with the rate of evaporation of refrigerant in the evaporator. The valve may operate either continuously or intermittently. The thing is that it maintains a constant level of liquid irrespective of the evaporator pressure and temperature.

In continuous operation of float valve, it regulates the flow so that the liquid level is maintained at all times. To do so it responses to very small changes of liquid level, while in intermittent type, the valve responds to only the minimum and maximum liquid levels. At these conditions, the valve will be either in fully opened or fully closed condition.

The low pressure float may be installed directly in the evaporator or accumulator in which it controls the liquid levels. Many a time it may be installed external to these units in a separate float chamber (Figure 13.15).

In Figure 13.15 a bypass line with hand-expansion valve is provided as part of safety precaution at the time of failure of float valve. This arrangement is very necessary for large refrigeration units. In the said figure one can see that there are two expansion valves, one on each side of the float valve. At a time, only one will be in operation and another may be idle or it may be in repairing or servicing condition.

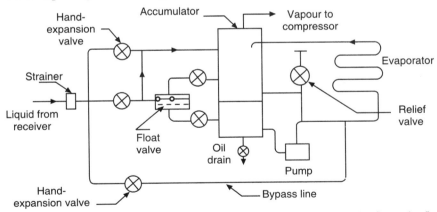

Figure 13.15 Low side float valve regulates the flow of liquid from the receiver. In the figure the flow maintains a constant level in the accumulator. The pump circulates liquid through the evaporator.

13.6.2 High Pressure Float Valve

The principle of high pressure float valve is similar to the low pressure float valve, i.e. it controls the refrigerant flow to the evaporator as per the load variation. The only difference is that it is located on the high pressure side of the system in a separate float chamber as indicated in

Figure 13.16. It controls the liquid level in this chamber. A weight valve is also provided after the float chamber but before the evaporator to present flash gas in the liquid line. This is necessary only when the float chamber is located at a considerable distance from the evaporator.

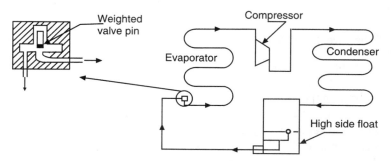

Figure 13.16 High side float valve. Float valve regulates flow into the evaporator at the rate refrigerant is condensed. The weighted valve prevents flash gas in the liquid line.

The working principle of this valve is very simple. The rate of condensation of refrigerant in the condenser is usually constant. Therefore, the condensate flows to the float chamber because of which the level of liquid in it rises and subsequently the float valve opens. The refrigerant flows through the opening to the evaporator. In this way only a small quantity of liquid refrigerant remains on the high pressure side, i.e. in float chamber and the rest is in evaporator.

This valve can be used either on a dry expansion system or on a flooded system. When the compressor stops, the high side float valve will also close, thus stopping the float of refrigerant to the evaporator. The high side float valve automatically and indirectly maintains a constant level of refrigerant in the evaporator.

EXERCISES

1. Explain the functions of expansion valve and name the expansion valves.
2. Explain with a neat sketch the hand-operated expansion valve.
3. Explain with a neat sketch the automatic expansion valve. Enlist the factors affecting the valve capacity.
4. Explain the working principle of thermostatic expansion valve with a neat sketch.
5. Why is an external equalizer required? Explain how does it work with TEV.
6. Write a short note on capillary tube.
7. Explain the procedure to get a balance point with the help of capillary tube.
8. Explain the working of float valves with the help of examples.

Chapter 14

Refrigerant Piping, Accessories and System Practices

14.1 INTRODUCTION

The successful operation of refrigeration plants is also dependent on its piping design and proper layout, etc. All refrigerant piping should be installed to provide the following functions.

1. A leak-proof path for the refrigerant to follow.
2. Proper feed of the refrigerant to the evaporator.
3. A proper oil return path to the compressor.
4. A properly-sized piping design to permit full capacity and highest efficiency of the system. To keep the pressure drop to a low level.
5. Prevent excessive amounts of lubricating oil from being trapped in any part of the system.
6. Prevent liquid refrigerant or oil slugs from entering the compressor during operation and idle time.
7. Maintain a clean and dry system.

14.2 MATERIALS

Minimum requirements for refrigerant piping in respect of material are given in the safety codes of BIS standards which must be followed. Piping materials may be steel, copper or steel pipes with coatings. The material selected must conform to the refrigerant used, the size of the job and the nature of the application.

Copper is suitable for halogeneted hydrocarbon refrigerants. Copper is not suitable for ammonia since in the presence of water, ammonia attacks nonferrous materials. Copper is either hard drawn or soft temper. It is measured in outside dimensions (OD) and inside dimensions (ID).

The copper joints are made by brazing and have the melting points above 600°C. Welding is the most common way of joining steel or iron piping. All piping must be properly supported with hangers or brackets. When piping passes through a wall or floor, a suitable sleeve must be provided. The noise and vibration have many undesirable effects but the important ones are: breaking of brazed joints which may result into loss of charge, transmission of noise through piping and building. So vibration eliminators need to be provided. On long runs of tubing, allowance must be provided in the piping for expansion and contraction.

Occasionally, discharge tubing will vibrate due to the pulsations in it caused by the reciprocating type compressor. This movement can break hangers on a discharge line to an air-cooled or evaporative condenser. Usually this situation is corrected by the use of discharge mufflers in the compressor.

14.3 SYSTEM PRACTICE FOR HCFC SYSTEMS REFRIGERANT LINE SIZING

The sizes of refrigerant lines should be kept as small as possible due to cost considerations. But many a time, it may not be a criterion for the selection of a proper size as far as the efficiency of the unit is concerned. Suction and discharge line pressure drops cause loss of compressor capacity and increase the power consumption. Excessive pressure drop may cause the flashing of liquid in the liquid line, resulting into faulty operation of the expansion valve. Therefore, line sizing is always done to keep the pressure drop to a minimum. Refrigerating line sizing involves liquid line, suction and discharge lines.

14.3.1 Pressure Drop

The pressure drop in a pipeline for single-phase flow is calculated by the Darcy's formula

$$\Delta p = \frac{\rho . f . L v^2}{2D} \tag{14.1}$$

where Δp is the pressure drop (N/m^2), ρ the density (kg/m^3), L the length (m), D the diameter (m), v the velocity (m/s) and f is the friction factor. For the friction factor in suction and discharge lines, a value of 0.003 may be taken, whereas for liquid lines, it may be taken as 0.004. These are approximate values. The accurate values may be obtained considering tube material, surface roughness, type of flow, etc. by referring to good books on fluid mechanics.

It is a usual practice to evaluate the pressure drop in terms of the number of degree change in the saturation temperature due to this loss.

14.3.2 Sizing of Liquid Lines

The pressure drop in liquid lines should not exceed the value corresponding to 0.5 K and 1 K. The liquid pressure losses for a change of 0.5 K in saturation temperature at 40°C condensing temperature are given in Table 14.1.

Table 14.1 Pressure drops

Refrigerant	Pressure drop (kPa)
R12	11.7
R134a	13.3
R123	2.5
R22	18.6
R502	19.4

Following are a few design considerations for liquid lines.

(i) Liquid lines should be designed so that a slightly subcooled liquid reaches the liquid feed device to prevent the formation of flash gas. The flash gas in the liquid causes an increase in the pressure drop and further flashing may lead to reduction in the capacity of the expansion device and very irregular control of liquid refrigerant entering the evaporator.

For sizing of liquid lines, suction lines and discharge lines, the data in Table 14.2 and Table 14.4 may be used.

Table 14.2 Suction, discharge and liquid line capacities in kW for Refrigerant 22 (single- or high-stage application)

Nominal line size (mm)	Suction lines ($\Delta T = 0.04$ K/m)					Discharge lines ($\Delta T = 0.02$ K/m, $\Delta p = 74.90$)			Liquid lines	
	Saturated suction temperature (°C)					Saturated suction temp. (°C)			Velocity = 0.5 m/s	$\Delta T = 0.02$ K/m
	−40	−30	−20	−5	+5	−40	−20	+5		
	Corresponding Δp (Pa/m)									$\Delta p = 74.9$
	196	277	378	572	731					
COPPER LINE										
12	0.32	0.50	0.75	1.28	1.76	2.30	2.44	2.60	7.08	11.24
15	0.61	0.95	1.43	2.45	3.37	4.37	4.65	4.95	11.49	21.54
18	1.06	1.66	2.49	4.26	5.85	7.59	8.06	8.59	17.41	37.49
22	1.88	2.93	4.39	7.51	10.31	13.32	14.15	15.07	26.66	66.18
28	3.73	5.82	8.71	14.83	20.34	26.24	27.89	29.70	44.57	131.0
35	6.87	10.70	15.99	27.22	37.31	48.03	51.05	54.37	70.52	240.7
42	11.44	17.80	26.56	45.17	61.84	79.50	84.52	90.00	103.4	399.3
54	22.81	35.49	52.81	89.69	122.7	157.3	167.2	178.1	174.1	794.2
67	40.81	63.34	94.08	159.5	218.3	279.4	297.0	316.3	269.9	1415.0
79	63.34	98.13	145.9	247.2	337.9	431.3	458.5	488.2	376.5	2190.9
105	136.0	210.3	312.2	527.8	721.9	919.7	977.6	1041.0	672.0	4697.0
STEEL LINE										
10	0.47	0.72	1.06	1.78	2.42	3.04	3.23	3.44	10.66	15.96
15	0.88	1.35	1.98	3.30	4.48	5.62	5.97	6.36	16.98	29.62
20	1.86	2.84	4.17	6.95	9.44	11.80	12.55	13.36	29.79	62.55
25	3.52	5.37	7.87	13.11	17.82	22.29	23.70	25.24	48.19	118.2
32	7.31	11.12	16.27	27.11	36.79	46.04	48.94	52.11	83.56	244.4
40	10.98	16.71	24.45	40.67	55.21	68.96	73.31	78.07	113.7	366.6
50	21.21	32.23	47.19	78.51	106.4	132.9	141.3	150.5	187.5	707.5
65	33.84	51.44	75.19	124.8	169.5	211.4	224.7	239.3	267.3	1127.3
80	59.88	90.95	132.8	220.8	299.5	373.6	397.1	422.9	412.7	1991.3
100	122.3	185.6	270.7	450.1	610.6	761.7	809.7	862.2	711.2	4063.2

Notes:

1. Table capacities are in kilowatts of refrigeration.

   Δp = pressure drop due to line friction (Pa/m)
   ΔT = change in saturation temperature (K/m)

2. Line capacity for other saturation temperatures ΔT and equivalent lengths L_e

   $$\text{Line capacity} = \text{Table capacity} \left(\frac{\text{Table } L_e}{\text{Actual } L_e} \times \frac{\text{Actual } \Delta T}{\text{Table } \Delta t} \right)^{0.55}$$

3. Saturation temperature ΔT for other capacities and equivalent lengths L_e

   $$\Delta T = \text{Table } \Delta T \, \frac{\text{Actual } L_e}{\text{Table } L_e} \left(\frac{\text{Actual capacity}}{\text{Table capacity}} \right)^{1.8}$$

4. Values in the table are based on 40°C condensing temperature. We have to multiply the table capacities by the following factors for other condensing temperatures:

Condensing temperature	Suction lines	Discharge lines
20	1.18	0.80
30	1.10	0.88
40	1.00	1.00
50	0.91	1.11

EXAMPLE 14.1 A 140 kW R22 refrigeration unit operates at 40°C condensing temperature and 5°C suction temperature. The equivalent length of the liquid line including fittings and accessories is 30 m. Determine the size and pressure drop equivalent (in degrees) for the liquid line. Also find the subcooling required.

Solution:

(a) From Table 14.2, 42 mm OD copper tube can provide a capacity of 103.4 kW. And 54 mm OD copper tube can provide a capacity of 174.1 kW. Both these are based on a pressure drop equivalent to 0.02 K/m. Therefore, 54 mm OD copper tube fulfils the requirement. **Ans.**

(b) The actual pressure loss in degrees

$$\Delta T = 0.02 \times 30 \left(\frac{140}{174.1} \right)^{1.8} = 0.4 \text{ K}$$

Since 0.4 K is below the recommended 1 K, let us recompute ΔT for the smaller tube (42 mm) for which $\Delta T > 1.0$ K. As this temperature drop exceeds 1 K, therefore, the 54 mm tube is recommended. **Ans.**

(c) The pressure drop = 74.9 Pa/m from Table 14.2.

For 30 m, $\Delta p = 74.9 \times 30 = 0.224$ bar.

The pressure at the condenser corresponding to 40°C is 15.34 bar.

The pressure at the refrigerant control is 15.1 bar (15.34 − 0.224) which corresponds to a saturation temperature of 39°C. The amount of subcooling required is therefore approximately 1°C. **Ans.**

14.3.3 Sizing of Suction Lines

Sizing of the suction line is more critical than any other refrigerant line. An undersize pipe will cause excessive pressure drop in the suction line. So the compressor has to operate at relatively low suction pressure increasing its power input. An oversize pipeline will result into a low refrigerant velocity which will be inadequate to return oil from the evaporator to the compressor. Suction lines are designed: (1) To keep the pressure drop from friction not greater than the equivalent of 1 K change in saturation temperature. The equivalent pressure loss at 5°C saturated suction temperature is in Table 14.3. (2) To return the oil to the compressor effectively, especially in haloginated hydrocarbon refrigerants.

Table 14.3 The pressure loss

Refrigerant	Suction loss (K)	Pressure loss
R12	1	11.3 kPa
R134a	1	12.23 kPa
R123	1	1.75 kPa
R22	1	18.1 kPa
R502	1	19.7 kPa

At suction temperatures lower than 5°C, the pressure drop equivalent to a given degree change decreases. For example, at –30°C suction with R22, the pressure drop equivalent to 1 K change in saturation temperature is about 6.5 kPa. Therefore, low temperature lines need to be sized for very small pressure drops. If suction and hot gas risers are used in the system they must be sized in such a way that oil is returned to the compressor in the best possible manner at minimum load condition.

How the sizing of suction line is made, is explained with the help of Example 14.2.

EXAMPLE 14.2 A 55 kW R22 refrigeration unit is operating at 5°C suction and 38°C condensing temperatures. The total equivalent suction length is 30 m (including elbows, joints, bends, etc). Determine the (a) line size and (b) pressure drop equivalent (in degrees) for the suction line.

Solution: Refer to Table 14.2. The corrected capacity for 38°C is to be determined first. The correction factor, by interpolation is 0.976. The corrected capacity = 55 × 0.976 = 53.68 kW.

(a) For 35 mm OD copper tube, the capacity is 37.31 kW. For 42 mm OD copper tube the capacity is 61.84 kW. Therefore, we select a line size of 42 mm OD. **Ans.**

(b) The actual pressure drop in degrees

$$\Delta T = 0.04 \times 30 \times \left(\frac{55}{61.84}\right)^{1.8} = 0.97 \text{ K}$$

Since 0.97 K is less than the recommended 1 K, let us recompute ΔT for the small diameter tube (35 mm) for which ΔT = 2.4 K. As this temperature drop is too large, therefore, the 42 mm tube is recommended. **Ans.**

Many refrigeration units employ suction riser when the evaporator is at a level lower than the compressor. The oil will be transported only along with the return gas. Therefore, the size, shape and orientation of suction risers is very significant. The principal criteria to determine the transport of oil are gas velocity, gas density and inside pipe diameter. Actually, the velocity of return gas dominates the other two in transportation of oil. Therefore, a greater velocity is required at subzero temperatures as the density decreases. The riser diameter should be such that it has sufficient velocity. Usually the suction risers are sized for minimum system capacity. The sizing of suction risers is explained in Chapter 3 of ASHRAE Handbook of Refrigeration (volume–1990). The readers may refer to the same.

14.3.4 Sizing of Discharge Lines

The discharge lines should be sized to provide the minimum pressure drop since pressure loss in hot gas lines increases the required compressor power per unit refrigeration. Not only this but the compressor capacity also decreases. The discharge line should be sized in such a way that it provides a sufficient velocity to carry the oil and also the pressure drop is kept to the minimum. For this, the pressure drop is normally designed such that it does not exceed the equivalent of 1 K change in saturation temperature.

14.4 PIPING LAYOUT

We have discussed the sizing of liquid line, suction line and discharge line based on the pressure drop. To provide a properly designed piping system, the layout of piping to provide a suitable arrangement to return the entrapped oil to the compressor is very necessary. The following descriptions and drawings therefore relate to the piping design. Three different areas are discussed—suction lines, hot gas lines and condenser to receiver lines.

14.4.1 Piping Layout for Suction Lines

1. The proper piping for a suction riser from a suction evaporator is shown in Figure 14.1. The sump or loop at the outlet of the evaporator coil will allow good oil drainage. If the return gas velocity is maintained between 5 m/s and 20 m/s through the riser, then the refrigerant oil can be returned to the compressor efficiently.
2. When the system operates over a wide capacity range or at full load, double suction risers may be used, which ensures proper oil return (Figure 14.1). For part load operation, a single riser will be sufficient to meet the capacity. At full load, both risers are put into use. The size of suction line to the compressor must be based on the sum of the cross-sections of the two risers. At minimum or part load, the trap is filled with oil which blocks the riser B, so riser A will be in operation (not shown in the figure).

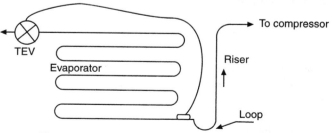

Figure 14.1 Suction riser with sump for oil return.

3. The compressor and evaporator are located at the same level or the compressor level is below that of the evaporator. A suction loop needs to be used to prevent liquid drainage from the evaporator to the compressor during shutdown (Figure 14.2). The loop must be as high as the top of the evaporator coil. If the system is provided with an automatic pump down which pumps the liquid out of the evaporator coil before shutdown, then the suction loop can be eliminated.

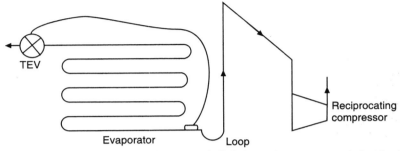

Figure 14.2 Suction loop prevents refrigerant and oil drainage to compressor during the off cycle.

14.4.2 Piping Layout for Hot Gas Lines

1. If the compressor is below the condenser, then it is essential to provide a trap at the outlet of compressor covering the full height of compressor in order to carry the oil to the condenser (Figure 14.3).
2. If the compressor is above the condenser, a simple piping arrangement is shown in Figure 14.4.
3. When the compressor is located in an area that can become colder than the condenser, it is better to use a check valve at the inlet to the condenser. This prevents refrigerant flow during the off cycle (Figure 14.5) to the compressor.

On halocarbon systems, if an oil separator is used it should be located between the compressor and the hot gas loop (Figure 14.6). This keeps any condensed liquid refrigerant out of the oil separator.

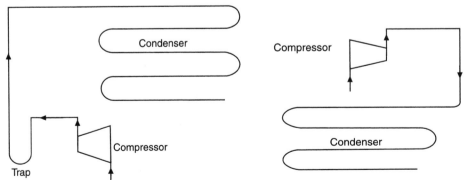

Figure 14.3 Hot gas line from compressor must turn downwards.

Figure 14.4 Simple connection between compressor and condenser when the compressor is above the condenser.

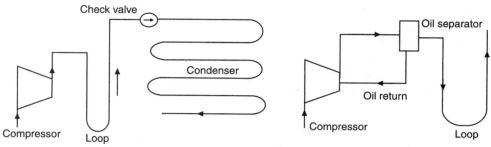

Figure 14.5 On discharge line piping, both check valve and loop are used to prevent liquid refrigerant flow to the compressor.

Figure 14.6 Where oil separator is used with halocarbon refrigerant, it should be placed close to the compressor and before the loop.

14.4.3 Condenser-to-Receiver Lines

A receiver tank stores the liquid refrigerant. It is located between the condenser and the expansion valve in the system. The receiver offers the following benefits:

1. It can pump down the storage capacity when another part of the system is being serviced or if the system has to be shut down for a long time.
2. A receiver is required to store the excess refrigerant charge that occurs with air-cooled condensers.
3. As per the loading condition, the charge in evaporator and condenser varies, and the receiver performs the task of accommodating the fluctuating charges.
4. On systems with multi-evaporators, the receiver holds the charge of one or more evaporators which are short-circuited due to partial load.
5. In air-cooled systems, the receiver provides sufficient refrigerant, thus keeping away the evaporator from the starving condition at the time of start-up.

Due to the number of benefits of receiver, one is required to give much importance to the pipeline between the receiver and the condenser. It should provide free drainage of the liquid from the condenser and permit the flow of vapour back to the condenser, all the time. If the pressure builds up in the receiver above that in the condenser, then liquid will not drain easily. This happens when the receiver is warmer than the condenser. Such situations arise in winter conditions, i.e. during the periods of reduced loading.

The size of line is such that the velocity of liquid flow is about 0.5 m/s or more and even the pipeline should have some slope.

There are two types of liquid receivers: (a) Through-flow type and (b) Surge type.

In the through-flow type arrangement, all the liquid from the condenser drains into the receiver before entering into the liquid line.

In a surge type receiver, only part of liquid from the condenser enters the receiver to remain in it. This part of liquid may not be required by the evaporator because of its low loading.

The piping between the condenser and the receiver can be equipped with a separate vent line to allow the receiver and condenser pressures to equalize. This vent line may or may not carry the check valve. The condensate line should be sized so that the velocity does not exceed 0.75 m/s. For large condensers and where the horizontal distance to the receiver is more than

2 m, an equalizer line between the inlet to the condenser and top of the receiver should be installed (Figure 14.7).

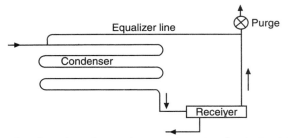

Figure 14.7 Use an equalizer line where the condenser and receiver are horizontally more than 2 m apart.

14.5 MULTIPLE SYSTEM PRACTICES FOR HCFC SYSTEMS

In this section, the arrangement for refrigerant lines for multiple compressors, evaporators and condensers has been discussed. Different practical arrangements have been shown instead of a full theoretical analysis. While discussing the practical arrangement of refrigerant lines, the following points must be considered.

1. Efficient oil return to the compressor
2. To provide protection to the compressor against 'slugging' with liquid
3. To provide protection to the compressor during the 'off' cycle

14.5.1 Hot Gas Lines

The best piping arrangement for connecting hot gas lines on multiple compressors is indicated in Figure 14.8. This arrangement does not permit the drainage of oil and refrigerant to the idle compressor. The convenient location for the header is at the floor level.

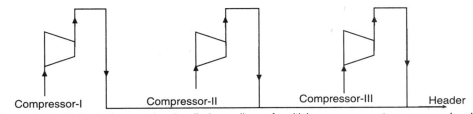

Figure 14.8 Method of connecting the discharge lines of multiple compressors to a common header.

Two or more condensers may be connected in series or parallel and can be used in a single refrigerant circuit. When these are connected in series, the pressure of each condenser is added. Usually condensers are connected in parallel (Figure 14.9). In the parallel connection, the pressure drop remains equal to any of the condensers.

14.5.2 Suction Lines

Suction to multiple compressors is given through a suction manifold as shown in Figure 14.10. In such an arrangement it is possible to prevent oil drain to the idle compressor by gravity. When

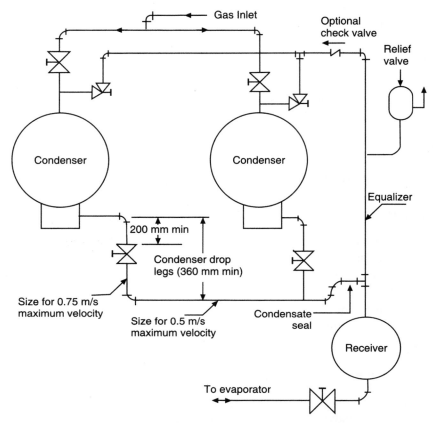

Figure 14.9 Parallel condensers with through-type receiver.

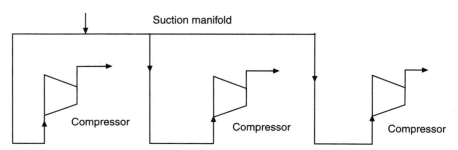

Figure 14.10 Suction line manifold should be arranged to the idle compressor.

the suction manifold (header) is below the compressors, the take off should be made to equalize the return of oil from each compressor (Figure 14.11).

14.6 SYSTEM PRACTICES FOR AMMONIA

The properties of ammonia have been covered in Chapter 3 and we know that ammonia has excellent thermodynamic properties but at low concentration levels of 35 to 50 mg/kg, it is

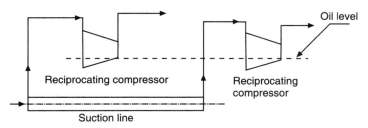

Figure 14.11 Method of connecting the suction header to multiple elevated compressors to provide equal oil return to each compressor.

considered toxic. It can burn and explode if present by 16 to 25% by volume in the presence of a flame. Cold liquid ammonia should not be confined between closed valves in a pipe where the liquid can get warm and expand, thus bursting piping components. Ammonia is a powerful solvent that can remove dirt, scale and sand or moisture remains in the pipes, valves and fittings during installation. These substances may enter the compressor along with the suction gas, and disturb the working of bearings, piston, cylinder walls, valves and lubricating oil. So a suction strainer is a must.

A damaging kind of wear of cylinder walls and rings may result if liquid droplets enter the compressor with return gas. Liquid entry to the compressor reduces the lubrication, resulting in rapid abrasion of cylinder walls and rings. Like these, there are still some more disadvantages of liquid entering the compressor.

During the pipe layout and installation for the ammonia system, the information given above may be remembered. Pipe sizing should be made carefully to provide low pressure drop and avoid capacity or power penalties caused by inadequate piping.

Copper and copper bearings are attacked by ammonia due to which they are not used for ammonia piping. Iron and steel piping fittings and valves are used for ammonia gas and liquid.

Tested union joints, gasket joints and thread joints for pipe connections can be used but utmost care must be taken to avoid leakage with the use of good sealing materials. The insulated pipelines should be located at least 2.3 m above the floor. There must be some space between the pipelines and the adjacent wall surfaces.

14.6.1 Pipe Sizing

While sizing the ammonia pipeline the theory discussed earlier in this chapter is also useful. Table 14.4 may be employed for the purpose. This table presents the practical suction line sizing data based on 0.02 K total pressure drop equivalent per metre equivalent length of pipe. Data on equivalent lengths of valves and pipefittings, etc. are given in ASHRAE Handbook—1990. The readers may refer to the same for details.

Table 14.4 Suction, discharge and liquid capacities for ammonia (single or high-stage applications)

Steel line size	Suction Lines $\Delta T = 0.02$ K/m					Discharge Lines $\Delta T = 0.02$ K/m, $\Delta p = 684.0$ Pa/m			Liquid Lines		
	Saturated suction temperature, (°C)								Steel line size		
Nominal (mm)	−40 $\Delta p = 76.9$	−30 $\Delta p = 116.3$	−20 $\Delta p = 168.8$	−5 $\Delta p = 276.6$	5 $\Delta p = 370.5$	−40	−20	+5	Nominal (mm)	Velocity = 0.5 m/s	$\Delta p = 450.0$
10	0.8	1.2	1.9	3.5	4.9	8.0	8.3	8.5	10	3.9	63.8
15	1.4	2.3	3.6	6.5	9.1	14.9	15.3	15.7	15	63.2	118.4
20	3.0	4.9	7.7	13.7	19.3	31.4	32.3	33.2	20	110.9	250.2
25	5.8	9.4	14.6	25.9	36.4	59.4	61.0	62.6	25	179.4	473.4
32	12.1	19.6	30.2	53.7	75.4	122.7	126.0	129.4	32	311.0	978.0
40	18.2	29.5	45.5	80.6	113.3	184.4	189.4	194.5	40	423.4	1469.4
50	35.4	57.2	88.1	155.7	218.6	355.2	364.9	374.7	50	697.8	2840.5
65	56.7	91.6	140.6	248.6	348.9	565.9	581.4	597.0	65	994.8	4524.8
80	101.0	162.4	249.0	439.8	616.9	1001.9	1029.3	1056.9	80	1536.3	8008.8
100	206.9	332.6	509.2	697.8	1258.6	2042.2	2098.2	2154.3	-	-	-
125	375.2	601.8	902.6	1622.0	2271.4	3682.1	3783.0	3884.2	-	-	-
150	608.7	975.6	1491.4	2625.4	3672.5	5954.2	6117.4	6281.0	-	-	-
200	1252.3	2003.3	3056.0	5382.5	7530.4	12195.3	12529.7	12864.8	-	-	-
250	2271.0	3625.9	5539.9	9733..7	13619.	22028.2	22632.2	23237.5	-	-	-
300	3640.5	5813.5	8873.4	15568.9	21787.	35239.7	36206.0	37174.3	-	-	-

Notes:

1. Table capacities are in kilowatts of refrigeration resulting in a line friction loss per unit equivalent pipe length with corresponding change in saturation temperature, where

   Δp = pressure drop due to line friction (Pa/m)
   ΔT = change in saturation temperature (K/m)

2. Line capacity for other saturation temperatures ΔT and equivalent lengths L_e

   $$\text{Line capacity} = \text{Table capacity} \left(\frac{\text{Table } L_e}{\text{Actual } L_e} \times \frac{\text{Actual } \Delta T}{\text{Table } \Delta T} \right)^{0.55}$$

3. Saturation temperature ΔT for other capacities and equivalent lengths L_e

   $$\Delta T = \text{Table } \Delta T \frac{\text{Actual } L_e}{\text{Table } L_e} \left(\frac{\text{Actual capacity}}{\text{Table capacity}} \right)^{1.8}$$

4. Values in the table are based on 30°C condensing temperature. Multiply the table capacities by the following factors for other condensing temperatures:

Condensing temperature	Suction lines	Discharge lines
20	1.04	0.86
30	1.00	1.00
40	0.96	1.24
50	0.91	1.43

5. Discharge and liquid line capacities are based on −5°C suction.

Just to introduce how the sizes of ammonia lines are decided, Example 14.3 has been worked out.

EXAMPLE 14.3 A 225 kW ammonia refrigeration plant is operating at 5°C suction and 38°C condensing temperatures. The total equivalent length is 30 m (including equivalent lengths of elbows, joints, bends, etc). Determine the line size and pressure drop equivalent (in degrees) for the suction line.

Solution: Refer to Table 14.4. The table is based on 30°C condensing temperature. Therefore, determine the correction factor by interpolation and it is found to be 1.192.

The corrected capacity = (225)(1.192) = 268.2 kW

The pipe size of 50 mm OD has a capacity of 218.6 kW at 5°C suction temperature, while for 65 mm OD it is 348.9 kW.

Therefore, we select a pipeline of size 65 mm OD.

The actual pressure drop in degrees

$$\Delta T = 0.02 \times 30 \times \left(\frac{268.2}{348.9}\right)^{1.8} = 0.37 \text{ K}$$

Since 0.37 K is less than the recommend 1 K, let us recompute ΔT for the smaller diameter (50 mm) tube, for which $\Delta T = 0.86$ K. This temperature drop is also less than 1 K. Instead of going for a higher diameter pipe, it is better to select a pipe size of 50 mm OD. **Ans.**

14.6.2 Ammonia Piping

Ammonia is a non-miscible refrigerant. The oil pumped along with ammonia into the discharge line will not be carried along all parts of the system with refrigerant. An oil separator is necessarily installed in the discharge line. The oil separated will be drained back to the crankcase.

Since oil is heavier than ammonia, it collects at the bottom of pressure vessels and at other points of the system. For this reason, oil sumps are provided at the bottom of all receivers, evaporators, accumulators and other vessels in the system containing ammonia liquid. Used oil must be removed from the system periodically and replaced in the compressor with new oil.

Since the lubricating oil does not circulate through the system piping, the minimum velocity criteria to sweep the oil with the refrigerant has no significance. Therefore, the piping is sized for a low pressure drop without much consideration about the refrigerant velocity. Pipeline layout is also of not much importance.

14.6.3 Compressor Piping

Figure 14.12 shows two compressors operating in parallel for the same suction main. Suction main should be in a position to return only clean, dry gas to the compressor. An oil separator should be provided away from the compressor as far as possible. Oil separates more easily at low temperatures. Figure 14.12 shows a common oil separator for the two compressors and a single oil receiver.

In Figure 14.12, check valves are also installed in the discharge line to each compressor. This minimizes the flow of hot gas back to the compressor during shutdown.

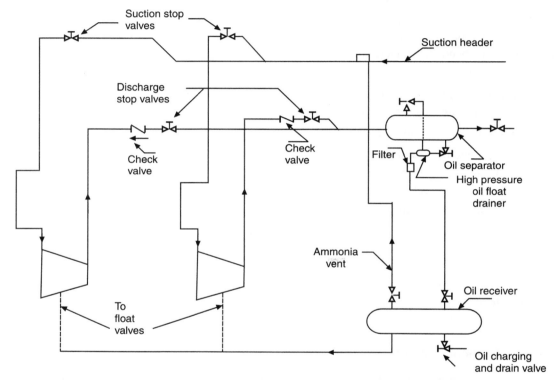

Figure 14.12 Parallel reciprocating compressors and common oil separator.

14.6.4 Condenser and Receiver Piping

Proper design of condenser and receiver piping is necessary for optimum utilization of condenser surface. This requires the condensate in the condenser to be drained to receiver effectively. Also, any non-condensable gases should be purged. The receiver should necessarily be located below the condenser so that the latter is not flooded with ammonia. The drain line between the condenser and the receiver is sized so that it allows free passage of liquid to receiver and vapour back from receiver to condenser. To satisfy this, velocity in this line is to be kept below 0.5 m/s. Still more ways of connecting condensers and receivers are possible, but it is not possible to cover all of them here.

14.6.5 Suction Traps

The liquid ammonia that enters the suction line of compressor, affects its working adversely as discussed earlier. So to ensure that the suction is free from liquid ammonia, suction traps are used (Figure 14.13). A properly designed trap will separate the liquid and ammonia vapours and ensure only the entry of vapour to the compressor.

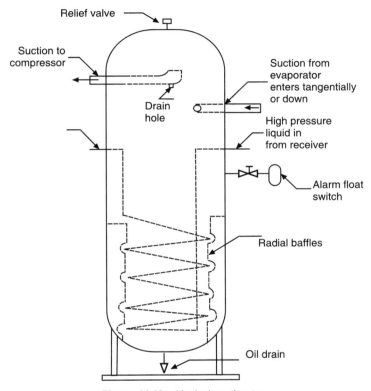

Figure 14.13 Vertical suction trap.

14.6.6 Oil Separators

Oil separators are preferred in systems where oil is non-miscible with the refrigerant. They are normally located in discharge line. Oil separators are also recommended for (i) low temperature systems, (ii) systems employing non-oil-returning evaporators, such as flooded liquid chillers, and (iii) any system where capacity control and/or long suction or discharge rise causes serious piping design problems.

Discharge line oil separators are of two types—impingement type and chiller type.

Impingement type oil separator consists of a wire mess, a screen or matrix. The wire matrix of different porosity is placed in a cylindrical shape container. The discharge gas line is connected to this. The oil particles due to higher momentum get separated in the matrix and collectively drained through a pipe to the compressor.

Chiller type oil separator is similar to the water-cooled condenser in construction. The water is circulated through the tubes while the discharge hot gas is circulated through the shell. The oil particles precipitate at the outer surface of cold water tubes and get collected at the bottom, and then drained to the compressor by any means. Utmost care should be taken so that the refrigerant does not condense. If it condenses, it may return in the liquid phase to the compressor crankcase. This means that the water temperature should not be too low so as to promote condensation.

14.7 INSTALLATION ARRANGEMENT

The installation of refrigeration systems is dependent on the following factors: (i) capacity, (ii) type of equipment used, (iii) refrigerant used and (iv) condenser type and unique features of applications. The installation must provide for a number of essential conditions, including:

- Proper location of equipment
- Piping arrangement
- Electrical connections
- Check test and start

It is also necessary to follow all the safety norms of the local bodies while installing the refrigeration plants.

14.7.1 Location of Equipment

While selecting a suitable location for the equipment, following points should be viewed.

Ventilation: The components of the system which require to dissipate heat should be located in such a place that they can do the job effectively. For example, air-cooled condenser, water-cooled condensers and evaporative condensers must be located outside the building where good free air circulation is available.

Weather conditions: Care must be taken not only in locating the equipment but also while selecting them because a particular equipment working in one climatic condition may not be so efficient in other climatic conditions. For example, water-circulating equipment may face the problem of freezing in winter months.

Space for installation and service: Provision should be made for space not only for installation of equipment but also for their servicing. If shell-and-tube condensers are to be installed in an equipment room, space needs to be provided for cleaning tubes and even replacing them if required.

Isolation of noise and vibration: Noise and vibrations of an operating machine are not completely eliminated but their intensity can be reduced to a certain extent. For example, if a compressor can be located on the ground floor on an isolation base, some of the problems related to vibration and sound transmission are solved. A compressor discharge line to an air-cooled condenser located on the roof should have a muffler installed in the discharge piping to reduce vibration. Also discharge lines should be securely bolted to solid areas of the structure.

Availability of utility services: The facilities such as electricity, water supply and drainage must also be considered. Any location of equipment that simplifies the connection to the existing facilities is helpful.

14.7.2 Piping Arrangement

For halocarbon refrigerants, copper piping is preferred and for ammonia, steel piping is needed. While joining the pipes either by brazing or welding process, care should be taken to prevent the formation of scale.

When brazing near a valve, care should be taken not to damage the mechanism by excessive heat. This can be done by either removing the internal parts or wrapping them with a wet rag. There are also certain compounds in the market that can be applied to the copper piping to isolate them from the heating effect.

For electrical connections and related study, the reader is required to refer the specialist books in this subject area.

14.8 DEHYDRATION

The excessive moisture in a refrigeration system may cause (i) capillary or expansion valve freeze-up, (ii) formation of sludges that are highly corrosive, (iii) flapper valve breakage, and (iv) hermetic motor burn-out.

Since all these effects, with the exception of freeze-up, cannot be easily detected during a standard factory test, it is necessary to use dehydration technique to reduce the level of moisture content to permissible limits. Permissible limits of moisture content can only be set when it is measured accurately. The acceptable limit of moisture content depends upon the application, type of refrigerant and size of the unit.

The hydrocarbon group of refrigerants such as ethane absorbs almost no water. It is heavier than refrigerant and shrinks to the bottom. The R22 absorbs a small amount of water while the synthetic lubricating oils used for R134a are highly hydroscopic in nature.

Refrigerant R717 (ammonia) absorbs water in any amount. For R717, moisture becomes a problem only when excessive amounts are absorbed. Generally, the problem is corrected by replacing the refrigerant.

Several satisfactory methods of dehydrating refrigeration systems and components are available. The choice of a method depends upon the size, moisture level, production quantities, cost and time.

Heat, vacuum and dry air are used for moisture removal either separately or in different combinations.

14.8.1 Dehydration by Heating

This method is used for dehydrating of open parts such as tubes, dryers, etc. The parts are heated for a long time in the oven while maintaining a high temperature, sufficient enough to vaporize the moisture. The relative humidity of air in the oven should be kept low. The oven should have an arrangement to exhaust the water vapour and provide supply of dry air. After removing the components from the oven, they must be capped or closed in order to prevent the moisture from entering them during cooling.

The disadvantages of this method are:

1. If the system or parts to be dehydrated are large, it is difficult to heat them to the required temperature and therefore either production is limited or a large operating area is required.
2. The system has a small opening for water vapour to escape, compared with its volume.

14.8.2 Dehydration by Vacuum

This method is employed only to the parts which can be closed and made leak tight. The boiling point of water is low under vacuum, for example, at 0.0031 bar the boiling point is 25°C. Therefore, the moisture can be easily converted to vapour and removed with the help of vacuum pump.

The disadvantages of this method are:

1. When the vacuum pump is not capable of drawing a deep vacuum, double or triple evacuation is required. So it takes a long time to reach the required level of vacuum.
2. To reach a low boiling point of the order of 25°C, a very high vacuum is required.

14.8.3 Dehydration by Dry Air

Dry air or nitrogen will be blown through the equipment which removes moisture by becoming totally or partially saturated.

Dry air or nitrogen sweeping is an effective way of removing moisture but

1. very high flow rates are required, and
2. moisture may remain in the pores of the metal due to surface viscosity.

Dry air is manufactured by several methods. In one of the methods, moisture is removed in refrigeration unit after which the same air is passed through the dryer unit charged with activated alumina, silica gel, etc.

The combination of two or three methods results in faster and more uniform dryness of the treated system.

14.8.4 Heat and Vacuum Method

The dehydrated units are heated in the oven and evacuated simultaneously. The heat being external force drives out the moisture from the surfaces and at the same time evacuation reduces the boiling point. Therefore, the combination of heating and evacuation is a faster method of dehydration. Oven temperature to be maintained is in the range of 120°C to 150°C.

14.8.5 Vacuum and Heat Method

In this method, parts are evacuated when cold and then heated in the oven followed by evacuation. The evacuation of the part while in cold removes the moisture from the inside volume. The heat applied in the oven drives out the moisture from the surfaces and brings in vapour state and further evacuation removes the vapour.

14.8.6 Heat and Dry Air Method

The heat applied to the parts allows the moisture trapped in the metal pores or blind pockets to boil off. This vapour is driven out by the sweeping dry air, which should have a dew point temperature of –40°C and –56°C. Where the entire system is to be dehydrated the combination of 'heat, dry air and vacuum' may be adapted to carry out dehydration at a faster rate.

A combination of oven and dry air has been used effectively for dehydration of hermetic compressors. The compressor before welding the shells is passed through an oven and swept with air with –4°C dew point. The oven heat would cause the moisture to vaporize and the air circulated across the unit absorbs it and carries it off. The air is dried, by passing it through silica gel or a similar drying material, before it enters the oven.

The advantages and disadvantages of the various combination methods depend upon the system components and the quality of dehydration.

14.9 CHARGING

The accuracy needed when charging refrigerant or oil into a unit depends on its size and application. While charging, the following points must be noted:

1. Careful handling of refrigerant and oil.
2. Where to charge and how much to charge?
3. If the oil is added after the refrigerant is already in the crankcase, excessive foaming and oil vapour may damage the bearings.
4. If the unit is charged before conducting the performance test, there is a possibility of liquid slugging in the crankcase of compressor.
5. Refrigerant lines must be dry and clean and all charging lines must be free of moisture and non-condensable gases.
6. Most of the lubricants are hygroscopic in nature. Care should be taken to avoid water contamination.
7. The oils must be tested before charging for the moisture content.
8. The charging lines must be purged and it must be made sure that all connections are tight so that air is not drawn in the system with the refrigerant.

The quality of refrigerant to be added to the system for initial charge or recharging depends on the size of the equipment and the amount of refrigerant to be circulated. In very large systems, it is a common practice to simply weigh the charge by placing the refrigerant cylinders on a suitable scale and observing the reduction in weight in kilograms. This method is fine for systems that have receivers or condenser volume ample enough to take a slight overcharge.

On smaller systems and particularly those that are self-contained packaged units without receivers, the system charge is critical to grams rather than kilograms. The refrigerant may be in either the liquid or the vapour form. Refrigerant is added in the vapour form through the suction valve when the unit is operating. Refrigerant may be added in the liquid form when the unit is off and in an evacuated condition through the liquid line service valve only.

14.9.1 Charging through Suction Valve

For simplicity, here we are showing only a refrigerant cylinder and assuming that the charge is weighed during operation.

(i) Install the gauge manifold.
(ii) The suction line and refrigerant cylinder are connected through valve B as shown in Figure 14.14.
(iii) The cylinder valve is opened and the flare out is loosened at the compressor end. This operation removes air from the line.
(iv) Turn the suction valve to close the suction line so that the compressor can draw gas directly from the cylinder.

(v) Observe the weight of the cylinder.
(vi) When the correct charge has been added, close the valve on the suction side and the valve on the refrigerant cylinder.
(vii) Back seat both suction and liquid lines service valves, remove hoses and cap ports.

If a weigh balance is not available, then the system is charged up to a certain pressure which is measured by the pressure gauge. The cylinder may be kept in warm water in order to increase the speed of charging.

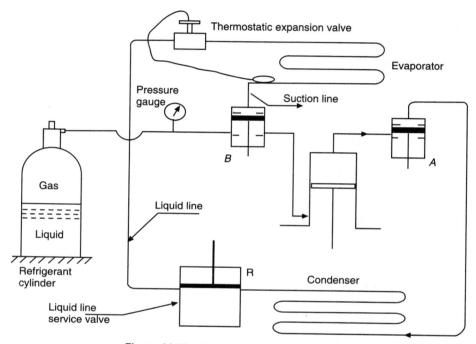

Figure 14.14 Charging through suction valve.

14.9.2 Charging through Charging Valve

The charging procedure for the liquid form (unit not operating and evacuated) is detailed below and shown in Figure 14.15.

(i) Install the gauge manifold.
(ii) Attach the refrigerant cylinder; invert the cylinder unless it is equipped with a liquid vapour valve, which permits liquid withdrawal in the upright position.
(iii) Open both suction and liquid service valves one turn-off back seat.
(iv) Open valve on the high side of the gauge manifold.
(v) Open valve on the refrigerant cylinder and add refrigerant.
(vi) After the correct charge is introduced, close the valve on the high side of manifold and close the valve on the refrigerant cylinder. Back-seat both the suction and liquid service valves.
(vii) Remove the gauge manifold.

Refrigerant Piping, Accessories and System Practices **479**

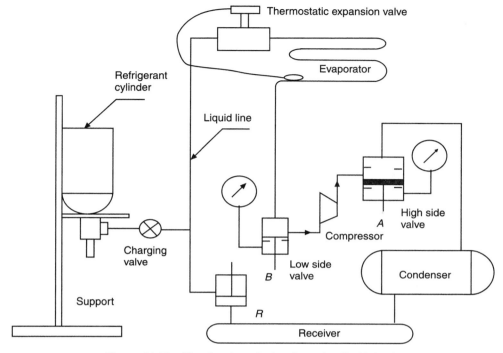

Figure 14.15 Charging through charging valve (liquid form).

14.9.3 Checking the Charge

Checking the charge of a new installation or of an existing unit is another function of the service manifold gauges. For example, the following procedure would be used for an air-cooled unit.

(i) Install the gauge manifold (suction and discharge side gauge).
(ii) Allow the system to operate until the pressure gauge readings stabilize (about 15 minutes).
(iii) While the unit is operating, record the following information.
    (a) High pressure gauge reading.
    (b) Dry-bulb temperature of air entering the condenser coil.
    (c) Wet-bulb temperature of air entering the evaporator coil (this is done with a wet wick thermometer).
(iv) A comparison of the above measurements with the head pressure charging table supplied with the unit will indicate if the system is adequately charged and operating properly.

14.10 TESTING FOR LEAKS

Testing for the leakage of refrigerant from the refrigeration units is very important. A very minute leakage may lead to the loss of refrigerant during the operation. Leakages on low-pressure side are still serious due to the air entering the system. Some leak detection methods are discussed here.

14.10.1 Water Submersion Test

It consists of a well-lit tank filled with clean water. It may be heated slightly to make it free from air. A component or unit under positive pressure of dry gas is placed in the water bath and observed for bubbles that indicate a leak. This method is simple and reliable.

The pressure of gas inside the test component and the time for which it is to be immersed in water will vary. Therefore, the sensitivity of the test is very difficult to define. The minimum leak rate that can be determined is 1/1000 cc per second. Many a time instead of clean water, bath soap solution can be applied at the joints. It is used for a large system in order to locate and repair the leak.

14.10.2 Pressure Testing

The unit or part is either pressurized or evacuated and then the decrease or rise in pressure is measured against time. It is necessary to maintain constant temperatures in order to eliminate pressure variations.

It is a simple, reliable and possible method for determining total leakage. However, it takes a long for pressure change to occur and the location of leak also cannot be determined.

14.10.3 Halide Leak Detector

The units or components charged with refrigerant containing a halogen may be tested for leaks with the help of the halide torch. Halide torch normally burns with almost colourless or faint blue colour.

The torch is held against the joint and if there is a leak, the colour of the flame turns greenish blue. The intensity of greenish blue is dependent on the rate of leakage.

This type of detector is capable of detecting leaks as small as 1/1000 cc per second. For most of the applications, this type of detector works satisfactorily. It is also available as a battery operated unit. Some units have a dual type of control, setting a single HI-LOW selector in addition to a balance control dial.

The electronic detectors are also used to detect small leaks. The pressure of refrigerant will cause the instrument to emit a loud noise.

If the refrigerant is ammonia, the location of the leak can be detected by burning a sulphur candle and passing the flame into the area of suspected leak. A safer method is the application of hydrochloric acid vapour used in the same manner. In either case, white smoke will form in the presence of ammonia. If ammonia has leaked into water or brine solution, Nessler's solution can be used to detect the presence of the refrigerant. Ammonia causes the solution to turn yellow.

EXERCISES

1. Explain the functions of refrigerant piping.
2. Discuss the factors involved in proper sizing of refrigerant lines.
3. Discuss the factors involved in proper layout of refrigerant lines.
4. What precautions are needed during pipe layout and installation for the ammonia system?
5. Write short notes on: (a) suction traps and (b) oil separators
6. Name the methods used for dehydrating refrigeration systems and components.
7. Explain any one method of testing for leaks of refrigerant.

Chapter 15

Air Distribution System and Duct Design

15.1 INTRODUCTION

Figure 15.1 shows the airflow arrangement in a typical air conditioning system. Essentially it consists of an air-conditioning unit, a fan, supply duct, supply air outlet, return air duct, air filters, etc.

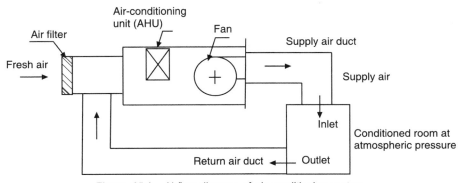

Figure 15.1 Airflow diagram of air conditioning system.

In this chapter we will focus on topics like air distribution, duct material and duct design.

The conditioned air from air handling unit (AHU) is carried to the room or air-conditioned space through the supply duct. For economical operation of the air conditioning system, the room air is returned through the return duct and mixed with fresh air in an appropriate proportion. This mixing is as per the quality of air to be supplied to the room. This requires an air-distribution system. There are four components of this system, namely (a) Supply air duct; (b) Supply air inlet; (c) Return air outlet; (d) Return air duct. The first two components have a major effect on air distribution.

The requirement of good air distribution is to create a proper combination of temperature, humidity and air motion in the conditioned room or zone at 1.8 m above the floor level. The temperature variation within a single room should not exceed 1°C and within rooms the variation should not be more than 2°C. The desired air velocity is about 9.0 m/min at the occupancy level. The proper air distribution is needed to mix the supply air with room air to achieve an air velocity of 9 m/min throughout the room.

15.2 CLASSIFICATION OF DUCTS

The purpose of the duct is to carry the conditioned air to the space, and room air back to the AHU. While doing so, one has to consider the power consumed by the fan, heat entering the air through the wall of a duct, the material of the duct, load on the duct, duct size, etc. The air flowing through the duct imposes two types of loads on duct—(1) The static load due to the mean static pressure differential across the duct wall. (2) The pulsating load due to turbulent air flow. This latter load is relatively small. The ducts are classified as:

Low pressure duct system: Average air velocity is less than 600 mpm and static pressure $\leq$ 5 cm of water gauge.

Medium pressure duct system: Average air velocity is less than 600 mpm or static pressure up to 15 cm of water gauge.

High pressure duct system: Average air velocity > 600 mpm or static pressure over 15 cm and up to 25 cm water gauge.

The low pressure system permits the use of simple forming methods. Majority of duct construction belongs to the low pressure type. Medium and high pressure ducts may be used to meet the space limitations. Design and data books include information about the thickness of duct wall as per the air pressure.

15.3 DUCT MATERIAL

The most common material used for ducting is the GI sheet. This being economical, it is acceptable in most of the applications.

The duct material for residential air conditioning may be galvanized iron, aluminium or material with UL standard 181 rating. Sheet metal ducts should be constructed of minimum thickness and installed with HVAC duct construction standards—Metal and Flexible (SMACNA 1985). The duct material and installation standards for residential, commercial and industrial purposes are different and air conditioning engineers have to follow them accordingly. These are not explained here.

Fibrous glass ducts are a composite of rigid fibre glass and a factory-applied coating (typically aluminium) which serves as the finish and vapour barrier.

Flexible ducts connect mixing boxes, light troffers, diffusers and other terminals to the air distribution system.

Aluminium is not recommended for installations handling abrasive materials. Thermoplastic and thermosetting plastic materials may be used in commercial and industrial installations.

It has been noticed that growth of bacteria takes place at rusted spots. GI sheets usually rust over a period, giving rise to bacterial growth. This is not of much concern as a normal healthy human being has adequate resistance to bacteria.

In hospitals, ducting may be made of aluminium to avoid rusting. However, the cost of aluminium makes it uneconomical in normal applications.

In certain corrosive atmospheres of chemical or similar factories, stainless steel can be a better alternative despite its high cost.

No comprehensive standards exist for underground air ducts. Coated steel, asbestos cement, plastic, tile, concrete and other materials have been used.

An air-conditioning engineer makes a choice of material after giving due consideration to all the aspects of a good duct design.

Shape of duct

The circular shape is the most compact shape, which requires minimum material and has the least frictional pressure drop. It is difficult to construct a circular duct at the site and it is architecturally unsuitable to pass through a space. So this shape is limited to factories or outside the space. Initially the ducts are sized for the required flow rate of air and velocity and pressure loss, etc. assuming circular shape. After finding the diameter it is converted into an equivalent rectangular cross-section with a proper aspect ratio.

We shall take up this topic later in this chapter, first we shall discuss the fundamentals of fluid flow.

15.4 CONTINUITY EQUATION

Water flowing through piping and air flowing through ducts in heating, ventilating and air-conditioning (HVAC) systems is usually under steady flow conditions. Steady flow means that the flow rate of the fluid at any point in a section of pipe or duct is equal to that at any other point in the same pipe or duct, regardless of the pipe or duct's shape or cross-section.

In HVAC systems, the density of the air or water flowing generally remains constant. When the density remains constant, the flow is called *incompressible*.

Let us consider the flow of air or water through a duct shown in Figure 15.2 in order to apply the continuity equation.

Volume flow rate of fluid in m³/s, $\text{VFR} = \rho_1 \times A_1 \times V_1 = \rho_2 \times A_2 \times V_2 = c$ \hfill (15.1)

A_1, A_2 = cross-sectional area of pipe or duct at any points 1 and 2, in m²

V_1, V_2 = velocity of fluid at any points 1 and 2, in m/s

Since density, $\rho_1 = \rho_2 = \rho$, then mass flow rate in kg/s = $\rho \times A \times V$

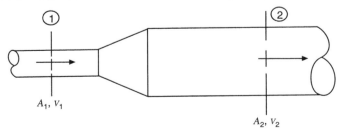

Figure 15.2 The continuity equation for steady flow of air or water through a duct or a pipe.

EXAMPLE 15.1 Air is flowing in a duct of 0.005 m² cross-sectional area at a velocity of 12 m/s. This high velocity results in a disturbing noise. The air-conditioning engineer wants to reduce the velocity to 6 m/s. What should be the size of the new duct?
Solution: Using Eq. (15.1),

$$A_2 = \frac{V_1}{V_2} \times A_1 = \frac{12}{6} \times 0.005 \text{ m}^2 = 0.01 \text{ m}^2 \qquad \textbf{Ans.}$$

15.5 ENERGY EQUATION FOR A PIPE FLOW

It is a very common practice to write the energy balance equation for different applications. Let us apply the energy equation to the pipe flow shown in Figure 15.3.

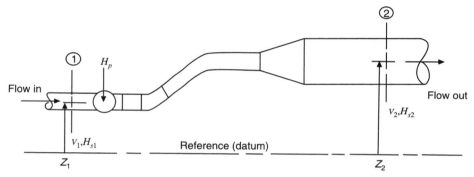

Figure 15.3 The energy equation applied to flow in a duct pipe.

$$E_1 + E_{add} = E_{lost} + E_2 \qquad (15.2)$$

where

E_1, E_2 = energy of fluid at points 1 and 2
E_{add} = energy added to fluid between points 1 and 2
E_{lost} = energy lost from fluid between points 1 and 2.

The energy of the fluid flowing through a closed conduit at any point is due to the pressure of fluid, velocity of fluid (kinetic energy) and elevation of fluid from datum (potential energy). The energy added to the fluid is from a pump or fan. The energy lost is due to friction. But energy would never be lost, it would appear in another form (e.g. a temperature change), but it is usually small and may be ignored.

$$E_1 = H_{s1} + \frac{V_1^2}{2} + gZ_1 \text{ and } E_2 = H_{s2} + \frac{V_2^2}{2} + gZ_2, \text{ and } E_{add} = H_p, E_{lost} = H_f$$

Therefore,

$$H_{s1} + \frac{V_1^2}{2} + gZ_1 + H_p = H_{s2} + \frac{V_2^2}{2} + gZ_2 + H_f \qquad (15.3)$$

where

H_{s1} and H_{s2} = static pressure of fluid at locations 1 and 2 respectively (m)
V_1 and V_2 = velocity of fluid at locations 1 and 2 respectively
g = gravitational constant (9.81 m/s²)
$\dfrac{V_1^2}{2}$ and $\dfrac{V_2^2}{2}$ = velocity pressure head at 1 and 2 (m)
Z_1 and Z_2 = elevation from datum (m)
H_p = pressure head imparted by pump or fan (m)
H_f = pressure head lost in piping or duct from friction (m)

Equation (15.3) is referred to as flow energy equation or generalized Bernoulli equation. It is useful in determining the pressure requirements of pumps and fans and in testing and balancing systems.

Equation (15.3) can be arranged in a useful form as:

$$H_p = (H_{s2} - H_{s1}) + \frac{(V_2^2 - V_1^2)}{2g} + (H_{e2} - H_{e1}) + H_f \qquad (15.4)$$

Note: Equations (15.2), (15.3) and (15.4) are valid for unit mass flow rate (kg/s).

EXAMPLE 15.2 A piping system delivers water from the basement to the roof storage tank, 50 m above. The friction loss in the piping, valves and fittings is 3.5 m. The water enters the pump at a gauge pressure of 3 m and is delivered at atmospheric pressure (all values are gauge pressure). The velocity at the pump suction is 2 m/s and at the piping exit is 3 m/s. What is the required pump pressure?

Solution: Using Eq. (15.4),

$$H_p = (H_{s2} - H_{s1}) + \frac{(V_2^2 - V_1^2)}{2g} + (H_{e2} - H_{e1}) + H_f$$

$$(H_{s2} - H_{s1}) = 0 - 3 \text{ m}$$
$$= -3 \text{ m (change in static pressure)}$$

$$\frac{(V_2^2 - V_1^2)}{2g} = \frac{((3)^2 - (2)^2)}{2 \times 9.81}$$

$$= 0.25 \text{ m (change in velocity pressure)}$$

$$(H_{e2} - H_{e1}) = 50 \text{ m (change in elevation)}$$

$$H_f = 3.5 \text{ m (friction pressure loss)}$$

$$H_p = -3 + 0.25 + 50 + 3.5$$

$$= 50.75 \text{ m of w.g.} \qquad \textbf{Ans.}$$

15.6 TOTAL, STATIC AND VELOCITY PRESSURE

The total pressure energy that a fluid has at any point consists of two parts, its static pressure energy and its velocity pressure energy. Thus, the total pressure H_t of a flowing fluid is defined as

$$H_t = H_s + H_v \tag{15.5}$$

where

H_t = total pressure of water column (m)
H_s = static pressure of water column (m)
H_v = velocity pressure of water column (m).

Here the pressure of a fluid is expressed in metre of water column, but it can also be expressed in N/m² if needed.

The static pressure is the pressure the fluid has at rest. But fluid particles are never at rest at definite temperature. Considering the activities of the fluid molecules at molecular level, fluid pressure is defined as the rate of change of momentum of the molecules per unit area.

The velocity pressure is defined as

$$H_v = \frac{v^2}{2g} \tag{15.6}$$

Thus, the velocity pressure concept is useful in measuring velocities and flow rates in piping and ducts. If the velocity pressure can be measured, the velocity can be found using Eq. (15.7)

$$v = \sqrt{2gH_v} \tag{15.7}$$

where

v = velocity (m/s)
g = gravitational constant (9.81 m/s²)
H_v = velocity head of fluid (m).

The static, total and velocity pressure can be measured using a U-tube manometer as shown in Figure 15.4(a), (b) and (c), respectively.

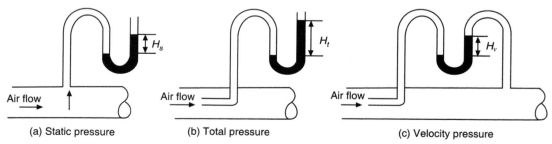

(a) Static pressure (b) Total pressure (c) Velocity pressure

Figure 15.4 Manometer arrangement to read static, total and velocity pressure.

15.7 STATIC REGAIN

During a flow in a duct or pipe the static pressure can increase in the direction of flow if the velocity decreases. This is due to the conversion of velocity energy to static (pressure) energy called *static regain*.

Consider the diverging air duct section in Figure 15.5. The difference in velocity between points 1 and 2 is

$$H_{v1} - H_{v2} = \frac{V_1^2}{2g} - \frac{V_2^2}{2g} \qquad (15.8)$$

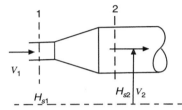

Figure 15.5 The flow through a pipe or duct with changing cross-section.

Applying the flow energy equation, assuming that there is no friction loss H_f and the change in elevation is negligible, then

$$H_{s2} - H_{s1} = H_{v1} - H_{v2} = \frac{V_1^2}{2g} - \frac{V_2^2}{2g} \qquad (15.9)$$

Equation (15.9) shows that if the velocity decreases in the direction of flow then the static pressure increases. Velocity energy has been converted to pressure energy. This effect is called *static pressure regain*.

Static regain is not as high as shown in Eq. (15.9) due to the presence of friction loss.

15.8 PRESSURE LOSS IN THE DUCT

There are two types of pressure losses when air flows through a duct.

(a) Loss of pressure due to friction between moving particles of fluid and the interior surface of duct. It occurs throughout the duct length.
(b) Dynamic loss of pressure occurs due to change in cross-section of the duct and also due to change in the direction of the duct. It does not occur in a straight duct.

Let us study both these types of duct losses.

15.8.1 Pressure Loss from Friction in Piping and Ducts

Friction is a resistance to flow resulting from fluid viscosity and from the walls of the pipe or duct. In previous examples, we have assumed values of friction pressure loss. Actually, we must be able to calculate it.

For the type of flow that usually exists in HVAC systems (called turbulent flow) the pressure loss or drop due to friction can be found from the following equation (called the Darcy–Weisbach relation):

$$H_f = f \frac{L}{D} \frac{v^2}{2g} \qquad (15.10)$$

where

H_f = pressure loss (drop) from friction in a straight pipe or duct (m)
f = a friction factor
L = length of pipe or duct (m)
D = diameter of pipe (m) (For any other cross-section it is hydraulic diameter, D_h)
v = velocity of fluid (m/s²).

Equation (15.10) can also be expressed in the following way:

$$\frac{\Delta p_f}{\rho} = gH_f = f \frac{L}{D} \frac{v^2}{2} \qquad (15.11)$$

where Δp_f is the frictional pressure drop in N/m².

The friction factor f depends on the roughness of the pipe or duct wall surface. A rough surface results into an increased frictional resistance. To keep energy loss minimum, surfaces of the pipe or wall should be smooth. Lower velocities and larger diameters reduce H_f and therefore result in lower energy consumption, although the cost of pipe or duct then increases.

H_f could be calculated each time from Eq. (15.10). Using Eq. (15.11) charts shown in Figures 15.6 and 15.7 have been developed that are much easier to use and show the same information for airflow. This chart shown in Figure 15.6 is for low velocities and Figure 15.7 is for high velocities. These charts are plotted for the volume flow rate in m³/s as a function of friction rate, i.e. the friction pressure drop per unit length. The other parameters are the duct diameter D and mean velocity v. The charts are valid for air at 20°C and 1.01325 bar and clean galvanised iron (GI) with joints and seams having good commercial practice. The following are the limitations of these charts:

1. For the duct materials like plastics, concrete and wood, they do not apply.
2. These charts are only for air.
3. These charts are based on standard air density.

Similar charts are available in open literature for other pipe materials as well as for water flow.

15.8.2 Friction Factor 'f'

The friction factor for smooth ducts may be obtained by the following formulae

$$f' = \frac{64}{R_N} \quad \text{for laminar flow}$$

$$f' = \frac{0.3164}{(R_N)^{0.25}} \quad \text{for turbulent flow}$$

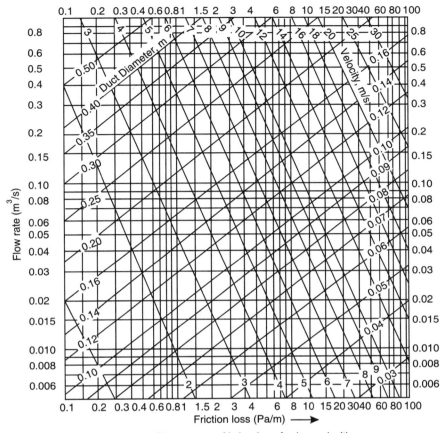

Figure 15.6 Flow rate vs. friction loss for low velocities.

where R_N is Reynold number. It is dimensionless and is defined as the ratio of inertial force to viscous force.

$$R_N = \frac{\rho_a D v}{\mu}$$

where ρ_a = density of air (kg/m³), D = diameter of duct (m), v = velocity of air (m/s) and μ = absolute viscosity (N/m²).

In the case of rough pipes or ducts, the friction factor depends upon the roughness factor e/D where 'e' is absolute roughness of the surface and D is diameter of the duct.

The friction factor may be calculated by the formula

$$f = \frac{1}{\left[1.74 - 2\log\left(\frac{2e}{D}\right)\right]^2}$$

or directly from Moody chart for different R_N and e/D.

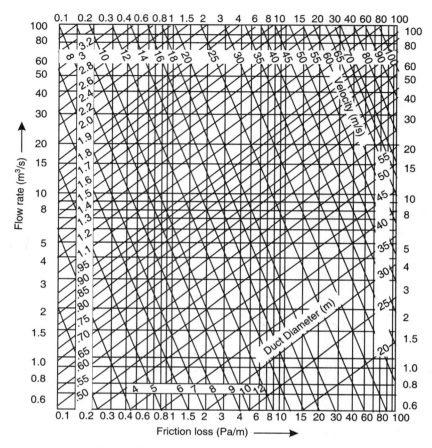

Figure 15.7 Duct friction chart for high velocities.

15.9 RECTANGULAR SECTIONS EQUIVALENT TO CIRCULAR SECTIONS

Ducts of rectangular sections are economical for carrying the air compared with circular sections. For this reason, ducts are designed on the basis of circular cross-section and then converted into an equivalent rectangular section. The frictional loss of the rectangular section is calculated after converting the rectangular section into an equivalent diameter.

There are two methods of finding out the equivalent diameter for a given cross-section.

The rectangular duct carrying the same quantity of air as the circular duct and the pressure drop per unit length of duct in both the cases is same.

The velocity of air in the duct is given by the equation

$$V = \frac{Q}{60\,A} \qquad (15.12)$$

where

Q = quantity of air (m³/min)
A = cross-sectional area (m²)
v = velocity of air (m/s).

Substituting the value of v from Eq. (15.12) into Eq. (15.11), we get

$$\frac{\Delta p_f}{\rho} = gH_f = f \frac{L}{D} \frac{v^2}{2}$$

or

$$H_f = f \times \frac{L}{D_h} \times \frac{1}{2g} \left(\frac{Q}{60A}\right)^2 = f \times \frac{L}{D} \times \frac{1}{2g} \left(\frac{Q}{60A}\right)^2$$

where hydraulic diameter $D_h = D$ for round duct.

$$\therefore \quad Q = \sqrt{\frac{2g(60)^2 \times H_f}{f \times L}} \sqrt{A^2 \times D}$$

$$= \sqrt{\frac{2g(60)^2 \times H_f}{f \times L}} \sqrt{\frac{A^3 \times 4}{P}} \quad \left[\because \frac{A^3}{A} \times D = \frac{A^3 \cdot D}{\frac{\pi}{4} \cdot D^2}\right] \quad (15.13)$$

where P is wetted permitted = πD.

If two ducts, of which one is circular and the other rectangular (different cross-sectional areas) carry the same quantity of air and have the same pressure drop, then the factor $\sqrt{\frac{A^3}{P}}$ in Eq. (15.13) must be same for both the ducts.

$$\sqrt{\frac{\pi^3 D^6}{(4)^3 \pi D}} = \sqrt{\frac{(ab)^3}{2(a+b)}}$$

where a and b are the two sides of the duct and D is the diameter of the circular duct.

$$\therefore \quad D = 1.265 \left[\frac{(ab)^3}{(a+b)}\right]^{1/5} \quad (15.14)$$

The major factor that influences the initial as well as the running cost for the flow of air through the duct is the *aspect ratio* of the duct which is defined as

$$\text{Aspect ratio} = \frac{a}{b} \text{ where } a > b.$$

For different value of (a/b), the values of (a/D) are calculated and represented in the form of graph as shown in Figure 15.8.

Air velocity through the rectangular duct is same as in circular duct and pressure drop per unit length of duct in both the cases is same.

Again using Eq. (15.12), we can write the velocity in terms of other factors as

$$V = \sqrt{\frac{2g \times H_f}{f \times L}} \sqrt{D_h} = \sqrt{\frac{2g \times H_f}{f \times L}} \sqrt{\frac{4 \times A}{P}} \quad (15.15)$$

For the given condition, we can equate the value of the term $\frac{A}{P}$ for circular and rectangular ducts.

$$\sqrt{\frac{\pi D^2}{4\pi D}} = \sqrt{\frac{(ab)}{2(a+b)}}$$

$$\therefore \quad D = \frac{2ab}{a+b} \quad (15.16)$$

Normally the duct sizes are worked out for the flow rates and air velocity in circular cross-section and then they are converted to rectangular sections considering the aspect ratio. The conversion charts and graphs are available for immediate calculations. One rough conversion chart is shown in Figure 15.8. Similar charts can be found in the open literature.

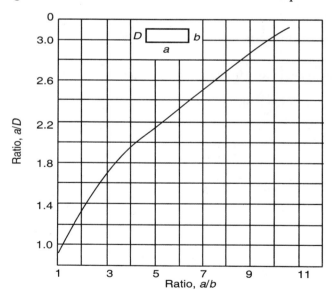

Figure 15.8 Equivalent duct sizes.

EXAMPLE 15.3 A rectangular duct 0.15 m × 0.12 m is 20 m long and carries standard air at the rate of 0.3 m³/s. Find the total pressure required at the inlet to the duct in order to maintain this air power. $f = 0.005$.

Solution:

$$\text{Velocity} = \frac{\text{Flow rate}}{\text{Area}} = \frac{0.3}{0.15 \times 0.12} = 16.7 \text{ m/s}$$

$$\text{Hydraulic mean depth, } R_h = \frac{\text{Area}}{\text{Perimeter}} = \frac{ab}{2(a+b)} = \frac{0.15 \times 0.12}{2(0.15+0.12)} = 0.033 \text{ m}$$

Assume air density, $\rho_a = 1.2$ kg/m³

Frictional pressure loss, $p_f = \dfrac{f L \rho_a v^2}{2 R_h}$

$$= \frac{0.005 \times 20 \times 1.2 \times (16.67)^2}{2 \times 0.033} = 505.3 \text{ N/m}^2$$

Velocity head, $P_v = \dfrac{\rho_a v^2}{2} = \dfrac{1.2 (16.67)^2}{2} = 166.7$ N/m²

Total pressure required $P_t = P_f + P_v$

$$= 505.25 + 166.73 = 671.98 \text{ N/m}^2 \qquad \textbf{Ans.}$$

Power required = $P_t \times$ discharge

$$= 671.98 \times 0.3 = 201.6 \text{ W} \qquad \textbf{Ans.}$$

EXAMPLE 15.4 A rectangular duct suction 500 mm × 350 mm size carries 1.25 m³/s of air having density of 1.15 kg/m³. Determine the equivalent diameter of the circular duct if (a) the quantity of air carried in both cases is same, (b) velocity of air in both the cases is same, and (c) if $f = 0.001$ for sheet metal, find the pressure loss per 100 m length of duct.

Solution: Given: $a = 0.5$ m, $\quad b = 0.35$ m
$Q = 1.25$ m³/s, $\quad \rho_a = 1.15$ kg/m³, $\quad f = 0.001$

(a) Equivalent diameter for the same flow rate is given by formula

$$D = 1.265 \left(\frac{a^3 b^3}{a+b}\right)^{1/5} = 1.265 \left[\frac{(0.5)^3 (0.35)^3}{0.5 + 0.35}\right]^{1/5}$$

$$= 1.265 \times 0.363 = 0.46 \text{ m} \qquad \textbf{Ans.}$$

(b) Equivalent diameter for the same velocity

$$D = \frac{2ab}{a+b} = \frac{2 \times 0.5 \times 0.35}{0.5 + 0.35} = 0.412 \text{ m} \qquad \textbf{Ans.}$$

(c) $\qquad$ Velocity of air $= \dfrac{Q}{\text{Area}} = \dfrac{1.25}{0.5 \times 0.35} = 7.143$ m/s

$$\text{Mean hydraulic depth} = \dfrac{A}{P} = \dfrac{0.5 \times 0.35}{2(0.5 + 0.35)} = 0.103 \text{ m}$$

$$\text{Friction pressure loss, } H_f = f \times \rho \left(\dfrac{L}{R_h}\right)\left(\dfrac{v^2}{2g}\right) = 1.15 \dfrac{0.001 \times 100 \times (7.143)^2}{2 \times 9.81 \times 0.103}$$

$$= 2.9 \text{ m of water} \qquad \textbf{Ans.}$$

15.10 DYNAMIC LOSSES IN DUCT

In a flow through duct, whenever there is a direction or velocity change, the pressure loss is greater than what it would be if there were uninterrupted flow. The additional loss, in excess of the *straight-duct friction loss*, is the dynamic pressure loss. The dynamic losses in duct are caused due to:

1. Change in direction, i.e. due to elbows, bends, etc.
2. Changes in area or velocity, i.e. due to enlargement, contraction, suction and discharge openings, dampers, etc.

This pressure loss is expressed in two ways. One is called the *equivalent length method*, while the other is called the *loss coefficient method*. Equivalent length method is used to work out the pressure loss due to fittings. For more details the readers may refer to a good book on fluid mechanics or ASHRAE handbooks.

15.10.1 Pressure Loss due to Sudden Enlargement

A sudden or abrupt enlargement in duct cross-section causes the formation of eddies in the corners of the enlarged section of the duct as shown in Figure 15.9. The loss of pressure is due to these eddies.

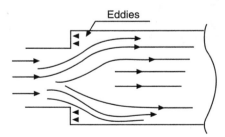

Figure 15.9 Sudden enlargement.

The pressure loss is given by the formula

$$(\Delta p_L)_{\max} = (0.5) \times \rho \times (V_1 - V_2)^2$$

$$= \left(1 - \frac{A_1}{A_2}\right)\frac{1}{2}\rho v_1^2 = (K)_{max}\, \rho v_2 \quad N/m^2$$

Here, $(K)_{max}$ is the maximum value of dynamic loss coefficient. The derivation part is not included here.

However, to avoid a large loss of pressure, abrupt enlargements are rarely done in the duct. In a gradual enlargement the pressure loss is given by the formula,

$$P_L = (C_r\, C_1 \times \rho \times v_1^2) \quad N/m^2$$

where C_r is the ratio of actual loss to loss in sudden enlargement. The C_r depends upon the angle of gradualness. The less this angle, the less is the C_r.

15.10.2 Pressure Loss Due to Contraction

An abrupt contraction in the cross-section of the duct causes the eddies' formation due to which there is added pressure drop. The pressure loss is strictly due to enlargement at vena-contracta.

The formula for loss of pressure is

$$P_L = C_2 \times \rho \times v_2^2$$

where C_2 is the constant of contraction and v_2 is the velocity in smaller cross-section.

In the case of gradual contraction, which is the norm in duct construction, the pressure loss is given by

$$P_L = C_r\, C_2 \times \rho \times v_2^2$$

where C_r is loss coefficient, which is smaller when the angle of contraction is lesser.

15.10.3 Pressure Loss at Entry or Exit from Duct

This loss is also as if due to contraction at entry and enlargement at exit of the duct. So the formula is given as

$$\text{Pressure loss } P_L = (C \times \rho \times v^2) \quad N/m^2$$

where v is velocity in duct and C is the constant based on contraction or enlargement.

15.10.4 Pressure Loss in Bends, Tees and Branch Offs

The pressure loss due to change in direction can be calculated by a formula similar to that used for the change in magnitude, i.e.

$$P_d = (C \times \rho \times v^2) \quad N/m^2$$

where C is the dynamic loss coefficient and v is the velocity of air.

15.10.5 Pressure Loss in Fittings

In addition to the pressure loss in straight lengths of duct, there is a pressure loss when the air flows through duct fittings (elbows, tees and transitions). These pressure losses are due to the turbulence and change in direction. Normally it is expressed in terms of equivalent length.

This data is available in reference data book (ASHRAE handbook, Fundamental volume). The dynamic loss should be equated to friction pressure loss, i.e.

$$P_d = \frac{fL_{eq}\, \rho_a V^2}{2D_h}$$

or

$$C \times \rho_a V^2 = \frac{fL_{eq}\, \rho_a V^2}{2m}$$

or

$$C = \frac{fL_{eq}}{2D_h}$$

$$\therefore \quad L_{eq} = C \times \frac{2D_h}{f}$$

This equivalent length obtained for dynamic loss in each fitting is added to the length of straight duct to find the total pressure loss.

15.11 METHODS OF DUCT DESIGN

In an air-conditioning distribution system, the velocity is maximum at the blower outlet and goes on reducing in subsequent sections. The maximum velocity permissible in the duct is as per permissible friction drop or by noise level. Generally, it is noticed that noise level limitations permit lower maximum velocity and therefore this becomes the decisive criterion. Table 15.1 gives the recommended and highest velocities permissible in different applications.

Table 15.1 Recommended and highest velocities permissible in different applications

Designation	Recommended velocities (mpm)			Maximum velocities (mpm)		
	Residences	Schools, theatres, public buildings	Industrial buildings	Residences	Schools, theatres, public buildings	Industrial buildings
Outside air intakes	152	152	152	244	275	365
Filters	76	91	107	91	107	107
Heating coils	137	152	183	152	183	214
Air washers	152	152	152	152	152	152
Suction connections	214	244	304	275	304	428
Fan outlets	304–487	394–608	487–732	518	457–670	518–853
Main ducts	214–275	304–394	365–548	244–365	335–488	394–608
Branch ducts	183	183–275	244–304	214–304	244–365	304–548
Branch risers	152	183–214	244	198–244	244–365	304–487

Three methods of sizing ducts will be explained here—equal friction method, static regain method and velocity reduction method.

15.11.1 Equal Friction Method

1. With this method, the same value of friction loss rate per length of duct is used to size each section of the duct in the system.
2. The friction loss rate is chosen to result in an economical balance between duct cost and energy cost. A higher friction loss results in smaller ducts but higher fan operating costs.
3. Duct systems for HVAC installations are losely classified into low velocity and high velocity groups, although these are not strictly separate categories.
4. Typical ranges of design equal friction loss rates used for low velocity systems are from 6.5 mm to 12.5 mm of water column per 100 m of duct. Maximum velocities in the main duct at the fan outlet are limited where noise generation is a problem (Table 15.2). However, sound attenuation devices and duct sound lining can be used, if needed.
5. High velocity ducts are designed with initial velocities from about 762 m/min to as high as about 1220 m/min. Corresponding friction loss rates may be as high as 50 mm to 100 m of water column.
6. High velocity duct systems are primarily used to reduce the overall duct sizes. In many large installations, space limitations (above hung ceilings in shafts) make it impossible to use the larger ducts resulting from low velocity systems.
7. The higher pressures result in certain special features of these systems. The ducts and fans must be constructed to withstand higher pressures. The noise produced at high velocities requires special sound attenuation.

The following example illustrates duct sizing by the equal friction method.

EXAMPLE 15.5 Find the size of each duct section for the system shown in Figure 15.10, using the equal friction design method. Use rectangular ducts. The system serves a public building.

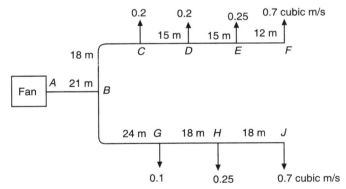

Figure 15.10 Duct system with flow rates—Example 15.5.

Solution:
1. Add up the flow rates (normally it is in CFM or cmm) backward from the last outlet, to find the flow rates in each duct section. The results are shown in Table 15.2.

2. Select a design velocity for the main from the fan using Table 15.1. A velocity of 300 m/min will be chosen that depends on the type of application.
3. From Figure 15.10, the friction loss rate for the main section AB is read as 0.4 Pa/m and the equivalent round duct diameter is read as 0.8 m.
4. The equivalent round duct diameter for each duct section is read from Figure 15.10 at the intersection of the design friction loss rate (0.4 Pa/m) and the flow rate for the section.
5. The rectangular duct sizes are calculated considering the aspect ratio. In actual installation, the duct proportions chosen would depend on space available.
6. The pressure loss in the system can be calculated as shown previously.

Table 15.2 Collection of results

Section	Flow (cubic m/s)	Velocity (m/s)	Friction loss (Pa/m)	Eq. duct diameter (m)	Rectangular duct size (m × m)
AB	2.4	5	0.4	0.8	One can find considering the aspect ratio and the availability of the space to lay the duct.
BC	1.35	4.7	0.4	0.60	
CD	1.15	4.3	0.4	0.55	
DE	0.95	4.2	0.4	0.49	
EF	0.7	4.0	0.4	0.425	
BG	1.05	4.4	0.4	0.52	
GH	0.95	4.2	0.4	0.49	
HJ	0.7	4.0	0.4	0.425	

Ans.

15.11.2 Static Regain Method

1. The static regain method of sizing ducts is most often used for high velocity systems with long round ducts, especially in large installations.
2. In this method, an initial velocity in the main duct leaving the fan is selected in the range of 762 m/min to 1220 m/min.
3. After the initial velocity is chosen, the velocities in each successive section of duct in the main run are reduced so that the resulting static pressure gain is enough to overcome the frictional losses in the next duct section.
4. The result is that the static pressure is the same at each junction in the main run. Because of this, there will not be extreme differences in the pressures among the branch outlets, so balancing is simplified.
5. One disadvantage of the static regain method of duct design is that it usually results in a system with some of the duct sections larger than those found by the equal friction method. For systems at high velocities, however, this method is recommended.

The following example illustrates duct sizing by the static pressure regain method.

EXAMPLE 15.6 Determine the duct sizes for the system shown in Figure 15.11, using the static regain method. Round ducts will be used.

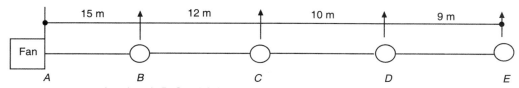

At points *A*, *B*, *C* and *D* there are 4 diffusers 15 cubic m/s each

Figure 15.11 Use of static regain method—Example 15.6.

Solution:
The results of the work are summarized in Table 15.3. The steps are as follows:
1. A velocity in the initial section is selected. (This system is a high velocity system, so the noise level will not determine the maximum velocity. Sound attenuating devices must be used.) An initial velocity of 14 m/s is chosen to explain the static regain method but it would be as per the application.
2. From Figure 15.10, the friction loss in section *AB* is determined. The duct size and static pressure loss due to friction loss per m are 0.8 m and 2.5 × 15 = 37.5 Pa.
3. The velocity must be reduced in section *BC* so that the static pressure regain will be equal to the friction loss in *BC*. There will not be a complete regain due to dynamic losses in the transition at *B*. We will assume a 75% regain factor for the fittings. A trial and error procedure is necessary to balance the regain against the friction loss. Let us try a velocity of 13 m/s in section *BC*. The friction loss is

$$\text{Loss in } BC = \frac{1.4 \text{ Pa}}{m} \times 12 \text{ m} = 16.8 \text{ N/m}^2$$

The static pressure regain available to overcome this loss is

$$\text{Regain at } B = 0.75 \left[\left(\frac{14}{4.04}\right)^2 - \left(\frac{13}{4.04}\right)^2 \right]$$

$$= 1.24 \text{ mm H}_2\text{O } (\rho_1 \times h_1 = \rho_2 \times h_2. \text{ Find } h_2 \text{ for the air column as } 1.033 \text{ m}, p = \rho \times g \times h = 1.2 \times 9.81 \times 1.033 = 12.16 \text{ N/m}^2$$

Actual pressure loss is 16.8 N/m² while static regain is 12.16 N/m² which is less than 16.8 N/m² and also very close to it. Otherwise one would require to conduct a trial and error method to achieve the same.

This trial is satisfactory. The regain at *B* is precise enough to overcome the loss in section *BC*. The duct size of *BC* is 0.75 m.

4. Continue the same procedure at transition *C*. Let us try a velocity of 9 m/s in *CD*.

$$\text{Loss in } CD = \frac{1.4 \text{ Pa}}{m} \times 10 \text{ m} = 14 \text{ Pa}$$

$$\text{Regain at } C = 0.75 \left[\left(\frac{10}{4.04}\right)^2 - \left(\frac{9}{4.04}\right)^2 \right] = 1.65 \text{ mm H}_2\text{O (i.e. } 13.5 \text{ N/m}^2)$$

The regain at point C is 13.5 N/m² which is very close to 14.0 N/m². The first guess is satisfactory. No further trial is needed. The duct size is 0.65 m.

5. Loss in section $DE = 1.4 \times 9$ m $= 12.6$ N/m².

$$\text{Regain at } D = 0.75 \left[\left(\frac{9}{4.04} \right)^2 - \left(\frac{8}{4.04} \right)^2 \right] = 1.04 \text{ mm H}_2\text{O (i.e. 10.2 N/m}^2)$$

Table 15.3 Collection of results

Section	Flow (cubic m/s)	Velocity (m/s)	Eq Diameter (m)	Velocity pressure (Pa/m)	Length (m)	Friction loss (Pa)	Static Pressure regain
AB	6	14	0.8	2.5	15	37.5	
B							12.2
BC	4.5	10.0	0.75	1.4	12	16.2	
C							13.5
CD	3	9.0	0.65	1.4	10	14.0	
D							10.2
DE	1.5	9	0.49	1.4	9	12.6	

Ans.

15.11.3 Velocity Reduction Method

1. This is the simplest and quickest method of duct design.
2. The designer reduces the velocity in subsequent sections based on his experience of the earlier designs.
3. Then the size is calculated by using the values of the velocity and discharge in the chart.
4. The major drawback of the method is that it is not based on any principle so it cannot be used by a new designer. However, even for an experienced designer it is difficult to use this method for a complex distribution system.

EXAMPLE 15.7 A typical duct system is shown in Figure 15.12.

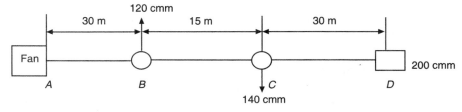

Figure 15.12 Use of equal friction method—Example 15.7.

Using the friction chart and equal friction pressure drop method, estimate the diameter and extra velocity pressure in *AB*, *BC*, *CD*, *BE*, and *CF*. State the assumptions, if any.

Solution:

Assume it is an office and the velocity at fan outlet as per noise level consideration is 400 m/min.

For *AB*, from friction chart (Figure 15.6) at 7.67 m³/s flow and 6.67 m/s velocity, the friction pressure drop/m length on abscissa is 0.038 N/m². The corresponding diameter of equivalent round duct is 1.2 m as obtained from the same chart.

For subsequent sections of the duct, the velocity and duct diameter are found out for the respective flow and friction pressure drop of 0.038 N/m² per metre length. The values are shown in Table 15.4.

Table 15.4 Collected results

Duct section	cmm	Flow (m³/s)	Velocity (m/s)	Pressure drop (N/m²)	Diameter (m)
AB	460	7.67	6.67	0.038	1.2
BC	340	5.67	6.0	0.038	1.1
CD	200	3.33	5.3	0.038	0.89
BE	120	2	4.6	0.038	0.69
CF	140	2.34	4.8	0.038	0.73

To find the velocity head in each section, assume air density ρ_a as 1.2 kg/m³.
Therefore, velocity head (dynamic head)

$$\text{In } AB = \left(\frac{V}{4.04}\right)^2 = \left(\frac{6.67}{4.04}\right)^2 = 2.72 \text{ mm of water}$$

$$\text{In } BC = \left(\frac{V}{4.04}\right)^2 = \left(\frac{6.0}{4.04}\right)^2 = 2.21 \text{ mm of water}$$

$$\text{In } CD = \left(\frac{V}{4.04}\right)^2 = \left(\frac{5.3}{4.04}\right)^2 = 1.72 \text{ mm of water}$$

$$\text{In } BE = \left(\frac{V}{4.04}\right)^2 = \left(\frac{4.6}{4.04}\right)^2 = 1.3 \text{ mm of water}$$

$$\text{In } CF = \left(\frac{V}{4.04}\right)^2 = \left(\frac{4.8}{4.04}\right)^2 = 1.41 \text{ mm of water}$$

Ans.

EXAMPLE 15.8 An air distribution system is shown in Figure 15.13. The air supply quantity and the equivalent lengths of duct are also shown. The velocity in *AB* is limited to 400 m/min as per noise level consideration. The ducts are of rectangular section and one side of each duct is 60 cm.

Find the size of the duct by the equal friction method, and the maximum pressure loss.

Solution: Assume it is an office; the velocity at fan outlet as per noise level consideration is 400 m/min.

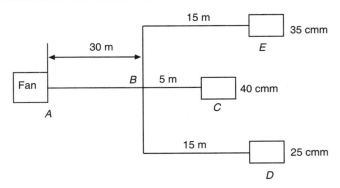

Figure 15.13 Use of equal friction method—Example 15.8.

For AB, from friction chart shown in Figure 15.6 and Figure 15.7 at 7.67 m³/s flow and 6.67 m/s velocity the friction pressure drop/m length on abscissa is 0.7 N/m². The corresponding diameter of equivalent round duct is 0.56 m obtained from the same chart.

The flow rate cmm is converted to m³/s, for example, 40 cmm = $\dfrac{40}{60}$ = 0.67 m³/s.

Table 15.5 Collection of results

Duct section	Length (m)	Discharge (m³/s)	Velocity (m/s)	Friction rate (Pa/m)	Diameter (m)
AB	15	1.67	6.67	0.7	0.56
BC	5	0.67	5.5	0.7	0.4
BD	15	0.42	4.8	0.7	0.34
BE	15	0.58	5.0	0.7	0.38

To find maximum pressure loss in the duct system, one has to multiply maximum length by friction rate.

AD or AE are the ducts of maximum length

Maximum pressure loss = (15 + 15) × 0.1 = 3 N/m² **Ans.**

15.12 DUCT ARRANGEMENT SYSTEMS

The requirement of cooling or that of creating comfort conditions in a residential space or in a factory space is met through an air conditioning system. The comfort conditions in a space are met through combinations of air temperature, humidity of air, velocity of supply air and uniform air distribution. The maximum temperature variation in an air conditioned space should not be more than 1°C. The stages of creating comfort conditions inside a residential space by air conditioning involve:

- Estimation of the cooling load.
- Selection of the air conditioning system such as the split AC, package AC, ductable AC, central AC, etc. in accordance with the estimated cooling load.
- Installation of the chosen AC units.

Only the ductable AC and the central AC systems need to have a duct system to carry the conditioned air from the air handling unit to the conditioned space and to return the air back to the conditioner. While supplying the conditioned air to a space through a duct arrangement, one must consider the following points.

1. To prevent any excessive pressure loss in the system.
2. To maintain a proper air velocity in the duct.
3. To keep the noise level as minimum as possible.
4. To keep the supply air from a supply outlet towards the direction of the occupants.
5. To keep the flow of supply air from top to bottom.
6. To prevent the supply air from being short-circuited to the doors of the conditioned space.

Keeping the above points in mind, the following are the few air distribution systems discussed in this chapter.

15.12.1 The Perimeter System

The perimeter system can be either of loop type as shown in Figure 15.14 or of radial type as shown in Figure 15.15. The AC system is usually located in the basement but as nearly as possible at the geometric centre of the building or all supply outlets of the AC. The supply outlets are placed close to the ceiling level at a height of about 2.5 m from the floor level. The ducts run through the basement, the building foundation slab, the floor and connect the air-conditioner to the outlet grills. The return grills are generally located on the bottom side of the inside wall. This arrangement is commonly used for residential systems.

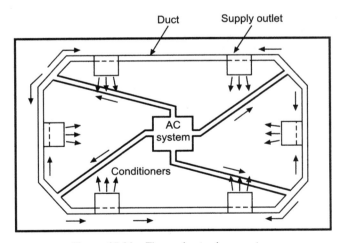

Figure 15.14 The perimeter loop system.

The advantages of the perimeter system are that there is proper distribution of the conditioned air. The used air can be easily brought back to the conditioner through return ducts.

The disadvantages are that the system needs planning at the time of construction of the building itself. The fresh air intake is thus possible with proper planning at the construction stage of the building.

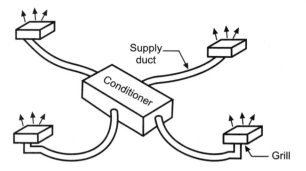

Figure 15.15 Radial perimeter system.

15.12.2 Extended Plenum System

The simple arrangement of ducts of this system is shown in Figure 15.16. The important advantage of this system is that the grills can be located at any required point as per the structural demands. The air conditioner unit may be located in the attic, in the basement or in any other convenient place. This system can be used either for residential or for commercial purpose. The main duct and the branch ducts are suspended at a height of about 2.5 to 3 m from the floor.

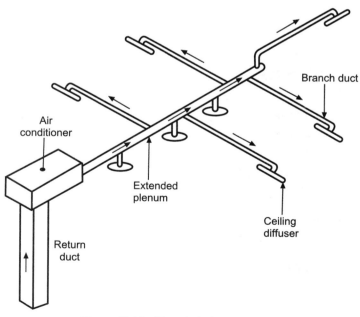

Figure 15.16 Extended plenum system.

A system arrangement of ducts in a commercial space is shown in Figure 15.17. Air is uniformly distributed in the space. Also, the return air is taken back to the air conditioner easily.

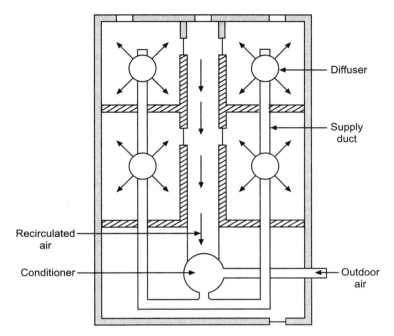

Figure 15.17 Arrangement of ducts in a commercial space.

The arrangement of the duct system on the basis of the number of ducts is discussed below.

Single duct system

This simple arrangement is shown in Figure 15.18. It consists of an air filter, a blower (to increase the pressure), a heating/cooling coil and a duct system to carry the air as shown. The conventional duct system operates at a low pressure with ordinary balancing dampers and booster coils or at a medium pressure with the more sophisticated air valves, booster coils on the low pressure side of the air valves and a low pressure distribution duct work. Such a system offers a high degree of comfort, ease of control, simple operation, and low cost.

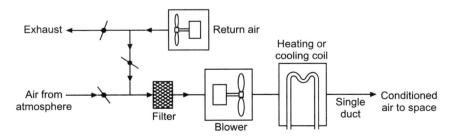

Figure 15.18 Single duct system.

The geographic areas where dehumidification is not a problem, the installation of a single duct re-cool system is more advantageous. In this system, air is supplied through a cooling coil of minimum capacity which cools the air in accordance with the setting of the room-thermostat. During winter, hot water is circulated through the same coil, because dehumidification is not required since condensation does not occur on the cooling coils. Hence, there is no problem of drain pans and eliminators. If this system is provided with a four-pipe system having hot and chilled water available at all times, a system with extremely flexible operating characteristics can be achieved.

Dual duct system

The availability of high pressure air handling units with their inherent stability characteristics, has permitted the development of the various dual duct designs.

A simple form of a dual duct system is shown in Figure 15.19. The cooling and/or heating coils are located on the discharge side of the supply fan. A return fan provides the static pressure necessary to vary the return and outside air. When the outside air temperature is less than the comfort temperature, then the outside air is appropriately mixed with the return air before passing it through an air filter and a blower for supplying the conditioned air to the space.

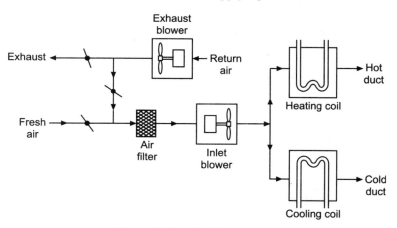

Figure 15.19 Dual duct system.

Similarly, when the outside temperature is much higher than the comfort temperature, the air is cooled by passing the same over a cooling coil before being supplied to the space. This is acutally done by varying the damper position. The quantity of air passing through the hot and cold ducts is controlled by the terminal units.

The dual duct system offers the following advantages:

- Since the cooling and heating needs of the conditioned air are achieved through separate heating or cooling coils in separate ducts, the system leads to energy saving.
- It is an efficient system for use in refrigeration because of absence of reheat.
- It is flexible and amenable to change even after construction.
- It can be used with many different air outlet arrangements such as the overhead system, under the window units and base board units.

This system is particularly economical and versatile where different conditions of air are required simultaneously such as those in hospitals.

Dual duct and induction system combined

A combined dual duct and induction system is shown in Figure 15.20. In this case, the room coil is used only for cooling. Both warm and cold primary air is available at all times. A thermostat controls the quantities of hot and cold air mixed in an induction unit to achieve the desired room temperature.

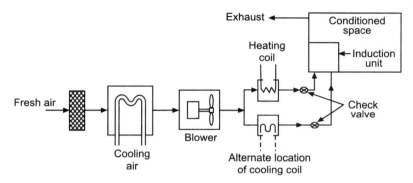

Figure 15.20 Combined dual duct and induction systems.

To achieve the proper temperature that the room needs, the hot and cold check valves can be operated either way. For example, if a more quantity of hot air is needed, then the check valve of the hot air line opens and at the same time the check valve of the cold air line closes. This arrangement is suitable for a hospitals since the air is re-circulated only within the room and the excess can be exhausted through outlets.

Return in duct system

The used air from the conditioned space is recirculated after treating in the conditioner. Therefore, in a junction of two air streams, the return air stream through the return duct is connected to the supply air duct. Four types of arrangements are illustrated with the help of simple sketches as shown in Figure 15.21.

For the straight angle inlet fitting as shown in Figure 15.21(a), the momentum of return air deflects the main stream, resulting in a contraction followed by an *abrupt expansion loss*. In addition, the branch stream is mixed with the main stream and accelerated, resulting in further loss.

The modification of the fitting as shown in Figure 15.21(b), results in less contraction of the stream and lower loss. Both types in Figure 15.21(a) and (b) result in a pressure loss which is a substantial part of the leaving velocity head. The pressure loss is usually several times as great as the friction loss.

A type of fitting that can be used in place of a damper in a return air riser is a shaft as shown in Figure 15.21(c). By proportioning the area of branch to the area of the main as the ratio of the air quantities of the branch to those of the main, the velocities of the merging streams are made equal and hence the mixing loss does not occur.

A type of fitting that combines the branch damper and also the duct size change with the fittings is shown in Figure 15.21(d). Since friction loss in the main upstream of the branch always occurs, the velocity of the branch air leaving the damper is higher than that of the air coming down the main, and a pressure rise always occurs in the fitting. This results in minimum pressure loss in the system and hence by a proper design, a substantially constant static pressure in the main can be achieved.

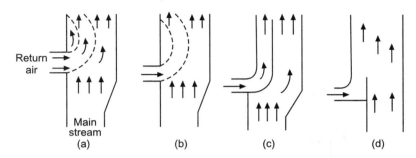

Figure 15.21 Four types of arrangements of return air duct branch fittings.

A return air fan is used in all cases, except in very small installations. It is typically selected to work against a static pressure of 0.75 to 1.5 inch W.G. Since the pressure resistance of most systems is much below this range, most of the pressure loss occurs in the branch fittings.

The pressure loss is less for the fittings shown in Figures 15.21(c) and (d) for the reasons mentioned above. The other advantages of these fittings are listed below:

- Much higher duct velocities with lower noise generation can be achieved because noise generation is dependent on the relative velocity of the streams merging in the fittings and also on the amount of turbulence generated.
- The smaller ducts (due to high velocity) result in savings in cost and shaft space and can serve a larger number of floors from one apparatus.
- The system balancing can be minimised because of less pressure drop along the duct and high velocity through the fittings.
- The need for a return air fan can be eliminated in many systems.
- The effect of outside temperature variations on the infiltrations and exhilarations of air at different levels of a tall building, can be minimised.

15.13 AIR DISTRIBUTION SYSTEMS

The aim of air distribution in a warm air heating and air conditioning system are:

1. To create the proper combination of temperature, humidity and air motion in the occupied zone of the conditioned room.
2. To establish comfort conditions within this zone as per acceptable limits established in respect of air temperature, air motion, relative humidity and their physiological effects on the human body. Any variation from accepted standards of one of these

parameters may result in discomfort to the occupants. Discomfort may also be caused by lack of uniformity of conditions within the space or by excessive fluctuations of conditions in the same part of the space. Such discomfort may rise due to excessive variations in room air temperature (horizontally, vertically or both), excessive air motion (draft), failure to deliver or distribute the air according to the load requirements at the different locations, or too rapid fluctuations of room temperature or that of air motion.

3. Conditioned air must be delivered to the air conditioned space and distributed at the desired temperature and velocity. The temperature variation should not be more than 2°C in the occupied zone of the room. The desired air movement around the bodies of the occupants is 7.5 m/min. The maximum allowable movement is 15 m/min when the occupants are sitting and 20 m/min when the occupants are moving.
4. To prevent oversupply of the cold conditioned air to a particular point so that there does not occur a cold spot in the conditioned space.
5. To prevent undersupply of the conditioned air so that there does not exist a hot spot/space at a higher temperature than needed.
6. To ensure that the supply air from the supply outlets is directed towards the face of the occupants.
7. To ensure that the air movement takes place from top to bottom.
8. To prevent short-circuiting of supply air to the doors of the conditioned space.
9. The temperature of the conditioned air may be above or below that of the air in the occupied zone.

The proper air distribution in the conditioned space requires entrainment of room air by the primary air stream outside of the zone of occupancy in order to ensure that air motion and temperature differences are reduced to acceptable limits before the air enters the occupied zone.

There are several methods of air distribution systems which can be successfully used for different purposes. The common air-distribution systems are explained with the help of Figures 15.22 to 15.27.

Simple type

The supply and exhaust grills are installed in the same wall. This system is simple in construction and is cheaper. The velocity at the outlet of the inlet grill must be sufficient to throw the air horizontally but not that high so as to cause objectionable downward reverse currents. The return on the far wall tends to create a stagnant space below the supply outlet. The system shown in Figure 15.22 is commonly used for many installations; it is, however, more suitable for the summer air-conditioning system than for the winter air-conditioning system.

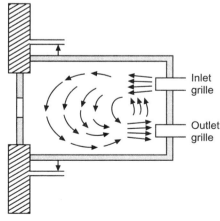

Figure 15.22 Air flow distribution patterns for cooling and heating.

Pan type

The pan type arrangement of the supply air shown in Figure 15.23 provides a uniform discharge around the perimeter of the conditioned space. It may be seen that the air distribution obtained in the room is more uniform than that of the arrangement shown in Figure 15.22. Another arrangement shown in Figure 15.24 assures an even more uniform distribution of air in the room than the arrangement shown in Figure 15.23.

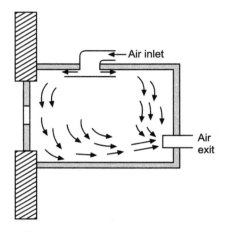

Figure 15.23 Pan type arrangement.

The arrangement shown in Figure 15.24 has a combined supply and return openings in a single unit. This method is used for heating as well as for cooling applications. If used for a cooling system, the effectiveness of air circulation is enhanced by natural upward convection currents. If used for comfort heating in cold climates, this arrangement will probably produce large ceiling-to-floor temperature differentials.

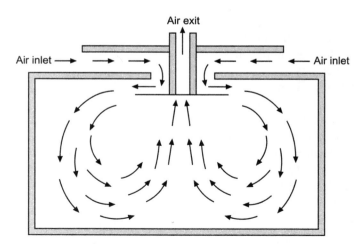

Figure 15.24 Pan type arrangement but with a much better uniform distribution of air in the room.

The system shown in Figure 15.24 is more advantageous when the ducts cannot run in the partitions or columns. The pan or plate may be ornamented to harmonise within the decorations in the room. This arrangement involves less capital and low running cost. It is a useful arrangement for single storey buildings.

Ventilation type

The inlets and outlets in the room may be located as shown in Figures 15.25 and 15.26 where only ventilation is required. The incoming air must have a low velocity to avoid uncomfortable drafts. The arrangement in Figure 15.25 is preferred for heating and that in Figure 15.26 for cooling.

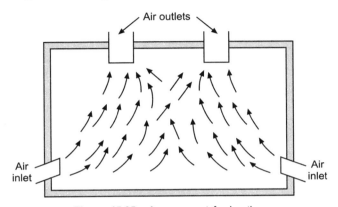

Figure 15.25 Arrangement for heating.

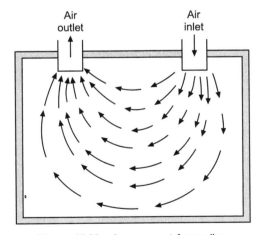

Figure 15.26 Arrangement for cooling.

Air-distribution system for auditoriums

The central air-distribution system is commonly used for auditoriums. The central air-distribution system is provided with a fan to blow air into the air-conditioned space through ducts and inlet openings.

The central air-distribution system is further divided into three groups according to the direction of air flow from the air inlet into the room.

1. **Upward flow system:** In this type of air-distribution system, the air is brought into the space through the inlets near the floor or through the pedestals of the chairs in the auditorium. The air flows upwards as shown in Figure 15.27.

The exhaust air outlets are located in the side walls near the ceiling or in the ceiling itself. This air-distribution system is used in rooms where there is a marked tendency of air being heated by the occupants. The upward flow carries with it the vitiating products from the bodies of the occupants.

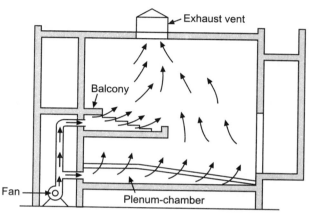

Figure 15.27 Upward air-distribution system for auditoriums.

The major difficulties faced in the design of this system are avoidance of drafts and non-uniform heating of the space. Hence this air-distribution system is not satisfactory for summer air-conditioning purposes where the supply air temperature has to be less than the room air temperature.

2. **Downward flow or overhead system:** This system is generally preferred when the ceiling is not free from obstructions. In this type of the air-distribution system, the air is introduced at the ceiling and over the balcony and is removed through the grills located on the main floor and from an exhaust chamber below the balcony seats as shown in Figure 15.28. This system is more suitable for installation wherein the ceiling is not free from obstructions. Such a downward air-distribution system is designed to spread the incoming air uniformly over the occupied zone to secure uniform conditions within the air-conditioned space. The downward air-distribution system is more satisfactory for theatres, auditoriums and movie halls. It is a much better system for summer air-conditioning than the upward air-distribution system.

3. **Ejector air-distribution system:** This system is generally adopted when the ceiling is free from obstructions. It is also a form of downward air-distribution system but the velocity of air at the inlet is considerably higher than the air velocity in the ordinary type of downward air-distribution system. The air is introduced at a high velocity through specially designed nozzles

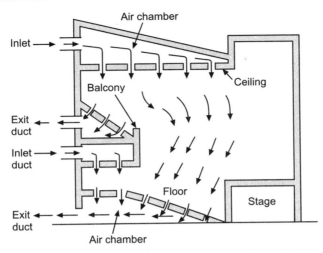

Figure 15.28 Overhead air-distribution system (for downward flow of air) for auditoriums.

at the rear of the building and is discharged well above the occupied zone as shown in Figure 15.29. The jet action of the incoming air causes a better diffusion of fresh air within the space. The outlets are located above the floor as shown in figure, so that the return air flows uniformly over the body of the occupants.

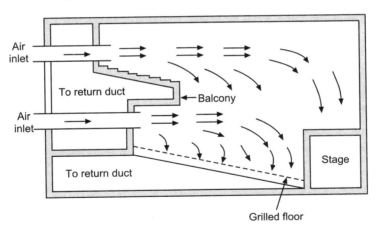

Figure 15.29 Ejector air-distribution system for auditoriums.

With any of the above three types of the systems used, supplementary supply of air from the wall openings must be provided as the space in auditoriums is very large.

The location of the return and exhaust inlets does not significantly affect the air motion, except possibly in the area near the inlet where the velocities may exceed the comfort limits. The inlets should be located in such a way so that the short-circuiting of the supply air is avoided.

The exhaust inlets in bars, kitchens, lavatories, dining rooms, club rooms, etc. should be located near the ceiling level so as to collect warm air, odours and fumes from such spaces.

An exhaust system provides vent-flues or ducts for the exhaust of air. The air flow through the exhaust system, is provided either by air pressure in the room or by the action of the exhaust fan. Exhaust ducts from the various points are connected to an attic space and then the air from the attic space is exhausted by a fan. The trunk system is also used for the exhaust of air where the ducts from the different rooms are connected to the main duct (trunk duct) and the air is then exhausted by an exhaust fan. The exhaust fan with a trunk duct system gives much better results.

It is possible to reduce the operating cost of an air-conditioning system by re-circulating part of the air supplied to the air-conditioned spaces. The proportions of fresh and re-circulated air mostly range from 20% to 30% fresh air and 70% to 80% re-circulated air. It is not desirable to re-circulate air from garages and workshops as it contains a high percentage of CO_2. Similarly, the air from kitchens, lavatories and dinning rooms must not be re-circulated as this exhaust air carries many objectionable fumes.

EXERCISES

1. Discuss the various materials used to construct ducts.
2. What is the aspect ratio of a rectangular duct? What are the advantages in keeping it low?
3. What are pressure losses in air distribution ducts? Explain them in brief.
4. What is equivalent diameter of rectangular duct? Prove that the equivalent diameter of rectangular duct for same air flow is $D_{eq} = 1.265 \left[\dfrac{a^3 b^3}{a+b}\right]^{0.2}$.
5. Describe the different methods of duct design.
6. Compare the duct design methods, i.e. equal friction and static regain methods.
7. Describe with a suitable sketch the duct arrangement of extended plenum type of system.
8. What is a dual duct system? Explain with the help of a suitable sketch.
9. What should be the design aims of an air distribution system?
10. Explain with the help of a sketch a suitable air distribution system for auditoriums.

Chapter 16

Cryogenics

16.1 INTRODUCTION

The term 'cryogenic' is derived from the Greek word *Kryos* meaning cold or frost, and is generally applied to very low temperature (below 125 K) applications. There is no definite demarcation between refrigeration and cryogenics. It means that refrigeration is treated up to sub-zero temperatures and below those temperatures it is assumed cryogenics. Normally, temperatures below 125 K are generally included in cryogenics, because boiling points of all permanent gases are below this limit. Table 16.1 gives the boiling points of the so-called permanent gases.

Table 16.1 Permanent gases in air near the earth's surface

Gas	Volume or mole%	Parts per million (ppm)	Boiling point (K)
Nitrogen (N_2)	78.08	—	77.36
Oxygen (O_2)	20.95	—	90.18
Argon (Ar)	0.93	—	87.28
Neon (Ne)	0.0018	18.2	27.09
Helium (He)	0.0005	5.2	4.21
Krypton (Kr)	0.000114	1.14	119.83
Xenon (Xe)	0.0000086	0.086	165.0
Hydrogen (H_2)	0.00005	0.5	20.3

The gases listed in the table remain unchanged in concentrations in the dry atmosphere.

Realistically cryogenic engineering is not a field in itself, but may well be regarded as an extension of many other fields of engineering into the extreme realm of low temperatures. The unusual phenomenon that occurs at extremely low temperatures and the special techniques that the engineers must employ tend to make cryogenic engineering a unique field.

The various methods used to liquefy air and separate its components, is a vast field which would be introduced in this chapter. Before that we shall recollect the applications of nitrogen, oxygen and argon either in liquid or gaseous state.

16.2 APPLICATIONS OF NITROGEN

Air is the main source of nitrogen. Nitrogen in its purest form can be economically produced in large scale by cryogenic air separation technology that finds widespread applications. The quality of nitrogen product required differs significantly from one industry to another.

(a) **Electronics industry:** The electronics industry demands the highest purity of 1 ppb O_2, dust level 1/std ft^3 and moisture level at 0.1 ppm. The US Semiconductor Industry Association's National Technology for Semiconductors forecasts that the N_2 purity requirement at a point of use (POU) would be 0.01 to 0.1 ppb in the year 2010.

(b) **Metals industry:** The purity of 2 to 100 ppm O_2 is sufficient for aluminium, rubber, glass, textile, chemicals and steel industries. Nitrogen is used in the manufacture of steel and other metals and as a shield gas in the heat treatment of iron, steel and other metals. It is also used as a process gas, together with other gases for reduction of carbonization and nitriding. 'Flash' or 'fins' on cast metal can be removed by cooling with liquid nitrogen, making them brittle. The purity of 10 ppm at a pressure of 300 to 400 bar (30 to 40 Mpa) will meet its requirements in petroleum industry.

(c) **Manufacturing:** Shrink fitting is an interesting alternative to traditional expansion fitting. Instead of heating the outer metal part, the inner part is cooled by liquid nitrogen so that the metal shrinks and can be inserted. When the metal returns to its normal temperature, it expands to its original size, leading to a very tight fit. Liquid nitrogen is used to cool concrete which leads to better cured properties.

(d) **Construction:** When construction operations must be done in soft, water-soaked ground such as tunnel construction underneath waterways, the ground can be frozen effectively with liquid nitrogen. Pipes are driven into the ground and liquid nitrogen is pumped through the pipes under the earth's surface. When the nitrogen exits into the soil, it vaporizes removing heat from the soil and freezing it.

(e) **Chemicals, pharmaceuticals and petroleum:** Refineries, petrochemical plants and marine tankers use nitrogen to purge equipment. Cold nitrogen gas is used to cool reactors filled with catalyst during maintenance work. The cooling time can be reduced substantially. Cooling reactors (and the materials inside) to low temperature allows better control of side reactions in complex reactions in the pharmaceutical industry. Liquid nitrogen is often used to provide the necessary refrigeration as it can produce rapid temperature reduction and easily maintain the required cold reaction temperatures. Reactor cooling and temperature control systems usually employ a circulating low temperature heat transfer fluid to transfer refrigeration produced by vaporizing liquid nitrogen to the shell of the reactor vessel.

(f) **Healthcare:** Nitrogen is used as a shield gas in the packaging of some medicines to prevent degradation by oxidation or moisture adsorption. Nitrogen is also used to freeze blood as well as viruses for vaccination.

  (i) Using liquid nitrogen, food can be frozen in few seconds thus preserving much of its original taste, colour and texture. It is reported that weight losses can be reduced considerably when food is frozen cryogenically rather than by any other means. The purity of LIN ranging from 95 to 98% is sufficient for freezing food.

(ii) Cryosurgery is a technique that destroys cancer cells by freezing. It has been used at some top medical centres for tumorous of the prostate, liver, lung, breast and brain as well as for cataracts, gyneacological problems and other diseases.

(iii) LIN is used in storing biological specimens, especially bull semen for the cattle industry.

(g) **Automobile industry:** The automobile tyres have been one of the most difficult items to recycle—or even worse—to discard. Cryogenics provides the necessary technology for the effective recovery, separation, and reuse of all materials used in the tyre. In fact, the use of LIN is the only known way to recover rubber from the steel radial tyre.

(h) **Grinding:** Mechanical breakdown of solids into smaller particles is known as grinding. Cryogenic grinding is a method of powdering materials at sub-zero temperatures. The materials are frozen with LIN when they are ground.

16.3 APPLICATIONS OF OXYGEN

Oxygen was one of the first atmospheric gases to be liquefied by Cailletet and Pictet in 1877. Later Polish scientists Olzewski and Wroblewski at Cracow in 1883 produced stable liquid oxygen in U-tube whose properties could be studied. Therefore, oxygen began its useful life in industry early in the twentieth century.

1. A huge quantity of oxygen is consumed in steel-making industry following LD or BOF process. These processes need 99.5% (conventional standard grade) purity of oxygen to accelerate the oxidation and for conversion of iron to steel.
2. The daily consumption amounts to several thousand tonnes and all modern steel plants, therefore, have tonnage oxygen plants.
3. One of the most common usages of oxygen is in the fabrication and cutting of metals using oxy-acetylene torch.
4. Another major use of oxygen is in the field of medicine to help patients breath. Oxygen is used in the preparation of chemicals. For example, manufacturing of ethylene oxide requires 40% oxygen while acetylene consumes almost 20% oxygen. Titanium dioxide, propylene oxide and vinyl acetate need 10–15% O_2 for their manufacture.
5. In glass manufacture, oxygen is added to enrich the combustion air in glass-melting furnaces. Jet aircraft for high altitude missions are equipped with oxygen systems for breathing purposes. Coal gasification is also one of the large consumers of gaseous oxygen.
6. Oxygen is increasingly becoming important as a bleaching chemical. In the manufacture of high-quality bleached pulp, the lignin in the pulp must be removed in a bleaching process. Chlorine has been used for this purpose but new processes using oxygen are found to reduce water pollution. Oxygen plus caustic soda can replace hypochlorite and chlorine dioxide in the bleaching process, resulting in lower costs. In a chemical pulp mill, oxygen added to the combustion air increases the production capacity of the soda recovery boiler and the lime-reburning kiln. The use of oxygen in black liquor oxidation reduces the discharge of sulphur pollutants into the atmosphere.

16.4 APPLICATIONS OF ARGON

Usually argon is obtained from air which contains 0.0093% of argon by volume. It is highly inert and finds its applications over a wide range of conditions, both at cryogenic and at high temperatures. Argon is relatively expensive and its use is limited to applications where its highly inert properties are essential.

1. The largest usage of argon worldwide is the argon-oxygen decarburization process for producing low carbon stainless steels.
2. MIG welding developed by Airco in the 1940s and TIG welding represent large markets for argon.
3. The lamp industry uses argon to fill incandescent bulbs. This gives longer life to the filament because argon does not react even at high temperatures.
4. Plasma-arc cutting and plasma-arc welding employ plasma gas (argon and hydrogen) to provide a very high temperature when used with a special torch. The air components like nitrogen, oxygen and argon are separated from air in their purest form through a cryogenic air separation plant. Therefore these are referred to as *cryogens*. Let us introduce the cryogenic distillation of air at this stage.

16.5 CRYOGENIC AIR SEPARATION PLANT

Most of the items of equipment involved in an air separation plant are dedicated to the liquefaction processes which involve compressors, equipment for separation of moisture, CO_2, hydrocarbons, heat exchangers, expansion devices, cryogenic pump, etc.

The compression block, consisting of a multistage centrifugal compressor with intercoolers and aftercoolers, is used to compress air to the pressure required in the high-pressure column (HP column).

Air is then processed through a purification block, which consists of a refrigeration/cooling system for air to be cooled to a suitable temperature for H_2O, CO_2 and hydrocarbons to be adsorbed in the molecular sieve adsorbers.

Air further passes through the heat exchanger block/consisting of plate-fin heat exchangers, where it is cooled to a temperature at which it is partially liquefied, using the cold from the return product and waste streams.

Partially liquefied air is sent to the double column system integrated with crude argon column (distillation block), where it is separated into O_2, N_2 and crude argon fractions. Product O_2 is taken from the bottom of the low-pressure column (LP column). While part of it may be taken as liquid product, the remaining is passed through the main heat exchanger (in the heat exchanger block) and collected as gaseous O_2 product after the recovery of its cold. Liquid nitrogen is drawn from the top of the high-pressure column. Gaseous N_2 is drawn from the top of the pure nitrogen column which is placed on the top of the low pressure column. The non-condensable gases are purged from the top of the high pressure column. The crude argon, normally with 97.5% argon content, is drawn from the top of the crude argon column and is processed in a purification unit to achieve the required purity level.

Recycle compressor and liquid production block are responsible for creating the liquid which is needed for distillation, to overcome heat in-leak and to produce liquid cryogens. Product gases or waste gases are pressurized to the required pressure in the recycle compressor and booster compressor, cooled in the refrigeration units and/or heat exchangers and are further

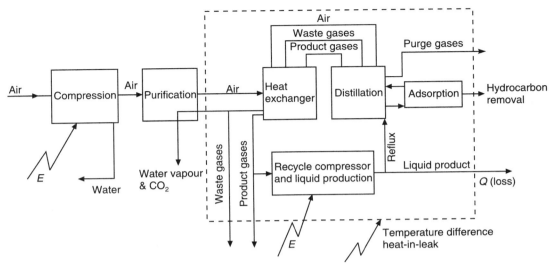

Figure 16.1 Block diagram of a cryogenic air separation plant.

expanded in the turbine to produce refrigeration, so that product gases can be liquefied. The output of the turbine is used to drive the booster compressor.

In the adsorption block, 'the hydrocarbons are removed from LOX and oxygen-rich bottom product of HP column by passing through the MS adsorbers.

Though the raw material is available absolutely free in nature, cryogenic air separation plants are energy intensive. Under the growing crunch of energy crisis, the air separation industry has no option but to seek improvement through optimized design and operation.

Cryogenic distillation involves complex processes and a large number of items of equipment. For the operation of plant at its best efficiency, each of the items of equipment has to work at its specified level. A slight deterioration in the performance of any one device affects the performance of the whole plant and undermines the good effects of other efficient components. Therefore, the design, operating and maintenance engineers need to understand how various design and operating parameters affect the working of individual items of equipment, which in turn affect the performance of the whole plant.

From the above discussion it is clear that air has to be liquefied first and then put into the distillation column for separation. Liquefaction of air needs cold or refrigeration to take the atmospheric air to its boiling point. Therefore, let us understand the cooling methods at this stage.

16.6 COOLING METHODS

There are three primary methods by which fluids may be cooled:
1. By transferring heat to a boiling liquid or cold gas stream, often but not always, using a heat exchanger. Note that the liquid or cold gas must, however, have been made in the first place by either method (2) or method (3).
2. By allowing the fluid to expand isenthalpically (a Joule–Thomson or J–T expansion), provided the starting temperature and pressure are low enough.
3. By causing the fluid to do external work using an expansion engine.

A bath of pure liquid boiling at constant pressure will give refrigeration at constant temperature. The use of a number of liquid baths to cool a gas to successively lower temperatures is known as the *cascade* method. It has been used since the earliest days of cryogenics but now with the development of reliable and efficient expansion machines, it is not employed to that extent.

The cascade process is the mixed refrigerant cycle, which is used extensively in processes for the liquefaction of natural gas.

Another way to achieve the cooling effect is through an isenthalpic expansion (or throttling) which may be discussed with reference to the Joule–Thomson coefficient $\mu = (\delta T/\delta p)_h$.

$$\mu = \left(\frac{\delta T}{\delta p}\right)_h = \frac{1}{c_p}\left\{T\left(\frac{\delta v}{\delta p}\right)_T - v\right\} = \frac{(T\beta - 1)v}{c_p} \quad (16.1)$$

where β is the volume expansion coefficient. Depending on the value of $T\beta$, the Joule–Thomson coefficient may be positive or negative, that is, heating or cooling may occur on throttling. The locus of points for which μ is zero is known as the *inversion curve* as shown in Figure 16.2, within which μ is positive. The highest temperature on the inversion curve is commonly known as the *inversion temperature,* and the highest pressure is known as the *inversion pressure*. The inversion temperatures and inversion pressures of some common gases are given in Table 16.2.

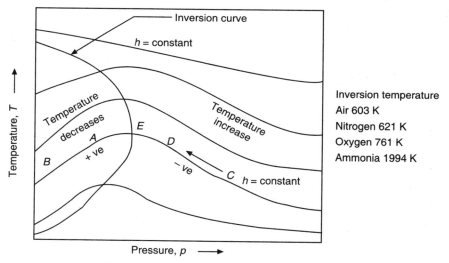

Figure 16.2 Isenthalpic expansion of a real gas.

It is clear from Figure 16.2 that if the upstream (*A*) and downstream (*B*) pressures lie within the inversion curve, cooling is obtained; if they are both outside it (*C, D*), the gas warms on expansion; and the greatest cooling is obtained if the initial pressure is on the inversion curve (*E*). There is clearly no point in compressing the gas to a point outside the inversion curve (*C*) and expanding to a point within it (*B*), since the same final temperature could have been reached by an expansion from a point (*A*) within the inversion curve with a consequent saving in compressor power. In practice, expansion usually begins from well within the inversion curve, as the decrease in compressor power outweighs the small loss in yield.

Throttling may by carried out by passing the fluid through an orifice or a short length of small diameter tube, but often a valve is used so that the pressure ratio can be controlled, and also in the event of blockage by solid matter, the valve can be manipulated to clear it. Although throttling is irreversible and therefore contributes to plant efficiency, it is widely used in the lowest temperature stages of a plant to enable liquid to be produced, mainly because of its simplicity, cheapness and reliability compared with an expansion engine.

Table 16.2 Inversion temperatures and pressures of common gases

Gas	Inversion temperature (K)	Inversion pressure (bar)	Highest temperature from which liquid can be made by throttling (K)
Methane	>500	533	260
Oxygen	742	570	220
Nitrogen	621	380	185
Neon	260	—	70
Hydrogen	200	164	50
Helium	43	39	7.7

There are two devices used for doing external work, a reciprocating engine and a turbine. The reciprocating engine, essentially a compressor operating in reverse, consists of a piston and cylinder with suitable inlet and exhaust valves to regulate the gas flow. The piston and cylinder are usually at low temperature, so that oil cannot be used for lubrication, leading to maintenance problems, which have now been largely overcome by the use of dry lubrication materials. Typical expansion engine efficiencies are 70–90%.

The other external work device is the turboexpander, in which the process gas drives a radial-flow turbine. These are essentially high-speed devices, running at thousands of revolutions per second; the bearings may be lubricated by oil or gas. Efficiencies are usually in the range 65–85%.

16.7 AIR LIQUEFACTION SYSTEM

There has been a continuous development of the air liquefaction systems. The liquefaction of air in the production of oxygen was the first engineering application of cryogenics. Even today, the production and sale of the liquefied gases is an important area in cryogenic engineering. We shall discuss several of the systems used to liquefy cryogenic fluids.

We are concerned with the performance of various systems, where performance is specified by the system performance parameters or payoff functions.

16.7.1 System Performance Parameters

There are three payoff functions which we might use to indicate the performance of a liquid system.

(i) Work required per unit mass of gas compressed $\left(\dfrac{w}{m}\right)$.

(ii) Work required per unit mass of gas liquefied $\left(\dfrac{w}{m_f}\right)$.

(iii) Fraction of the total flow of gas that is liquefied, $y = \dfrac{m_f}{m}$.

Theses payoff functions are different for different gases; therefore we should also need another performance parameter that allows the comparison of the same system using different fluids. *The figure of merit* (FOM) for a liquefaction system is such a parameter. It is defined as the theoretical minimum work requirement divided by the actual work requirement for the system:

$$\text{FOM} = \frac{w_i}{w} = \frac{-w_i/m_f}{-w/m_f}$$

The figure of merit is a number between 0 and 1. It gives a measure of how closely the actual system approaches the ideal system performance.

There are several performance parameters that apply to the components of real systems. These include compressor and expander adiabatic efficiencies, compressor and expander mechanical efficiencies, heat exchanger effectiveness, pressure drop through piping, heat exchangers, heat transfer to the system from ambient surroundings, and so on.

The gas to be liquefied is compressed reversibly and isothermally from ambient conditions to a very high pressure. This high pressure is selected so that the gas will become saturated liquid upon reversible isentropic expansion through the expander. The final condition at point f (Figure 16.3) is taken at the same pressure as the initial pressure at point 1. The pressure attained at the end of the isothermal compression is extremely high, of the order of 70 Gpa or 80 Gpa for nitrogen. In the analysis of each of the liquefaction systems, we shall apply the first law of thermodynamics for steady flow which may be written in general as in Eq. (16.2).

$$Q_{net} - W_{net} = \Sigma m \left(h + \frac{v^2}{2} + gz \right)_{out} - \Sigma m \left(h + \frac{v^2}{2} + gz \right)_{in} \qquad (16.2)$$

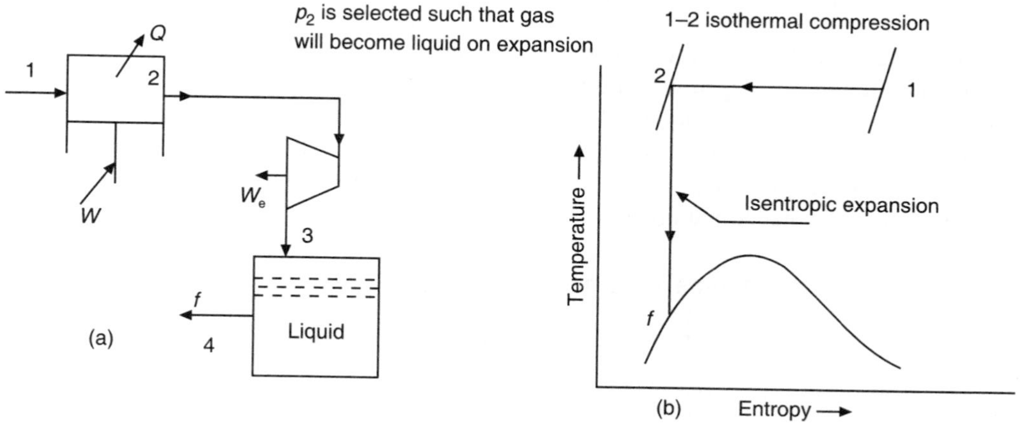

Figure 16.3 Thermodynamically ideal liquefaction cycle.

The summation signs in Eq. (16.2) imply that we add the enthalpy terms, kinetic energy terms and potential energy terms for all the inlets and outlets of the systems. A system might be consisting of several different streams, which would result in more than one inlet and one outlet. In all our system analyses, we shall assume that the kinetic and potential energy changes are much smaller than the enthalpy changes, and these energy terms may be neglected. Thus in our special case, the first law for steady flow may be written as in Eq. (16.3).

$$Q_{net} - W_{net} = \Sigma m(h)_{out} - \Sigma m(h)_{in} \tag{16.3}$$

Applying the first law to the system as shown in Figure 16.3,

$$Q_r - W_i = m(h_f - h_1) = -m(h_1 - h_f) \tag{16.4}$$

The heat transfer process is reversible and isothermal in the Carnot cycle. Thus from the second law of thermodynamics,

$$Q_r = -mT_1(s_1 - s_2) = -mT_1(s_1 - s_f) \tag{16.5}$$

Because the process from point 2 to point f is isentropic where $s_2 = s_1$ (s is the entropy of the liquid), we get the following equation

$$\frac{-W_{net}}{m} = -T_1(s_1 - s_2) - (h_1 - h_f) \tag{16.6}$$

In the ideal system, 100 percent of the gas compressed is liquefied, or $m = m_f$ so that $y = 1$. Notice that a liquefaction system is a work absorbing system; therefore, the net work requirement is negative and the term $-W_i/m$ is a positive number.

Equation (16.2) gives the minimum work required to liquefy a gas so this is the value we should try to approach in any practical system. Because we have set the final pressure at point 1 and point f is on the saturated liquid curve, the ideal work requirement depends only on the pressure and temperature at point 1 and type of the liquefied. Ordinarily, we take point 1 at the ambient conditions.

16.8 SIMPLE LINDE CYCLE

The simple Linde or Joule–Thomson (J–T) expansion cycle is shown schematically in Figure 16.4(a) and (b).

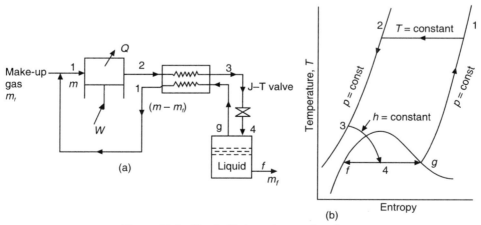

Figure 16.4 Simple Linde cycle as a liquefier.

The various processes involved in the cycle are given as follows.
(a) **Isothermal compression (1–2):** In the ideal process, the air is compressed isothermally at ambient temperature, rejecting heat to a coolant. The compression is carried out in multi-stages with intercoolers and aftercooler.
(b) **Constant pressure cooling (2–3):** The compressed air is cooled before it reaches the throttling valve in a heat exchanger by the stream returning to the compressor intake. The cooling of the air is such that after expansion in the J–T valve it converts into liquid.
(c) **Expansion in J–T valve (3–4):** The sufficiently cooled air is isenthalpically expanded in the J–T valve at the end of which air will be in mixed phase (liquid + vapour). Liquid air is collected in the insulated container while the vapour part is recirculated or passed through heat exchanger to cool the incoming air.
(d) Unliquefied fraction and the vapour formed by liquid evaporation from the absorbed heat, Q are warmed in the heat exchanger as they are returned to the compressor intake.

An energy balance around the heat exchanger, expansion valve and liquid reservoir now results in Eq. (16.7) as follows:

$$mh_2 = (m - m_f)h_1 + m_f h_f \tag{16.7}$$

Because of the unbalanced flow in the liquefaction system, if we define the fraction liquefied in a liquefier as $y = \dfrac{m_f}{m}$, then we can solve Eq. (16.7) for the fraction liquefied in a simple Linde cycle as:

$$y = \frac{(h_1 - h_2)}{(h_1 - h_f)} \tag{16.8}$$

The simple Linde cycle may also be used as a liquefier for fluids that have an inversion temperature above the ambient temperature. Under such circumstances, the refrigeration duty Q is replaced by a draw—offstream of mass m_f representing the liquefied mass of fluid that is continuously withdrawn, mass of fluid is warmed in the counter-current heat exchanger and returned to the compressor, where h_f is the specific enthalpy of the liquid being withdrawn. Note that liquefaction is maximized when the difference h_1 and h_2 is maximized. Since h_1 and h_f are generally fixed, this means that h_2 must be minimized. Mathematically, since $T_2 = T_1$, this means that

$$\left(\frac{\partial h}{\partial p}\right)_{T=T_1} = 0 \tag{16.9}$$

This is equivalent to saying that the high pressure p_2 which minimizes h_2 is the pressure at which the Joule–Thomson coefficient is zero for temperature T_1. In other words, for maximum liquid yield, point 2 in Figure 16.4(b) should occur at the intersection of T_1 and the inversion curve of the fluid at pressure p_2.

To account for heat inleak q_L into the system, the relation in Eq. (16.7) needs to be modified to

$$y = \frac{(h_1 - h_2 - q_L)}{(h_1 - h_f)} \tag{16.10}$$

16.9 CLAUDE CYCLE

The main difference between this cycle and the Linde Cycle is in the generation of refrigeration effect or cold produced. The air after compression is cooled to point 3 (Figure 16.5) and part of it is expanded in the expansion device which helps in producing more refrigeration effect than that produced in Linde Cycle. Therefore, the liquid yield per unit energy input is greater in this cycle. In large machines, the work produced during expansion is conserved. In small refrigerators, the energy from the expansion is usually expended in an energy-absorbing process or device.

A schematic of Claude cycle is shown in Figure 16.5 and the corresponding T–s cycle is shown in Figure 16.6.

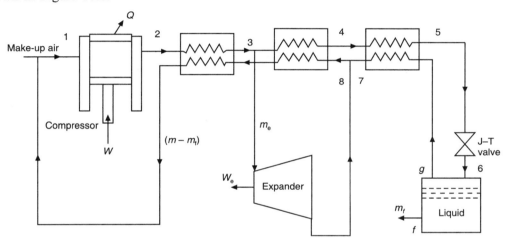

Figure 16.5 Block diagram of Claude cycle.

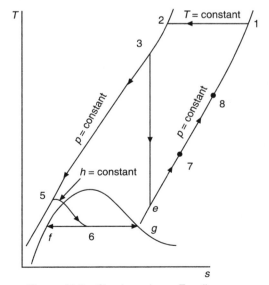

Figure 16.6 Claude cycle on T–s diagram.

The various processes involved in the cycle are given as follows:

1. **Isothermal compression (1–2):** In the ideal process, the air is compressed isothermally at ambient temperature, rejecting heat to a coolant. The compression is carried out in multi-stages with intercoolers and aftercooler.
2. **Constant pressure cooling (2–3):** The compressed air is cooled before it reaches the throttling valve in a heat exchanger by the stream returning to the compressor intake. The cooling of the air is such that after expansion in the J–T valve it converts into liquid.
3. **Expansion in expander (3–e):** A fraction of main stream air is expanded in the turbo-expander, very cold air at the outlet of the turbo-expander is mixed with the return gas and is helpful to cool the main stream air further to point 5 (in Figure 16.6). Expansion of air from this temperature through J–T valve would yield more liquid.
4. **Expansion in J–T valve (5–6):** The sufficiently cooled air is isenthalpically expanded in the J–T valve at the end of which air will be in mixed phase (liquid + vapour). Liquid air is collected in the insulated container while the vapour part is recirculated or passed through heat exchanger to cool the incoming air.
5. Unliquefied fraction and the vapour formed by liquid evaporation from the absorbed heat Q are warmed in the heat exchanger as they are returned to the compressor intake.

16.10 SMALL-TO-MEDIUM-SIZED HYDROGEN LIQUEFIERS

The properties of hydrogen are given in Table 16.3.

Table 16.3 Properties of hydrogen

Properties	Values
Atomic weight	2.02
Boiling point	–252.77°C at 101.3 kPa
Specific volume	11.967 m^3/kg at 20°C, 101.3 kPa
Critical temperature	–239.77°C
Critical pressure	1298 kPa
Heat capacity	1432 J/kg-K at 25°C, 101.3 kPa

The normal boiling point of hydrogen is very low at –252.77°C. Therefore to liquefy it, one has to produce such a low temperature.

Figure 16.7 shows a simplified flowsheet for a conventional Linde liquefier. This type of unit was constructed for capacities in the range of 14 L/h to 60 L/h.

In this process, pure hydrogen is taken either from bottles T_2 or from a gasholder T_1, and compressed in C_1 to a high pressure, typically 13000 kPa, and dried in column V_1 before being admitted to the cryogenic section. Here the hydrogen is cooled in heat exchangers, E_1, E_2 and E_3 by returning vapour hydrogen and liquid nitrogen, which is admitted into vessel V_2. V_2 contains E_2, which provides the majority of refrigeration needed at above –280°C (the temperature of the liquid N_2). The cold hydrogen emanating from E_3 is flashed across JT_3 to yield two phases. The liquid product H_2 is drawn off for use in the laboratory while the vapour is reheated to provide some refrigeration in E_3 and E_1. Vacuum pump C_2 is used to ensure that the liquid N_2 boils as cold as possible. Vacuum pump C_3 is used to ensure that the cold box is maintained at a good vacuum to minimized heat inleak, which is a significant factor for such small Joule–Thomson liquefiers.

Modest rocketry research programs and low-temperature physics laboratories use the small- to medium-sized hydrogen liquefiers, which are usually inexpensive, simple, and easy to run. The capacity range is from about 5L/h. This corresponds to a range of 0.0085–0.85 tonnes per day liquid H_2.

This size of liquefier usually generates hydrogen for immediate use, and only a small storage capacity (Dewar storage containers) is usually provided in any establishment. Therefore it is unlikely to be economic to convert ortho H_2 to para H_2, and the design of small-to medium-sized H_2 liquefiers does not provide for ortho to para conversion.

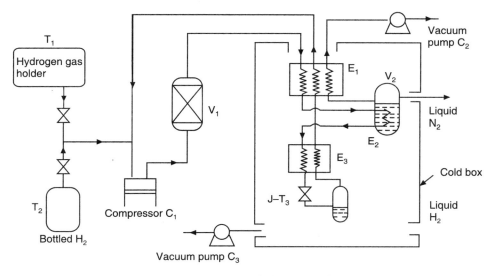

Figure 16.7 A small hydrogen liquefier.

16.11 SIMON HELIUM LIQUEFIER

The properties of helium are given in Table 16.4.

Table 16.4 Properties of helium

Property	Value
Atomic weight	4.0
Boiling point	−268/93°C at 101.3 kPa
Specific volume	6.03 m³/kg at 20°C, 101.3 kPa
Critical temperature	−267.95°C
Critical pressure	228 kPa
Heat capacity	5238 J/kg-K at 25°C, 101.3 kPa

The normal boiling point of helium is very low at –268.93°C. Therefore to liquefy it, one has to produce that temperature. This was first achieved by scientist Simon through a cascade system and is explained here.

A special application of the isentropic expansion concept is utilized in the Simon helium liquefier to produce small batch quantities of liquid helium. A simplified equipment arrangement for this system is shown in Figure 16.8. The process path can be described by following the temperature-entropy diagram shown in Figure 16.9. The steps are as follows:

Process 1–2: The heavy-walled container is filled in with high pressure helium gas, at about 15 Mpa and ambient temperature.

Process step 2–3: The container and the high pressure helium gas are cooled to 77 K with liquid nitrogen. Helium gas is added at 15 Mpa to maintain a constant pressure in the inner container during this cooling process.

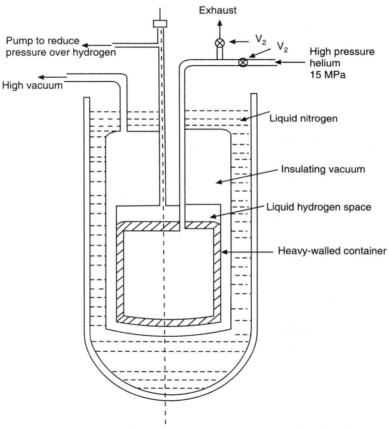

Figure 16.8 Simplified equipment arrangement for the Simon helium liquefaction system.

Process step 3–4: The helium gas originally in the vacuum space is removed once liquid nitrogen temperature has been achieved, while the helium gas in the liquid hydrogen space is replaced by liquid hydrogen. This permits the inner container and its contents to be cooled to 20.4 K, the temperature of liquid hydrogen.

Process step 4–5: Temperature of the inner container and contents at 15 Mpa are further lowered to near 10–12 K by pumping on the liquid hydrogen and reducing its pressure below the triple point.

Process step 5–6: Finally, the helium exhaust valve is opened which permits a reduction in pressure from 15 to 0.101 Mpa. The helium, which remains in the container, does work by displacing a small quantity of helium vapour. The expansion of the helium which is essentially isentropic during this step liquefies 80 to 100% of the helium in the container.

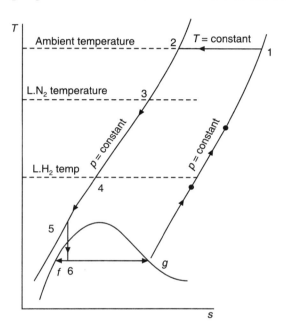

Figure 16.9 Claude cycle on T–s diagram

EXERCISES

1. What do you understand by cryogenics?
2. Explain the applications of cryogen "Nitrogen".
3. Explain the applications of cryogen "Oxygen".
4. Explain the applications of cryogen "Argon".
5. Explain a typical cryogenic air separation plant with the help of a block diagram.
6. Explain the ideal liquefaction cycle for a gas.
7. Explain the terms FOM and liquid yield.
8. Discuss the working of Linde cycle for liquefaction of air with the help of a block diagram and T–s diagram.

Chapter 17

Food Preservation

17.1 INTRODUCTION

The preservation of perishable commodities, particularly foodstuffs, is one of the most common uses of mechanical refrigeration. Therefore, it is one of the comprehensive study areas of refrigeration.

The food preservation is the process of treating and handling of perishable foodstuffs to stop or slow down spoilage and thus allow for longer periods of storage. The spoilage of food means loss of quality, edibility or nutritional value.

Preservation involves preventing the growth of bacteria, yeasts, fungi and other micro-organisms and further retarding the oxidation of fats which cause rancidity.

There are a large number of food preservation methods. Fruits are preserved by turning them into jams. For example, fruits are boiled to reduce their moisture content and to kill the bacteria, yeasts, etc., followed by sugaring to prevent their regrowth and subsequently sealing into air-tight jars. During food preservation, maintaining the nutritional value, texture and flavour are some of the important aspects. In many cases, such changes have now come to be seen as desirable qualities—cheese, yogurt and pickled onions being common examples.

17.2 FOOD DETERIORATION AND SPOILAGE

It is important to first understand the causes of food deterioration or spoilage so that one can study the food preservation methods accordingly.

Food, either in the form of plant or animal, contains basically three major components—carbohydrates, proteins and fats. The destruction of any one of these ingredients causes the spoilage of food. The spoilage period varies from one food to another based on the food—solid, liquid, semi-solid, etc. The spoilage of food comes in the form of bad odour, uncommon colour, bad taste and bad physical appearance. The result of food spoilage may lead to loss of weight, softening, souring, rotting, wilting or maybe any of these in a combined form.

Few examples of food spoilage are listed below:

1. Spoilage of milk results into bad smell and taste.
2. The inside gaur of apples turn to a blackish colour.

3. Spoilage of grapes makes them blackish with a smelly liquid oozing out from them.
4. Spoiled eggs lose weight and give out a very bad odour.
5. A slimy coating is observed on spoiled meat.
6. Spoiled butter and cheese turn black and give out a bad odour.

17.3 FACTORS OF FOOD DETERIORATION AND SPOILAGE

As defined earlier, the preservation of food means maintaining the nutritional value, the colour, and the taste throughout the storage period. The deterioration of food may start either by internal agents or by external agents and/or simultaneously by both the agents, making it a combined effect.

The spoilage of perishable food is caused by a series of complex chemical changes which take place in the foodstuff after harvesting or killing. These chemical changes are due to internal and external agents.

The internal agents, i.e. *natural enzymes* are inherent in all organic materials. The external agents, i.e. micro-organisms grow in and out on the surface of the foodstuffs.

In conclusion, spoilage agents are classified into two groups:

1. Enzymes—internal agents.
2. Micro-organisms—bacteria, yeast and moulds.

17.3.1 Enzymes

These are complex, protein-like chemical substances, essential in the chemistry of all living processes and hence are normally present in all organic materials (both in plants and animals, both living and dead). The other important points to know about enzymes are:

- Enzymes are best described as chemical catalytic agents which bring chemical changes in organic materials.
- There are many enzymes, each produces only one specific chemical reaction.
- One enzyme, namely *lactase* which acts to convert lactose (milk sugar) to lactic acid, is responsible for souring of milk.
- Enzymes are manufactured by all living cells to carry on the living activities of the cell such as respiratory sprouting of seeds, growth of plants and animals, the ripening of fruits and the digestive processes of animals.
- Enzymes are catabolic as well as anabolic. That is they act to destroy dead cell tissues as well as to maintain live cell tissue.
- Whether the enzymes action is catabolic or anabolic, they are always destructive to perishable foods, and, therefore their activities in perishable food need to be controlled.
- Enzymes are very sensitive to the surrounding temperature.
- Enzymes are eliminated at temperatures above 71°C. Hence cooking a food substance completely destroys the enzymes contained in it.
- The chemical activity of enzymes is greatly reduced below 0°C. Their activity is maximum in the presence of free oxygen. Hence their activity will drastically decrease with the diminishing supply of oxygen.

17.3.2 Micro-organisms

The term 'micro-organism' is used to cover the complete group of minute plants and animals of microscopic and sub-microscopic size. However, the following three types are of particular interest in the context of food spoilage/food preservation: (i) Bacteria, (ii) Yeasts and (iii) Moulds.

These tiny organisms are found in large numbers anywhere—in the air, in the ground, in water, in and on bodies of plants and animals, and in every other place where favourable temperature and humidity conditions are present. The growth of micro-organisms in and on the surface of perishable foods causes complex chemical changes in the food substance which results in alterations in the taste, the odour and the appearance of food, making such food unfit for consumption. On the other hand, micro-organisms have many useful and necessary functions. If it were not for the work of micro-organisms, life of any kind would not be possible. Decay and decomposition of all dead animal tissues are essential to make space available for new life and growth.

Micro-organisms provide us with the following important benefits:

(i) They play an important role in the food chain, by helping to keep the essential materials in circulation.

(ii) Micro-organisms are essential in processing of certain fermented foods and other commodities. For example, bacteria are responsible for the fermentation needed in the processing of pickles, olives, coco, and certain milk products such as butter, cheese, yogurt etc. Micro-organisms are essential in the production of vinegar from various alcohols. Yeasts are responsible for the alcoholic fermentation of products manufactured by the brewing and wine-making industries, and also baking industry.

17.3.3 Bacteria

Bacteria are made up of one single living cell. Reproduction of bacteria is accomplished by cell division. On reaching maturity, the bacterium divides into two separate and equal cells. The lifecycle of bacteria is relatively short, being a matter of minutes or hours. The rate at which bacteria grow and reproduce depends upon such environmental conditions as temperature, light and degrees of acidity and alkalinity and upon the availability of oxygen and adequate supply of soluble food. There are varieties of bacteria which reproduce and grow in different environment and conditions.

Some bacteria are 'free living' and feed on only animal wastes and on the dead tissues of animals. Some, however, are parasites and require a living host. Some bacteria need moisture and soluble food, however, some do not. Some bacteria need a slightly acidic and some other need a slightly alkaline environment for reproduction and growth. Bacteria grow in indirect sunlight but not in direct sunlight. For each species of bacteria, there is an optimum temperature at which bacteria will grow at the highest rate. The optimum temperature for most saprophytes (bacteria which live on animal waste and dead tissue) is usually between 24°C and 30°C. The optimum temperature for parasites (bacteria which live on living host) is around 37°C. It has been found that foods subjected to 120°C will kill all types of bacteria and their activity will be stopped when foods are subjected to temperatures below $-20°C$.

17.3.4 Yeasts

Yeasts are simple, one-cell plants of the fungus family. Yeasts are somewhat larger and more complex than bacteria cells. The reproduction of yeasts is by budding. The buddings separate from mother yeasts to form new ones.

Like bacteria, yeasts are the agents of fermentation and decay. They bring about chemical changes in the food upon which they grow. Yeasts are widespread in nature and yeast spores are invariably found in air and on the surfaces of fruits and berries. Yeasts also need air, food and moisture for growth and are sensitive to temperature and degrees of acidity and alkalinity in the environment.

17.3.5 Moulds

Similar to yeasts, moulds are simple plants of the fungi family. Moulds are much more complex in structure than either bacteria or yeast plants. Mould plant is made up of a number of cells which are positioned end to end to form long threadlike fibre called hyphase. The network formed by these threadlike fibres is called the mycelium and is visible to the naked eye. Moulds reproduce by spore formation. Mould spores are actually seeds and will germinate and produce mould growth on any food substance under certain environment conditions.

Moulds are less resistant to high temperatures than are bacteria and they are more tolerant to low temperature (close to zero degree centigrade). All mould growth ceases at temperatures of $-12°C$ and below. Mostly all moulds can be destroyed with heating above $60°C$.

17.4 FOOD PRESERVATION PROCESSES

Various types of foods are processed and preserved in a number of ways, which are briefly described below.

17.4.1 Drying

This is one of the most ancient food preservation techniques. Drying of food reduces the moisture content in the foodstuff, which in turn decrease the activity of micro-organisms. Insects do not attack the grains free of moisture.

17.4.2 Refrigerated Storage

Refrigerated storage is divided into three groups: (i) short-term storage, (ii) long-term storage, and (iii) frozen storage.

In short-term storage, the food products are chilled and stored at some temperature above the freezing point. This method is normally followed by retail establishments where rapid turnover of the product is expected. Depending upon the product, the storage period ranges one to two days.

In long-term storage, the food products are chilled and stored at some temperature above the freezing point. The storage period depends upon the type of product stored and the condition of the food product entering the storage. Maximum storage-periods for long-term storage range from seven to ten days like ripe tomatoes. Some products are stored for more than six months like onions and some smoked meat.

When perishable food items are to be stored for longer periods, they should be frozen and placed in frozen storage. For example, fish are stored in frozen condition.

Storage conditions

The optimum storage condition for a product held in either short or long duration depend upon the nature of the product, length of time period the product is kept in storage and whether the product is in packed or unpacked condition. ASHRAE fundamental volume lists the various food products, storage temperature, humidity, air velocity, etc.

Storage temperature

The optimum storage temperature for most of the products is one that is slightly above the freezing point of the product. There are, however, notable exceptions. The incorrect storage temperature generally lowers the product quality and shortens the storage life. Some fruits are very much sensitive to the storage temperature. For example, citrus fruits frequently develop rind pitting when stored at relatively high temperatures. On the other hand, they are subject to *scald* and watery breakdown when stored at temperatures below their critical temperature. Bananas suffer peel injury when stored below 13°C.

Humidity and air-motion in storages

The perishable food items stored in cold-storages need a specific range of humidity and air-motion apart from temperature control. Foodstuff such as meat, fish, fruit, vegetables when stored in unpacked condition lose moisture by evaporation at the surface. This is known as dehydration of the foodstuff; due to which there is a weight and vitamin loss. Therefore, it is very much necessary to maintain the required humidity and air-motion in storages.

17.5 MIXED STORAGE

There are various foodstuffs which require specific storage conditions in order to maintain the optimum quality and also the storage life. It is not economical to establish separate storages to satisfy the storage conditions if these foodstuffs are in small quantities. Financially, economical considerations often demand that a number of refrigerated products be placed in a common storage. Naturally, the differences in the storage conditions required by the various products pose a problem with regard to the conditions to be maintained in a space designed for common storage. Generally, storage conditions in such spaces represent a compromise in the required storage conditions. In such mixed storages, the products which require higher storage temperature may be subjected to low temperature and vice versa.

The product condition is also important when it enters into the storage. Many a time the products are precooled to a certain temperature before these are brought into the cold storages.

17.6 FREEZING AND FROZEN STORAGE

When foodstuffs are to be preserved in their original fresh state for relatively long periods, these are usually frozen and stored at approximately $-18°C$ or below. The foods which are normally frozen for preservation at fresh state include fruits, fruit juices, berries, meat, sea foods, etc.

The following factors govern the ultimate quality and storage life of any frozen product.
1. The nature and composition of the foodstuff to be frozen.
2. The precautions taken in selecting, handling and preparing the foodstuff for freezing.
3. The freezing method.
4. The storage conditions.

There are various food freezing methods, the important ones blast freezing, indirect-contact freezing, immersion freezing. There are many more methods of food preservation, which are not included here in this book.

17.6.1 Advantages of Food Preservation

1. The seasonal availability of many foods increases.
2. The transportation of delicate, perishable foods over long distances becomes possible.
3. Foods become safer to eat as micro-organisms are either killed or their activity towards food spoilage is reduced.
4. Modern supermarkets/big bazars become possible with modern food processing techniques.
5. There is more consistency in marketing and distribution of food products.
6. It becomes possible to alleviate food shortages and improve the overall nutrition value of the food.
7. Food processing reduces the food borne diseases.
8. Extremely modern food diet becomes possible on a wide scale because of food processing and preservation.
9. The modern techniques of food processing often improve the taste of food.
10. It takes less time to prepare food from the processed food items than that from raw food ingredients.
11. It improves the quality of life for people with allergies, diabetes, etc.

17.6.2 Disadvantages of Food Preservation

1. Any processed food has a slight effect on the nutritional density. Vitamin C is destroyed by heat and therefore canned fruits have a lower content of vitamin C.
2. The processed and preserved foods lose their nutritional value by about 5 to 20%.
3. The food additives used in processing may often pose a health risk.
4. Foods may get contaminated as these have to go through mixing, grinding, chopping and emulsifying equipment during processing.
5. The processed foods may lose the original taste and colour.

17.7 COLD STORAGE

It is a building space in which the temperature, humidity and air circulation are maintained to satisfy the storage conditions of the commodities. The atmosphere so controlled and maintained in the cold storage is known as controlled atmosphere and maintained atmosphere (CAMA).

The commodities, for example, include grapes, raisins, potatoes, onions, seafood, etc. These commodities are required to be stored at different climatic conditions.

As per the present day practice the cold stores can be classified as follows:

(a) *Cold stores based on a seasonal basis:* Stores used for storing potatoes, grapes, chillies, apples, etc.
(b) *Stores for storing the commodities round the year:* These cold stores are designed to store a variety of commodities. Such cold stores are called multipurpose cold stores. The commodities to be stored include fruits, vegetables, dry fruits, spices, pulses, milk products, etc. These stores are usually located near the consuming centres.
(c) *Cold stores for temporary storage of commodities:* There will be a precooling facility for fresh fruits and vegetables mainly for export-oriented items like grapes, etc. These are located near grapes growing areas like Tasgaon, Nashik, etc.
(d) *Frozen food stores:* The items like fish, meat, poultry, dairy products, etc. are first frozen and then stored in such stores. There is a large potential in this sector for creating facilities in India.
(e) *Small stores:* These include mini cold store units or walk-in-cold stores located in hotels, restaurants, malls, supermarkets, etc.
(f) Cold stores mainly used for bananas and mangoes.

17.7.1 Cold Chain

A large number of cold stores created for storing a variety of commodities in different parts of a country is known as a 'Cold chain'. India ranks number one in milk production, number two in fruits and vegetables production and has a substantial potential for marine, meat and poultry products. There is a sufficient scope for the creation of "cold chain" facilities for these commodities.

The cold storage sector is undergoing a continuous upgradation for improving their energy efficiency. Realising the importance of the cold chain industry, the Government of India has taken initiatives, to establish standards for the cold stores. Efforts are also being made to evolve a new concept called Green-cold-chain.

17.7.2 Construction of Cold Stores

Cold stores are built in a number ways using a variety of construction materials and methods such as:

(a) Conventional cold storages made in RCC frames, brick walls, RCC slabs or truss type sheet roofs, steel frames with wooden or steel grating.

(b) Cold storage buildings with single floor structure designed for mechanised loading and unloading of products.

(c) Pre-engineered building structures designed with cold chambers constructed from insulated PUF panels. Cold stores may have many chambers in a single floor construction with heights varying from 5 to 12 m or higher. These may have mechanised loading and unloading facilities for goods.

In the case of medium and large cold stores the facilities provided include:

(a) Loading/unloading areas
(b) Ante rooms
(c) Cold storage chambers
(d) Staircases and lifts
(e) Machine room
(f) Office and toilet blocks

It is advisable to have multi-commodity cold stores for better capacity utilisation. For example, a potato cold storage may be designed for the storage of seed potato and table potato or processing potato, etc. Similarly, for energy efficiency, cold stores have to be multi-chamber. Therefore, a cold storage should have at least two chambers and the chamber size should be of the capacity range of 1000 MT to 1500 MT for proper capacity utilisation and energy efficiency.

EXERCISES

1. Define food preservation. What are the advantages and disadvantages of food preservation?
2. What are the parameters needed to be considered for storing the food items in cold stores?
3. Explain the various methods of food preservation.
4. Why do we need food preservation systems?
5. Explain the "Micro-organisms" and Enzymes responsible for food spoilage.
6. What is the meaning of Cold chain?
7. What is CAMA?

Food Preservation

17.1 INTRODUCTION

The storage of most perishable commodities, particularly foodstuffs, is one of the most common applications of mechanical refrigeration. Therefore, it is one of the comprehensive study areas of refrigeration.

Food preservation is the process of treating and handling of perishable foodstuffs to stop or slow down the spoilage and thus allow for longer periods of storage. The spoilage of food means loss of quality, edibility or nutritional value.

Preservation entails preventing the growth of bacteria, yeasts, fungi and other micro-organisms, as well as retarding the oxidation of fats which cause rancidity.

There are a number of food preservation methods. Fruits are preserved by turning them into jam, for example, fruits are boiled to reduce their moisture content and to kill the bacteria, etc., sweetened by sugaring to prevent their regrowth and subsequently sealing within an airtight jar. Because food preservation, maintaining the nutritional value, texture and flavour is one of the important aspects. In many cases, such changes have now come to be regarded as desirable qualities—cheese, yogurt and pickled onions being common examples.

17.2 FOOD DETERIORATION AND SPOILAGE

It is important to understand the causes of food deterioration or spoilage so that one can study the various preservation methods.

The food of an organism contains basically three major components — carbohydrates, proteins and fats. The deterioration in any one of these ingredients causes the spoilage of food. The spoilage varies from one food to another based on the food — solid, liquid, semi-solid, etc. The spoilage of food comes in the form of bad odour, uncommon colour, bad taste and bad feel. In addition, as a result of food spoilage, may lead to loss of weight, softening, souring, etc. Some of the types of food spoilages in general form:

Some of the spoilages are listed below:
- (i) Milk products produce bad smell and taste.
- (ii) Peeled or cut apples turn to a brownish colour.

Appendix A

Table A1 Dry saturated steam (pressure-based)

Pressure, p (bar)	Saturation temperature, T (°C)	Specific volume of steam, v_g (m³/kg)	Specific enthalpy, in kJ/kg			Specific entropy, in kJ/kg-K		
			Water (h_f)	Latent heat (h_{fg})	Dry steam (h_g)	Water (s_f)	Latent heat (s_{fg})	Dry steam (s_g)
0.00611	0.0000	206.16	0.8	2501.6	2501.6	0.000	9.158	9.158
0.02	17.51	67.012	73.5	2460.2	2533.7	0.261	8.464	8.725
0.04	28.98	34.803	121.4	2433.1	2554.5	0.423	8.053	8.476
0.06	36.18	23.741	151.5	2416.0	2567.5	0.521	7.810	8.331
0.08	41.54	18.104	173.9	2403.2	2577.1	0.593	7.637	8.230
0.10	45.83	14.674	191.8	2392.9	2584.8	0.649	7.502	8.151
0.12	49.45	12.361	206.9	2384.2	2591.2	0.696	7.391	8.087
0.14	52.58	10.693	220.0	2376.7	2596.7	0.737	7.297	8.033
0.16	55.34	9.4325	231.6	2370.0	2601.6	0.772	7.215	7.987
0.18	57.83	8.4446	242.0	2363.9	2605.9	0.804	7.142	7.946
0.20	60.09	7.6492	251.5	2358.4	2609.9	0.832	7.077	7.909
0.25	64.99	6.2040	272.0	2346.4	2618.4	0.893	6.939	7.832
0.30	69.13	5.229	289.3	2336.1	2625.4	0.944	6.825	7.769
0.35	72.71	4.5255	304.3	2327.2	2631.5	0.988	6.729	7.717
0.40	75.89	3.9932	317.7	2319.2	2636.9	1.026	6.645	7.671
0.45	78.74	3.5761	329.6	2312.0	2641.6	1.060	6.570	7.630
0.50	81.35	3.2401	340.6	2305.4	2646.0	1.091	6.504	7.595
0.60	85.95	2.7317	359.9	2293.6	2653.5	1.146	6.387	7.533
0.70	89.96	2.3647	376.8	2283.3	2660.1	1.192	6.288	7.480
0.80	93.51	2.0869	391.7	2274.0	2665.7	1.233	6.202	7.435
0.90	96.71	1.8691	405.2	2265.6	2670.9	1.270	6.126	7.395
1.00	99.63	1.6937	417.5	2257.9	2675.4	1.303	6.057	7.360
1.013	100.00	1.6730	419.1	2256.9	2676.0	1.307	6.048	7.355
1.20	104.8	1.4281	439.4	2244.1	2683.4	1.361	5.937	7.298
1.40	109.3	1.2363	458.4	2231.9	2690.3	1.411	5.836	7.247
1.60	113.3	1.0911	475.4	2220.8	2696.2	1.455	5.747	7.202
1.80	116.9	0.9772	490.7	2210.8	2701.5	1.494	5.668	7.162
2.00	120.2	0.8854	504.7	2201.6	2706.3	1.530	5.597	7.127
2.50	127.4	0.7184	535.4	2181.0	2716.4	1.607	5.445	7.052

(Contd.)

Table A1 Dry saturated steam (pressure-based) *(Contd.)*

Pressure, p (bar)	Saturation temperature, T (°C)	Specific volume of steam, v_g (m³/kg)	Specific enthalpy, in kJ/kg			Specific entropy, in kJ/kg-K		
			Water (h_f)	Latent heat (h_{fg})	Dry steam (h_g)	Water (s_f)	Latent heat (s_{fg})	Dry steam (s_g)
3.00	133.5	0.6055	561.4	2163.2	2724.7	1.672	5.319	6.991
3.50	138.9	0.5240	584.3	2147.3	2731.6	1.727	5.212	6.939
4.0	143.6	0.4622	604.7	2132.9	2737.6	1.726	5.118	6.894
4.5	147.9	0.4137	623.2	2119.7	2742.9	1.820	5.034	6.855
5.0	151.9	0.3747	640.1	2107.4	2747.5	1.860	4.959	6.819
6.0	158.8	0.3155	670.4	2085.1	2755.5	1.931	4.827	6.758
7.0	165.0	0.2727	697.1	2064.9	2762.0	1.992	4.713	6.705
8.0	170.4	0.2403	720.9	2046.6	2767.5	2.046	4.614	6.660
9.0	175.4	0.2148	742.6	2029.5	2772.1	2.094	4.525	6.619
10.0	179.9	0.1943	762.6	2013.6	2776.2	2.138	4.445	6.583
12.0	188.0	0.1632	798.4	1984.3	2782.7	2.216	4.303	6.519
14.0	195.0	0.1407	830.1	1957.7	2787.8	2.284	4.181	6.465
16.0	201.4	0.1237	858.5	1933.2	2791.1	2.344	4.074	6.418
18.0	207.1	0.1103	884.5	1910.3	2794.8	2.398	3.978	6.375
20.0	212.4	0.0995	908.6	1888.7	2797.3	2.447	3.890	6.337
22.0	217.2	0.0907	930.9	1868.2	2799.1	2.492	3.809	6.301
24.0	221.8	0.0832	951.9	1848.5	2800.4	2.534	3.735	6.269
26.0	226.0	0.0768	971.7	1829.7	2801.4	2.574	3.665	6.239
28.0	230.0	0.0714	990.5	1811.5	2802.0	2.611	3.600	6.211
30.0	23.8	0.0666	1008.3	1794.0	2802.3	2.646	3.538	6.184
32.0	237.4	0.0624	1025.4	1776.9	2802.3	2.679	3.480	6.159
34.0	240.9	0.0587	1041.8	1760.3	2802.1	2.710	3.424	6.134
36.0	244.2	0.0554	1057.5	1744.2	2801.7	2.740	3.371	6.111
38.0	247.3	0.0524	1072.7	1728.4	2801.1	2.769	3.321	6.090
40.0	250.3	0.0497	1087.4	1712.9	2800.3	2.797	3.272	6.069
42.0	253.2	0.0473	1101.6	1697.8	2799.4	2.823	3.225	6.048
44.0	256.1	0.0451	1115.4	1682.9	2798.3	2.849	3.180	6.029
46.0	258.8	0.0430	1128.8	1668.3	2797.1	2.874	3.136	6.010
48.0	261.4	0.0411	1141.8	1653.9	2795.7	2.897	3.094	5.991
50.0	263.9	0.0394	1154.5	1639.7	2794.2	2.921	3.053	5.974
52.0	266.4	0.0378	1166.9	1625.7	2792.6	2.943	3.013	5.956
54.0	268.8	0.0363	1179.0	1611.8	2790.8	2.965	2.974	5.939
56.0	271.1	0.0349	1190.8	1598.2	2789.0	2.986	2.936	5.923
58.0	273.4	0.0336	1202.4	1584.6	2787.0	3.007	2.899	5.906
60.0	275.6	0.0324	1213.7	1571.3	2785.0	3.027	2.863	5.891
62.0	277.7	0.0313	1224.9	1558.0	2782.9	3.047	2.828	5.875
64.0	279.8	0.0302	1235.8	1544.8	2780.6	3.066	2.794	5.860
66.0	281.9	0.0292	1246.5	1531.8	2778.3	3.085	2.760	5.845
68.0	283.9	0.0283	1257.1	1518.8	2775.9	3.104	2.727	5.831
70.0	285.8	0.0274	1267.5	1506.5	2773.4	3.122	2.694	5.816
72.0	287.7	0.0265	1277.7	1493.2	2770.9	3.140	2.662	5.802
74.0	289.6	0.0257	1287.8	1480.5	2768.2	3.157	2.631	5.788
76.0	291.4	0.0249	1297.7	1467.8	2765.5	3.174	2.600	5.774

(Contd.)

Table A1 Dry Saturated Steam (pressure-based) *(Contd.)*

Pressure, p (bar)	Saturation temperature, T (°C)	Specific volume of steam, v_g (m³/kg)	Specific enthalpy, in kJ/kg			Specific entropy, in kJ/kg-K		
			Water (h_f)	Latent heat (h_{fg})	Dry steam (h_g)	Water (s_f)	Latent heat (s_{fg})	Dry steam (s_g)
78.0	293.2	0.0242	1307.5	1455.3	2762.7	3.191	2.569	5.760
80.0	295.0	0.0235	1317.2	1442.7	2759.9	3.208	2.539	5.747
82.0	296.7	0.0228	1326.7	1430.3	2757.0	3.224	2.510	5.734
84.0	298.4	0.0222	1336.2	1417.8	2754.0	3.240	2.481	5.721
86.0	300.1	0.0216	1345.4	1405.5	2750.9	3.256	2.452	5.708
88.0	301.7	0.0210	1354.7	1393.1	2747.8	3.271	2.423	5.694
90.0	303.3	0.0205	1363.8	1380.8	2744.6	3.287	2.395	5.682
92.0	304.9	0.0200	1372.8	1368.5	2741.3	3.302	2.367	5.669
94.0	306.5	0.0195	1381.7	1356.3	2738.0	3.317	2.340	5.657
96.0	308.0	0.0190	1390.6	1344.1	2734.7	3.332	2.313	5.645
98.0	309.5	0.0185	1399.4	1331.9	2731.3	3.346	2.286	5.632
100.0	311.0	0.0180	1408.1	1319.7	2727.8	3.361	2.259	5.620
110.0	318.0	0.0160	1450.6	1258.8	2709.4	3.431	2.129	5.560
120.0	324.6	0.0143	1491.7	1197.5	2698.2	3.497	2.003	5.500
130.0	330.8	0.0128	1531.9	1135.1	2667.0	3.561	1.880	5.441
140.0	336.6	0.0115	1571.5	1070.9	2642.4	3.624	1.756	5.380
150.0	342.1	0.0103	1610.9	1004.2	2615.1	3.686	1.632	5.318
160.0	347.3	0.0093	1650.4	934.5	2584.9	3.747	1.506	5.253
170.0	352.3	0.0084	1691.6	860.0	2551.6	3.811	1.375	5.186
180.0	357.0	0.0075	1734.8	779.0	2513.8	3.877	1.236	5.113
190.0	361.4	0.0067	1778.7	691.8	2470.5	3.943	1.090	5.033
200.0	365.7	0.0059	1826.6	591.6	2418.2	4.015	0.926	4.941
210.0	369.8	0.0050	1886.3	461.2	2347.5	4.105	0.717	4.822
220.0	373.7	0.0037	2010.3	186.3	2196.6	4.293	0.288	4.581
221.2	374.15	0.0032	2107.4	000.0	2107.4	4.443	0.000	4.443

Table A2 Dry saturated steam (temperature-based)

Saturation temperature, T (°C)	Pressure, p (bar)	Specific volume of steam, v_g (m³/kg)	Specific enthalpy, in kJ/kg			Specific entropy in kJ/kg-K		
			Water (h_f)	Evaporation (h_{fg})	Steam (h_g)	Water (s_f)	Evaporation (s_{fg})	Steam (s_g)
0	0.00611	206.16	0.0	2501.6	2501.6	0.000	9.158	9.158
1	0.00657	192.61	4.2	2499.2	2503.4	0.015	9.116	9.131
2	0.00706	179.92	8.4	2496.8	2505.2	0.031	9.074	9.105
3	0.00758	168.17	12.6	2494.5	2507.1	0.046	9.033	9.079
4	0.00813	157.27	16.8	2492.1	2508.9	0.061	8.992	9.053
5	0.00872	147.16	21.0	2489.7	2510.7	0.076	8.951	9.027
6	0.00935	137.78	25.2	2487.4	2512.6	0.091	8.910	9.001
7	0.01001	129.06	29.4	2485.0	2514.4	0.106	8.870	8.976
8	0.01072	120.97	33.6	2482.6	2516.2	0.121	8.830	8.951

(Contd.)

Table A2 Dry saturated steam (temperature-based) *(Contd.)*

Saturation temperature, T (°C)	Pressure, p (bar)	Specific volume of steam, v_g (m³/kg)	Specific enthalpy, in kJ/kg			Specific entropy in kJ/kg-K		
			Water (h_f)	Evaporation (h_{fg})	Steam (h_g)	Water (s_f)	Evaporation (s_{fg})	Steam (s_g)
9	0.01147	113.44	37.8	2480.3	2518.1	0.136	8.790	8.926
10	0.01227	106.43	42.0	2477.9	2519.9	0.151	8.751	8.902
11	0.01312	99.909	46.2	2475.5	2521.7	0.166	8.712	8.878
12	0.01401	93.835	50.4	2473.2	2523.6	0.181	8.673	8.854
13	0.01497	88.176	54.6	2470.8	2525.4	0.195	8.635	8.830
14	0.01597	82.900	58.8	2468.5	2527.2	0.210	8.596	8.806
15	0.01704	77.978	62.9	2466.1	2529.1	0.224	8.558	8.8782
16	0.01817	73.384	67.1	2463.8	2530.9	0.239	8.520	8.759
17	0.01936	69.095	71.3	2461.4	2532.7	0.253	8.483	8.736
18	0.02062	65.087	75.5	2459.0	2534.5	0.268	8.446	8.714
19	0.02196	61.341	79.7	2456.7	2536.4	0.282	8.409	8.691
20	0.02337	57.838	83.9	2454.3	2538.2	0.296	8.372	8.668
22	0.02642	51.492	92.2	2449.6	2541.8	0.325	8.299	8.624
24	0.02982	45.926	100.6	2444.9	2545.5	0.353	8.228	8.581
26	0.03360	41.034	108.9	2440.2	2549.1	0.381	8.157	8.538
28	0.03778	36.728	117.3	2435.4	2552.7	0.409	8.087	8.496
30	0.04242	32.929	125.7	2430.7	2556.4	0.437	8.018	8.455
32	0.04753	29.572	134.1	2425.9	2560.0	0.464	7.950	8.414
34	0.05318	26.601	142.4	2421.2	2563.6	0.491	7.883	8.374
36	0.05940	23.967	150.7	2416.4	2567.1	0.518	7.816	8.334
38	0.06624	21.627	159.1	2411.7	2570.8	0.545	7.751	8.296
40	0.07375	19.546	167.5	2406.9	2574.4	0.572	7.686	8.258
42	0.08199	17.692	175.8	2402.1	2577.9	0.599	7.622	8.221
44	0.09100	16.036	184.2	2397.3	2581.5	0.625	7.559	8.184
48	0.1116	13.233	200.9	2387.7	2588.6	0.678	7.435	8.113
50	0.1234	12.046	209.3	2382.9	2592.2	0.704	7.374	8.078
52	0.1361	10.890	217.6	2378.1	2595.7	0.729	7.314	8.043
54	0.1500	10.022	226.0	2373.2	2599.2	0.755	7.254	8.009
56	0.1651	9.1587	234.4	2368.4	2602.8	0.780	7.196	7.976
58	0.1815	8.3808	242.7	2363.5	2606.8	0.806	7.137	7.943
60	0.1992	7.6785	251.1	2358.6	2609.7	0.831	7.080	7.911
62	0.2184	7.0437	259.5	2353.7	2613.2	0.856	7.023	7.879
64	0.2391	6.4690	267.8	2348.8	2616.6	0.881	6.967	7.848
66	0.2615	5.9482	276.2	2343.9	2620.1	0.906	6.911	7.817
68	0.2856	5.4756	284.6	2338.9	2623.5	0.930	6.856	7.786
70	0.3116	5.0463	293.0	2334.0	2627.0	0.955	6.802	7.757
72	0.3396	4.6557	301.4	2329.0	2630.3	0.979	6.748	7.727
74	0.3696	4.3000	309.7	2324.0	2633.7	1.003	6.695	7.698
76	0.4019	3.9757	318.2	2318.9	237.1	1.027	6.42	7.669
78	0.4365	3.6796	326.5	2313.9	2640.4	1.051	6.590	7.641
80	0.4736	3.4091	334.9	2308.8	2643.7	1.075	6.538	7.613
82	0.5133	3.1616	343.3	2303.8	2647.1	1.099	6.487	7.586

(Contd.)

Table A2 Dry saturated steam (temperature-based) *(Contd.)*

Saturation temperature, T (°C)	Pressure, p (bar)	Specific volume of steam, v_g (m³/kg)	Specific enthalpy, in kJ/kg			Specific entropy in kJ/kg-K		
			Water (h_f)	Evaporation (h_{fg})	Steam (h_g)	Water (s_f)	Evaporation (s_{fg})	Steam (s_g)
84	0.5557	2.9350	351.7	2298.6	2650.4	1.123	6.436	7.559
86	0.6011	2.7272	360.1	2293.5	2653.6	1.146	6.386	7.532
88	0.6495	2.5365	368.5	2288.4	2656.9	1.169	6.337	7.506
90	0.7011	2.3613	376.9	2283.2	2660.1	1.193	6.287	7.480
92	0.7561	2.2002	385.4	2278.0	2663.4	1.216	6.239	7.454
94	0.8146	2.0519	393.8	2272.8	2666.6	1.239	6.190	7.429
96	0.8769	1.9153	402.2	2267.5	2669.7	1.261	6.143	7.404
98	0.9430	1.7893	410.6	2262.2	2672.8	1.285	6.095	7.380
100	1.013	1.6730	419.1	2256.9	2676.0	1.307	6.048	7.355
110	1.433	1.2099	461.3	2230.0	2691.3	1.419	5.820	7.239
120	1.985	0.89152	503.7	2202.3	2706.0	1.528	5.601	7.129
130	2.701	0.66814	546.3	2173.6	2719.9	1.634	5.392	7.026
140	3.614	0.50850	589.1	2144.0	2733.1	1.739	5.189	6.928
150	4.760	0.39245	632.1	2113.2	2745.3	1.842	4.994	6.836
160	6.181	0.30676	675.5	2081.3	2756.7	1.943	4.805	6.748
170	7.920	0.24255	719.2	2047.9	2667.1	2.042	4.621	6.663
180	10.027	0.19380	763.1	2013.2	2776.3	2.139	4.443	6.582
190	12.551	0.15632	807.5	1976.7	2784.2	2.236	4.268	6.504
200	15.549	0.12716	852.3	1938.6	2790.9	2.331	4.097	6.428
210	19.077	0.10424	897.7	1898.5	2796.2	2.425	3.929	6.354
220	23.198	0.08604	943.7	1856.2	2799.9	2.518	3.764	6.282
230	27.976	0.07145	990.3	1811.7	2802.0	2.610	3.601	6.211
240	33.478	0.05965	1037.6	1764.4	2802.2	2.702	3.439	6.141
250	39.776	0.05004	1085.8	1714.7	2800.5	2.794	3.277	6.071
260	46.943	0.04213	1134.9	1661.5	2796.4	2.885	3.116	6.001
270	55.058	0.03559	1185.2	1604.6	2789.8	2.976	2.954	5.930
280	64.202	0.03013	1236.8	1543.6	2780.4	3.068	2.790	5.858
290	74.461	0.02554	1290.0	1477.6	2767.6	3.161	2.624	5.785
300	85.927	0.02165	1345.1	1406.0	2751.1	3.255	2.453	5.708
310	98.700	0.01833	1402.4	1327.6	2730.0	3.351	2.277	5.628
320	112.89	0.01548	1462.6	1241.1	2703.7	3.450	2.092	5.542
330	128.63	0.01299	1526.5	1143.6	2670.1	3.553	1.896	5.449
340	146.05	0.01078	1595.5	1030.7	2626.2	3.662	1.681	5.343
350	165.35	0.00880	1671.9	895.7	2567.6	3.780	1.438	5.218
360	186.75	0.00694	1764.2	721.3	2485.5	3.921	1.139	5.060
370	210.54	0.00497	1890.2	452.6	2342.8	4.111	0.704	4.814
374.15	221.20	0.00317	2107.4	0.0	2107.4	4.443	0.000	4.443

Table A3 Saturated trichloromonofluoromethane (CCl₃F), R11 datum at −40°C, $h_f = 0$, $s_f = 0$

Saturation temperature, T (°C)	Pressure, p (bar)	Specific volume of steam, in m³/kg		Specific enthalpy in kJ/kg		Specific entropy, in kJ/kg-K		
		Liquid (v_f)	Vapour (v_g)	Liquid (h_f)	Vapour (h_g)	Latent (h_{fg})	Liquid (s_f)	Vapour (s_g)
−40	0.05096	0.000617	2.7625	0.00	203.48	203.48	0.0000	0.8730
−38	0.5942	0.000619	2.5360	1.65	204.48	202.83	0.0068	0.8701
−36	0.06787	0.000621	2.3094	3.30	205.48	202.18	0.0137	0.8671
−34	0.07633	0.000623	2.0827	4.95	206.48	201.53	0.0206	0.8642
−32	0.08478	0.000624	1.8562	6.60	207.48	200.88	0.0274	0.8613
−30	0.09323	0.000626	1.6269	8.25	208.48	200.23	0.0343	0.8583
−28	0.10345	0.000627	1.4448	9.90	209.48	199.58	0.0410	0.8554
−26	0.11586	0.000629	1.3123	11.55	210.49	198.94	0.0477	0.8533
−24	0.12828	0.000631	1.1798	13.20	211.49	198.29	0.0544	0.8508
−22	0.14292	0.000633	1.0722	14.85	212.50	197.65	0.0611	0.8483
−20	0.15869	0.000634	0.9770	16.51	213.50	196.99	0.0678	0.8462
−18	0.17414	0.000636	0.8793	18.19	214.52	196.33	0.0741	0.8441
−16	0.19280	0.000638	0.8045	19.82	215.53	195.71	0.0808	0.8420
−14	0.21260	0.000639	0.7345	21.49	216.54	195.05	0.0875	0.8403
−12	0.23390	0.000641	0.6713	23.17	217.55	194.38	0.0938	0.8387
−10	0.25848	0.000643	0.6191	24.83	218.56	193.73	0.1009	0.8366
−8	0.28306	0.000645	0.5671	26.50	219.57	193.07	0.1063	0.8349
−6	0.30280	0.000647	0.5194	28.17	220.59	192.42	0.1126	0.8332
−4	0.33967	0.000649	0.4801	29.85	221.61	191.76	0.1189	0.8315
−2	0.36983	0.000651	0.4409	31.54	222.63	191.09	0.1252	0.8303
0	0.40360	0.000653	0.4069	33.22	223.64	190.42	0.1315	0.8286
1	0.42190	0.000653	0.3920	34.06	224.14	190.08	0.1344	0.8280
2	0.44021	0.000654	0.3770	34.90	224.65	189.75	0.1374	0.8274
3	0.45852	0.000655	0.3620	35.74	225.15	189.41	0.1405	0.8267
4	0.47683	0.000656	0.3471	36.59	225.65	189.06	0.1436	0.8261
5	0.49720	0.000657	0.3340	37.43	226.16	188.73	0.1465	0.8254
6	0.51917	0.000658	0.3225	38.28	226.66	188.38	0.1498	0.8248
7	0.54118	0.000659	0.3109	39.13	227.17	188.04	0.1526	0.8242
8	0.56317	0.000660	0.2994	39.98	227.68	187.70	0.1562	0.8236
9	0.58517	0.000661	0.2878	40.83	228.18	187.35	0.1586	0.8229
10	0.60717	0.000662	0.2763	41.68	228.69	187.01	0.1616	0.8223
11	0.63320	0.000663	0.2673	42.53	229.20	186.67	0.1646	0.8219
12	0.65921	0.000664	0.2584	43.39	229.71	186.32	0.1675	0.8215
13	0.68523	0.000665	0.2495	44.24	230.21	185.97	0.1706	0.8210
14	0.71124	0.000666	0.2405	45.10	230.72	185.62	0.1738	0.8206
15	0.73727	0.000667	0.2316	45.95	231.22	185.27	0.1767	0.8200
16	0.76552	0.000669	0.2235	46.81	231.73	184.92	0.1796	0.8194
17	0.79655	0.000670	0.2163	47.67	232.23	184.56	0.1825	0.8170
18	0.82759	0.000671	0.2092	48.54	232.73	184.19	0.1855	0.8186
19	0.85862	0.000672	0.2021	49.40	233.24	183.84	0.1885	0.8180
20	0.88966	0.000673	0.1950	50.26	233.74	183.48	0.1913	0.8173
21	0.92070	0.000674	0.1878	51.12	234.24	183.12	0.1943	0.8169
22	0.95625	0.000675	0.1821	51.99	234.74	182.75	0.1972	0.8165

(Contd.)

Table A3 Saturated trichloromonofluoromethane (CCl$_3$F), R11 datum at $-40°C$, $h_f = 0$, $s_f = 0$
(Contd.)

Saturation temperature, T (°C)	Pressure, p (bar)	Specific volume of steam, in m^3/kg		Specific enthalpy in kJ/kg		Specific entropy, in kJ/kg-K		
		Liquid (v_f)	Vapour (v_g)	Liquid (h_f)	Vapour (h_g)	Latent (h_{fg})	Liquid (s_f)	Vapour (s_g)
23	0.99237	0.000676	0.1764	52.85	235.24	182.39	0.2003	0.8163
24	1.02496	0.000677	0.1708	53.72	235.75	182.03	0.2031	0.8160
25	1.06117	0.000678	0.1651	54.59	236.25	181.66	0.2061	0.8157
26	1.10074	0.000679	0.1595	55.46	236.75	181.29	0.2089	0.8152
27	1.13880	0.000680	0.1542	56.32	237.25	180.93	0.2120	0.8150
28	1.17917	0.000681	0.1495	57.20	237.76	180.56	0.2148	0.8148
29	1.22273	0.000682	0.1452	58.07	238.25	180.18	0.2177	0.8144
30	1.26070	0.000684	0.1401	58.94	238.76	179.82	0.2206	0.8140
31	1.30665	0.000685	0.1362	59.82	239.25	179.43	0.2234	0.8135
32	1.34821	0.000686	0.1316	60.70	239.75	179.05	0.2261	0.8131
33	1.39568	0.000687	0.1278	61.57	240.24	178.67	0.2290	0.8128
34	1.44414	0.000689	0.1242	62.45	240.73	178.28	0.2320	0.8126
35	1.49276	0.000690	0.1205	63.32	241.23	177.91	0.2349	0.8123
36	1.54130	0.000691	0.1168	64.20	241.72	177.52	0.2378	0.8119
37	1.58983	0.000692	0.1131	65.07	242.22	177.15	0.2406	0.8116
38	1.64000	0.000693	0.1096	65.95	242.72	176.77	0.2433	0.8114
39	1.69571	0.000694	0.1066	66.84	243.20	176.36	0.2462	0.8112
40	1.75145	0.000696	0.1037	67.74	243.69	175.95	0.2490	0.8110
41	1.80718	0.000697	0.1007	68.63	244.18	175.55	0.2518	0.8108
42	1.86292	0.000698	0.0977	69.52	244.67	175.15	0.2546	0.8106
43	1.91866	0.000699	0.0947	70.41	245.16	174.75	0.2574	0.8104
44	1.97952	0.000700	0.0921	71.31	245.65	174.34	0.2600	0.8102
45	2.04300	0.000701	0.0896	72.20	246.14	173.94	0.2029	0.8100
46	2.10641	0.000702	0.0871	73.08	246.63	173.55	0.2658	0.8098
47	2.16980	0.000704	0.0847	73.98	247.12	173.14	0.2686	0.8096
48	2.23310	0.000705	0.0823	74.87	247.60	172.73	0.2713	0.8094
49	2.29670	0.000706	0.0798	75.76	248.07	172.31	0.2741	0.8092
50	2.36000	0.000707	0.0774	76.67	248.53	171.86	0.2768	0.8090

Table A4 Saturated dichlorodifluoromethane (CCl$_2$F$_2$), R12 datum at $-40°C$, $h_f = 0$, $s_f = 0$

Saturation temperature, T (°C)	Pressure, p (bar)	Specific volume of steam, in m^3/kg		Specific enthalpy in kJ/kg		Specific entropy, in kJ/kg-K		
		Liquid (v_f)	Vapour (v_g)	Liquid (h_f)	Vapour (h_g)	Latent (h_{fg})	Liquid (s_f)	Vapour (s_g)
−100	0.01185	0.000600	10.1951	−51.84	142.00	193.84	−0.2567	0.8628
−95	0.01864	0.000604	6.6231	−47.56	144.22	191.78	−0.2323	0.8442
−90	0.02843	0.000608	4.4206	−43.28	146.46	189.74	−0.2086	0.8273
−85	0.04254	0.000613	3.0531	−39.00	148.73	187.73	−0.1856	0.8122
−80	0.06200	0.000617	2.1519	−34.72	151.02	185.74	−0.1631	0.7985

(Contd.)

Table A4 Saturated dichlorodifluoromethane (CCl_2F_2), R12 datum at $-40°C$, $h_f = 0$, $s_f = 0$

(Contd.)

Saturation temperature, T (°C)	Pressure, p (bar)	Specific volume of steam, in m³/kg		Specific enthalpy in kJ/kg		Specific entropy, in kJ/kg-K		
		Liquid (v_f)	Vapour (v_g)	Liquid (h_f)	Vapour (h_g)	Latent (h_{fg})	Liquid (s_f)	Vapour (s_g)
−75	0.08826	0.000622	1.5462	−30.42	153.32	183.74	−0.1412	0.7861
−70	0.12298	0.000627	1.1314	−26.12	155.63	181.75	−0.1198	0.7665
−65	0.16807	0.000632	0.8421	−21.81	157.96	179.77	−0.0988	0.7648
−60	0.22665	0.000637	0.6401	−17.48	160.29	177.77	−0.0783	0.7558
−55	0.30052	0.000643	0.4930	−13.14	162.62	175.76	−0.0581	0.7475
−50	0.39237	0.000648	0.3845	−8.78	164.95	173.73	−0.0384	0.7401
−45	0.50512	0.000654	0.3035	−4.39	167.27	171.66	−0.0190	0.7334
−40	0.64190	0.000660	0.2422	0.00	169.60	169.60	0.0000	0.7274
−38	0.70460	0.000663	0.2221	1.76	170.52	168.76	0.0075	0.7251
−36	0.77196	0.000665	0.2040	3.53	171.44	167.91	0.0149	0.7230
−34	0.84421	0.000667	0.1877	5.31	172.36	167.05	0.0224	0.7209
−32	0.92776	0.000670	0.1729	7.08	173.28	166.20	0.0298	0.7190
−30	1.00441	0.000673	0.1596	8.86	174.20	165.34	0.0371	0.7171
−28	1.09311	0.000676	0.1475	10.64	175.11	164.47	0.0444	0.7153
−26	1.18778	0.000678	0.1364	12.43	176.02	163.59	0.0516	0.7135
−24	1.28858	0.000681	0.1265	14.22	176.93	162.71	0.0588	0.7118
−22	1.39581	0.000683	0.1173	16.02	177.83	161.81	0.0660	0.7102
−20	1.50972	0.000686	0.1090	17.82	178.73	160.91	0.0731	0.7087
−18	1.63104	0.000689	0.1090	19.62	179.63	160.01	0.0801	0.7073
−16	1.75963	0.000692	0.1014	21.43	180.53	159.10	0.0871	0.7059
−14	1.89575	0.000695	0.0944	23.23	181.42	159.19	0.0941	0.7045
−12	2.04605	0.000698	0.0880	25.05	182.31	157.26	0.1010	0.7032
−10	2.19172	0.000701	0.0821	26.87	183.19	156.32	0.1080	0.7019
−8	0.35272	0.000704	0.0767	28.70	184.06	155.36	0.1148	0.7007
−6	2.52244	0.000707	0.0717	30.53	184.94	154.41	0.1217	0.6996
−4	2.70116	0.000710	0.0672	32.37	185.80	153.43	0.1285	0.6986
−2	2.88921	0.000713	0.0630	34.20	186.67	152.47	0.1352	0.6975
0	3.08690	0.000717	0.0591	36.05	187.53	151.48	0.1420	0.6965
1	3.18974	0.000719	0.0555	36.98	187.95	150.97	0.1453	0.6961
2	3.29513	0.000720	0.0538	37.90	188.38	150.48	0.1489	0.6956
3	3.40310	0.000721	0.0521	38.83	188.81	149.98	0.1521	0.6951
4	3.51367	0.000723	0.0505	39.76	189.23	149.47	0.1553	0.6947
5	3.62690	0.000725	0.0490	40.69	189.65	148.96	0.1586	0.6943
6	3.74280	0.000727	0.0475	41.62	190.07	148.45	0.1620	0.6938
7	3.86141	0.000729	0.0461	42.56	190.49	147.93	0.1653	0.6933
8	3.98283	0.000730	0.0447	43.50	190.91	147.41	0.1686	0.6929
9	4.10702	0.000732	0.0434	44.43	191.32	146.89	0.1719	0.6925
10	4.23407	0.000734	0.0422	45.37	191.74	146.37	0.1752	0.6921
11	4.36442	0.000736	0.0410	46.31	192.15	145.84	0.1784	0.6917
12	4.49763	0.000738	0.0398	47.26	192.56	145.30	0.1817	0.6913
13	4.63386	0.000739	0.0386	48.20	192.97	144.77	0.1850	0.6909
14	4.77312	0.000741	0.0375	49.15	193.38	144.23	0.1883	0.6906
15	4.91545	0.000743	0.0355	50.10	193.79	143.69	0.1915	0.6902

(Contd.)

Table A4 Saturated dichlorodifluoromethane (CCl_2F_2), R12 datum at $-40°C$, $h_f = 0$, $s_f = 0$

(Contd.)

Saturation temperature, T (°C)	Pressure, p (bar)	Specific volume of steam, in m³/kg		Specific enthalpy in kJ/kg			Specific entropy, in kJ/kg-K		
		Liquid (v_f)	Vapour (v_g)	Liquid (h_f)	Vapour (h_g)	Latent (h_{fg})	Liquid (s_f)	Vapour (s_g)	
16	5.06087	0.000745	0.0345	51.05	194.19	143.14	0.1948	0.6898	
17	5.20942	0.000747	0.0335	52.00	194.59	142.59	0.1981	0.6894	
18	5.36117	0.000749	0.0326	52.95	194.99	142.04	0.2013	0.6891	
19	5.51614	0.000751	0.0317	53.91	195.38	141.47	0.2046	0.6888	
20	5.67441	0.000753	0.0308	54.87	195.78	140.91	0.2078	0.6884	
21	5.83635	0.000756	0.0300	55.83	196.17	140.34	0.2110	0.6881	
22	6.00171	0.000758	0.0292	56.79	196.56	139.77	0.2143	0.6878	
23	6.17050	0.000759	0.0284	57.75	196.96	139.21	0.2174	0.6875	
24	6.34269	0.000761	0.0276	58.73	197.34	138.61	0.2207	0.6872	
25	6.51840	0.000764	0.0269	59.70	197.73	138.03	0.2239	0.6868	
26	6.69765	0.000766	0.0262	60.67	198.11	137.44	0.2271	0.6865	
27	6.88048	0.000768	0.0255	61.65	198.50	136.85	0.2303	0.6862	
28	7.06704	0.000770	0.0248	62.63	198.87	136.24	0.2335	0.6859	
29	7.25738	0.000772	0.0241	63.61	199.25	135.64	0.2368	0.6856	
30	7.45103	0.000775	0.0235	64.59	199.62	135.03	0.2400	0.6853	
31	7.64903	0.000777	0.0230	65.58	199.99	134.41	0.2431	0.6850	
32	7.85089	0.000779	0.0225	66.57	200.36	133.79	0.2463	0.6847	
33	8.05662	0.000782	0.0218	67.56	200.73	133.17	0.2495	0.6845	
34	8.26621	0.000784	0.0212	68.56	201.09	132.53	0.2527	0.6842	
35	8.48000	0.000786	0.0207	69.56	201.45	131.89	0.2559	0.6839	
36	8.69766	0.000789	0.0202	70.55	201.80	131.25	0.2591	0.6836	
37	8.91904	0.000792	0.0196	71.55	202.16	130.61	0.2623	0.6833	
38	9.14483	0.000794	0.0191	72.56	202.51	129.95	0.2654	0.6830	
39	9.37497	0.000796	0.0186	73.57	202.86	129.29	0.2685	0.6828	
40	9.60897	0.000799	0.0182	74.59	203.20	128.61	0.2718	0.6825	
41	9.84793	0.000802	0.0177	75.61	203.54	127.93	0.2750	0.6822	
42	10.09131	0.000804	0.0173	76.62	203.87	127.25	0.2782	0.6820	
43	10.33862	0.000807	0.0168	77.65	204.21	126.56	0.2814	0.6817	
44	10.59021	0.000810	0.0164	78.68	204.55	125.87	0.2846	0.6814	
45	10.84655	0.000813	0.0160	79.71	204.87	125.16	0.2878	0.6811	
46	11.10758	0.000815	0.0156	80.75	205.19	124.44	0.2909	0.6808	
47	11.37304	0.000818	0.0153	81.79	205.51	123.72	0.2941	0.6805	
48	11.64290	0.000821	0.0149	82.83	205.83	123.00	0.2973	0.6802	
49	11.91724	0.000824	0.0146	83.88	206.14	122.26	0.3005	0.6800	
50	12.19655	0.000827	0.0142	84.94	206.45	121.51	0.3037	0.6797	

Table A5 Saturated monochlorodifluoromethane (CHClF$_2$), R22 datum at $-40°C$, $h_f = 0$, $s_f = 0$

Saturation temperature, T (°C)	Saturation pressure, p (bar)	Specific volume of steam, in m^3/kg		Specific enthalpy in kJ/kg			Specific entropy, in kJ/kg-K		
		Liquid (v_f)	Vapour (v_g)	Liquid (h_f)	Vapour (h_g)	Latent (h_{fg})	Liquid (s_f)	Vapour (s_g)
−100	0.02009	0.000643	8.3412	−63.45	203.73	267.18	−0.3144	1.2293
−95	0.03150	0.000647	5.4344	−58.14	206.19	264.33	−0.2843	1.2004
−90	0.04792	0.000652	3.6381	−52.87	208.64	261.51	−0.2550	1.1736
−85	0.07731	0.000656	2.5204	−47.61	211.11	258.72	−0.2269	1.1489
−80	0.10393	0.000661	1.7816	−42.40	213.60	256.00	−0.1989	1.1267
−75	0.14759	0.000666	1.2842	−37.17	216.11	253.28	−0.1721	1.1066
−70	0.20517	0.000672	0.9420	−31.93	218.62	250.55	−0.1461	1.0874
−65	0.27965	0.000677	0.7037	−26.68	221.16	247.84	−0.1206	1.0702
−60	0.37448	0.000683	0.5351	−21.42	223.67	245.09	−0.0959	1.0543
−55	0.49621	0.000689	0.4131	−16.13	226.18	242.31	−0.0712	1.0396
−50	0.64758	0.000696	0.3229	−10.81	228.69	239.50	−0.0473	1.0262
−45	0.83241	0.000702	0.2556	−5.40	231.20	236.60	−0.0234	1.0137
−40	1.05586	0.000709	0.2049	0.00	233.67	233.67	0.0000	1.0024
−38	1.15862	0.000712	0.1882	2.20	234.65	232.45	0.0096	0.9982
−36	1.26910	0.000715	0.1728	4.40	235.65	231.25	0.0188	0.9940
−34	1.38731	0.000719	0.1590	6.58	236.60	230.02	0.0280	0.9902
−32	1.51324	0.000721	0.1465	8.83	237.60	228.77	0.0373	0.9860
−30	1.64690	0.000724	0.1353	11.05	238.55	227.50	0.0464	0.9283
−28	1.78938	0.000727	0.1253	13.29	239.52	226.23	0.0557	0.9785
−26	1.94069	0.000731	0.1161	15.58	240.48	224.90	0.0645	0.9747
−24	2.10207	0.000734	0.1077	17.77	241.41	223.64	0.0733	0.9710
−22	2.27448	0.000738	0.1000	19.99	242.33	222.34	0.0821	0.9676
−20	2.45793	0.000741	0.0930	22.21	243.25	221.04	0.0908	0.9638
−18	2.65310	0.000744	0.0865	24.47	244.17	219.70	0.0996	0.9605
−16	2.85903	0.000748	0.0806	26.72	245.08	218.36	0.01080	0.9576
−14	3.07876	0.000752	0.0752	28.94	245.96	217.02	0.1164	0.9538
−12	3.31172	0.000756	0.0701	31.16	246.84	215.68	0.1248	0.9508
−10	3.55793	0.000759	0.0655	33.40	247.72	214.32	0.1336	0.9479
−8	3.81321	0.000763	0.0612	35.66	248.60	212.94	0.1419	0.9454
−6	4.09172	0.000767	0.0573	37.92	249.46	211.54	0.1503	0.9425
−4	4.37972	0.000771	0.0536	40.19	250.30	210.11	0.1591	0.9396
−2	4.68317	0.000775	0.0503	42.52	251.14	208.62	0.1675	0.9370
0	5.00207	0.000779	0.0471	44.94	251.97	207.03	0.1763	0.9345
1	5.16841	0.000781	0.0457	46.16	252.37	206.21	0.1809	0.9333
2	5.33917	0.000783	0.0443	47.38	252.77	205.39	0.1855	0.9320
3	5.51434	0.000785	0.0429	48.64	253.17	204.53	0.1901	0.9303
4	5.69352	0.000787	0.0416	49.91	253.56	203.65	0.1943	0.9291
5	5.87621	0.000790	0.0403	51.16	253.95	202.79	0.1989	0.9278
6	6.06276	0.000792	0.0391	52.41	254.33	201.92	0.2035	0.9266
7	6.25393	0.000794	0.0379	53.68	254.71	201.03	0.2077	0.9253
8	6.44993	0.000797	0.0368	54.94	255.08	200.14	0.2123	0.9241

(Contd.)

Table A5 Saturated monochlorodifluoromethane ($CHClF_2$), R22 datum at $-40°C$, $h_f = 0$, $s_f = 0$

(Contd.)

Saturation temperature, T (°C)	Saturation pressure, p (bar)	Specific volume of steam, in m³/kg		Specific enthalpy in kJ/kg			Specific entropy, in kJ/kg-K	
		Liquid (v_f)	Vapour (v_g)	Liquid (h_f)	Vapour (h_g)	Latent (h_{fg})	Liquid (s_f)	Vapour (s_g)
9	6.65034	0.000799	0.0357	56.22	255.44	199.22	0.2169	0.9228
10	6.85517	0.000801	0.0346	57.52	255.81	198.29	0.2211	0.9216
11	7.06621	0.000803	0.0336	58.80	256.15	197.35	0.2257	0.9203
12	7.27724	0.000805	0.0326	60.07	256.48	196.41	0.2303	0.9190
13	7.49793	0.000808	0.0316	61.38	256.83	195.45	0.2345	0.9178
14	7.72552	0.000810	0.0307	62.72	257.17	194.45	0.2391	0.9165
15	7.95517	0.000813	0.0298	64.02	257.50	193.48	0.2437	0.9152
16	8.18758	0.000815	0.0289	65.32	257.81	192.48	0.2483	0.9140
17	8.42552	0.000817	0.0281	66.63	258.13	191.50	0.2529	0.9127
18	8.63241	0.000820	0.0273	67.95	258.44	190.49	0.2571	0.9115
19	8.91793	0.000822	0.0266	69.27	258.73	189.46	0.2617	0.9102
20	9.17241	0.000825	0.0258	70.59	259.00	188.41	0.2663	0.9089
21	9.43310	0.000827	0.0251	71.93	259.29	187.36	0.2709	0.9077
22	9.69931	0.000830	0.0244	73.31	259.58	186.27	0.2755	0.9065
23	9.97103	0.000833	0.0237	74.66	259.85	185.19	0.2801	0.9052
24	10.24828	0.000835	0.0230	76.04	260.11	184.07	0.2847	0.9039
25	10.53103	0.000838	0.0224	77.39	260.38	182.99	0.2889	0.9027
26	10.81931	0.000841	0.0218	78.79	260.64	181.85	0.2935	0.9014
27	11.11310	0.000844	0.0212	80.16	260.89	180.73	0.2981	0.9002
28	11.41241	0.000846	0.0206	81.54	261.12	179.58	0.3023	0.8989
29	11.69034	0.000849	0.0200	82.96	261.37	178.41	0.3069	0.8977
30	12.03448	0.000852	0.0194	84.38	261.60	177.22	0.3115	0.8964
31	12.35103	0.000855	0.0189	85.77	261.81	176.04	0.3161	0.8948
32	12.67310	0.000858	0.0184	87.17	262.02	174.85	0.3207	0.8935
33	13.00070	0.000861	0.0179	88.57	262.22	173.65	0.3249	0.8922
34	13.33793	0.000864	0.0174	90.00	262.40	172.40	0.3295	0.8909
35	13.68276	0.000867	0.0169	91.43	262.58	171.15	0.3337	0.8893
36	14.03034	0.000870	0.0165	92.85	262.74	169.89	0.3383	0.8880
37	14.38207	0.000874	0.0161	94.24	262.88	168.64	0.3429	0.8868
38	14.74345	0.000877	0.0156	95.63	263.00	167.37	0.3471	0.8851
39	15.11103	0.000881	0.0152	97.03	263.13	166.10	0.3517	0.8838
40	15.48965	0.000884	0.0148	98.44	263.21	164.77	0.3563	0.8822
41	15.86827	0.000887	0.0144	99.82	263.29	163.47	0.3601	0.8809
42	16.25793	0.000891	0.0140	101.24	263.38	162.14	0.3651	0.8797
43	16.65380	0.000894	0.0137	102.68	263.48	160.80	0.3693	0.8780
44	17.05517	0.000898	0.0133	104.12	263.58	159.46	0.3739	0.8767
45	17.46552	0.000902	0.0130	105.58	263.67	158.09	0.3785	0.8755
46	17.88414	0.000906	0.0126	107.04	263.76	156.72	0.3827	0.8738
47	18.30827	0.000910	0.0123	108.51	263.87	155.36	0.3871	0.8723
48	18.73793	0.000914	0.0119	109.98	263.93	153.95	0.3915	0.8709
49	19.12414	0.000918	0.0116	111.42	263.99	152.57	0.3959	0.8694
50	19.61380	0.000922	0.0113	112.86	264.05	151.19	0.4003	0.8680

Table A6 Saturated ammonia (NH$_3$), R717 datum at $-40°C$, $h_f = 0$, $s_f = 0$

Saturation temperature, T (°C)	Saturation pressure, p (bar)	Specific volume of steam, in m³/kg		Specific enthalpy in kJ/kg			Specific entropy, in kJ/kg-K	
		Liquid (v_f)	Vapour (v_g)	Liquid (h_f)	Vapour (h_g)	Latent (h_{fg})	Liquid (s_f)	Vapour (s_g)
−50	0.40896	0.001426	2.6281	−44.43	1373.27	1417.70	−0.1943	6.1603
−48	0.45972	0.001431	2.3565	−35.44	1376.80	1412.24	−0.1551	6.1192
−46	0.51600	0.001436	2.1177	−26.60	1380.20	1406.80	−0.1157	6.0789
−44	0.57710	0.001441	1.9062	−17.81	1383.31	1401.12	−0.0769	6.0394
−42	0.64455	0.001446	1.7196	−8.97	1386.67	1395.64	−0.0384	6.0008
−40	0.71793	0.001451	1.5537	0.00	1390.02	1390.02	0.0000	5.9631
−38	0.79384	0.001456	1.4077	8.97	1393.13	1384.16	0.0381	5.9262
−36	0.88607	0.001462	1.2775	17.81	1396.35	1378.54	0.0758	5.8900
−34	0.98096	0.001467	1.1614	26.84	1399.51	1372.67	0.1131	5.8545
−32	1.08165	0.001472	1.0574	35.68	1402.48	1366.80	0.1506	5.8198
−30	1.19586	0.001477	0.9644	44.66	1405.60	1360.94	0.1876	5.7856
−28	1.31724	0.001483	0.8820	53.68	1408.53	1354.85	0.2242	5.7521
−26	1.44790	0.001488	0.8069	62.61	1411.45	1348.84	0.2607	5.7195
−24	1.58841	0.001494	0.7397	71.73	1414.39	1342.66	0.2970	5.6872
−22	1.73986	0.001500	0.6793	80.76	1417.28	1336.52	0.3330	5.6556
−20	1.90276	0.001505	0.6244	89.78	1420.02	1330.24	0.3984	5.6244
−18	2.07807	0.001511	0.5750	98.76	1422.72	1323.96	0.4043	5.5939
−16	2.26551	0.001517	0.5303	107.83	1425.28	1317.45	0.4397	5.5639
−14	2.46634	0.001523	0.4896	116.95	1427.88	1310.93	0.4747	5.5356
−12	2.68069	0.001529	0.4526	126.16	1430.54	1304.38	0.5096	5.5055
−10	2.90896	0.001536	0.4189	135.37	1433.05	1297.68	0.5443	5.4770
−8	3.15365	0.001541	0.3884	144.35	1435.33	1290.98	0.5789	5.4487
−6	3.41380	0.00548	0.3604	153.56	1437.93	1284.37	0.6139	5.4210
−4	3.69062	0.001554	0.3348	162.77	1440.02	1277.25	0.6473	5.3940
−2	3.98427	0.001561	0.3113	171.98	1442.17	1270.19	0.6812	5.3670
0	4.29586	0.001567	0.2898	181.20	1444.45	1263.25	0.7151	5.3405
1	4.45848	0.001571	0.2798	185.80	1445.49	1259.69	0.7321	5.3277
2	4.62662	0.001574	0.2702	190.40	1446.54	1256.14	0.7487	5.3145
3	4.79931	0.001578	0.2610	195.17	1447.59	1252.42	0.7653	5.3017
4	4.97682	0.001582	0.2521	199.85	1448.63	1248.78	0.7818	5.2888
5	5.15862	0.001585	0.2436	204.46	1449.56	1245.10	0.7989	5.2765
6	5.34745	0.001589	0.2354	209.06	1450.49	1241.43	0.8154	5.2638
7	5.54007	0.001592	0.2276	213.73	1451.54	1237.81	0.8320	5.2513
8	5.73820	0.001595	0.2201	218.50	1452.54	1234.04	0.8487	5.2389
9	5.94186	0.001598	0.2128	223.11	1453.39	1230.28	0.8652	5.2266
10	6.15103	0.001603	0.2060	227.72	1454.22	1226.50	0.8814	5.2141
11	6.36641	0.001607	0.1992	232.53	1455.30	1222.77	0.8979	5.2020
12	6.58731	0.001610	0.1928	237.15	1456.11	1218.96	0.9142	5.1900
13	6.81420	0.001614	0.1866	241.92	1456.96	1215.04	0.9307	5.1780
14	7.04717	0.001617	0.1807	246.60	1457.80	1211.20	0.9470	5.1659
15	7.28276	0.001621	0.1751	251.44	1458.63	1207.19	0.9634	5.1542

(Contd.)

Table A6 Saturated ammonia (NH$_3$), R717 Datum at $-40°C$, $h_f = 0$, $s_f = 0$ *(Contd.)*

Saturation temperature, T (°C)	Saturation pressure, p (bar)	Specific volume of steam, in m³/kg		Specific enthalpy in kJ/kg		Specific entropy, in kJ/kg-K		
		Liquid (v_f)	Vapour (v_g)	Liquid (h_f)	Vapour (h_g)	Latent (h_{fg})	Liquid (s_f)	Vapour (s_g)
16	7.53104	0.001624	0.1696	256.14	1459.47	1203.33	0.9794	5.1421
17	7.78138	0.001628	0.1643	259.88	1460.24	1200.36	0.9956	5.1302
18	8.03862	0.001633	0.1592	265.54	1460.92	1195.38	1.0118	5.1186
19	8.30551	0.001637	0.1543	270.35	1461.75	1191.40	1.0280	5.1073
20	8.57241	0.001641	0.1496	275.16	1462.60	1187.44	1.0442	5.0956
21	8.85172	0.001645	0.1450	279.77	1463.21	1183.44	1.0604	5.0843
22	9.13655	0.001648	0.1407	284.56	1463.84	1179.28	1.0763	5.0729
23	9.42690	0.001652	0.1365	289.37	1464.63	1175.26	1.0924	5.0616
24	9.72690	0.001656	0.1324	294.19	1465.33	1171.14	1.1083	5.0503
25	10.02760	0.001661	0.1284	298.90	1465.84	1166.94	1.1242	5.0391
26	10.34069	0.001665	0.1246	303.82	1466.59	1162.77	1.1402	5.0279
27	10.66137	0.001669	0.1210	308.63	1467.22	1158.59	1.1563	5.0170
28	10.99172	0.001673	0.1174	313.45	1467.85	1154.40	1.1721	5.0061
29	11.32758	0.001678	0.1140	318.26	1468.45	1150.19	1.1879	4.9951
30	11.66896	0.001682	0.1107	323.08	1468.87	1145.79	1.2037	4.9842
31	12.01655	0.001686	0.1075	327.89	1469.50	1141.61	1.2195	4.9733
32	12.37517	0.001691	0.1045	332.71	1469.94	1137.23	1.2350	4.9624
33	12.74482	0.001695	0.1015	337.52	1470.36	1132.84	1.2508	4.9517
34	13.12137	0.001700	0.0987	342.48	1470.92	1128.44	1.2664	4.9409
35	13.50345	0.001704	0.0960	347.50	1471.43	1123.93	1.2821	4.9302
36	13.89379	0.001709	0.0932	352.29	1471.70	1119.41	1.2978	4.9196
37	14.29517	0.001713	0.0907	357.25	1472.19	1114.94	1.3135	4.9091
38	14.70620	0.001718	0.0882	362.12	1472.45	1110.33	1.3290	4.8985
39	15.12276	0.001723	0.0857	367.11	1472.87	1105.76	1.3445	4.8885
40	15.54483	0.001727	0.0834	371.93	1473.30	1101.37	1.3600	4.8774
41	15.98551	0.001732	0.0811	376.95	1473.50	1096.55	1.3754	4.8669
42	16.43172	0.001738	0.0789	381.79	1473.70	1091.91	1.3908	4.8563
43	16.88344	0.001742	0.0768	386.76	1473.91	1087.15	1.4065	4.8461
44	17.34482	0.001747	0.0747	391.79	1474.12	1082.33	1.4221	4.8357
45	17.81724	0.001752	0.0727	396.81	1474.33	1077.52	1.4374	4.8251
46	18.30070	0.001757	0.0708	401.84	1474.54	1072.70	1.4528	4.8147
47	18.79518	0.001763	0.0689	406.86	1474.68	1067.82	1.4683	4.8045
48	19.29931	0.001768	0.0670	411.89	1474.76	1062.87	1.4835	4.7938
49	19.80965	0.001773	0.0652	416.91	1474.84	1057.93	1.4990	4.7834
50	20.33103	0.001779	0.0636	421.94	1474.92	1052.98	1.5148	4.7732

Table A7 Saturated carbon dioxide (CO_2), R744 datum at $-40°C$, $h_f = 0$, $s_f = 0$

Saturation temperature, T (°C)	Saturation pressure, p (bar)	Specific volume of steam, in m³/kg		Specific enthalpy in kJ/kg			Specific entropy, in kJ/kg-K	
		Liquid (v_f)	Vapour (v_g)	Liquid (h_f)	Vapour (h_g)	Latent (h_{fg})	Liquid (s_f)	Vapour (s_g)
−40	10.05517	0.000898	0.03821	0.00	320.52	320.52	0.0000	1.3754
−38	10.84965	0.000904	0.03577	3.77	320.86	317.09	0.0161	1.3654
−36	11.64413	0.000911	0.03332	7.54	321.19	313.65	0.0322	1.3553
−34	12.46676	0.000918	0.03100	11.32	321.51	310.19	0.0483	1.3453
−32	13.38786	0.000925	0.02906	15.17	321.76	306.59	0.0641	1.3356
−30	14.30896	0.000932	0.02712	19.02	322.01	302.99	0.0800	1.3259
−28	15.28855	0.000939	0.02535	22.95	322.22	299.27	0.0956	1.3163
−26	16.34124	0.000947	0.02379	26.97	322.39	295.42	0.1109	1.3067
−24	17.39393	0.000955	0.02222	30.99	322.56	291.57	0.1264	1.2971
−22	18.54262	0.000963	0.02084	35.06	322.72	287.66	0.1419	1.2879
−20	19.73931	0.000971	0.01957	39.17	322.89	283.72	0.1574	1.2787
−18	20.93600	0.000980	0.01830	43.27	323.06	279.79	0.1733	1.2692
−16	22.22758	0.000990	0.01716	47.45	322.89	275.44	0.1888	1.2605
−14	23.58345	0.00100	0.01611	51.73	322.61	270.888	0.2060	1.2523
−12	25.01131	0.001010	0.01513	56.32	322.58	266.26	0.2260	1.2464
−10	26.54069	0.001021	0.01426	60.85	322.24	261.39	0.2432	1.2372
−8	28.07007	0.001032	0.01338	65.37	321.91	256.54	0.2600	1.2280
−6	29.66069	0.001043	0.01256	70.06	321.54	251.48	0.2771	1.2192
−4	31.37379	0.001056	0.01182	75.08	321.12	246.04	0.2954	1.2101
−2	33.08689	0.001069	0.01110	80.10	320.71	240.61	0.3135	1.2010
0	34.91034	0.001082	0.01041	85.27	320.01	234.74	0.3312	1.1909
1	35.86620	0.001090	0.01010	87.91	319.55	231.65	0.3400	1.1854
2	36.82207	0.001098	0.00979	90.54	319.09	228.55	0.3488	1.1799
3	37.77800	0.001106	0.00947	93.18	318.63	225.45	0.3575	1.1744
4	38.73380	0.001113	0.00916	95.82	318.11	222.35	0.3662	1.1690
5	39.75034	0.001122	0.00887	98.55	317.57	219.02	0.3754	1.1633
6	40.81545	0.001131	0.00860	101.36	316.86	215.50	0.3849	1.1575
7	41.88055	0.001140	0.00834	104.16	316.14	211.98	0.3944	1.1516
8	42.94565	0.001149	0.00807	106.97	315.43	208.46	0.4039	1.1458
9	44.01076	0.001158	0.00780	109.77	314.72	204.95	0.4134	1.1400
10	45.07586	0.001167	0.00753	112.58	314.01	201.43	0.4229	1.1342
11	46.25517	0.001178	0.00729	115.55	312.79	197.24	0.4329	1.1274
12	47.43448	0.001190	0.00705	118.52	311.58	193.06	0.4430	1.1205
13	48.61379	0.001202	0.00682	121.49	310.36	188.87	0.4534	1.1137
14	49.79310	0.001213	0.00658	124.47	309.15	184.68	0.4635	1.1068
15	50.97241	0.001225	0.00634	127.44	307.94	180.50	0.4735	1.1000
16	52.20580	0.001238	0.00611	130.68	306.41	175.79	0.4843	1.0921
17	53.50676	0.001254	0.00590	134.05	304.48	170.43	0.4954	1.0830
18	54.80772	0.001270	0.00569	137.48	302.55	165.07	0.5065	1.0738
19	56.10870	0.001286	0.00547	140.92	300.63	159.71	0.5176	1.0647
20	57.40965	0.001302	0.00526	144.35	298.70	154.35	0.5286	1.0556

(Contd.)

Table A7 Saturated carbon dioxide (CO$_2$), R744 datum at $-40°C$, $h_f = 0$, $s_f = 0$ *(Contd.)*

Saturation temperature, T (°C)	Saturation pressure, p (bar)	Specific volume of steam, in m³/kg		Specific enthalpy in kJ/kg		Specific entropy, in kJ/kg-K		
		Liquid (v_f)	Vapour (v_g)	Liquid (h_f)	Vapour (h_g)	Latent (h_{fg})	Liquid (s_f)	Vapour (s_g)
21	58.71062	0.001318	0.00505	147.78	296.78	149.00	0.5397	1.0465
22	60.12745	0.001345	0.00482	151.96	293.29	141.33	0.5539	1.0332
23	61.55876	0.001374	0.00460	156.23	289.60	133.37	0.5681	1.0194
24	62.99007	0.001404	0.00438	160.50	285.92	125.42	0.5832	1.0056
25	64.42138	0.001433	0.00416	164.77	282.23	117.46	0.5978	0.9918
26	65.85270	0.001462	0.00394	169.04	278.55	109.51	0.6124	0.9781
27	67.31930	0.001501	0.00371	174.08	274.16	100.08	0.6289	0.9617
28	68.85655	0.001560	0.00347	180.64	268.37	87.73	0.6490	0.9402
29	70.39380	0.001619	0.00323	187.19	262.58	75.39	0.6691	0.9187
30	71.93103	0.001678	0.00299	193.75	256.79	63.04	0.6892	0.8972
31	73.53103	0.002160	0.00216	225.62	225.62	0.00	0.7913	0.7913

Table A8 Saturated sulphur dioxide (SO$_2$), R764 datum at $-40°C$, $h_f = 0$, $s_f = 0$

Saturation temperature, T (°C)	Saturation pressure, p (bar)	Specific volume, in m³/kg		Specific enthalpy in kJ/kg		Specific entropy, in kJ/kg-K		
		Liquid (v_f)	Vapour (v_g)	Liquid (h_f)	Vapour (h_g)	Latent (h_{fg})	Liquid (s_f)	Vapour (s_g)
−40	0.21628	0.000652	1.4012	0.00	415.44	415.44	0.0000	1.7820
−38	0.24594	0.000654	1.2694	2.45	416.52	414.07	0.0101	1.7715
−36	0.27561	0.000656	1.1375	4.90	417.61	412.71	0.0203	10.7610
−34	0.30731	0.000659	1.0143	7.38	418.68	411.30	0.0305	1.7505
−32	0.34578	0.000661	0.9211	9.93	419.64	409.71	0.0409	1.7404
−30	0.38432	0.000663	0.8280	12.49	420.62	408.13	0.0514	1.7303
−28	0.42757	0.000665	0.7462	15.09	421.56	406.47	0.0619	1.7203
−26	0.47673	0.000667	0.6794	17.75	422.45	404.70	0.0726	1.7106
−24	0.52590	0.000669	0.6125	20.42	423.33	402.91	0.0833	1.7008
−22	0.58345	0.000671	0.5575	23.13	424.17	401.04	0.0941	1.6912
−20	0.64517	0.000673	0.5090	25.88	424.93	399.05	0.1049	1.6818
−18	0.70693	0.000676	0.4606	28.63	425.73	397.10	0.1158	1.6723
−16	0.77683	0.000678	0.4199	31.44	426.46	395.02	0.1267	1.6630
−14	0.85324	0.000680	0.3841	34.24	427.13	392.89	0.1376	1.6538
−12	0.93517	0.000682	0.3518	37.06	427.74	390.68	0.1485	1.6447
−10	1.02345	0.000685	0.3227	39.89	428.31	388.42	0.1594	1.6356
−8	1.11820	0.000687	0.2966	42.76	428.86	386.10	0.1703	1.6267
−6	1.21972	0.000689	0.2729	45.63	429.36	383.73	0.1812	1.6177
−4	1.32758	0.000692	0.2509	40.50	429.80	381.30	0.1919	1.6088
−2	1.44289	0.000695	0.2319	51.39	430.21	378.82	0.2028	1.6000
0	1.56620	0.000697	0.2148	54.26	430.54	376.28	0.2135	1.5912
1	1.63076	0.000699	0.2061	55.73	430.71	374.28	0.2189	1.5868
2	1.69782	0.000701	0.1983	57.19	430.86	373.67	0.2242	1.5824
3	1.76783	0.000702	0.1910	58.65	430.99	372.34	0.2295	1.5781

(Contd.)

Table A8 Saturated Sulphur dioxide (SO$_2$), R764 Datum at $-40°C$, $h_f = 0$, $s_f = 0$ *(Contd.)*

Saturation temperature, T (°C)	Saturation pressure, p (bar)	Specific volume, in m^3/kg		Specific enthalpy in kJ/kg			Specific entropy, in kJ/kg-K	
		Liquid (v_f)	Vapour (v_g)	Liquid (h_f)	Vapour (h_g)	Latent (h_{fg})	Liquid (s_f)	Vapour (s_g)
4	1.83785	0.000703	0.1837	60.11	431.11	371.00	0.2349	1.5737
5	1.91103	0.000704	0.1768	61.56	431.22	369.66	0.2402	1.5694
6	1.98676	0.000705	0.1704	63.02	431.31	368.29	0.2455	1.5651
7	2.06248	0.000707	0.1640	64.48	431.40	366.92	0.2509	1.5608
8	2.14276	0.000708	0.1581	65.94	431.48	365.54	0.2561	1.5564
9	2.22496	0.000710	0.1524	67.40	431.54	364.14	0.2614	1.5522
10	2.30689	0.000711	0.1467	68.87	431.61	362.74	0.2667	1.5479
11	2.39627	0.000712	0.1417	70.34	431.65	361.31	0.2719	1.5436
12	2.48565	0.000714	0.1366	71.80	431.68	359.88	0.2771	1.5393
13	2.57655	0.000715	0.1317	73.27	431.70	358.43	0.2823	1.5350
14	2.67289	0.000716	0.1273	74.72	431.69	356.97	0.2875	1.5308
15	2.67924	0.000718	0.1228	76.18	431.68	355.50	0.2927	1.5266
16	2.86910	0.000719	0.1186	77.64	431.66	354.02	0.2978	1.5224
17	2.97338	0.000720	0.1146	79.09	431.62	352.53	0.3030	1.5182
18	3.07765	0.000722	0.1106	80.55	431.58	351.03	0.3082	1.5140
19	3.18673	0.000723	0.1069	82.01	431.52	349.51	0.3132	1.5097
20	3.29820	0.000725	0.1033	83.46	431.46	348.00	0.3183	1.5055
21	3.40965	0.000726	0.0997	84.92	431.39	346.47	0.3233	1.5013
22	3.52910	0.000728	0.0965	86.38	431.29	344.91	0.3283	1.4971
23	3.64951	0.000729	0.0933	87.83	431.19	343.36	0.3334	1.4929
24	3.77092	0.000731	0.0902	89.29	431.08	341.79	0.3384	1.4887
25	3.90027	0.000732	0.0873	90.74	430.94	340.20	0.3434	1.4846
26	4.02965	0.000734	0.0844	92.19	430.80	338.61	0.3484	1.4805
27	4.16207	0.000735	0.0817	93.64	430.65	337.01	0.3532	1.4764
28	4.30124	0.000737	0.0791	95.08	430.48	335.40	0.3580	1.4721
29	4.44027	0.000738	0.0765	96.53	430.31	333.78	0.3629	1.4677
30	4.58276	0.000740	0.0740	97.97	430.12	332.15	0.3677	1.4635
31	4.73172	0.000741	0.0718	99.41	429.91	330.50	0.3725	1.4593
32	4.88069	0.000743	0.0695	100.86	429.70	328.84	0.3773	1.4551
33	5.03641	0.000745	0.0673	102.29	429.47	327.18	0.3821	1.4509
34	5.19407	0.000746	0.0653	103.71	429.22	325.51	0.3867	1.4467
35	5.35172	0.000748	0.0632	105.13	428.98	323.85	0.3914	1.4425
36	5.52353	0.000749	0.0612	106.56	428.70	322.14	0.3961	1.4383
37	5.69534	0.000751	0.0593	107.98	428.43	320.45	0.4007	1.4341
38	5.87100	0.000753	0.0575	109.40	428.14	318.74	0.4053	1.4299
39	6.06019	0.000754	0.0558	110.81	427.81	317.00	0.4098	1.4256
40	6.24938	0.000756	0.0542	112.21	427.48	315.27	0.4143	1.4214
41	6.43856	0.000758	0.0526	113.26	427.15	313.53	0.4188	1.4172
42	6.62775	0.000759	0.0510	115.03	426.82	311.79	0.4233	1.4129
43	6.81693	0.000761	0.0493	116.43	426.46	310.06	0.4278	1.4087
44	7.05520	0.000763	0.0478	117.83	426.11	308.28	0.4322	1.4044
45	7.31793	0.000765	0.0465	119.22	425.71	306.49	0.4365	1.4002
46	7.58080	0.000767	0.0451	120.61	425.32	304.71	0.4409	1.3959
47	7.84360	0.000769	0.0438	122.00	424.92	302.92	0.4452	1.3916
48	8.10640	0.000771	0.0424	123.39	424.52	301.13	0.4495	1.3874
49	8.36598	0.000773	0.0411	124.78	424.11	299.33	0.4538	1.3831
50	8.59986	0.000775	0.0401	126.13	423.60	297.47	0.4578	1.3787

Table A9 Refrigerant 13 (chlorotrifluoromethane) properties of saturated liquid and saturated vapour

Saturation temperature, T (°C)	Saturation pressure, p (bar)	Volume Vapour, v_g (m³/kg)	Density Liquid, v_f (kg/m³)	Specific enthalpy in kJ/kg Liquid, h_f (kJ/kg)	Vapour, h_g (kJ/kg)	Specific entropy in kJ/kg-K Liquid, h_{fg} (kJ/(kg-K))	Vapour, s_g (kJ/(kg-K))
−120	0.006988	1.7339	1661.6	84.310	250.08	0.45943	1.5418
−115	0.010751	1.1611	1644.2	88.240	252.17	0.48467	1.5212
−110	0.016055	0.79969	1626.5	92.233	254.28	0.50951	1.5027
−105	0.023338	0.56491	1608.7	96.294	256.38	0.53400	1.4861
−100	0.033107	0.40825	1590.7	100.43	258.48	0.55818	1.4710
−98	0.037837	0.36061	1583.4	102.10	259.32	0.56778	1.4654
−96	0.043095	0.31954	1576.1	103.79	260.15	0.57734	1.4600
−94	0.048922	0.28400	1568.7	105.49	260.99	0.58686	1.4548
−92	0.055362	0.25314	1561.3	107.20	261.82	0.59634	1.4499
−90	0.062458	0.22627	1553.8	108.93	262.64	0.60578	1.4451
−88	0.070257	0.20279	1546.3	110.66	263.46	0.61519	1.4405
−86	0.078805	0.18221	1538.8	112.41	264.28	0.62457	1.4360
−84	0.088151	0.16412	1531.2	114.18	265.09	0.63391	1.4318
−82	0.098346	0.14818	1523.6	115.95	265.90	0.64322	1.4277
−81.45	0.101325	0.14410	1521.4	116.45	266.12	0.64579	1.4266
−80	0.10944	0.13409	1515.9	117.74	266.70	0.65249	1.4237
−78	0.12148	0.12160	1508.1	119.55	267.50	0.66173	1.4199
−76	0.13453	0.11051	1500.3	121.36	268.29	0.67094	1.4162
−74	0.14864	0.10063	14920.4	123.19	269.07	0.68012	1.4126
−72	0.16389	0.09182	1484.5	125.03	269.84	0.68926	1.4092
−70	0.18025	0.08393	1476.5	126.88	270.61	0.69838	1.4059
−68	0.19788	0.07686	1468.5	128.75	271.37	0.70746	1.4027
−66	0.21678	0.07050	1460.4	130.63	272.12	0.71651	1.3996
−64	0.23703	0.06477	1452.2	132.52	272.86	0.72552	1.3966
−62	0.25868	0.05961	1443.9	134.42	273.59	0.73451	1.3936
−60	0.28180	0.05494	1435.6	136.33	274.32	0.74346	1.3908
−58	0.30644	0.05071	1427.2	138.26	275.03	0.75238	1.3881
−56	0.33267	0.04687	1418.6	140.20	275.73	0.76126	1.3854
−54	0.36054	0.04338	1410.1	142.15	276.42	0.77012	1.3828
−52	0.39012	0.04020	1401.4	144.12	277.10	0.77894	1.3803
−50	0.42147	0.03730	1392.6	146.09	277.76	0.78773	1.3778
−48	0.45465	0.03465	1383.7	148.08	278.42	0.7949	1.3754
−46	0.48973	0.03223	1374.7	150.08	279.06	0.80521	1.3730
−44	0.52677	0.03001	1365.6	152.09	279.69	0.81390	1.3707
−42	0.56584	0.02797	1356.4	154.11	280.30	0.82257	1.3685
−40	0.60699	0.02609	1347.1	156.15	280.90	0.83121	1.3663
−38	0.65030	0.02436	1337.6	158.19	281.48	0.83981	1.3641
−36	0.69583	0.02277	1328.0	160.25	282.04	0.84838	1.3620
−34	0.74364	0.02130	1318.3	162.32	282.59	0.85693	1.3598
−32	0.79380	0.01994	1308.4	164.41	283.13	0.86545	1.3578
−30	0.84637	0.01868	1298.3	166.50	283.64	0.87394	1.3557

(Contd.)

Table A9 Refrigerant 13 (chlorotrifluoromethane) properties of saturated liquid and saturated vapour *(Contd.)*

Saturation temperature, T (°C)	Saturation pressure, p (bar)	Volume Vapour, v_g (m³/kg)	Density Liquid, v_f (kg/m³)	Specific enthalpy in kJ/kg Liquid, h_f (kJ/kg)	Vapour, h_g (kJ/kg)	Specific entropy in kJ/kg-K Liquid, h_{fg} (kJ/(kg-K))	Vapour, s_g (kJ/(kg-K))
−28	0.90143	0.01751	1288.1	168.61	284.14	0.88241	1.3536
−26	0.95904	0.01643	1277.7	170.74	284.61	0.89085	1.3516
−24	1.0193	0.01542	1267.1	172.88	285.06	0.89928	1.3496
−22	1.0822	0.01448	1256.4	175.03	285.50	0.90768	1.3475
−20	1.1479	0.01361	1245.3	177.20	285.90	0.91607	1.3455
−18	1.2164	0.01279	1234.1	179.38	286.29	0.92445	1.3434
−16	1.2878	0.01203	1222.6	181.58	286.64	0.93281	1.3414
−14	1.3622	0.01132	1210.8	183.80	286.97	0.94117	1.3393
−12	1.4396	0.01065	1198.8	186.04	287.27	0.94953	1.3372
−10	1.5202	0.01002	1186.4	188.30	287.54	0.95789	1.3350
−8	1.6040	0.009434	1173.7	190.59	287.77	0.96627	1.3328
−6	1.6911	0.008881	1160.6	192.90	287.97	0.97465	1.3305
−4	1.7815	0.008360	1147.1	195.23	288.13	0.98306	1.3282
−2	1.8754	0.007869	1133.1	197.60	288.24	0.99151	1.3258
0	1.9729	0.007405	1118.7	200.00	288.31	1.0000	1.3233
2	2.0741	0.006965	1103.6	202.44	288.32	1.0085	1.3207
4	2.1790	0.006549	1087.9	204.92	288.27	1.0172	1.3179
6	2.2878	0.006153	1071.5	207.45	288.16	1.0259	1.3150
8	2.4006	0.005775	1054.3	210.04	287.98	1.0347	1.3119
10	2.5176	0.005415	1036.1	212.70	287.71	1.0437	1.3086
12	2.6389	0.002069	1016.7	215.43	287.34	1.0529	1.3051
14	2.7646	0.004735	996.01	218.27	286.86	1.0623	1.3012
16	2.8950	0.004412	973.57	221.24	286.23	1.0721	1.2969
18	3.0303	0.004097	948.97	224.36	285.42	1.0823	1.2920
20	3.1708	0.003785	921.51	227.71	284.38	1.0932	1.2865
22	3.3168	0.003472	890.04	231.38	283.02	1.1050	1.2799
24	3.4690	0.003149	852.38	235.56	281.18	1.1183	1.2718
26	3.6280	0.002793	803.49	240.74	278.47	1.1347	1.2608
28	3.7955	0.002322	723.19	248.98	273.40	1.1605	1.2415
*28.9	3.870	0.00173	578	264.7	264.7	1.209	1.209

Appendix B

Table B1 Conversion factors (MKS to SI Units and vice versa)

Physical quantity	Symbol	SI to MKS	MKS to SI
Force	F	1 N = 0.1020 kg_f	1 kg_f = 9.807 N
Pressure	P	1 kN/m^2 = 0.0102 kg_f/cm^2	1 kg_f/cm^2 = 98.07 kN/m^2
		1 N/m^2 = 9.869 × 10^{-3} atmosphere	1 atm = 101.325 kN/m^2
		1 kN/m^2 = 0.0102 ata (Tech. atm)	1 ata = 98.07 kN/m^2
		1 kN/m^2 = 7.501 mm of Hg	1 mm of Hg = 0.1333 kN/m^2
Energy, Work	W	1 J = 0.1020 kg_f metre	1 kg_f metre = 9.807 J
Energy, heat	Q	1 J = 0.2388 cal	1 cal = 4.187 J
Energy, electrical	E	1 MJ = 0.2778 kWh	1 kWh = 3.6 MJ
Energy, heat	Q	1 W = 0.0143 kcal/min	1 kcal/min = 69.77 W
Horsepower (metric)	hp	1 kW = 1.36 hp	1 hp = 0.7355 W
Heat flow rate	Q	1 W = 0.86 kcal/h	1 kcal/h = 1.163 W
Heat flux	$q = Q/A$	1 W/m^2 = 0.86 kcal/h-m^2	1 kcal/h-m^2 = 1.163 W/m^2
Heat generation per unit volume	Q	1 W/m^3 = 0.86 kcal/h-m^3	1 kcal/h.m^3 = 1.163 W/m^3
Specific heat	C	1 kJ/kg °C = 0.2388 kcal/kg-°C	1 kcal/kg °C = 4.187 kJ/kg-°C
Thermal conductivity	k	1 W/m-K = 0.86 kcal/m-h-°C	1 kcal/m.h °C = 1.163 W/m-K
Heat transfer coefficient	h	1 W/m^2-K = 0.86 kcal/m^2-h-°C	1 kcal/m^2.h°C = 1.163 W/m^2-K
Dynamic viscosity	μ	1 kg/m-s = 10 Poise (gm/cm-s) or 1 $N-s/m^2$ = 0.120 $kg_f/s-m^2$	1 Poise (gm/cm-s) = 0.1 kg/m-s or 1 $kg_f/m^2 s$ = 9.807 $N-s/m^2$
Viscosity, thermal diffusivity	ν, α	1 m^2/s = 10^4 strokes (cm^2/s)	1 stroke = 10^{-4} m^2/s
Temperature	T	1 K = °C + 273.15	1 °C = K − 273.15

Appendix C

Table C1 Properties of insulating, building and other materials

Substance	Temperature, °C	k, W/m-°C	ρ, kg/m^3	C, kJ/kg-°C	α, m^2/s × 10^7
Structural and heat-resistant materials					
Asphalt	20–55	0.74–0.76			
Building brick, common	20	0.69	1600	0.84	5.2
Carborundum brick	600	18.5			
	1400	11.1			
Chrome brick	200	2.32	3000	0.84	9.2
	550	2.47			9.8
	900	1.99			7.9
Diatomaceous earth, moulded and fired	200	0.24	2000	0.96	5.4
	870	0.31			
Fireclay brick, burnt 1330°C	500	1.04			
	800	1.07			
	1100	1.09			
Burnt 1450°C	500	1.28	2300	0.96	5.8
	800	1.37			
Missouri	200	1.00	2600	0.96	4.0
	600	1.47			
	1400	1.77			
Magnesite	200	3.81		1.13	
	650	2.77			
	1200	1.90			
Cement, Portland		0.29	1500		
Mortar	23	1.16			
Concrete, cinder	23	0.76			
Stone 1-2-4 mix	20	1.37	1900–2300	0.88	8.2–6.8
Glass, window	20	0.78 (avg)	2700	0.84	3.4
Corosilicate	30–75	1.09	2200		
Plaster, gypsum	20	0.48	1440	0.84	4.0
Metal lath	20	0.47			
Wood lath	20	0.28			
Stone					
Granite		1.73–3.98	2640	0.82	8–18
Limestone	100–300	1.26–1.33	2500	0.90	5.6–5.9
Marble		2.07–2.94	2500–2700	0.80	10–13.6
Sandstone	40	1.83	2160–2300	0.71	11.2–11.9
Wood (across the grain)					

(Contd.)

Table C1 Properties of insulating, building and other materials *(Contd.)*

Substance	Temperature, °C	k, W/m-°C	ρ, kg/m^3	C, kJ/kg-°C	α, m^2/s $\times 10^7$
Balsa	30	0.055	140		
Cypress	30	0.097	460		
Fir	23	0.11	420	2.72	0.96
Maple or oak	30	0.166	540	2.4	1.28
Yellow pine	23	0.147	640	2.8	0.82
White pine	30	0.112	430		
	Insulating materials				
Asbestos					
Loosely packed	−45	0.149			
	0	0.154	470–570	0.816	3.3–4
	100	0.161			
Asbestos–cement boards	20	0.74			
Sheets	51	0.166			
Felt,	38	0.057			
16 laminations/cm	150	0.099			
	260	0.083			
8 laminations/cm	38	0.078			
	150	0.095			
	260	0.112			
Corrugated,	38	0.087			
150 plies/m	93	0.100			
	150	0.119			
Asbestos cement	—	2.08			
Balsam wool,	32	0.04	35		
Cardboard, corrugated	—	0.064			
Celotex	32	0.048	160		
Cork, regranulated	32	0.045	45–120	1.88	2–5.3
Ground	32	0.043	150		
Diatomaceous earth (Sil–0–cel)	0	0.061	320		
Felt, hair	30	0.036	130–200		
Wool	30	0.052	330		
Fibre, insulating board	20	0.048	240		
Glass wool,	23	0.038	24	0.7	22.6
Insulex, dry	32	0.064			
		0.144			
Kapok	30	0.035			
Magnesia, 85%	38	0.067	270		
	93	0.071			
	150	0.074			
	204	0.080			
Rock wool	32	0.040	160		
Loosely packed	150	0.067	64		
	260	0.087			
Sawdust	23	0.059			
Silica aerogel	32	0.024	140		
Wood shavings	23	0.059			

Appendix D

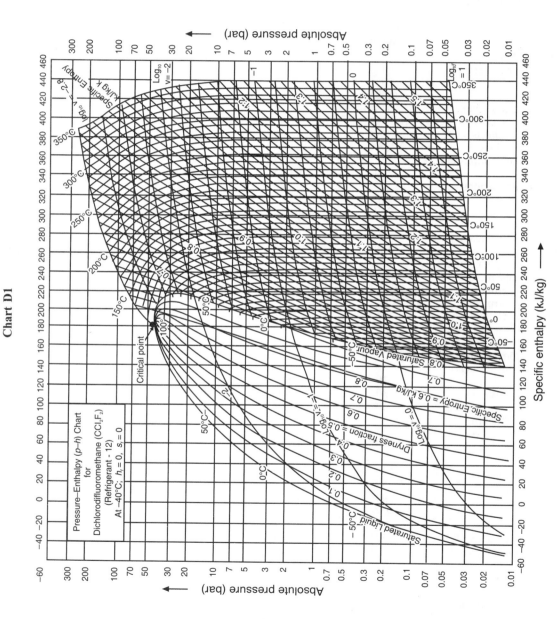

Chart D1

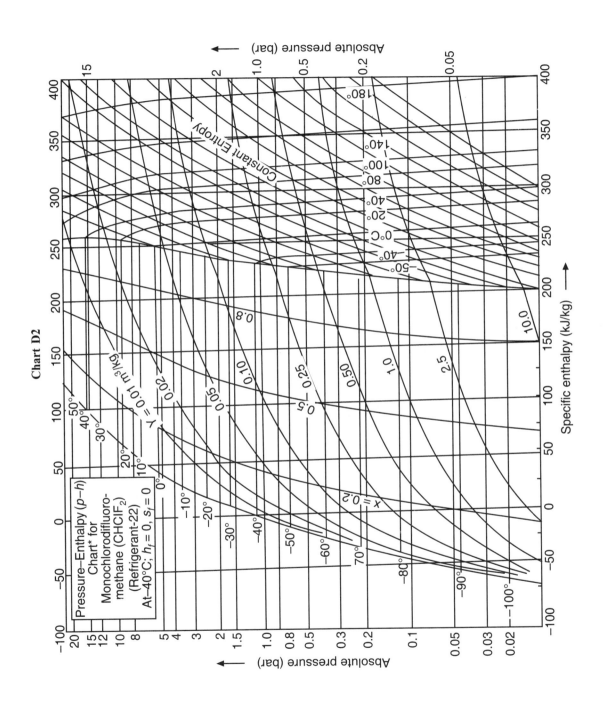

Chart D2 Pressure–Enthalpy (p–h) Chart* for Monochlorodifluoromethane (CHClF$_2$) (Refrigerant-22) At $-40°C$; $h_f = 0$, $s_f = 0$

Chart D3

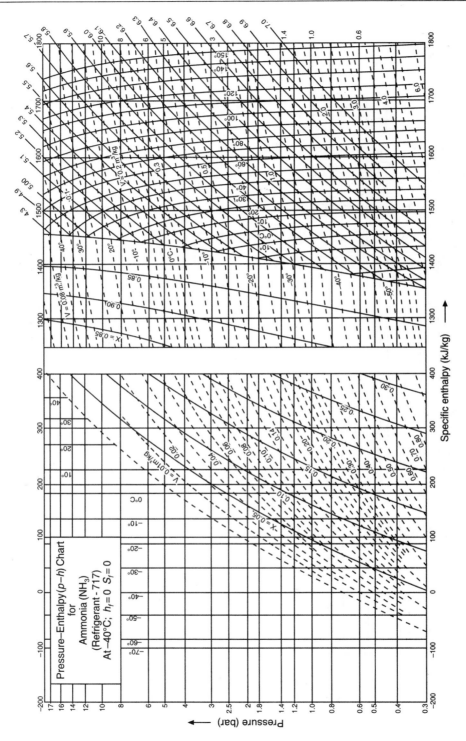

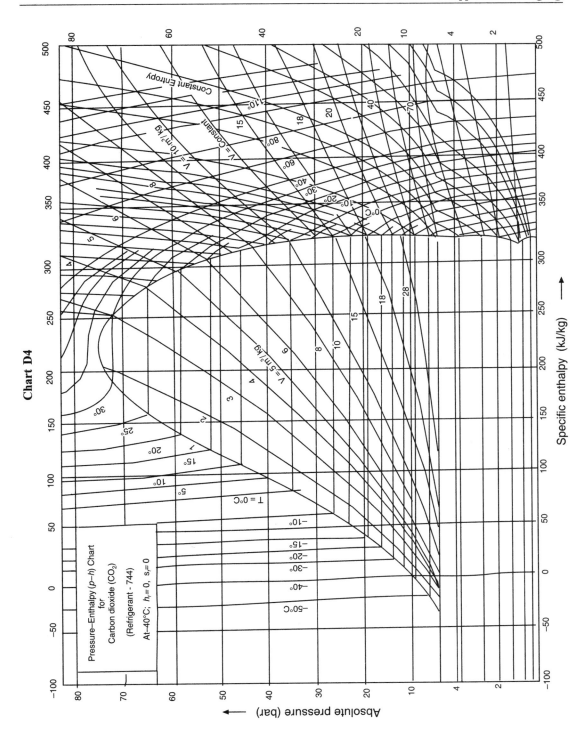

Chart D4 Pressure–Enthalpy (p–h) Chart for Carbon dioxide (CO_2) (Refrigerant - 744) At $-40°C$; $h_f = 0$, $s_f = 0$

Chart D5

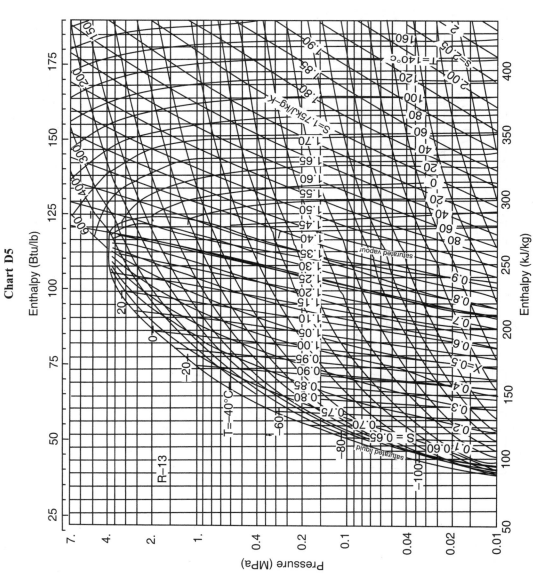

Pressure–enthalpy diagram for Refrigerant 13

Chart D6

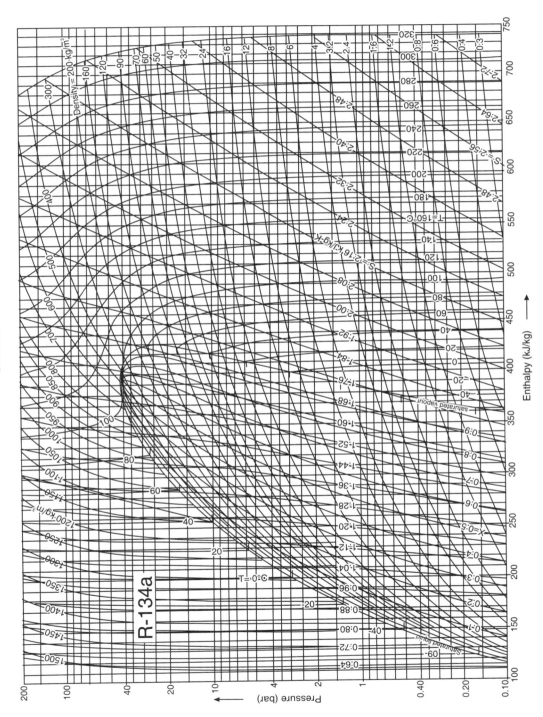

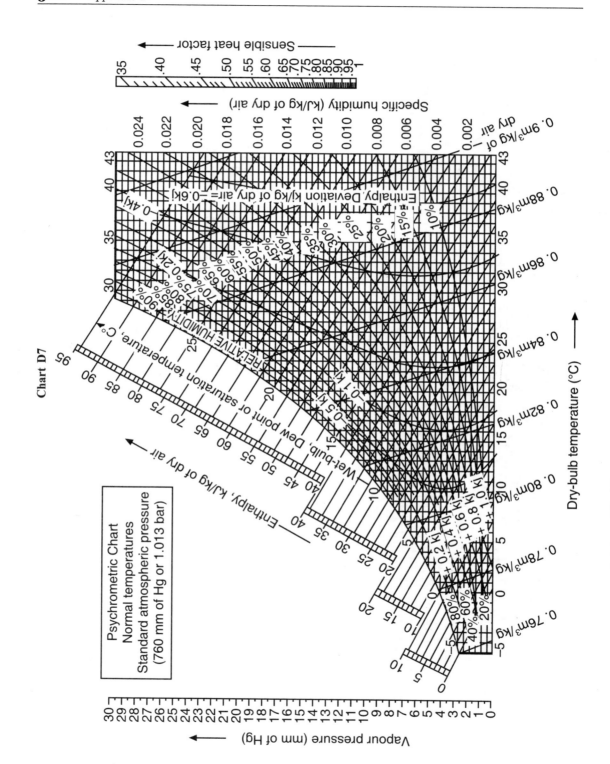

Index

Abrupt expansion loss, 507
Absolute humidity, 280
Absolute pressure, 4
Acceleration due to gravity, 2
Actual Bell–Colemann cycle, 67
Actual performance, 397
Adiabatic, 8
   cooling, 305
   demagnetisation, 49
   magnetisation, 48
   mixing of air streams, 309
   process, 20
   saturation, 285
Air change method, 330
Air cleaning devices, 372
Air conditioner
   split, 362
   window, limitations of, 361, 362
Air conditioning equipment, 370
Air conditioning systems, 360
   classification of, 360
Air distribution systems, 481, 503, 508
Air liquefaction system, 521
Air refrigeration
   advantages of, 115
   limitations of, 115
Air refrigeration cycle, 49
Air refrigeration systems, 53, 57
   method of, 68
Air washer, 312, 376
Air-cooled condensers, 437
Aircraft refrigeration, application of, 68
Air-distribution system for auditoriums, 511
Air-water combination systems, 360
Air-water systems, 369
All-air systems, 360
All-water (hydronic) systems, 360, 368

Altitude, 328
Ammonia (NH_3), 185
   characteristic of, 262
   piping, 471
Anabolic, 531
Analyzer, 264
Angle of incidence, 328
Aniline point, 196
Appliances, 325, 333
Argon, applications of, 38, 518
Aspect ratio, 491
Atmospheric pressure, 4
Automatic expansion valve (AEV), 443, 444
Automobile industry, 517
Automobile tyres, 38
Axial fans, 377
Azeotropes, 180

Backward curved blades, 378
Bacteria, yeast, 531, 532
Bakery, 277
Baudelot cooler, 431
Bell–Colemann, 63
Blower power, 326, 334
Blueprint, 334
Body heat loss, 275
Boiling point (BP), 182
Boot-strap air cooling system, 88
Boot-strap air evaporative cooling system, 100
Brazed plate, 435
   condensers, 436
Brines, 193
Building peak cooling load, 335
Bypass factor (BF), 300
Bypass method, 400

Capacity control of reciprocating compressor, 399
Capacity of a compressor, 398
Capillary tube, 444, 453
    selection of, 456
Carbon dioxide, 552
Carbon dioxide (R744), 186
Carnot COP, 59
Carnot refrigerator, 57
Catabolic, 531
Central air conditioning systems, 362
Central or built-up systems, 361
Centrifugal fans, 377, 379
CFCs, 179
Charging, 477
    through charging valve, 478
    through suction valve, 477
Checking the charge, 479
Chemical dehumidification or sorbent dehumidification, 313
Chemical properties of refrigerants, 183
Chemical stability, 195, 196
Chemicals, pharmaceuticals and petroleum, 516
Chiller type oil separator, 473
Claude cycle, 525
Clausius statement, 27, 28
Clearance volumetric efficiency, 390
CLTD method, 327
Coefficient of performance (COP), 6, 56
Coffeemaker, 334
Coil bypass factor, 300
Coil sensible heat factor (CSHF), 303
Cold chain, 536
Cold storage, 536
Cold stores based on a seasonal basis, 536
Cold stores for temporary storage of commodities, 536
Comfort air conditioning, 275
Comfort chart, 323
Comfortable in summer, 320
Comparison of various air cooling systems used for aircraft, 111
Compression process, 384
Compressor, 383
    piping, 472
    slugging, 443
Computer devices, 334
Condenser and receiver piping, 472
Condenser pressure, effect of, 154
Condensers, 432
Condenser-to-receiver lines, 466
Condensing pressure, 183
Condensing unit off-time, 428

Conduction, 416
    through exterior structures, 326
Consequences of ozone depletion, 199
Constant pressure, 17
Constant temperature process, 18
Constant volume, 17
Construction, 516
    of cold stores, 536
Contaminants in the air, 375
Continuity equation, 483
Control volume, 23
Convection heat transfer, 420
Convective heat loss, 321
Cooling and dehumidification, 312
    of moist air, 302
Cooling coil, 363, 370
    load, 335
Cooling load (condenser load), 435
Cooling load estimation, 327
Cooling methods, 519
Cooling with humidification, 305
Copiers/Typesetters, 334
Copies (large), 334
Corrosiveness, 184
Cost and availability, 185
Crack method, 329
Critical point, 30
Critical pressure, 30
Critical temperature and pressure, 183
Cross-charged valves, 451
Cryogenic air separation plant, 518
Cryogenic distillation, 519
Cryogenic grinding, 38
Cryogenics, 515
Cryogens, 37
Cryosurgery, 37
Cycle, 8
Cyclic
    heat engine, 27
    process, 8
    refrigeration systems, 39

Dampers, 494
Day of the year and time of the day, 328
DBT lines, 296
Defrost, method of, 427
Degree of saturation, 281
Dehydration, 475
    by dry air, 476
    by heating, 475
    by vacuum, 476

Index

Density (ρ), 1
Desiccants, 313
Design features of a reciprocating compressor, 394
Desirable properties, 181
  of (absorber) solvent, 261
Destruction of ozone, 198
Determination of compressor motor power, 395
Dew point temperature, 279
Dielectric strength, 185
Direct expansion type, 430
Discharge line oil separators, 473
Discharge pressure, effect of, 397
Disposable air filters, 374
Domestic Electrolux (ammonia–hydrogen) refrigerator, 269
Downward flow or overhead system, 512
DPT lines, 297
Dry air, 278
Dry and flooded evaporators, 423
Dry compression, 135
Dry saturated steam, 539
Dry-bulb temperature (DBT), 277, 279
Drying, 533
Dry-type filters, 373
Dual duct and induction system combined, 507
Dual duct system, 366, 506
Duct air leakage, 326, 334
Duct arrangement systems, 502
Duct, classification of, 482
Duct design, 481
  method of, 496
Duct furnaces, 372
Duct heat gain, 326, 334
Duct material, 482
Ductwork, 363
Dynamic losses in duct, 494
Dynamic pressure loss, 494

EER, 6
Effective room latent cooling load, 342
Effective room sensible cooling load, 342
Effective room sensible heat factor, 342
Effective temperature, 322
Efficient oil return, 467
Ejector air-distribution system, 512
Elbows, 494
Electric current, 2
Electric defrost, 428
Electric heaters, 372
Electronic air cleaners, 374

Electronic devices, 325
Electronics industry, 516
Energy, 4
  efficiency ratio, 6
  equation for a pipe flow, 484
Enthalpy, 5
  of air, 283
Entropy, 5
Enzymes, 531
Equal friction method, 497
Equilibrium state, 7
Equipment, 325
Equivalent length method, 494
Estimation, method of, 328
Ethylene glycol, 193
Evaporating pressure, 183
Evaporative
  condensers, 440
  cooling, 305
  heat loss, 321
  refrigeration, 40
Evaporator, 423
  and condensers, 415
  pressure, effect of, 152
Expansion devices, 443
Exposure of the glass, 328
Extended plenum system, 504
Extended surface (Fin), 426
Extensive, 7
External equalizer, 448
External sources, 325

Factors affecting solar radiation, 328
Factors of food deterioration, 531
Fadeout point, 452
Fan, 376
  laws, 381
  optimum conditions of, 381
  ratings, 379
  selection, 379
  system interaction, 380
  types of, 376
Figure of merit (FOM), 522
Filter, 374
Filters/cleaners, types of, 373
Fin density, 370
Finding the infiltration rate, 329
First law, 25
Flammability, 183
Flash point, 184, 196

Float valves, 456
Flooded shell-and-tube coolers, 431
Flow work, 14
Fluid pressure, 3
Food deterioration and spoilage, 530
Food preservation, 530
    advantages of, 535
    disadvantages of, 535
    processes, 533
Force, 3
Forced convection, 421
Formation, 198
Forward curved blades, 378
Freezing and frozen storage, 535
Freezing point, 182
Friction factor 'f', 488
Frosting and defrosting of coolers, 427
Frozen food stores, 536

Gas-charged TEV, 450
Gauge pressure, 4
Global warming potential, 199
Grand sensible heat factor, 338, 339
Grashoff number, 421, 422
Grinding, 517

Halide leak detector, 480
Halocarbon compounds, 179, 194
Handling and maintenance, 185
Hand-operated expansion device, 443
Hand-operated expansion valve, 444
Haze, 328
HCFC–R123, 201
Healthcare, 516
Heat, 11
    and dry air method, 476
    and vacuum method, 476
    engine, 55
    exchange loss, effect of, 390
    exchangers, 264
    gain, 157
    gain from equipment, 333
    gain from human beings, 332
    gain through glass, 327
    load from people, 332
    load on the evaporator, 425
    loss, 157
    pump, 56
    transfer, 434
    transfer and pressure drop, 438
    treatment of metals, 38
Heating and humidification, 307, 312
    by steam injection, 308
Heating load, 54
HEPA, 374
Hermetic compressors, 384
Hermetically sealed compressor, 413
HFC-134a as replacement for R12, 202
HFC–R245, 201
High pressure duct system, 482
High pressure float valve, 444, 457
Hot gas bypass, 400
Hot gas lines, 467
Hot pipes and tanks, 326
Hot water coil, 372
h–s diagram, 31
Human comfort chart, 323
Humidification, 312
Humidifier, 375
    duty, 306
Humidity, 328
    and air-motion in storages, 534
    ratio (specific humidity), 279
    relative humidity, 277
Hydrocarbons, 180
    propane (R290), 203
Hydronic systems, 368

Ice refrigeration, 39
Ideal compressor, performance of, 396
Ideal gases, 17
Ideal refrigerant, 185
Ideal vapour absorption refrigeration system, 265
Index of compression process, 183
Indicator diagrams, 388
Indoor air quality, 276
Induction unit, 369
Industrial air conditioning, 276
Infiltration, 329
Inorganic compounds, 180
Installation arrangement, 474
Integral two-stage, 383
Intensive, 7
Internal energy, 9
Internal sources, 325
Irreversibility, 8
Isentropic, 386
    process, 141
Isobars, 29

Isobutane (R600a), 203
Isolation of noise and vibration, 474
Isomagnetic enthalpic heat transfer, 48
Isomagnetic entropic transfer, 49
Isothermal, 386
    compression process, 385
    process, 19

Joule cycle, 63
Joule's equivalent, 15

Kelvin and Planck, 55
    statement, 27
Kinetic energy, 8
Kyoto protocol, 204

Lactase, 531
Laminar flow, 421
Latent heat
    defrost, 428
    of evaporation, 181
    of fusion, 29
    loss effect of infiltration air, 329
    of vaporisation, 29
Leak detection, 184
Leakage loss, 391
Lights, 325
Liquid coolers, 430
Liquid gas refrigeration, 42
Liquid line, 157
Liquid nitrogen, 37
Liquid subcooling, 141
Liquid-suction heat exchanger, 151
Lithium bromide absorption, 271
Location and space allotment, 435
Location of equipment, 474
Loss coefficient method, 494
Losses in vapour compression refrigeration system, 156
Low pressure duct system, 482
Low pressure float valve, 444, 457
Lubricating oils, 195
Luminous intensity, 3

Macroscopic, 8
Magnetocaloric effect, 48

Maintenance cost considerations, 435
Manufacturing, 516
Mass, 1
Materials, 459
Mathematical model, 321
Maximum COP, 268
Mechanical solar refrigeration system, 46
Medium pressure duct system, 482
Metals industry, 516
Microcomputer/word processor, 334
Micro-organisms, 531, 532
Microscopic, 8
Microwave oven, 334
Milk and milk products, 35
Minicomputer, 334
Miscellaneous, 334
Miscibility with oil, 184
Mixed storage, 534
Modified reversed Carnot cycle, 61
Moist air, 278
Montreal protocol, 200
Moulds, 531, 533
MS adsorbers, 519
Multiple system practices for HCFC systems, 467
Multizone system, 365
Natural convection, 421

Neutralization number, 196, 197
New refrigerants, 189
Nitrogen, applications of, 37, 516
Non-condensable gases, 436
Non-cyclic refrigeration systems, 39
Nusselt number, 422

Occupancy, 325
Oil, 195
    separators, 473
Oil-injected compressor, 404
Oil-injection free compressor, 405
On-off control, 400
Open-type, 383
Outside air load, 331
Outside design conditions, 324
Overall heat transfer coefficient, 326, 419
Overall volumetric efficiency, 388
Oxygen, applications of, 38, 517
Ozone depletion, 197
    potential (ODP), 199

Pan humidifier, 376
Pan type, 510
Partial pressure of water vapour, 282
Path, 7
Peltier effect, 44
Performance characteristics, 378
Performance criteria, 375
Perimeter system, 503
Permanent, 374
  gases, 278, 515
p–h diagram, 31, 32
Photovoltaic operated refrigeration system, 46
Pipe sizing, 469
Piping arrangement, 475
Piping layout, 464
  for hot gas lines, 465
  for suction lines, 464
PMM-I, 25
Polytropic
  compression of vapour, 158
  with cooling, 386
  with friction, 386
  process, 21
Positive displacement, 383
Potential energy, 8
Pour and/or floc point, 195, 196
Power, 5
  fluid, 451
  input, 411
  requirement, 386
Practical vapour absorption system, 263
Prandtl number, 421
Preheat coil, 363
Pressure, 3, 282
  drop, 426, 460
  drops at the compressor delivery valve, 158
  drops in the condenser, 158
  drops in the delivery line, 158
  drops in the evaporator, 158
  process, 18
  testing, 480
Pressure loss, 141
  in bends, tees and branch offs, 495
  due to contraction, 495
  in the duct, 487
  at entry or exit from duct, 495
  in fittings, 495
  from friction in piping and ducts, 487
  due to sudden enlargement, 494
Primary air, 369
Primary area, 370

Primary refrigerants, 179
Printer (laser), 334
Printer (line, high-speed), 334
Printing, 277
Products brought in, 325
Propylene glycol, 193
Psychrometric analysis of the air conditioning, 335
Psychrometric chart, 296
Psychrometry, 277
Pure substance, 28
p–v diagram, 31

Quality and temperature of available water, 435
Quasi-static process, 12

R11 (CCl3F), 187
R12 (CCl2F2), 187
R12, R134a, 30
R134a, 190
R22 (CHClF2), 187
R407, 202
R410A, 202
R500 (CCl2F2/CH3CHF2), 187
Radiation heat loss, 321
Ramming process, 69
Rating and selection
  of air-cooled condensers, 439
  of TEV, 453
  of water-cooled condenser, 437
Recommended inside design conditions—summer and winter, 276
Recommended velocities, 496
Rectangular sections equivalent to circular, 490
Rectifier, 264
Reduced ambient air cooling system, 101
Refrigerant, 13, 178, 435, 555, 556
  classification of, 179
  designation of, 180
  relationship, 195
  secondary, 180, 191
  selection of, 187
Refrigerant-charged expansion valves, 450
Refrigerant–solvent combination, 261
Refrigerant–solvent properties, 261
Refrigerated storage, 533
Refrigerating effect, 5
Refrigeration, 35
  applications of, 35
  by dry ice, 41

load, 54
magnetic, 48
systems, 39
Refrigerator, 55
Regenerative air cooling system, 106
Reheat system, 365
Reheating coil, 363
Relative COP, 134
Relative humidity, 282
Renewable air filters, 374
Required operating pressure, 435
Return air fan, 363
Return in duct system, 507
Reversed Carnot cycle, limitations of, 61
Reversible process, 12
Reynolds number, 421
RH lines, 296
Rolling piston, 401
Room cooling load, 335
Room latent cooling load, 336
Room peak load, 335
Room sensible heat factor line, 337
Room total cooling load (RTCL), 336, 337
Room total heat, 337
Rotary compressors, 401
Rotary vane type compressor, 402
Rotating vane, 401
Rutherford, 12

Safe handling of refrigerants, 194
Saturated
   air, 278
   ammonia, 550
   dichlorodifluoromethane, 546, 547
   monochlorodifluoromethane, 548
   trichloromonofluoromethane, 544, 545
Saturation pressure of water vapour, 282
Screw compressors, 401, 403
   design, effect of, 407
Second law of thermodynamics, 27
Secondary area, 370
Seebeck effect, 44
Seer, 7
Semi-hermetic compressors, 384
Sensible cooling, 312
   of moist air, 301
Sensible heat factor (SHF), 303
Sensible heat loss effect of infiltration air, 329
Sensible heating of air, 298
Shape of duct, 483

Shell and coil, 435
   condensers, 435
   cooler, 431
Shell-and-tube type, 435
   condensers, 435
Simon helium liquefier, 527
Simple air evaporative cooling system, 84
Simple air-cooling system, 68
Simple Linde cycle, 523
Simple type, 509
Simple vapour absorption system, 262
Single duct system, 505
Single screw compressor, 403
Single-stage, 383
Single zone or multiple zone systems, 360
Single-zone central air conditioning, 363
Sink, 26
Size of various contaminants, 375
Sizing of discharge lines, 464
Sizing of liquid lines, 460
Sizing of suction lines, 463
Small stores, 536
Small-to-medium-sized hydrogen liquefiers, 526
Soft drinks, 36
Solar refrigeration, 46
Solar vapour absorption system, 46
Sorbent dehumidification, 313
Source, 26
Space for installation and service, 474
Specific enthalpy lines, 297
Specific gravity, 2, 195, 196
Specific heat, 5
Specific humidity lines, 296
Specific volume, 1, 184, 284
   lines, 297
Speed control method, 399
Spray chamber, 376
Stability, 184
Static regain, 487
   method, 498
Stationary, 374
Steady state conduction
   through a composite cylinder, 418
   through a composite wall, 417
Steady state heat conduction through a slab, 416
Steady state heat flow through a cylindrical wall, 417
Steam coils, 371
Steam jet refrigeration, 43
Steam type humidifier, 376
Storage conditions, 534
Storage of vegetables and food products, 36

Storage temperature, 534
Stored energy, 8
Stores for storing the commodities round the year, 536
Straight-duct friction loss, 494
Subcooling, effect of, 146
Sublimates, 30
Substitutes for CFC, R12 refrigerant, 202
Suction
    gas superheating, 141, 143
    lines, 467
    pressure control, 428
    pressure, effect of, 396
    traps, 473
Sulphur dioxide, 553
Summer air conditioning system, 338
Superheating the suction vapour, 142
Supply air conditions, 335
Surging effect, 412
System characteristics, 380
System heat gain, 326, 334
System performance parameters, 521
System practice for HCFC systems refrigerant line, 460

Temperature, 3
    of adiabatic saturation, 284
Testing for leaks, 479
Textile, 277
Thermal conductivity, 185
Thermal energy reservoir, 26
Thermodynamic properties, 181
Thermodynamic wet-bulb temperature, 284, 285
Thermodynamics of human body, 321
Thermoelectric refrigeration, 44
Thermostatic expansion, 443
Thermostatic expansion valve (TEV), 446
Time clock initiation
    and suction pressure termina, 428
    and temperature termination, 428
    and termination, 428
Total equivalent warming impact (TEWI), 204
Total, static, 486
Toxicity, 184, 194
TR, 54
Triple point, 30

T–s diagram, 31
Tube-in-tube condensers, 435
Twin or double helical screw compressor, 406
Typical air conditioning processes, 298
Typical samples of inside design conditions, 277

Unitary or central systems, 360
Unitary system, 360, 361
Unitary vs. central systems, 369
Unsaturated organic compounds, 180
Upward flow system, 512

Vacuum and heat method, 476
Valve lifting method, 399
Valve pressure drops, effect of, 391
Vapour compression refrigeration cycle, 50
Variable air volume (VAV) system, 367
Variable substances in air, 278
VC cycle, 116
Velocity, 2
    pressure, 486
    reduction method, 500
Ventilation, 330, 474
    type, 511
Viscosity, 184, 196, 197
    index, 196, 197
Viscous impingement filters, 373
Vortex tube, 44

Water, 193
    cooler, 334
    submersion test, 480
Water-cooled condenser types, 433, 435
Weather conditions, 474
Wet compression, 135
Wet-bulb temperature (WBT), 277, 279
    lines, 296
Wetted pack humidifier, 376
Winter air conditioning, 340
Winter season the comfort conditions, 320
Work, 10
    done by impeller, 410

Yeasts, 533